T0122369

Differential-Algebraic Equations Forum

Differential-Algebraic Equations Forum

The series *Differential-Algebraic Equations Forum* is concerned with analytical, algebraic, control theoretic and numerical aspects of differential algebraic equations, as well as their applications in science and engineering. It is aimed to contain survey and mathematically rigorous articles, research monographs and textbooks. Proposals are assigned to a Publishing Editor, who recommends publication on the basis of a detailed and careful evaluation by at least two referees. The appraisals will be based on the substance and quality of the exposition.

More information about this series at http://www.springer.com/series/11221

Timo Reis • Sara Grundel • Sebastian Schöps
Editors

Progress in Differential-Algebraic Equations II

 Springer

Editors
Timo Reis
Fachbereich Mathematik
Universität Hamburg
Hamburg, Germany

Sara Grundel
Max-Planck-Institut für Dynamik
komplexer technischer Systeme
Magdeburg, Sachsen-Anhalt
Germany

Sebastian Schöps
Technische Universität Darmstadt
Darmstadt, Germany

ISSN 2199-7497   ISSN 2199-840X   (electronic)
Differential-Algebraic Equations Forum
ISBN 978-3-030-53904-7   ISBN 978-3-030-53905-4   (eBook)
https://doi.org/10.1007/978-3-030-53905-4

This Springer imprint is published by the registered company Springer Nature Switzerland AG.
The registered company address is: Gewerbestrasse 11, 6330 Cham, Switzerland

Preface

The "9th Workshop on Descriptor Systems" took place on March 17–20, 2019 in Paderborn, Germany. Following the tradition of the preceding workshops organized by Prof. Peter C. Müller between 1992–2005 and 2013, the workshop brought together more than 40 mathematicians and engineers from various fields, such as numerical and functional analysis, control theory, mechanics and electromagnetic field theory. The participants focused on the theoretical and numerical treatment of "descriptor" systems, i.e., differential-algebraic equations (DAEs).

This book contains the proceedings of this workshop. It discusses the wide range of current research topics in descriptor systems, including mathematical modeling, index analysis, stability, stabilization, well-posedness of problems, stiffness and different timescales, co-simulation and splitting methods, and convergence analysis. In addition, it also presents applications from the automotive and circuit industries that show that descriptor systems are challenging problems from the point of view of theory and practice.

This book is organized into three parts with the first part covering analysis. It features a contribution by Diana Estévez Schwarz and René Lamour that discusses orthogonal transformations for decoupling of DAEs with a higher index. Different types of the so-called higher-index components with regard to the explicit and hidden constraints are characterized. This also results into a straightforward possibility for linear DAEs to determine an orthogonally projected explicit ODE. The second chapter consists of an article by Michael Hanke and Roswitha März on an operator-theoretic view to linear DAEs with constant and non-constant coefficients. Conditions concerning basic characteristics such as normal solvability (closed range), Fredholmness, etc. are presented. In particular, it is proven that actually the operators having tractability index zero and one constitute the class of normally solvable differential-algebraic operators.

The second part of this book covers numerical analysis and model order reduction. It consists of five contributions: The article by Andreas Bartel and Michael Günther contains a complete convergence theory for inter/extrapolation-based multirate schemes for both ODEs and DAEs of index one, along with a convergence analysis based on linking these schemes to multirate dynamic iteration. The second contribution in this part by Michael Hanke and Roswitha März is on overdetermined polynomial least-squares collocation for two-point boundary value problems for higher-index DAEs. Basic properties, such as convergence properties, of this method for initial value problems by a windowing technique are proven. The article by Robert Altmann and Christoph Zimmer is devoted to the construction of exponential integrators of the first and second order for the time discretization of constrained parabolic partial-differential-algebraic systems. Exponential integrators for unconstrained systems are combined with the solution of certain saddle point problems in order to meet the constraints throughout the integration process, along with a convergence analysis. The succeeding contribution by Gerd Steinebach is about an improvement of the known Rosenbrock–Wanner method `rodasp`, which results in a less distinctive drop of the convergence order. The contribution by Thanos Antoulas, Ion Victor Gosea, and Matthias Heinkenschloss treats the computation of reduced-order models for a class of semi-explicit DAEs, which includes the semi-discretized linearized Navier–Stokes and Oseen equations, by a data-driven Loewner framework.

The third part is devoted to control aspects of DAEs. The article by Thomas Berger, Lê Huy Hoàng, and Timo Reis treats adaptive control for a large class of multiple-input multiple-output DAEs by a novel funnel controller. To this end, a generalization of the concept of vector relative degree is presented. The second contribution in this part by Thomas Berger and Lukas Lanza is about state estimation for nonlinear DAE systems with inputs and outputs. The presented observer unifies earlier approaches and extends the standard Luenberger type observer design. The succeeding contribution by Matthias Gerdts and Björn Martens covers implicit Euler discretization for linear-quadratic optimal control problems with index two DAEs. The discretized problem is reformulated such that an approximation of an index reduced problem with suitable necessary conditions is obtained. Under some additional circumstances, it is shown that the controls converge with an order of $\frac{1}{2}$ in the L_1-norm. These error estimates are further improved with slightly stronger smoothness conditions of the problem data and switching function, which results into a convergence order of one.

The fourth part contains four articles on applications of DAEs. The article by Sarah-Alexa Hauschild, Nicole Marheineke, Volker Mehrmann, Jan Mohring, Arbi Moses Badlyan, Markus Rein, and Martin Schmidt is about modeling of district heating network in the port-Hamiltonian framework. By introducing a model hierarchy of flow equations on the network, a thermodynamically consistent port-Hamiltonian embedding of the resulting partial differential-algebraic systems is presented. It is further shown that a spatially discretized network model describing the advection of the internal energy density with respect to an underlying incompressible stationary Euler-type hydrodynamics can be considered as a parameter-dependent finite-dimensional port-Hamiltonian system. Moreover, an infinite-dimensional port-Hamiltonian formulation for a compressible instationary thermodynamic fluid flow in a pipe is presented. The subject of the article by Steffen Plunder and Bernd Simeon is a coupled system composed of a linear DAE and a linear large-scale system of ODEs, where the latter stands for the dynamics of numerous identical particles. Such systems, for instance, arise in mathematical models for muscle tissue where the macroscopic behavior is governed by the equations of continuum mechanics. Replacing the discrete particles by a kinetic equation for a particle density, the mean-field limit results into a new class of partially kinetic systems. The influence of constraints on those systems is investigated. As a main result, Dobrushin's stability estimate for systems of this type is presented. The estimate implies convergence of the mean-field limit and provides a rigorous link between the particle dynamics and their kinetic description. In the article by Idoia Cortes Garcia, Sebastian Schöps, Christian Strohm, and Caren Tischendorf, a definition of generalized circuit elements which may, for example, contain additional internal degrees of freedom, such that those elements still behave structurally like resistances, inductances, and capacitances, is presented. Several complex examples demonstrate the relevance of those definitions. Finally, the article by Diana Estévez Schwarz, René Lamour, and Roswitha März focuses on a classical benchmark problem for higher-index DAEs, namely a robotic arm resulting from a tracking problem in mechanical engineering. The difficulty of this problem is the appearance of certain singularities, whose thorough analysis is the subject of this article. To this end, different methodologies are elaborated, such as the projector-based analysis of the derivative array and the direct projector-based DAE analysis associated with the tractability index. As a result, with both approaches, the same kinds of singularities are identified. Some of them are obvious, but others are unexpected.

We would like to take the opportunity to thank all the individuals who contributed to the workshop and to this volume. Special gratitude goes to Prof. Dr. Peter C. Müller from BU Wuppertal for encouraging us to pursue his traditional and distinguished workshop series. We hope to see you at many of the future workshops on descriptor systems!

Hamburg, Germany   Timo Reis
Magdeburg, Germany   Sara Grundel
Darmstadt, Germany   Sebastian Schöps
May 2020

Contents

Part I
Analysis and Decoupling

A Projector Based Decoupling of DAEs Obtained from the Derivative Array

Diana Estévez Schwarz and René Lamour

Abstract The solution vector of a differential-algebraic equation contains different types of components, that can be analyzed with regard to various properties. In this paper, we particularly present an orthogonal decoupling that, for higher-index DAEs, describes in which context these orthogonal components appear in the derivative array. In this sense, we characterize different types of so-called "higher-index" components with regard to the explicit and hidden constraints. As a consequence, for linear DAEs we obtain a straightforward possibility to determine an orthogonally projected explicit ODE and compare it with the so-called inherent regular ODE related to the projector-based decoupling associated with the tractability matrix sequence. By several examples we illustrate the differences of these two projector-based approaches and discuss their relationship.

Keywords DAE · Differential-algebraic equation · Index · Derivative array · Projector based analysis · Constraints · Orthogonal decoupling · Tractability · MNA

Mathematics Subject Classification (2010) MSC 34A09, MSC 65L05, MSC 65L80

D. Estévez Schwarz (✉)
Beuth Hochschule für Technik Berlin, Berlin, Germany
e-mail: estevez@beuth-hochschule.de

R. Lamour
Institut für Mathematik, Humboldt Universität zu Berlin, Berlin, Germany
e-mail: lamour@math.hu-berlin.de

© The Editor(s) (if applicable) and The Author(s), under exclusive licence
to Springer Nature Switzerland AG 2020
T. Reis et al. (eds.), *Progress in Differential-Algebraic Equations II*,
Differential-Algebraic Equations Forum,
https://doi.org/10.1007/978-3-030-53905-4_1

1   Introduction

Higher-index differential-algebraic equations (DAEs) present explicit and hidden constraints that restrict the choice of consistent initial values. In fact, the dynamics can be characterized by lower-dimensional ODEs that might be not unique.

Example 1.1 Let us consider a well-understood higher-index example from [21, 22, 24]:

$$\begin{pmatrix} 1&0&0&0&0 \\ 0&0&1&0&0 \\ 0&0&0&1&0 \\ 0&0&0&0&1 \\ 0&0&0&0&0 \end{pmatrix} x' + \begin{pmatrix} -\alpha&-1&0&0&0 \\ 0&1&0&0&0 \\ 0&0&1&0&0 \\ 0&0&0&1&0 \\ 0&0&0&0&1 \end{pmatrix} x = \begin{pmatrix} q_1 \\ q_{2,1} \\ q_{2,2} \\ q_{2,3} \\ q_{2,4} \end{pmatrix}.$$

The explicit constraint reads

$$x_5 = q_{2,4}$$

and the hidden constraints result to be

$$x_4 = q_{2,3} - q'_{2,4},$$
$$x_3 = q_{2,2} - (q_{2,3} - q'_{2,4})',$$
$$x_2 = q_{2,1} - (q_{2,2} - (q_{2,3} - q'_{2,4})')'.$$

Therefore, the degree of freedom d results to be one. To characterize the one-dimensional dynamics, there are different possibilities. On the one hand, the explicit scalar ODE

$$x'_1 - \alpha x_1 = q_1 + q_{2,1} - q'_{2,2} + q''_{2,3} - q'''_{2,4}, \tag{1.1}$$

that depends on derivatives of the right-hand side q, could be considered. On the other hand, for

$$u_e := x_1 + x_3 - \alpha x_4 + \alpha^2 x_5$$

the explicit scalar ODE

$$u'_e - \alpha u_e = q_1 + q_{2,1} - \alpha q_{2,2} + \alpha^2 q_{2,3} - \alpha^3 q_{2,4}, \tag{1.2}$$

that does not depend on derivatives of q, could be considered. For the initialization, this means that if we consider (1.1), then an initial value is prescribed for $x_1(t_0)$. In contrast, if (1.2) is considered, then an initial value is prescribed for $u_e(t_0)$. In both cases, $x_2(t_0), \ldots, x_5(t_0)$ are determined by the explicit and hidden constraints and cannot be prescribed.

In terms of the projectors that we will introduce in forthcoming sections, we will decouple x in different orthogonal components, essentially

$$
\begin{pmatrix} 1 & & & \\ & 0 & & \\ & & 0 & \\ & & & 0 \\ & & & & 0 \end{pmatrix} x, \quad \begin{pmatrix} 0 & & & \\ & 1 & & \\ & & 1 & \\ & & & 1 \\ & & & & 0 \end{pmatrix} x, \quad \begin{pmatrix} 0 & & & \\ & 0 & & \\ & & 0 & \\ & & & 0 \\ & & & & 1 \end{pmatrix} x,
$$

where

- the left-hand side component, which corresponds to x_1, appears, together with its derivative x_1', in the original DAE and in the ODE (1.1),
- the component in the middle corresponds to x_2, x_3, x_4 which are determined by constraints, although the derivatives x_2', x_3', x_4' appear in the original DAE,
- the right-hand side component, i.e. x_5, is determined by constraints while the derivative x_5' does not appear in the original DAE.

While a general projector based characterization of ODEs associated to a DAE that does not involve derivatives (like (1.2)) can be found in [21] and the related work, such a general projector based description has not been developed so far for ODEs associated to a DAE with orthogonality properties like (1.1). Such ODEs will depend, in general, on derivatives of parts of the original DAE, i.e., parts of the so-called derivative array.

In this setting, our starting point is a projector based decoupling of the solution vector x into the derivative and the non-derivative part. Our goal is to provide the framework of a projector based analysis of DAEs for approaches that are based on the consideration of the derivative array, where the associated ODE may depend on derivatives of the right-hand side, like (1.1). Further, all considered projectors will be orthogonal.

In a first step into this direction, a new approach to compute consistent initial values for higher-index DAEs using the derivative array and a projector based approach was recently developed in [10, 12]. Starting from these results, in this paper we address an orthogonal decoupling of the solution vector and a corresponding decoupling of the equations of the DAE. In order to facilitate the readability, we start defining the basic concepts briefly again.

We consider DAEs of the form

$$
f(x'(t), x(t), t) = 0, \tag{1.3}
$$

for $f : \mathcal{G}_f \to \mathbb{R}^n$, $\mathcal{G}_f \subset \mathbb{R}^n \times \mathbb{R}^n \times \mathbb{R}$, where the partial Jacobian $f_{z_1}(z_1, z_0, t)$ is singular. We assume that

$$\ker f_{z_1}(z_1, z_0, t)$$

does not depend on (z_1, z_0) and that a continuously differentiable orthogonal projector $Q = Q(t)$ onto $\ker f_{z_1}$ exists. On the basis of the complementary projector $P = P(t) := I - Q(t)$ we can then reformulate the DAE as

$$f(x', x, t) = f(Px', x, t) = f((Px)' - P'x, x, t) = 0, \qquad (1.4)$$

as already introduced in [18], where we drop the argument t for the sake of simplicity. In this sense, we will use the notation:

- Px for the differentiated component,
- Qx for the undifferentiated component,

since, for the decoupling $x' = (Px)' + (Qx)'$, there is a function φ_1 such that $(Px)' = \varphi_1(x, t)$ is implicitly given, cf. [10, 12] and Sect. 2. In [10] we presented an orthogonal decoupling of Qx with regard to the explicit and hidden constraints.

In this article, we complete this approach by decoupling Px analogously, thus obtaining an orthogonal decoupling of the complete vector $x = Px + Qx$. The paper is organized as follows.

In Sect. 2 we summarize some definitions and the notations introduced in [10, 12]. Based on that, the orthogonal projectors used for the decoupling of x are defined in Sect. 3. In particular, the projector Π is defined, which is analyzed in more detail in Sect. 4.

Section 5 presents an extensive discussion of linear DAEs. For linear DAEs, Π turns out to deliver a description of an associated explicit ODE. We show and illustrate with examples the differences between the introduced orthogonal decoupling and the projector based decoupling associated with the tractability matrix sequence. In particular, we extensively analyze two illustrative classes of linear DAEs with constant coefficients

The computation of the projectors for the Modified Nodal Analysis (MNA) is briefly presented in Sect. 6 in order to show that the new orthogonal decoupling is a direct generalization of a result presented already in [5].

In the Appendix, we provide some required results from linear algebra.

2 Reinterpretation of the Differentiation Index

The conventional definition of the differentiation index is targeted on a representation of the so-called completion ODE or underlying ODE (see Sect. 5.1).

Definition 2.1 ([2]) The differentiation index is the smallest integer ν such that

$$f(x', x, t) = 0,$$

$$\frac{d}{dt} f(x', x, t) = 0,$$

$$\vdots$$

$$\frac{d^\nu}{dt^\nu} f(x', x, t) = 0,$$

uniquely determines x' as a continuous function of (x, t).

In order to allow for the differentiations, we consider

$$F_j(x^{(j+1)}, x^{(j)}, \ldots, x', x, t) := \frac{d^j}{dt^j} f(x', x, t),$$

and define for $k \in \mathbb{N}$, $k \geq 1$, $z_i \in \mathbb{R}^n$, $i = 0, \ldots, k$,

$$g^{[k]}(z_0, z_1, \ldots, z_k, t) := \begin{pmatrix} f(z_1, z_0, t) \\ F_1(z_2, z_1, z_0, t) \\ \vdots \\ F_{k-1}(z_k, \ldots, z_0, t) \end{pmatrix}, \tag{2.1}$$

which corresponds to the derivative array [2]. Let us further denote by

$$G^{[k]}_{(z_1, \ldots, z_k)}(z_0, z_1, \ldots, z_k, t) \in \mathbb{R}^{nk \times nk}$$

the Jacobian matrix of $g^{[k]}(z_0, z_1, \ldots, z_k, t)$ with respect to $(z_1, \ldots, z_k)$.

In practice, the index ν from Definition 2.1 at $(z_0^*, z_1^*, \ldots, z_k^*, z_{k+1}^*, t^*)$ can be determined by a rank check, verifying for $k = 1, \ldots$ whether the matrix

$$\mathcal{A}^{[k+1]} := G^{[k+1]}_{(z_1, \ldots, z_k, z_{k+1})}(z_0^*, z_1^*, \ldots, z_k^*, z_{k+1}^*, t^*) \in \mathbb{R}^{n(k+1) \times n(k+1)}$$

fulfills

$$\ker \mathcal{A}^{[k+1]} \subseteq \left\{ \begin{pmatrix} s_1 \\ s_2 \end{pmatrix} : s_1 \in \mathbb{R}^n, \ s_1 = 0, \ s_2 \in \mathbb{R}^{nk} \right\}. \tag{2.2}$$

This means that the matrix $\mathcal{A}^{[k+1]}$ is 1-full, with respect to the first n columns, cf. [4, 12, 20]. Therefore, at $(z_0^*, z_1^*, \ldots, z_k^*, z_{k+1}^*, t^*)$ the index is ν, if ν is the smallest integer for which $\mathcal{A}^{[\nu+1]}$ is 1-full in a neighborhood of $(z_0^*, \ldots, z_{k+1}^*, t^*)$.

With the decoupling $x = Px + Qx$ in mind, we will use the following definition of the differentiation index, which was introduced in [10, 12] focusing on the computation of consistent initial values and the characterization of singularities. Roughly speaking, if we formulate this index characterization in an analogous manner to Definition 2.1, it reads:

Definition 2.2 The differentiation index is the smallest integer μ such that

$$f(x', x, t) = 0,$$

$$\frac{d}{dt} f(x', x, t) = 0,$$

$$\vdots$$

$$\frac{d^{\mu-1}}{dt^{\mu-1}} f(x', x, t) = 0,$$

uniquely determines Qx as a function of (Px, t), provided that the rank conditions[1] introduced in Definition 2 of [10] are given.

Due to (1.4), we assume that there exists a function φ_1 such that, locally,

$$(Px)' = \varphi_1(x, t)$$

holds. If, according to Definition 2.2, there exists another function φ_2 such that

$$Qx = \varphi_2(Px, t),$$

and sufficient smoothness is given, then one further differentiation provides

$$(Qx)' = \varphi_3((Px)', Px, t) = \tilde{\varphi}_3(x, t).$$

Consequently, if μ is the differentiation index according to Definition 2.2 and sufficient smoothness is given, then the conventional differentiation index ν results to be μ as well.

To compute the index μ in this context, for $z_i \in \mathbb{R}^n$, $i = 0, \ldots, k$, we denote by

$$G_{(z_0)}^{[k]}(z_0, z_1, \ldots, z_k, t) \in \mathbb{R}^{nk \times n}$$

---

[1]These rank conditions correspond to the assumption that the projectors we introduce later on in Sect. 3 to characterize the different types of components have all constant rank in a so-called regularity region, cf. [10].

the Jacobian matrix of $g^{[k]}(z_0, z_1, \ldots, z_k, t)$ with respect to z_0, and consider at $(z_0^*, z_1^*, \ldots, z_k^*, t^*)$ the matrix

$$
\mathcal{B}^{[k]} := \begin{pmatrix} P(t^*) & 0 \\ G_{(z_0)}^{[k]}(z_0^*, z_1^*, \ldots, z_k^*, t^*) & G_{(z_1, \ldots, z_k)}^{[k]}(z_0^*, z_1^*, \ldots, z_k^*, t^*) \end{pmatrix}.
$$

According to [12], we check if the matrices $\mathcal{B}^{[k]} \in \mathbb{R}^{n(k+1) \times n(k+1)}$ are 1-full with respect to the first n columns for $k = 1, 2, \ldots$, i.e., whether

$$
\ker \mathcal{B}^{[k]} \subseteq \left\{ \begin{pmatrix} s_0 \\ s_1 \end{pmatrix} : s_0 \in \mathbb{R}^n, \ s_0 = 0, \ s_1 \in \mathbb{R}^{nk} \right\}. \tag{2.3}
$$

We conclude that at $(z_0^*, z_1^*, \ldots, z_k^*, t^*)$ the index is μ, if the constant rank assumptions are given in a neighborhood of $(z_0^*, z_1^*, \ldots, z_k^*, t^*)$ and μ is the smallest integer for which $\mathcal{B}^{[\mu]}$ is 1-full. We emphasize that $g^{[\mu]}$ consists of $f, F_1, \ldots, F_{\mu-1}$, such that no μ-th differentiation is needed.

Recall further that the rank conditions from Definition 2 of [10] were introduced for linear DAEs. For nonlinear DAEs we consider the linearization, following the reasoning that the nonlinear DAE has index μ iff the linearized DAE has it, cf. [21].

3 Defining Projectors with the Derivative Array

In order to characterize the different components, for

$$
G_L^{[k]} := G_{(z_0)}^{[k]}(z_0^*, z_1^*, \ldots, z_k^*, t^*)
$$
$$
G_R^{[k]} := G_{(z_1, \ldots, z_k)}^{[k]}(z_0^*, z_1^*, \ldots, z_k^*, t^*)
$$

we have a closer look onto the matrix

$$
G^{[k]} := \begin{pmatrix} G_L^{[k]} & G_R^{[k]} \end{pmatrix}, \tag{3.1}
$$

where L and R stand for left- and right-hand side, respectively. For the sake of simplicity, in the following we also drop the argument t^* for the projectors and Q and P.

– To decouple the undifferentiated component Qx for $k = 1, \ldots, \mu$, we consider an orthogonal basis[2] $B_R^{[k]}$ with the property $\ker B_R^{[k]} = \operatorname{im} G_R^{[k]}$ and define the projector T_k as the orthogonal projector onto

$$\ker \begin{pmatrix} P \\ B_R^{[k]} G_L^{[k]} \end{pmatrix} =: \operatorname{im} T_k.$$

Consequently, $T_k x$ corresponds to the part of the undifferentiated component Qx that, after k-1 differentiations, cannot yet be represented as a function of (Px, t). Note that, by definition, $T_k \neq 0$ for $k < \mu$ and $T_\mu = 0$, cf. [10].

– To characterize the different parts of the differentiated component Px, we further decouple $G^{[k]}$ in each step k into $G_L^{[k]} P$ and $G_L^{[k]} Q$ and consider

$$\begin{pmatrix} Q & 0 & 0 \\ G_L^{[k]} P & G_L^{[k]} Q & G_R^{[k]} \end{pmatrix}.$$

With this decoupling from [11] in mind, we consider an orthogonal basis[3] $B_{LQ-R}^{[k]}$ with

$$\ker B_{LQ-R}^{[k]} = \operatorname{im} \begin{pmatrix} G_L^{[k]} Q & G_R^{[k]} \end{pmatrix}$$

and finally define the orthogonal projector V_k onto

$$\ker \begin{pmatrix} Q \\ B_{LQ-R}^{[k]} G_L^{[k]} \end{pmatrix} =: \operatorname{im} V_k.$$

Then $V_k x$ represents the part of the differentiated components Px that is not determined by the constraints resulting after k-1 differentiations. By construction, the degree of freedom d is rank V_μ. In accordance with our previous work we define

$$\Pi := V_\mu.$$

---

[2] Instead of a basis, any matrix $W_R^{[k]}$ with $\ker W_R^{[k]} = \operatorname{im} G_R^{[k]}$ could be used in this context, especially a projector. According to our implementation in InitDAE [8, 13] , we consider a basis here.

[3] Again, instead of a basis, any matrix $W_{LQ-R}^{[k]}$ with $\ker W_{LQ-R}^{[k]} = \operatorname{im} G_{LQ-R}^{[k]}$ could be used in this context, analogously as for $B_R^{[k]}$.

Note that, also by construction, we have $QV_k = 0$ for all k and, hence, $Z_k = P - V_k$ results to be a projector:

$$Z_k \cdot Z_k = (P - V_k)(P - V_k) = P - 2 \cdot PV_k + V_k = P - V_k = Z_k.$$

Consequently, $Z_k x$ describes the differentiated components that are determined by constraints resulting after k-1 differentiation and, in particular, $(P - \Pi)x = Z_\mu x$ corresponds to the differentiated components that are determined by constraints after μ-1 differentiations.

According to Theorem 1 in [10], it holds

$$T_k = Q_0 T_k = T_k Q_0 = T_{k-1} T_k = T_k T_{k-1}, \tag{3.2}$$

and it can be proved analogously that

$$V_k = P_0 V_k = V_k P_0 = V_{k-1} V_k = V_k V_{k-1}. \tag{3.3}$$

Therefore, for $Z_k = P - V_k$, $U_k := Q - T_k$, $x = Px + Qx$ we can consider the decoupling

$$Px = PZ_1 x + V_1 Z_2 x + V_2 Z_3 x + \ldots + V_{\mu-2} Z_{\mu-1} x + \Pi x, \tag{3.4}$$

$$Qx = Q_0 U_1 x + T_1 U_2 x + T_2 U_3 x + \ldots + T_{\mu-2} U_{\mu-1} x + T_{\mu-1} x. \tag{3.5}$$

Example 3.1 Let us consider the DAE resulting from the exothermic reactor model (cf. [25]), also described in [2]:

$$C' = K_1(C_0 - C) - R,$$

$$T' = K_1(T_0 - T) + K_2 R - K_3(T - T_C),$$

$$0 = R - K_3 e^{-\frac{K_4}{T}} C,$$

$$0 = C - u,$$

where K_1, K_2, K_3, K_4 are constants, C_0 and T_0 are the feed reactant concentration and feed temperature (assumed to be known functions). The variables C and T are the corresponding quantities in the product, $u(t)$ is an input function prescribing C, R is the reaction rate per unit volume, and T_C is the temperature of the cooling medium. The corresponding projectors can be found in Table 1. The index is three and since $\Pi = 0$, the degree of freedom is zero and no initial values can be prescribed in this case.

Note that although for this nonlinear example all the projectors from Table 1 are constant, in general they may depend on $(z_0^*, z_1^*, \ldots, z_k^*, t^*)$.

Table 1 Projectors associated with the derivative array analysis for the exothermic reactor model (Example 3.1)

$x = (C, T, R, T_C)$		
A	$Q = \begin{pmatrix} 0&0&0&0 \\ 0&0&0&0 \\ 0&0&1&0 \\ 0&0&0&1 \end{pmatrix},$	$P = \begin{pmatrix} 1&0&0&0 \\ 0&1&0&0 \\ 0&0&0&0 \\ 0&0&0&0 \end{pmatrix}$
$G^{[1]}$	$T_1 = \begin{pmatrix} 0&0&0&0 \\ 0&0&0&0 \\ 0&0&0&0 \\ 0&0&0&1 \end{pmatrix},$	$V_1 = \begin{pmatrix} 0&0&0&0 \\ 0&1&0&0 \\ 0&0&0&0 \\ 0&0&0&0 \end{pmatrix}$
$G^{[2]}$	$T_2 = \begin{pmatrix} 0&0&0&0 \\ 0&0&0&0 \\ 0&0&0&0 \\ 0&0&0&1 \end{pmatrix},$	$V_2 = \begin{pmatrix} 0&0&0&0 \\ 0&0&0&0 \\ 0&0&0&0 \\ 0&0&0&0 \end{pmatrix}$
$G^{[3]}$	$T_3 = \begin{pmatrix} 0&0&0&0 \\ 0&0&0&0 \\ 0&0&0&0 \\ 0&0&0&0 \end{pmatrix},$	$V_3 = \begin{pmatrix} 0&0&0&0 \\ 0&0&0&0 \\ 0&0&0&0 \\ 0&0&0&0 \end{pmatrix} =: \Pi$

4 Properties of the Orthogonal Projector Π

To simplify the notation, we introduce matrices N and W fulfilling

$$N := B_R^{[\mu]} G_L^{[\mu]}, \quad \ker W = \operatorname{im} NQ, \tag{4.1}$$

where W can be an arbitrary matrix (e.g. an orthogonal basis or projector). Consequently, the orthogonal projector Π fulfills

$$\ker \begin{pmatrix} Q \\ WN \end{pmatrix} = \ker Q \cap \ker WN = \operatorname{im} \Pi.$$

According to the index definition, it further holds

$$\ker \begin{pmatrix} P \\ N \end{pmatrix} = \ker \begin{pmatrix} \Pi \\ N \end{pmatrix} = \{0\}.$$

Consequently, Lemma A.1 from the Appendix implies that there exists a function φ_4 such that

$$(I - \Pi)x = \varphi_4(\Pi x, t). \tag{4.2}$$

In [11, 12] we have shown that, under suitable assumptions, the constrained optimization problem

$$\min \quad \left\| P(z_0 - \alpha) \right\|_2 \tag{4.3}$$

$$\text{subject to} \quad g^{[\mu]}(z_0, z_1, \ldots, z_\mu, t_0) = 0, \tag{4.4}$$

turns out to compute consistent initial values fulfilling

$$\Pi(z_0 - \alpha) = 0.$$

In this sense, the consistent initialization computed by (4.3)–(4.4) corresponds to

$$z_0 = \Pi\alpha + \varphi_4(\Pi\alpha, t_0).$$

In the following, we pursue this idea for linear DAEs in order to obtain an associated ODE explicitly.

5 Linear DAEs

In this section we consider linear DAEs with constant or time-dependent coefficient matrices of the form

$$A(t)x' + B(t)x = q(t), \tag{5.1}$$

which are regular on an open finite interval $\mathcal{I}$ according to the definition introduced in [10]. This implies that all the projectors introduced in the above sections have constant rank on $\mathcal{I}$ and can be interpreted as projector functions in dependence of t. Recall further that therefore this regularity assumption also excludes so-called harmless critical points like the one described in Example 2.71 from [21]. With the notation from (4.1), the explicit and hidden constraints can then be described in terms of

$$N(t)x = s(t) := B_R^{[\mu]}(t) \begin{pmatrix} q(t) \\ q'(t) \\ \vdots \\ q^{(\mu-1)}(t) \end{pmatrix}. \tag{5.2}$$

Recall further that for linear DAEs

$$W(t)N(t)x = W(t)N(t)P(t)x = W(t)s(t)$$

represents the constraints that restrict $P(t)x$. For simplicity, we will drop the argument t in the following.

For our purposes, we basically consider the orthogonal splittings

$$P = P\Pi + P(I - \Pi) = \Pi + (P - \Pi), \quad I = \Pi + (I - \Pi)$$

and assume that the coefficients A and B of (5.1) are as smooth as needed for the pseudo-inverses used below. For a detailed discussion on the properties of time-dependent pseudo-inverses in an analogous context we refer to [21, Proposition A.17], [23].

Note that there are some relations between Π and the projector $\Pi_{\mu-1}$ from Chapter 2.4.2 in [21]. In fact, by definition, for index-2 DAEs $I - \Pi$ results to be the orthogonal projector along im $\Pi_1 = \ker(I - \Pi_1)$, i.e., $\ker(I - \Pi) = \ker(I - \Pi_1)$.

In the case that the index is greater than two, the relationship between Π and $\Pi_{\mu-1}$ seems to be more complex. For a better appraisal, we start comparing the definitions of explicit ODEs related to a DAE that result from the different concepts.

5.1 On Explicit ODEs Associated with a DAE

In the literature, there are several explicit ODEs that are associated with DAEs, in particular:

- The completion ODE, or underlying ODE, is an explicit ODE for the complete vector x that is associated with the differential index concept. It can be extracted from the derivative array (cf., e.g., [2, 19] and the references therein) and depends on the derivatives of q up to the order μ:

$$x' = \varphi_c(x, q, q', \ldots, q^{(\mu)}),$$

for a suitable function φ_c.
- The inherent explicit regular ODE (IERODE) is closely related to the tractability index concept. It is formulated for $u_i := \Pi_{\mu-1}x \in \mathbb{R}^n$, where $\Pi_{\mu-1}$ is a suitably defined projector fulfilling rank $\Pi_{\mu-1} = d$. It lives in $\mathbb{R}^n$, $n \geq d$ and is unique in the scope of fine decoupling (see [21, 24] and the references therein). The projector $\Pi_{\mu-1}$ is precisely chosen such that the IERODE does not depend on derivatives of q, i.e.,

$$(\Pi_{\mu-1}x)' = \varphi_i(\Pi_{\mu-1}x, q), \quad \text{or,} \quad u_i' = \varphi_i(u_i, q),$$

for $u_i : \mathcal{I} \to \mathbb{R}^n$ and a suitable function φ_i.

– An essential underlying ODEs (EUODEs) has minimal size d (cf. [1, 24] and the references therein). There may be several EUODEs living in a transformed space with dimension d. EUODEs are also free of derivatives of q and can be considered a condensed IERODE, cf. [24]. We will represent EUODEs in terms of

$$u'_e = \varphi_e(u_e, q)$$

for $u_e : \mathcal{I} \to \mathbb{R}^d$ and a suitable function φ_e.

In this section, we consider a closely related definition of explicit ODEs:

– An orthogonally projected explicit ODE (OPE-ODE) obtained from the derivative array of a DAE is the explicit ODE formulated for $u_p := \Pi x \in \mathbb{R}^n$ for the orthogonal projector Π discussed in Sect. 4. An OPE-ODE lives in $\mathbb{R}^n$, $n \geq \operatorname{rank} \Pi = d$ and may depend on derivatives of q up to the order μ:

$$(\Pi x)' = \varphi_p(\Pi x, q, q', \dots, q^{(\mu)}),$$

or

$$u'_p = \varphi_p(u_p, q, q', \dots, q^{(\mu)}),$$

for $u_p : \mathcal{I} \to \mathbb{R}^n$ and a suitable function φ_p.
– Essential orthogonally projected explicit ODEs (EOPE-ODEs) are corresponding condensed OPE-ODEs with minimal size rank Π. They can also depend on derivatives of q in general. We will represent EOPE-ODEs in terms of

$$u'_{ep} = \varphi_{ep}(u_{ep}, q, q', \dots, q^{(\mu)})$$

for $u_{ep} : \mathcal{I} \to \mathbb{R}^d$ for $d = \operatorname{rank} \Pi$ and a suitable function φ_{ep}.

The following Lemma generalizes Lemma 2.27 from [21], that is formulated there for IERODEs i.e. for $\Pi_p = \Pi_{\mu-1}$, in a more general manner, such that we can apply it also for OPE-ODEs, i.e. for $\Pi_p = \Pi$:

Lemma 5.1 *Let Π_p be a projector with $\Pi_p \in C^1(\mathcal{I}, \mathbb{R}^{n \times n})$ and $u \in C^1(\mathcal{I}, \mathbb{R}^n)$ be a solution of an ODE of the form*

$$u' - \Pi'_p u + \Pi_p C(t) u = \Pi_p c(t) \tag{5.3}$$

for suitable $C(t), c(t), t \in \mathcal{I}$. Then the subspace im Π_p *is an invariant subspace for the ODE (5.3), i.e., the following assertion is valid for the solutions $u \in C^1(\mathcal{I}, \mathbb{R}^n)$:*

$u(t_*) \in$ im $\Pi_p(t_*)$, *with a certain $t_* \in \mathcal{I} \Leftrightarrow u(t) \in$ im $\Pi_p(t)$ for all $t \in \mathcal{I}$.*

Proof This proof follows the steps of Lemma 2.27 in [21], which traces back to [18]. Let $\bar{u} \in \mathcal{C}^1(\mathcal{I}, \mathbb{R}^n)$ denote the unique solution of

$$\bar{u}' - \Pi_p'(t)\bar{u} + \Pi_p(t)C(t)\bar{u} = \Pi_p(t)c(t), \tag{5.4}$$

$$\bar{u}(t_*) = \Pi_p(t_*)\alpha \tag{5.5}$$

for an arbitrary $\alpha \in \mathbb{R}^n$. If we multiply (5.4) and (5.5) by $(I - \Pi_p(t))$ and $(I - \Pi_p(t_*))$, respectively, then we obtain

$$(I - \Pi_p(t))\bar{u}' - (I - \Pi_p(t))\Pi_p'(t)\bar{u} = 0,$$

$$(I - \Pi_p(t_*))\bar{u}(t_*) = 0.$$

For the function $\bar{v} := (I - \Pi_p)\bar{u} \in \mathcal{C}^1(\mathcal{I}, \mathbb{R}^n)$ with

$$\bar{v}' = (I - \Pi_p)'\bar{u} + (I - \Pi_p)\Pi_p'\bar{u}'$$

then

$$0 = \bar{v}' - (I - \Pi_p)'\bar{u} - \underbrace{(I - \Pi_p)\Pi_p'}_{-(I-\Pi_p)'\Pi_p}\bar{u} = \bar{v}' - (I - \Pi_p)'(I - \Pi_p)\bar{u}$$

and, therefore, $\bar{v}' - (I - \Pi_p)'\bar{v} = 0$ and $\bar{v}(t_*) = 0$ hold. Consequently, $\bar{v}$ vanishes identically, implying $\bar{u} = \Pi_p u(t)$.   □

In the following, we focus in OPE-ODEs since they are specially relevant for the analysis of the Taylor series method discussed in [13]. Since automatic differentiation is used there, the higher order derivatives can perfectly be handled for sufficiently smooth DAEs. This is a fundamental difference to other integration schemes, which require a special treatment of these derivatives in general.

5.2   A Closer Look at the Constraints

With the results from the Appendix, the constraints (5.2) can be split into different parts with regard to $P(t)$ and $\Pi(t)$. Again, we assume that all pseudo-inverses are as smooth as needed and drop the argument t for the sake of simplicity.

– On the one hand, we consider the constraints for Px

$$WNx = Ws, \tag{5.6}$$

which leads to

$$(P - \Pi)x = (WN)^+(WN)x = (WN)^+Ws, \tag{5.7}$$

where $(\ldots)^+$ denotes the Moore-Penrose inverse.
- On the other hand, we reformulate (5.2), obtaining

$$N(I - \Pi)x = s - N\Pi x.$$

According to Corollary A.5 from the Appendix, the multiplication by $(N(I - \Pi))^+$ provides the representation

$$(I - \Pi)x = \left(N(I - \Pi)\right)^+ (s - N\Pi x). \tag{5.8}$$

Note that this particularly yields

$$Qx = Q\left(N(I - \Pi)\right)^+ (s - N\Pi x).$$

Therefore, if Πx is known, then $(I - \Pi)x$ can be computed accordingly. On that account, we deduce a projected explicit ODE for Πx in the following.

5.3 Obtaining an Orthogonally Projected Explicit ODE for $u = \Pi x$

Now we show how to obtain a orthogonally projected explicit ODE (OPE-ODE) for Πx in four steps.

(i) *Reformulation of the derivative with the projector P*
   For this step, we suppose that there exists a matrix-valued function $\hat{A}(t)$ such that $\hat{A}(t)A(t) = P(t)$. To construct such a matrix different approaches can may be possible:

   - If A is constant, for $r = \operatorname{rank} A$ and the SVD

   $$A = U\operatorname{diag}(\sigma_1, \ldots, \sigma_r, 0, \ldots, 0)V^T,$$

   the nonsingular matrix $\hat{A}$ can be defined by

   $$\hat{A} = V\operatorname{diag}(\frac{1}{\sigma_1}, \ldots, \frac{1}{\sigma_r}, 1, \ldots, 1)U^T,$$

   since then the property $\hat{A}A = P$ is given by construction. If $A(t)$ is a time-dependent matrix with constant rank whose elements are analytic functions of t, then $\hat{A}(t)$ may be constructed analogously as above using the analytic SVD, cf. [3].

- In case that

$$A(t) = P(t)C(t)P(t)$$

holds for a positive definite matrix $C(t)$, then also the nonsingular matrix

$$\hat{A}(t) := \left(P(t)C(t)P(t) + Q(t)\right)^{-1}$$

can be considered.

- In general, a singular matrix $\hat{A}(t)$ can be defined by

$$(A^T(t)A(t) + Q(t))^{-1}A^T(t),$$

since

$$\begin{aligned}P(t) &= (A^T(t)A(t) + Q(t))^{-1}(A^T(t)A(t) + Q(t))P(t)\\ &= (A^T(t)A(t) + Q(t))^{-1}A^T(t)A(t).\end{aligned}$$

By definition, the multiplication of (5.1) by $\hat{A}(t)$ leads to

$$(Px)' + B_{(i)}x = q_{(i)} \tag{5.9}$$

for

$$B_{(i)} := \hat{A}B + P', \quad q_{(i)} := \hat{A}q,$$

where we dropped again the arguments for readability, and drop them also in the next steps.

Note that for a nonsingular matrix $\hat{A}$, (5.9) is a DAE and that for a singular matrix $\hat{A}$ it is only the part of a DAE which is required for our forthcoming considerations.

(ii) *Reformulation of the derivative with the projector Π*

If we use Eq. (5.7) and the splitting

$$(Px)' = (\Pi x)' + ((P - \Pi)x)',$$

then Eq. (5.9) leads to

$$(\Pi x)' + B_{(i)}x = q_{(ii)} \tag{5.10}$$

for

$$q_{(ii)} = q_{(i)} - ((WN)^+ Ws)',$$

provided that $((WN)^+ Ws)$ is differentiable.

(iii) *Formulation of an ODE in terms of Πx*

With Eq. (5.8), in Eq. (5.10) we consider the splitting

$$B_{(i)}x = B_{(i)}(\Pi x + \underbrace{\left(N(I - \Pi)\right)^+ (s - N\Pi x)}_{=(I-\Pi)x}).$$

Consequently, for

$$B_{(iii)} := B_{(i)}(I - \left(N(I - \Pi)\right)^+ N)\Pi,$$

$$q_{(iii)} := q_{(ii)} - B_{(i)}\left(N(I - \Pi)\right)^+ s,$$

we obtain the ODE

$$(\Pi x)' + B_{(iii)}(\Pi x) = q_{(iii)}. \tag{5.11}$$

(iv) *Formulation of an invariant ODE for $u = \Pi x$*

If we finally multiply (5.11) by $\Pi(t)$, suppose that Π is differentiable and use $(\Pi x)' = (\Pi \Pi x)' = \Pi'(\Pi x) + \Pi(\Pi x)'$, the orthogonally projected explicit ODE (OPE-ODE)

$$(\Pi x)' - \Pi'(\Pi x) + \Pi C(t)(\Pi x) = \Pi c(t) \tag{5.12}$$

results for

$$C(t) = B_{(iii)} = (\hat{A}B + P')(I - (N(I - \Pi))^+ N)\Pi \tag{5.13}$$

$$c(t) = q_{(iii)} = \hat{A}q - ((WN)^+ Ws)' - (\hat{A}B + P')(N(I - \Pi))^+ s \tag{5.14}$$

in the invariant subspace im Π, cf. Lemma 5.1.

Summarizing, we have proved the following result:

Theorem 5.2 *Let the DAE (5.1) be regular with index μ such that the constraints can be described by (5.2). Let us further assume that the coefficients A and B of (5.1) are as smooth as needed for the used pseudo-inverses, leading to sufficiently smooth expressions, in particular to differentiable $((WN)^+ Ws)$, and smoothly differentiable P and Π. Then a solution $x = \Pi x + (I - \Pi)x$ of the DAE can be determined*

– *considering an initial value problem for the ODE (5.12) in the invariant subspace* im Π *in order to obtain Πx, and*
– *computing $(I - \Pi)x$ afterwards according to (5.8).*

Remark 5.1 In general we allow for (5.6) that

$$WNx = Ws = \phi(q, \ldots, q^{(\mu-1)})$$

and, therefore,

$$c(t) = \hat{c}(t, q, q', \ldots, q^{(\mu)}),$$

such that φ_p and φ_{ep} may depend on derivatives of q up to order μ. However, for the classes of DAEs inspected rigorously in [9, 10] we obtain $V_\mu = V_{\mu-1}$, consequently $Z_\mu = Z_{\mu-1}$, and therefore

$$WNx = Ws = \phi(q, \ldots, q^{(\mu-2)})$$

and

$$c(t) = \hat{c}(t, q, q', \ldots, q^{(\mu-1)}).$$

This holds particularly for properly stated linear DAEs of index $\mu \leq 2$ and linear DAEs with constant coefficient matrices with an arbitrary index. Consequently, for these classes of DAEs, φ_p and φ_{ep} depend on derivatives of q up to order μ-1.

5.4 Illustrative Examples

Example 5.1 We start illustrating our approach with a small index-2 example, which is slightly more general than the one discussed in [10].

$$\underbrace{\begin{pmatrix} 1 & 1 & 0 \\ 1 & 2 & 0 \\ 0 & 0 & 0 \end{pmatrix}}_{A} \begin{pmatrix} x_1 \\ x_2 \\ x_3 \end{pmatrix}' + \underbrace{\begin{pmatrix} 1 & 0 & a \\ 1 & 1 & 1 \\ 1 & 2 & 0 \end{pmatrix}}_{B} \begin{pmatrix} x_1 \\ x_2 \\ x_3 \end{pmatrix} = \begin{pmatrix} q_1 \\ q_2 \\ q_3 \end{pmatrix} \tag{5.15}$$

for functions $q_1(t)$, $q_2(t)$, $q_3(t)$ and a parameter a. According to the analysis shown in Table 2, the differentiation index is 2 and the constraints can be described by

$$\underbrace{\begin{pmatrix} 1 & 2 & 0 \\ 1 & 1 & 1 \end{pmatrix}}_{=:N} \begin{pmatrix} x_1 \\ x_2 \\ x_3 \end{pmatrix} = \underbrace{\begin{pmatrix} q_3 \\ q_2 - q_3' \end{pmatrix}}_{=:s}.$$

Table 2 Projectors associated with the derivative array analysis for Example 5.1

	$x = (x_1, x_2, x_3)$	
A	$Q = \begin{pmatrix} 0 & 0 & 0 \\ 0 & 0 & 0 \\ 0 & 0 & 1 \end{pmatrix}$,	$P = \begin{pmatrix} 1 & 0 & 0 \\ 0 & 1 & 0 \\ 0 & 0 & 0 \end{pmatrix}$
$G^{[1]}$	$T_1 = \begin{pmatrix} 0 & 0 & 0 \\ 0 & 0 & 0 \\ 0 & 0 & 1 \end{pmatrix}$,	$V_1 = \frac{1}{5}\begin{pmatrix} 4 & -2 & 0 \\ -2 & 1 & 0 \\ 0 & 0 & 0 \end{pmatrix}$
$G^{[2]}$	$T_2 = \begin{pmatrix} 0 & 0 & 0 \\ 0 & 0 & 0 \\ 0 & 0 & 0 \end{pmatrix}$,	$V_2 = \frac{1}{5}\begin{pmatrix} 4 & -2 & 0 \\ -2 & 1 & 0 \\ 0 & 0 & 0 \end{pmatrix} =: \Pi$

Consequently,

$$NQ = \begin{pmatrix} 0 & 0 & 0 \\ 0 & 0 & 1 \end{pmatrix}, \quad W = \begin{pmatrix} 1 & 0 \end{pmatrix}, \quad WN = WNP = \begin{pmatrix} 1 & 2 & 0 \end{pmatrix},$$

$$(WN)^+ WN = \frac{1}{5}\begin{pmatrix} 1 \\ 2 \\ 0 \end{pmatrix}\begin{pmatrix} 1 & 2 & 0 \end{pmatrix} = \frac{1}{5}\begin{pmatrix} 1 & 2 & 0 \\ 2 & 4 & 0 \\ 0 & 0 & 0 \end{pmatrix} = (P - \Pi),$$

$$N(I - \Pi) = \begin{pmatrix} 1 & 2 & 0 \\ \frac{3}{5} & \frac{6}{5} & 1 \end{pmatrix}, \quad (N(I - \Pi))^+ = \begin{pmatrix} \frac{1}{5} & 0 \\ \frac{2}{5} & 0 \\ -\frac{3}{5} & 1 \end{pmatrix},$$

$$(N(I - \Pi))^+ N(I - \Pi) = \begin{pmatrix} \frac{1}{5} & \frac{2}{5} & 0 \\ \frac{2}{5} & \frac{4}{5} & 0 \\ 0 & 0 & 1 \end{pmatrix} = (I - \Pi),$$

$$(I - (N(I - \Pi))^+ N) = \frac{1}{5}\begin{pmatrix} 4 & -2 & 0 \\ -2 & 1 & 0 \\ -2 & 1 & 0 \end{pmatrix}.$$

With

$$\hat{A} = \begin{pmatrix} 2 & -1 & 0 \\ -1 & 1 & 0 \\ 0 & 0 & 1 \end{pmatrix}, \quad \hat{A}B = \begin{pmatrix} 1 & -1 & 2a - 1 \\ 0 & 1 & 1 - a \\ 1 & 2 & 0 \end{pmatrix}$$

the OPE-ODE described by Eq. (5.12) reads

$$
\left(\frac{1}{5} \begin{pmatrix} 4 & -2 & 0 \\ -2 & 1 & 0 \\ 0 & 0 & 0 \end{pmatrix} x \right)' + (2-a)\frac{1}{5} \begin{pmatrix} 4 & -2 & 0 \\ -2 & 1 & 0 \\ 0 & 0 & 0 \end{pmatrix} x = \begin{pmatrix} r_1 \\ r_2 \\ r_3 \end{pmatrix}
$$

for

$$
\begin{pmatrix} r_1 \\ r_2 \\ r_3 \end{pmatrix} = \begin{pmatrix} 2q_1 - 2aq_2 - \frac{2}{5}(1-3a)q_3 \\ -q_1 + aq_2 + \frac{1}{5}(1-3a)q_3 \\ 0 \end{pmatrix} + \begin{pmatrix} (2a - \frac{6}{5}) \\ -(\frac{3}{5} - a) \\ 0 \end{pmatrix} q_3'.
$$

Hence, an EOPE-ODE can be formulated for $u_{ep} := 2x_1 - x_2$:

$$
u'_{ep} + (2-a)u_{ep} = -5r_2. \tag{5.16}
$$

Once this ODE is solved, the solution of the original DAE can be computed using

$$
(I - \Pi)x = \frac{1}{5} \begin{pmatrix} x_1 + 2x_2 \\ 2x_1 + 4x_2 \\ 5x_3 \end{pmatrix} = \frac{1}{5} \begin{pmatrix} q_3 \\ 2q_3 \\ q_2 - q_3' - 3q_3 - (2x_1 - x_2) \end{pmatrix} = \varphi_4(\Pi x, t).
$$

For this example, the matrices defined in [21], page 23 ff., which are part of the tractability matrix sequence, read:

$$
G_2 = \begin{pmatrix} 2a & 4a-1 & a \\ a+1 & 2a+2 & 1 \\ 1 & 2 & 0 \end{pmatrix}, \quad G_2^{-1} = \begin{pmatrix} 2 & -2a & 2a^2 - 2a + 1 \\ -1 & a & a - a^2 \\ 0 & 1 & -a - 1 \end{pmatrix},
$$

$$
\Pi_1 = \begin{pmatrix} 2 - 2a & 2 - 4a & 0 \\ a-1 & 2a-1 & 0 \\ 0 & 0 & 0 \end{pmatrix}.
$$

Consequently, for $u_i = \Pi_1 x$ the IERODE reads:

$$
u'_i + \begin{pmatrix} 2a^2 - 6a + 4 & 4a^2 - 10a + 4 & 0 \\ -a^2 + 3a - 2 & -2a^2 + 5a - 2 & 0 \\ 0 & 0 & 0 \end{pmatrix} u_i = \begin{pmatrix} 2 & -2a & 2a^2 - 2a + \frac{4}{5} \\ -1 & a & -a^2 + a - \frac{2}{5} \\ 0 & 0 & 0 \end{pmatrix} \begin{pmatrix} q_1 \\ q_2 \\ q_3 \end{pmatrix}
$$

Since it holds for $u_e = (a - 1)x_1 + (2a - 1)x_2$ that

$$\Pi_1 x = u_i = \begin{pmatrix} -2u_e \\ u_e \\ 0 \end{pmatrix},$$

it suffices to consider the EUODE

$$u'_e + (-a^2 + 3a - 2)(-2u_e) + (-2a^2 + 5a - 2)u_e = -q_1 + aq_2 + (-a^2 + a - \tfrac{2}{5})q_3,$$

i.e.,

$$u'_e + (2 - a)u_e = -q_1 + aq_2 + (-a^2 + a - \tfrac{2}{5})q_3.$$

Note that in contrast to (5.16), this ODE does not depend on derivatives of the right-hand side.

Example 5.2 Let us consider again Example 1.1. Since $T_3 \neq 0$ and $T_4 = 0$, the index is 4 and we obtain $\Pi = V_4$, cf. Table 3. Consequently, the associated EOPE-ODE we obtain coincides with the one discussed in [22, 24]:

$$x'_1 - \alpha x_1 = q_1 + q_{2,1} - (q_{2,2} - (q_{2,4} - q'_{2,4})')'.$$

In contrast, according to [22, 24], with

$$G_4 = \begin{pmatrix} 1 & -1 & \alpha & -\alpha^2 & \alpha^3 \\ 0 & 1 & 1 & 0 & 0 \\ 0 & 0 & 1 & 1 & 0 \\ 0 & 0 & 0 & 1 & 1 \\ 0 & 0 & 0 & 0 & 1 \end{pmatrix}, \quad \Pi_3 := \begin{pmatrix} 1 & 0 & 1 & -\alpha & \alpha^2 \\ 0 & 0 & 0 & 0 & 0 \\ 0 & 0 & 0 & 0 & 0 \\ 0 & 0 & 0 & 0 & 0 \\ 0 & 0 & 0 & 0 & 0 \end{pmatrix}$$

the EUODE (without derivatives of q) results to be

$$u'_e - \alpha u_e = q_1 + q_{2,1} - \alpha q_{2,2} + \alpha^2 q_{2,3} - \alpha^3 q_{2,4}$$

for

$$u_e = x_1 + x_3 - \alpha x_4 + \alpha^2 x_5.$$

Remark 5.2 Observe that, as expected, in Examples 5.1 and 5.2 the spectra of the EUODE and the EOPE-ODE coincide. This has to be given due to stability reasons. Indeed, for $q_1 \equiv q_2 \equiv q_3 \equiv q_4 \equiv q_5 \equiv 0$ we obtain $x_2 \equiv x_3 \equiv x_4 \equiv x_5 \equiv 0$ and therefore $u_e(t) = x_1(t)$ for all t.

A more general class of linear DAEs with constant coefficients that includes Example 5.2 is discussed in the next Sect. 5.5, see Example 5.3.

Table 3 Projectors associated with the derivative array analysis for Example 5.2

		$x = (x_1, x_2, x_3, x_4, x_5)$
A		$Q = \begin{pmatrix} 0\,0\,0\,0\,0 \\ 0\,1\,0\,0\,0 \\ 0\,0\,0\,0\,0 \\ 0\,0\,0\,0\,0 \\ 0\,0\,0\,0\,0 \end{pmatrix}, \; P = \begin{pmatrix} 1\,0\,0\,0\,0 \\ 0\,0\,0\,0\,0 \\ 0\,0\,1\,0\,0 \\ 0\,0\,0\,1\,0 \\ 0\,0\,0\,0\,1 \end{pmatrix}$
$G^{[1]}$	$T_1 = \begin{pmatrix} 0\,0\,0\,0\,0 \\ 0\,1\,0\,0\,0 \\ 0\,0\,0\,0\,0 \\ 0\,0\,0\,0\,0 \\ 0\,0\,0\,0\,0 \end{pmatrix}$	$V_1 = \begin{pmatrix} 1\,0\,0\,0\,0 \\ 0\,0\,0\,0\,0 \\ 0\,0\,1\,0\,0 \\ 0\,0\,0\,1\,0 \\ 0\,0\,0\,0\,0 \end{pmatrix}$
$G^{[2]}$	$T_2 = \begin{pmatrix} 0\,0\,0\,0\,0 \\ 0\,1\,0\,0\,0 \\ 0\,0\,0\,0\,0 \\ 0\,0\,0\,0\,0 \\ 0\,0\,0\,0\,0 \end{pmatrix}$	$V_2 = \begin{pmatrix} 1\,0\,0\,0\,0 \\ 0\,0\,0\,0\,0 \\ 0\,0\,1\,0\,0 \\ 0\,0\,0\,0\,0 \\ 0\,0\,0\,0\,0 \end{pmatrix}$
$G^{[3]}$	$T_3 = \begin{pmatrix} 0\,0\,0\,0\,0 \\ 0\,1\,0\,0\,0 \\ 0\,0\,0\,0\,0 \\ 0\,0\,0\,0\,0 \\ 0\,0\,0\,0\,0 \end{pmatrix}$	$V_3 = \begin{pmatrix} 1\,0\,0\,0\,0 \\ 0\,0\,0\,0\,0 \\ 0\,0\,0\,0\,0 \\ 0\,0\,0\,0\,0 \\ 0\,0\,0\,0\,0 \end{pmatrix}$
$G^{[4]}$	$T_4 = \begin{pmatrix} 0\,0\,0\,0\,0 \\ 0\,0\,0\,0\,0 \\ 0\,0\,0\,0\,0 \\ 0\,0\,0\,0\,0 \\ 0\,0\,0\,0\,0 \end{pmatrix}$	$V_4 = \begin{pmatrix} 1\,0\,0\,0\,0 \\ 0\,0\,0\,0\,0 \\ 0\,0\,0\,0\,0 \\ 0\,0\,0\,0\,0 \\ 0\,0\,0\,0\,0 \end{pmatrix} =: \Pi$

5.5 *Examples for Linear DAEs*

To facilitate the understanding of our approach, we show the differences between

- the introduced orthogonal decoupling, leading to an OEPE-ODE that involves derivatives of the right-hand side, and
- a decoupling leading to an IERODE that precisely does not involve any derivatives of the right-hand side

for the Kronecker Canonical Form and a slightly more general class of DAEs, which particularly includes Examples 1.1 and 5.2.

5.5.1   Kronecker Canonical Form (KCF)

A linear differential-algebraic equation with constant coefficients and regular matrix pair can be transformed by a premultiplication of a nonsingular matrix and a linear coordinate change into a DAE in Kronecker canonical form (KCF), i.e., a DAE of the form

$$\begin{pmatrix} I_{n_1} & 0 \\ 0 & \mathcal{N} \end{pmatrix} x' + \begin{pmatrix} \mathcal{W} & 0 \\ 0 & I_{n_2} \end{pmatrix} x = q(t) \tag{5.17}$$

for $x(t) \in \mathbb{R}^n$, an arbitrary $\mathcal{W} \in \mathbb{R}^{n_1 \times n_1}$, a nilpotent matrix $\mathcal{N} \in \mathbb{R}^{n_2 \times n_2}$ with nilpotency-index μ, i.e., $\mathcal{N}^{\mu-1} \neq 0$, $\mathcal{N}^{\mu} = 0$, $n = n_1 + n_2$, and identity matrices $I_{n_1} \in \mathbb{R}^{n_1 \times n_1}$ and $I_{n_2} \in \mathbb{R}^{n_2 \times n_2}$, cf. [17]. Rewriting the equations as

$$x_1' + \mathcal{W}x_1 = q_1(t), \tag{5.18}$$

$$\mathcal{N}x_2' + x_2 = q_2(t), \tag{5.19}$$

for $x_1(t) \in \mathbb{R}^{n_1}$, $x_2(t) \in \mathbb{R}^{n_2}$, Eq. (5.18) corresponds to the inherent ODE and, by a recursive approach, the so-called pure DAE corresponding to Eq. (5.19) leads to the constraints

$$x_2 = q_2(t) - \mathcal{N}x_2' = q_2(t) - \mathcal{N}(q_2'(t) - \mathcal{N}x_2') = \cdots = \sum_{j=0}^{\mu-1} (-1)^j \mathcal{N}^j q_2^{(j)}(t).$$

5.5.2 Π for DAEs in KCF

We consider $Q_{\mathcal{N}} := I - \mathcal{N}^+\mathcal{N}$, $P_{\mathcal{N}} = I - Q_{\mathcal{N}}$ and obtain the projectors

$$Q = \begin{pmatrix} 0 & \\ & Q_{\mathcal{N}} \end{pmatrix}, \quad P = \begin{pmatrix} I & \\ & P_{\mathcal{N}} \end{pmatrix}.$$

The Jacobian matrix (3.1) of the derivative array reads

$$G^{[k]} = \begin{pmatrix} \mathcal{W} & I & & & & & \\ & I & \mathcal{N} & & & & \\ & & \mathcal{W} & I & & & \\ & & & I & \mathcal{N} & & \\ & & & & \ddots & \ddots & \\ & & & & & \mathcal{W} & I \\ & & & & & I & \mathcal{N} \end{pmatrix} = \begin{pmatrix} G_L^{[k]} & G_R^{[k]} \end{pmatrix}.$$

For index μ DAEs, i.e., $\mathcal{N}^\mu = 0$, a basis $B_R^{[\mu]}$ with ker $B_R^{[\mu]} = $ im $G_R^{[\mu]}$ is given by

$$B_R^{[\mu]} = \begin{pmatrix} 0 & I & 0 & -\mathcal{N} & 0 & \mathcal{N}^2 & 0 & -\mathcal{N}^3 & \cdots & (-1)^{\mu-1}\mathcal{N}^{\mu-1} \end{pmatrix}.$$

Therefore, according to (4.1), $N := B_R^{[\mu]} G_L^{[\mu]} = \begin{pmatrix} 0 & I \end{pmatrix}$ and

$$\Pi = \begin{pmatrix} I \\ 0 \end{pmatrix}, \quad B(I - (N(I - \Pi))^+ N)\Pi = B\Pi = \begin{pmatrix} \mathcal{W} \\ 0 \end{pmatrix},$$

as expected. The corresponding projectors for the tractability index concept can be found in Section 1.2.6 from [21]. In this particular case, Π and $\Pi_{\mu-1}$ coincide and the OPE-ODE for Πx is the IERODE as well.

5.5.3 ODEs for Slightly More General DAEs

Consider the DAE

$$\begin{pmatrix} I_{n_1} & 0 \\ 0 & \mathcal{N} \end{pmatrix} x' + \begin{pmatrix} \mathcal{W}_1 & \mathcal{W}_2 \\ 0 & I_{n_2} \end{pmatrix} x = q(t). \tag{5.20}$$

Analogously as above, we obtain

$$x_2 = \sum_{j=0}^{\mu-1} (-1)^j \mathcal{N}^j q_2^{(j)}(t).$$

Obtaining the OPE-ODE for Πx corresponds to substituting this into the first block of equations, i.e.,

$$x_1' + \mathcal{W}_1 x_1 = -\mathcal{W}_2 \sum_{j=0}^{\mu-1} (-1)^j \mathcal{N}^j q_2^{(j)}(t) + q.$$

In fact, it holds

$$G^{[k]} = \begin{pmatrix} \mathcal{W}_1 & \mathcal{W}_2 & I & & & & & \\ & I & & \mathcal{N} & & & & \\ & & \mathcal{W}_1 & \mathcal{W}_2 & I & & & \\ & & & I & & \mathcal{N} & & \\ & & & & \ddots & & \ddots & \\ & & & & & \mathcal{W}_1 & \mathcal{W}_2 & I \\ & & & & & & I & \mathcal{N} \end{pmatrix} = \begin{pmatrix} G_L^{[k]} & G_R^{[k]} \end{pmatrix}.$$

Therefore, for the index μ DAEs, a basis $B_R^{[\mu]}$ with $\ker B_R^{[\mu]} = \operatorname{im} G_R^{[\mu]}$ is given again by

$$B_R^{[\mu]} = \begin{pmatrix} 0 & I & 0 & -\mathcal{N} & 0 & \mathcal{N}^2 & 0 & -\mathcal{N}^3 & \cdots & (-1)^{\mu-1}\mathcal{N}^{\mu-1} \end{pmatrix}.$$

Consequently, N, Π and $B\Pi$ are the same as above for DAEs in KCF, as expected. However, the projectors related to the tractability index concept, are different, since the OPE-ODE for Πx is not an IERODE.

For illustrative reasons, we show how the IERODE can be obtained for this particular class of DAEs without the tractability index sequence. We start noticing that we can substitute

$$x_2 = -\mathcal{N}x_2' + q_2$$

into the first block of equations, which leads to

$$\begin{pmatrix} I_{n_1} & -\mathcal{W}_2\mathcal{N} \\ 0 & \mathcal{N} \end{pmatrix} x' + \begin{pmatrix} \mathcal{W}_1 & 0 \\ 0 & I_{n_2} \end{pmatrix} x = \begin{pmatrix} q_1 - \mathcal{W}_2 q_2 \\ q_2 \end{pmatrix}.$$

This corresponds to a multiplication from the left-hand side by

$$\begin{pmatrix} I_{n_1} & -\mathcal{W}_2 \\ 0 & I_{n_2} \end{pmatrix}.$$

If we now define x_{1p} as follows

$$x = \begin{pmatrix} I & \mathcal{W}_2\mathcal{N} \\ 0 & I \end{pmatrix} \underbrace{\begin{pmatrix} I & -\mathcal{W}_2\mathcal{N} \\ 0 & I \end{pmatrix} x}_{=:x_{p1}} = \begin{pmatrix} I & \mathcal{W}_2\mathcal{N} \\ 0 & I \end{pmatrix} x_{p1},$$

then we obtain

$$\begin{pmatrix} I_{n_1} & -\mathcal{W}_2\mathcal{N} \\ 0 & \mathcal{N} \end{pmatrix} x' = \begin{pmatrix} I_{n_1} & 0 \\ 0 & \mathcal{N} \end{pmatrix} (x_{p1})'$$

and thus

$$\begin{pmatrix} I_{n_1} & 0 \\ 0 & \mathcal{N} \end{pmatrix} (x_{p1})' + \begin{pmatrix} \mathcal{W}_1 & \mathcal{W}_1\mathcal{W}_2\mathcal{N} \\ 0 & I_{n_2} \end{pmatrix} x_{1p} = \begin{pmatrix} q_1 - \mathcal{W}_2 q_2 \\ q_2 \end{pmatrix}.$$

This procedure can be repeated if we multiply from the left-hand side by

$$\begin{pmatrix} I_{n_1} & -\mathcal{W}_1\mathcal{W}_2\mathcal{N} \\ 0 & I_{n_2} \end{pmatrix}$$

to obtain

$$\begin{pmatrix} I_{n_1} & -\mathcal{W}_1\mathcal{W}_2\mathcal{N}^2 \\ 0 & \mathcal{N} \end{pmatrix}(x_{1p})' + \begin{pmatrix} \mathcal{W}_1 & 0 \\ 0 & I_{n_2} \end{pmatrix} x_{1p} = \begin{pmatrix} q_1 - \mathcal{W}_2 q_2 - \mathcal{W}_1\mathcal{W}_2\mathcal{N} q_2 \\ q_2 \end{pmatrix}.$$

If we repeat this analogously until the nilpotency index is reached, then we obtain

$$\begin{pmatrix} I_{n_1} & 0 \\ 0 & \mathcal{N} \end{pmatrix}(x_{p(\mu-1)})' + \begin{pmatrix} \mathcal{W}_1 & 0 \\ 0 & I_{n_2} \end{pmatrix} x_{p(\mu-1)} = \begin{pmatrix} q_1 - \sum_{j=0}^{\mu-1}(\mathcal{W}_1)^j\mathcal{W}_2\mathcal{N}^j q_2 \\ q_2 \end{pmatrix}$$

for

$$x_{p(\mu-1)} = \prod_{j=1}^{\mu-1} \begin{pmatrix} I_{n_1} & \mathcal{W}_1^{j-1}\mathcal{W}_2\mathcal{N}^j \\ 0 & I_{n_2} \end{pmatrix} x = \begin{pmatrix} I_{n_1} & \sum_{j=1}^{\mu-1} \mathcal{W}_1^{j-1}\mathcal{W}_2\mathcal{N}^j \\ 0 & I_{n_2} \end{pmatrix} x.$$

Example 5.3 For the Examples 1.1 and 5.2 this means

$$\mathcal{W}_1 = (-\alpha), \quad \mathcal{W}_2 = \begin{pmatrix} -1 & 0 & 0 & 0 \end{pmatrix}, \quad \mathcal{N} = \begin{pmatrix} 0 & 1 & 0 & 0 \\ 0 & 0 & 1 & 0 \\ 0 & 0 & 0 & 1 \\ 0 & 0 & 0 & 0 \end{pmatrix}, \quad \begin{pmatrix} q_1 \\ q_2 \end{pmatrix} = \begin{pmatrix} q_1 \\ q_{2,1} \\ q_{2,2} \\ q_{2,3} \\ q_{2,4} \end{pmatrix},$$

and therefore it holds

$$\left(1 \quad \sum_{j=1}^{\mu-1} \mathcal{W}_1^{j-1}\mathcal{W}_2\mathcal{N}^j\right) x = \begin{pmatrix} 1 & 0 & 1 & -\alpha & \alpha^2 \end{pmatrix} x = x_1 + x_3 - \alpha x_4 + \alpha^2 x_5,$$

$$-\sum_{j=0}^{\mu-1}(\mathcal{W}_1)^j\mathcal{W}_2\mathcal{N}^j q_2 = \begin{pmatrix} 1 & -\alpha & \alpha^2 & -\alpha^3 \end{pmatrix} q_2 = q_{2,1} - \alpha q_{2,2} + \alpha^2 q_{2,3} - \alpha^3 q_{2,4}.$$

Consequently, we obtain the IERODE and EUODE that are not a OPE-ODE or EOPE-ODE for Πx, respectively.

6 Modified Nodal Analysis (MNA)

For the equations resulting in circuit simulation with the conventional MNA, Lemma A.6 permits an easy interpretation of the representation of Π described already in [5] and the projector PQ_1 given in [16].

Using the same notation as in [5, 16], the conventional MNA for circuits without controlled sources leads to equations of the form

$$A_C C(A_C^T e, t) A_C e' + A_R r(A_R^T e, t) + A_L j_L + A_V j_V + A_I i(t) = 0,$$

$$L(j_L, t) j_L' - A_L^T e = 0,$$

$$A_V^T - v(t) = 0,$$

for incidence matrices A_C, A_R, A_V, A_L, A_I, suitable given functions C, L, r, v, i, and the unknown functions (e, j_L, j_V). If we suppose that $C(A_C^T e, t)$, $L(j_L, t)$ and $G(u, t) := \frac{\partial r(u,t)}{\partial u}$ are positive definite, in [5] it was shown that the projector Π is constant and depends only on the topological properties of the network.

For the description, we merely require projectors with

$$\text{im } Q_C = \ker A_C^T, \quad \text{im } Q_{CRV} = \ker (A_C A_R A_V)^T, \quad \text{im } \bar{Q}_{V-C} = \text{im } A_V^T Q_C.$$

Analogously to [6] we define

$$Q := \begin{pmatrix} Q_C & 0 & 0 \\ 0 & 0 & 0 \\ 0 & 0 & I \end{pmatrix}, \quad T = T_1 := \begin{pmatrix} Q_{CRV} & 0 & 0 \\ 0 & 0 & 0 \\ 0 & 0 & \bar{Q}_{V-C} \end{pmatrix},$$

but assume now that these projectors are orthogonal. Let us focus on the index-2 case, i.e. $T_1 \neq 0$, see Table 4. Due to the symmetry of the equations we can further define

$$\bar{H}_1 := \begin{pmatrix} A_C A_C^T & 0 & 0 \\ 0 & I & 0 \\ 0 & 0 & 0 \end{pmatrix} + Q$$

$$H_1(A_C^T e, j_L, t) := \begin{pmatrix} A_C C(A_C^T e, t) A_C^T & 0 & 0 \\ 0 & L(j_L, t) & 0 \\ 0 & 0 & 0 \end{pmatrix} + Q$$

$$WN := \begin{pmatrix} 0 & Q_{CRV}^T A_L & 0 \\ 0 & 0 & 0 \\ \bar{Q}_{V-C}^T A_V^T & 0 & 0 \end{pmatrix},$$

$$\bar{H}_2 := (WN)(WN)^T + (I - T)$$

Table 4 Projectors associated with the derivative array analysis for the conventional MNA equations

	$x = (e, j_L, jv)$
Index 1	
A	$Q = \begin{pmatrix} Q_C & 0 & 0 \\ 0 & 0 & 0 \\ 0 & 0 & I \end{pmatrix}, \; P = \begin{pmatrix} P_C & 0 & 0 \\ 0 & I & 0 \\ 0 & 0 & 0 \end{pmatrix}$
$G^{[1]}$	$T_1 = \begin{pmatrix} 0 & 0 & 0 \\ 0 & 0 & 0 \\ 0 & 0 & 0 \end{pmatrix}, \; V_1 = P = \begin{pmatrix} P_C & 0 & 0 \\ 0 & I & 0 \\ 0 & 0 & 0 \end{pmatrix} =: \Pi$
Index 2	
A	$Q = \begin{pmatrix} Q_C & 0 & 0 \\ 0 & 0 & 0 \\ 0 & 0 & I \end{pmatrix}, \; P = \begin{pmatrix} P_C & 0 & 0 \\ 0 & I & 0 \\ 0 & 0 & 0 \end{pmatrix}$
$G^{[1]}$	$T_1 = \begin{pmatrix} Q_{CRV} & 0 & 0 \\ 0 & 0 & 0 \\ 0 & 0 & \bar{Q}_{V-C} \end{pmatrix}, \; V_1 = P - (WN)^T \bar{H}_2^{-1}(WN)$
$G^{[2]}$	$T_2 = \begin{pmatrix} 0 & 0 & 0 \\ 0 & 0 & 0 \\ 0 & 0 & 0 \end{pmatrix}, \; V_2 = P - (WN)^T \bar{H}_2^{-1}(WN) =: \Pi$

Recall that the index is 1, iff $T_1 = 0$, i.e. $Q_{CRV} = 0$ (if there is no cut-set consisting of inductances and/or current sources only) and $\bar{Q}_{V-C} = 0$ (there is no loop consisting of capacitances and voltage sources). Oterwise, the index is 2, cf. [16]

$$= \begin{pmatrix} Q_{CRV}^T A_L A_L^T Q_{CRV} + P_{CRV} & 0 & 0 \\ 0 & I & 0 \\ 0 & 0 & \bar{Q}_{V-C}^T A_V^T A_V \bar{Q}_{V-C} + \bar{P}_{V-C} \end{pmatrix}$$

$$=: \begin{pmatrix} (\bar{H}_2)_{(1,1)} & 0 & 0 \\ 0 & I & 0 \\ 0 & 0 & (\bar{H}_2)_{(3,3)} \end{pmatrix},$$

$$H_2(A_C^T e, j_L, t) := (WN) H_1^{-1}(A_C^T e, j_L, t)(WN)^T + (I - T).$$

By construction, these matrices are nonsingular and $\bar{H}_2$ is symmetric such that the projector Π described already in [5] results to be the orthogonal projector Π, since

$$Z_2 = Z_1 = (WN)^+(WN) = (WN)^T \bar{H}_2^{-1}(WN)$$

$$= \begin{pmatrix} A_V \bar{Q}_{V-C} \left((\bar{H}_2)_{(3,3)}\right)^{-1} \bar{Q}_{V-C} A_V^T & 0 & 0 \\ 0 & A_L^T Q_{CRV} \left((\bar{H}_2)_{(1,1)}\right)^{-1} Q_{CRV}^T A_L\ 0 \\ 0 & 0 & 0 \end{pmatrix},$$

and therefore the orthogonal projector

$$\Pi = P - (WN)^T \bar{H}_2^{-1}(WN)$$

$$= \begin{pmatrix} P_C - A_V \bar{Q}_{V-C} \left((\bar{H}_2)_{(3,3)}\right)^{-1} \bar{Q}_{V-C} A_V^T & 0 & 0 \\ 0 & I - A_L^T Q_{CRV} \left((\bar{H}_2)_{(1,1)}\right)^{-1} Q_{CRV}^T A_L\ 0 \\ 0 & 0 & 0 \end{pmatrix}$$

results to be constant. In contrast, in [16] it was shown that

$$\Pi_1(A_C^T e, j_L, t) = P - \left(H_1(A_C^T e, j_L, t)\right)^{-1} (WN)^T \left(H_2(A_C^T e, j_L, t)\right)^{-1} (WN).$$

This projector is neither orthogonal nor constant in general. However, by construction it holds that ker Π = ker Π_1, cf. Lemma A.6.

7 Summary

In the present paper, we developed a new decoupling of DAEs that was obtained with orthogonal projectors and the derivative array.

The discussed projectors characterize the dependence of the different components on derivatives of the right-hand side. Moreover, they turned out to be constant for several examples from applications. Consequently, the components can be described easily and the verification of beneficial structural properties in the equations becomes simple. In fact, often higher-index components $T_k x$ appear only linearly, cf. [6], or in a restricted nonlinear form [7].

The presented decoupling of linear DAEs provides an orthogonally projected explicit ODE (OPE-ODE) that is described in terms of a specific orthogonal projector. The consideration of this particular OPE-ODE permits a better understanding of projected integration methods, in particular the Taylor series method described in [13] and [14].

The approach was applied to several examples, in particular to the equations from the exothermic reactor model discussed in [25], the MNA equations and DAEs in Kronecker canonical form. An application to the well-known index-5 DAE of the robotic arm can be found in [15] in this volume. Altogether, we illustrated that the introduced decoupling presents a valuable tool to analyze the structure of DAEs from various fields of applications. The algorithms for the computation were implemented in Python and are available online, cf. [8, 13].

Appendix: Linear Algebra Toolbox

In this appendix, we summarize some results concerning the relationship of (orthogonal) projectors and constraints.

Lemma A.1 ([10]) *Consider a pair of projectors $P, Q \in \mathbb{R}^{n \times n}$, $P = I - Q$.*

1. For a matrix $N \in \mathbb{R}^{m \times n}$ and a vector $b \in$ im N, the linear system of equations

$$Nz = b$$

uniquely determines Qz as a linear function of Pz and b iff

$$\ker \begin{pmatrix} P \\ N \end{pmatrix} = \{0\}. \tag{A.1}$$

2. For $G_L \in \mathbb{R}^{m_G \times n}$, $G_R \in \mathbb{R}^{m_G \times p}$, a projector W_R along im G_R, and for $b \in$ im $(G_L \ G_R)$, the linear system of equations

$$\begin{pmatrix} G_L & G_R \end{pmatrix} \begin{pmatrix} z_1 \\ z_2 \end{pmatrix} = b, \ z_1 \in \mathbb{R}^n, z_2 \in \mathbb{R}^p$$

uniquely determines Qz_1 as a linear function of Pz_1 and b iff, for $N := W_R G_L$,

$$\ker \begin{pmatrix} P \\ N \end{pmatrix} = \{0\}. \tag{A.2}$$

A proof can be found in [10] (Lemma 1).

Theorem A.2 ([11]) *Suppose that an arbitrary matrix $N \in \mathbb{R}^{m \times n}$ and complementary projectors Q, $P := I - Q \in \mathbb{R}^{n \times n}$ fulfilling*

$$\ker \begin{pmatrix} P \\ N \end{pmatrix} = \{0\}$$

are given, and that W is an arbitrary matrix with the property $\ker W = \operatorname{im} NQ$ *such that* $WN = WNP$. *Then all projectors* Π *onto*

$$\ker \begin{pmatrix} Q \\ WN \end{pmatrix} = \ker Q \cap \ker WN$$

fulfill

$$\ker \begin{pmatrix} \Pi \\ N \end{pmatrix} = \{0\}.$$

A proof that is based on the SVD can be found in [11], cf. Theorem 3.

Lemma A.3 *Consider an arbitrary matrix* $N \in \mathbb{R}^{m \times n}$ *and a pair of complementary orthogonal projectors* Q, $P := I - Q \in \mathbb{R}^{n \times n}$. *Then it holds*

$$\begin{pmatrix} P \\ NQ \end{pmatrix}^+ = \left(P (NQ)^+ \right).$$

Proof For $r := \operatorname{rank}(NQ)$, the singular value decomposition $NQ = U\Sigma V^T$ leads to

$$NQ = NQ \cdot Q = U\Sigma V^T \cdot Q = U\Sigma \begin{pmatrix} I_r & 0 \\ 0 & 0 \end{pmatrix} V^T Q = U\Sigma \begin{pmatrix} I_r & 0 \\ 0 & 0 \end{pmatrix} V^T.$$

Hence,

$$\begin{pmatrix} I_r & 0 \\ 0 & 0 \end{pmatrix} V^T = \begin{pmatrix} I_r & 0 \\ 0 & 0 \end{pmatrix} V^T Q \quad \text{and} \quad V \begin{pmatrix} I_r & 0 \\ 0 & 0 \end{pmatrix} = QV \begin{pmatrix} I_r & 0 \\ 0 & 0 \end{pmatrix}$$

and

$$(NQ)^+ = V\Sigma^+ U^T = Q \cdot V\Sigma^+ U^T = Q(NQ)^+,$$

such that

$$P \cdot (NQ)^+ = 0, \quad \left((NQ)^+ \right)^T \cdot P = 0. \tag{A.3}$$

With the properties (A.3), the four Moore-Penrose conditions for

$$A := \begin{pmatrix} P \\ NQ \end{pmatrix}$$

can be verified easily:

1.

$$AA^+A = \begin{pmatrix} P \\ NQ(NQ)^+NQ \end{pmatrix} = A.$$

2.

$$A^+AA^+ = \left(P \ (NQ)^+NQ(NQ)^+ \right) = A^+.$$

3.

$$AA^+ = \begin{pmatrix} P & 0 \\ 0 & (NQ)(NQ)^+ \end{pmatrix} = (AA^+)^T.$$

4.

$$A^+A = P + (NQ)^+(NQ) = P^T + ((NQ)^+(NQ))^T = (A^+A)^T. \qquad \square$$

Corollary A.4 *If, additionally to the assumptions of Lemma A.3, the property*

$$\ker \begin{pmatrix} P \\ N \end{pmatrix} = \{0\}$$

is given, then

$$(NQ)^+NQ = Q$$

holds.

Proof From

$$\{0\} = \ker \begin{pmatrix} P \\ N \end{pmatrix} = \ker \begin{pmatrix} P \\ NQ \end{pmatrix}$$

it follows that, in the proof of Lemma A.3, we have

$$I = A^+A = P + (NQ)^+(NQ)$$

such that $(NQ)^+(NQ) = Q$ must hold. $\qquad \square$

Corollary A.5 *If the assumptions of Corollary A.4 are given and we consider an arbitrary matrix W with the property* $\ker W = \mathrm{im}\ NQ$, *then, for the orthogonal* *projector Π fulfilling*

$$\ker \begin{pmatrix} Q \\ WN \end{pmatrix} = \ker Q \cap \ker WN = \operatorname{im} \Pi,$$

we have

$$I - \Pi = (N(I - \Pi))^+ N(I - \Pi) = (I - \Pi)(N(I - \Pi))^+ N(I - \Pi) \quad (A.4)$$

and

$$I - \Pi = Q + (WN)^+(WN), \tag{A.5}$$

where the latter representation implies

$$P - \Pi = (WN)^+(WN).$$

Proof Since

$$\ker \begin{pmatrix} P \\ N \end{pmatrix} = \ker \begin{pmatrix} \Pi \\ N \end{pmatrix} = \ker \begin{pmatrix} \Pi \\ N(I - \Pi) \end{pmatrix} = \{0\},$$

property (A.4) follows directly from Corollary A.4. Moreover, by the definition of Π, Lemma A.4 implies

$$I - \Pi = \begin{pmatrix} Q \\ WN \end{pmatrix}^+ \begin{pmatrix} Q \\ WN \end{pmatrix} = Q + (WN)^+(WN). \qquad \square$$

Let us now focus on some relationships used in Sect. 6.

Lemma A.6

1. *If A is an arbitrary matrix, Q is the orthogonal projector onto $\ker A$, then, for any positive definite matrix C, the matrix*

$$H_1 := A^T C A + Q$$

   is nonsingular and positive definite.
2. *We assume further that N is a matrix fulfilling*

$$\ker \begin{pmatrix} P \\ N \end{pmatrix} = \{0\},$$

W *is a matrix with* $\ker W = \operatorname{im} NQ$, *and*

$$P - \Pi = (WN)^+(WN).$$

Let further $\tilde{Q}$ *be an orthogonal projector onto* $\ker (WN)^T$. *Then the matrix*

$$H_2 = (WN)H_1^{-1}(WN)^T + \tilde{Q}$$

is nonsingular and positive definite.
3. *Under these assumptions, the matrix*

$$\Psi := H_1^{-1}(WN)^T H_2^{-1}(WN)$$

is a projector fulfilling $\Psi = \Psi \cdot P$ *and*

$$\Psi \cdot (P - \Pi) = \Psi, \quad (P - \Pi) \cdot \Psi = (P - \Pi),$$

i.e., $\ker \Psi = \ker (P - \Pi)$ *and therefore* $\Psi^+\Psi = (P - \Pi)$.
4. *Finally, the above equations lead to*

$$Q + \Psi^+\Psi = I - \Pi$$

and

$$(WN)\Psi = WN.$$

Proof

1. A slightly weaker form of this lemma was proved in [16] for a specific application. For completeness, we give a general proof here. Let z be an element of $\ker H$. Then we have

$$(A^T CA + Q)z = 0.$$

If we multiply this equation by Q, it results that $Qz = 0$. Hence,

$$A^T CAz = 0$$

holds. From the positive definiteness of C it follows that $Az = 0$, and therefore $Pz = 0$. Finally, the positive definiteness of H_1 follows from

$$H_1 = \begin{pmatrix} A^T & Q \end{pmatrix} \underbrace{\begin{pmatrix} C & 0 \\ 0 & I \end{pmatrix}}_{\text{positive definite}} \begin{pmatrix} A \\ Q \end{pmatrix} \quad \text{and} \quad \ker \begin{pmatrix} A \\ Q \end{pmatrix} = \ker A \cap \ker Q = \{0\}.$$

2. The second assertion results directly for $A = (WN)^T$, $C = H_1^{-1}$.
3. We focus now on the properties of Ψ:

(a) Let us first show that Ψ is a projector using $\tilde{P} := I - \tilde{Q}$

$$\Psi \cdot \Psi = H_1^{-1}(WN)^T H_2^{-1} \underbrace{(WN) \cdot H_1^{-1}(WN)^T}_{\tilde{P}H_2 = H_2 \tilde{P}} H_2^{-1}(WN)$$

$$= H_1^{-1}(WN)^T H_2^{-1}(WN) = \Psi.$$

(b) We finally show

$$\Psi \cdot (P - \Pi) = H_1^{-1}(WN)^T H_2^{-1}(WN) \cdot (WN)^+(WN)$$

$$= H_1^{-1}(WN)^T H_2^{-1}(WN) = \Psi,$$

$$(P - \Pi) \cdot \Psi = (WN)^+ \underbrace{(WN) \cdot H_1^{-1}(WN)^T}_{=\tilde{P}H_2} H_2^{-1}(WN)$$

$$= (WN)^+(WN) = (P - \Pi).$$

4. The last assertions follow directly form the above representation. □

With the notation of Lemma A.6 and

$$\Pi_\Psi := P - \Psi$$

we obtain the relations

$$\Pi_\Psi \Pi = \Pi, \quad \Pi \, \Pi_\Psi = \Pi_\Psi.$$

Note that in Sect. 6 we have shown that, for the considered index-2 DAEs, the projector Π_1 of the tractability index results to be a projector Π_Ψ with these properties.

References

1. Ascher, U.M., Petzold, L.R.: Projected implicit Runge-Kutta methods for differential-algebraic equations. SIAM J. Numer. Anal. **28**(4), 1097–1120 (1991)
2. Brenan, K., Campbell, S., Petzold, L.: Numerical Solution of Initial-Value Problems in Differential-Algebraic Equations. Classics in Applied Mathematics, vol. 14. Unabridged, Corrected Republication. SIAM, Society for Industrial and Applied Mathematics, Philadelphia (1996)
3. Bunse-Gerstner, A., Byers, R., Mehrmann, V., Nichols, N.K.: Numerical computation of an analytic singular value decomposition of a matrix valued function. Numer. Math. **60**(1), 1–39 (1991)

4. Campbell, S.L.: The numerical solution of higher index linear time varying singular systems of differential equations. SIAM J. Sci. Stat. Comput. **6**, 334–348 (1985)
5. Estévez Schwarz, D.: Topological analysis for consistent initialization in cicuit simulation. Technical Report 3, Institut für Mathematik, Humboldt-Universität zu Berlin (1999)
6. Estévez Schwarz, D.: Consistent initialization for index-2 differential algebraic equations and its application to circuit simulation. Ph.D. Thesis, Humboldt-University, Mathematisch-Naturwissenschaftliche Fakultät II, Berlin (2000). http://edoc.hu-berlin.de/docviews/abstract.php?id=10218
7. Estévez Schwarz, D.: Consistent initialization for DAEs in Hessenberg form. Numer. Algorithms **52**(4), 629–648 (2009). https://doi.org/10.1007/s11075-009-9304-1
8. Estévez Schwarz, D., Lamour, R.: InitDAE's documentation. https://www.mathematik.hu-berlin.de/~lamour/software/python/InitDAE/html/
9. Estévez Schwarz, D., Lamour, R.: Diagnosis of singular points of properly stated DAEs using automatic differentiation. Numer. Algorithms **70**(4), 777–805 (2015)
10. Estévez Schwarz, D., Lamour, R.: A new projector based decoupling of linear DAEs for monitoring singularities. Numer. Algorithms **73**(2), 535–565 (2016)
11. Estévez Schwarz, D., Lamour, R.: Consistent initialization for higher-index DAEs using a projector based minimum-norm specification. Technical Report 1, Institut für Mathematik, Humboldt-Universität zu Berlin (2016)
12. Estévez Schwarz, D., Lamour, R.: A new approach for computing consistent initial values and Taylor coefficients for DAEs using projector-based constrained optimization. Numer. Algorithms **78**(2), 355–377 (2018)
13. Estévez Schwarz, D., Lamour, R.: InitDAE: Computation of consistent values, index determination and diagnosis of singularities of DAEs using automatic differentiation in Python. J. Comput. Appl. Math. (2019). https://doi.org/10.1016/j.cam.2019.112486
14. Estévez Schwarz, D., Lamour, R.: Projected explicit and implicit Taylor series methods for DAEs. Technical Report, Institut für Mathematik, Humboldt-Universität zu Berlin (2019)
15. Estévez Schwarz, D., Lamour, R., März, R.: Singularities of the Robotic Arm DAE. In: Progress in Differential-Algebraic Equations II. Differential-Algebraic Equations Forum (DAE-F). Springer, Berlin (2020)
16. Estévez Schwarz, D., Tischendorf, C.: Structural analysis of electric circuits and consequences for the MNA. Int. J. Circuit Theory Appl. **28**(2), 131–162 (2000)
17. Gantmacher, F.: The Theory of Matrices. Chelsea House, Philadelphia (1960)
18. Griepentrog, E., März, R.: Differential-Algebraic Equations and Their Numerical Treatment. Teubner-Texte zur Mathematik, vol. 88. B.G. Teubner Verlagsgesellschaft, Leipzig (1986)
19. Hairer, E., Wanner, G.: Solving Ordinary Differential Equations II. Springer, Berlin (1996)
20. Kunkel, P., Mehrmann, V.: Differential-Algebraic Equations - Analysis and Numerical Solution. EMS Publishing House, Zürich (2006)
21. Lamour, R., März, R., Tischendorf, C.: Differential-Algebraic Equations: A Projector Based Analysis. Differential-Algebraic Equations Forum, no. 1. Springer, Berlin (2013)
22. Linh, V.H., März, R.: Adjoint pairs of differential-algebraic equations and their Lyapunov exponents. J. Dyn. Differ. Eq. **29**(2), 655–684 (2017)
23. März, R.: Differential-algebraic equations from a functional-analytic viewpoint: a survey. In: Surveys in Differential-Algebraic Equations II, pp. 163–285. Springer, Cham (2015)
24. März, R.: New answers to an old question in the theory of differential-algebraic equations: essential underlying ODE versus inherent ODE. J. Comput. Appl. Math. **316**, 271–286 (2017)
25. Pantelides, C.: The consistent initialization of differential-algebraic systems. SIAM J. Sci. Stat. Comput. **9**(2), 213–231 (1988)

Basic Characteristics of Differential-Algebraic Operators

Michael Hanke and Roswitha März

Abstract We investgate differential-algebraic operators, first with constant and then with variable coefficients, which act in Lebesgue spaces. We provide conditions concerning basic characteristics such as normal solvability (closed range), Fredholmness et cetera. In particular, we prove that actually the operators having tractability index zero and one constitute the class of normally solvable differential-algebraic operators.

Keywords Differential-algebraic operator · Closed operator · Closed range · Normal solvability · Tractability index · Fredholmness

Mathematics Subject Classification (2010) 34A09, 34L99, 47A05, 47E05

1 Introduction

This paper addresses differential-algebraic operators (DA operators) associated with differential-algebraic equations (DAEs) in standard form,

$$E(t)x'(t) - F(t)x(t) = q(t), \quad t \in [a, b], \tag{1.1}$$

M. Hanke (✉)
KTH Royal Institute of Technology, School of Engineering Sciences,
Department of Mathematics, Stockholm, Sweden
e-mail: hanke@nada.kth.se

R. März
Institute of Mathematics, Humboldt-University of Berlin, Berlin, Germany
e-mail: maerz@math.hu-berlin.de

T. Reis et al. (eds.), *Progress in Differential-Algebraic Equations II*,
Differential-Algebraic Equations Forum,
https://doi.org/10.1007/978-3-030-53905-4_2

with sufficiently smooth coefficient functions $E, F : [a, b] \to \mathbb{R}^{k \times m}$. In particular, we provide basic properties of DA operators,

$$T : \operatorname{dom} T \subset L^2((a, b), \mathbb{R}^m) \to L^2((a, b), \mathbb{R}^k), \tag{1.2}$$

$$\operatorname{dom} T = \{x \in L^2((a, b), \mathbb{R}^m) | Ex \in H^1((a, b), \mathbb{R}^k), (Ex)(a) = 0\}, \tag{1.3}$$

$$Tx = (Ex)' - (F + E')x, \quad x \in \operatorname{dom} T. \tag{1.4}$$

and their adjoint counterparts T^*. Specifically, we ask for closedness of the operators, describe their ranges and nullspaces, provide closed range conditions and conditions for them to be Fredholm. Having in mind, that the Moore-Penrose inverse T^+ is bounded, if and only if $\operatorname{im} T$ is closed, our main interest is directed to the closed range property, or equivalently, the *normal solvability* of T.[1]

We refer to the early papers [7, 17] for first findings concerning closedness and normal solvability of DA operators acting in spaces of integrable functions. Further contributions to DA operators acting in various function spaces are surveyed in [12]. Quite recently, a capacious analysis of operators T with constant matrix-coefficients E and F has been elaborated in [13, 14] by applying the *Quasi–Kronecker form* for matrix pencils, which decouples the matrix pencil into an underdetermined part, a regular part and an overdetermined part, [3]. We revisit some questions in this regard in Sect. 2.

Note that in [13, 14] operators acting in Lebesgue spaces L^p, $1 \leq p < \infty$, are considered. Nevertheless, the corresponding criteria result as conditions in terms of the given matrices E and F. That is why we here confine the presentation to the Hilbert space L^2 only.

The condition

$$\operatorname{im} F(t) \subseteq \operatorname{im} E(t) + F(t) \ker E(t), \quad t \in [a, b], \tag{A}$$

plays its role in several issues, e.g., [2, 12, 13]. It characterizes the class of strangeness free DAEs, see [12, Pages 192–193] and one could readily consider this condition as necessary and sufficient for normal solvability. Actually, Condition (A) is sufficient for normal solvability, but not necessary, [12, Theorem 3.2, Example 3.8], see also Example 1.2 below.

Several sufficient conditions for normal solvability are provided in [12], each of which characterizes a special class of DAEs with tractability index zero and one in the sense of [10, Definition 10.2]. It is conjectured, [12, Remark 3.2], *that all DAEs*

---

[1]We emphasize that we are interested in coefficient functions being as smooth as necessary but, on the other hand, as nonsmooth as possible. If E and F are real-analytic, and T acts from $C^\infty([a, b], \mathbb{R}^m)$ to $C^\infty([a, b], \mathbb{R}^k)$, then the range of T is simply always closed and each regular DA operator is surjective and has a finite-dimensional nullspace, and hence, it is Fredholm, see [5, Section 3.6], also [12, Section 2.4].

with tractability index zero and one yield normally solvable operators. In Sect. 3 we
will verify this conjecture.

The paper is organized as follows. In Sect. 2, we discuss different index notions
for matrix pencils in view of operator properties and provide then with Theorem 2.3
a new version of the results from [13, 14] in terms of matrix sequences originally
given in terms of the Quasi-Kronecker form. At the same time, the matrix sequences
serve as an easy introduction to deal with matrix function sequences later on. On
this background, the new statements in Sect. 3 concerning time-varying coefficients
become much more perspicuous. Section 3 provides basic properties of the corresponding DA operators, in particular, Theorem 3.4 on normal solvability. We add
some ideas concerning modifications and generalizations in Sect. 4. For an easier
reading we collect some material from [10] concerning the projector based analysis
and the tractability index of general possibly nonregular DAEs in the appendix.

To gain a first insight, we finish this section by considering operators associated
with simplest singular constant coefficient DAEs.

Example 1.1 For the operator T associated with the flat pencil $sE - F$ of size 2×3
given by

$$E = \begin{bmatrix} 1 & 0 & 0 \\ 0 & 1 & 0 \end{bmatrix}, \quad F = \begin{bmatrix} 0 & 1 & 0 \\ 0 & 0 & 1 \end{bmatrix}, \quad Tx = (Ex)' - Fx = \begin{bmatrix} x_1' - x_2 \\ x_2' - x_3 \end{bmatrix}, \quad x \in \text{dom } T,$$

$$\text{dom } T = \{x \in L^2((a,b), \mathbb{R}^3) | \, x_1, x_2 \in H^1((a,b), \mathbb{R}), x_1(a) = 0, x_2(a) = 0\},$$

the leading matrix E has full row-rank such that Condition (A) is trivially valid.
Writing $Tx = q$ as

$$\begin{bmatrix} x_1' \\ x_2' \end{bmatrix} = \begin{bmatrix} 0 & 1 \\ 0 & 0 \end{bmatrix} \begin{bmatrix} x_1 \\ x_2 \end{bmatrix} + \begin{bmatrix} q_1 \\ q_2 + x_3 \end{bmatrix}, \quad x_1(a) = 0, x_2(a) = 0,$$

we immediately conclude that T is surjective, im $T = L^2((a,b), \mathbb{R}^2)$, and

$$\ker T = \{x \in \text{dom } T | x_1' = x_2, x_2' = x_3\}$$

$$= \{x \in \text{dom } T | x_2 = \int_a x_3(s) ds, x_1 = \int_a \int_a^s x_3(\tau) d\tau ds\}.$$

Since $\ker T$ is infinite-dimensional, even though the DA operator T is normally
solvable it fails to be Fredholm. □

Example 1.2 For the operator associated with a steep pencil of size 3×2 given by

$$Tx = (\begin{bmatrix} 1 & 0 \\ 0 & 1 \\ 0 & 0 \end{bmatrix} x)' - \begin{bmatrix} 0 & 0 \\ 1 & 0 \\ 0 & 1 \end{bmatrix} x = \begin{bmatrix} x_1' \\ x_2' - x_1 \\ -x_2 \end{bmatrix}, \quad \text{dom } T = \{x \in H^1((a,b), \mathbb{R}^2) | x(a) = 0\},$$

the leading matrix E has full column-rank and Condition (A) is not valid. Observe that T is injective, that is, $\ker T = \{0\}$.

Writing $Tx = q$ as

$$x_1' = q_1,$$

$$x_2' - x_1 = q_2, \quad x_1(a) = 0,$$

$$-x_2 = q_3, \quad x_2(a) = 0,$$

we see that

$$\operatorname{im} T = \{q \in L^2((a,b), \mathbb{R}^3) | q_3 \in H^1((a,b), \mathbb{R}), q_3(a) = 0,$$

$$q_3' + q_2 \in H^1((a,b), \mathbb{R}), (q_3' + q_2)(a) = 0, \ q_1 = -(q_2 + q_3')'\}.$$

By straightforward computation we show that $\operatorname{im} T$ is closed. For that, let $q_* \in L^2((a,b), \mathbb{R}^3)$ be given as well as a sequence $q_n \in \operatorname{im} T$, $n \in \mathbb{N}$, tending to q_* in L^2. Denote $w_n = q_{n,3}' + q_{n,2}$ such that $w_n \in H^1((a,b), \mathbb{R})$, $w_n(a) = 0$, and further $w_n' = -q_{n,1}$, $w_n = -\int_a q_{n,1}(s)ds$. It follows that $w_n \to w_* := \int_a q_{*,1}(s)ds$ in $H^1((a,b), \mathbb{R})$ and $w_*(a) = 0$. Next we observe that $q_{n,3} \to q_{*,3}$, $q_{n,3}' = w_n - q_{n,2} \to w_* - q_{*,2}$ in L^2. This yields $q_{*,3} \in H^1((a,b), \mathbb{R})$, $q_{*,3}' = w_* - q_{*,2}$ and $0 = w_n' + q_{n,1} \to w_*' + q_{*,1}$, thus $q_{*,1} + (q_{*,3}' + q_{*,2})' = 0$. Finally, owing to the continuous embedding $H^1 \hookrightarrow C$ we obtain $|q_{*,3}(a)| = |q_{*,3}(a) - q_{n,3}(a)| \to 0$, thus $|q_{*,3}(a)| = 0$ which completes the proof that q_* belongs to $\operatorname{im} T$, thus $\operatorname{im} T$ is closed.[2]   □

Example 1.3 For the operator S associated with the singular pair $(-E^T, F^T)$, with E, F from Example 1.1,

$$Sy = -(E^T y)' - F^T y = -\left(\begin{bmatrix} 1 & 0 \\ 0 & 1 \\ 0 & 0 \end{bmatrix} y\right)' - \begin{bmatrix} 0 & 0 \\ 1 & 0 \\ 0 & 1 \end{bmatrix} y = \begin{bmatrix} -y_1' \\ -y_2' - y_1 \\ -y_2 \end{bmatrix},$$

$$\operatorname{dom} S = \{y \in H^1((a,b), \mathbb{R}^2) | y(b) = 0\},$$

---

[2]Since Condition (A) is not valid here, it fails to be a necessary condition of normal solvability, and we are confronted with a counterexample to [13, Theorem 1(iii)] claiming that T is normally solvable if and only if Condition (A) is valid. It should be noted that [13, Theorem 1] has already been corrected by the authors in [14].

the leading matrix has full column-rank and Condition (A) is not valid again. S is injective. Writing $Sy = p$ as

$$-y_1' = p_1,$$
$$-y_2' - y_1 = p_2, \quad y_1(b) = 0,$$
$$-y_2 = p_3, \quad y_2(b) = 0,$$

we see that

$$\operatorname{im} S = \{p \in L^2((a,b), \mathbb{R}^3) | p_3 \in H^1((a,b), \mathbb{R}), \ p_3(b) = 0,$$
$$p_3' - p_2 \in H^1((a,b), \mathbb{R}), \ p_3'(b) - p_2(b) = 0, \ p_1 + (p_3' - p_2)' = 0\},$$

which is closed by analogous arguments as used in Example 1.2. Moreover, with the flat operator T from Example 1.1, regarding $E^T = E^T E E^+$ and the boundary conditions, we have[3]

$$(Tx, y) = ((Ex)' - Fx, y) = (EE^+(Ex)' - Fx, y) = ((Ex)', EE^+y) - (x, F^T y)$$
$$= -(Ex, (EE^+y)') - (x, F^T y) = -(x, E^T(EE^+y)') - (x, F^T y)$$
$$= -(x, (E^T y)') - (x, F^T y) = (x, Sy), \quad x \in \operatorname{dom} T, \ y \in \operatorname{dom} S.$$

Both operators, T and S are densely defined and closed, and they form an adjoint pair, i.e., $S = T^*$. As an adjoint pair of such operators, T and $S = T^*$ have simultaneously a closed image or not. We know from Example 1.1 that T is surjective and, hence, im S is also closed. Therefore, Condition (A) is not necessary for the normal solvability of S, too. It is also easy to directly check the now expected relations

$$\operatorname{im} T = (\ker S)^\perp, \quad \ker T = (\operatorname{im} S)^\perp. \qquad \Box$$

2 Constant-Coefficient DA Operators, Matrix Pencils and Different Index Notions

For any given matrices $E, F \in \mathbb{R}^{k \times m}$, E singular but nontrivial, the ordered pair (E, F) stands for the pencil $sE - F$, $s \in \mathbb{R}$. If $k = m$ and the polynomial in s, $\det(sE - F)$, does not vanish identically, then the pencil is called regular, and otherwise singular.

---

[3] Here and in the following, $(\cdot, \cdot)$ denotes the scalar product in $L^2((a,b), \mathbb{R}^n)$ for any n.

There exist nonsingular matrices $\mathscr{L} \in \mathbb{R}^{k \times k}$ and $\mathscr{K} \in \mathbb{R}^{m \times m}$ transforming the pencil (E, F) into Quasi-Kronecker form, e.g., [2, 3], such that for all s,

$$s\mathscr{L}E\mathscr{K} - \mathscr{L}F\mathscr{K} = \mathscr{L}(sE - F)\mathscr{K} = \text{diag}(sE_{reg} - F_{reg}, \ sE_{sing} - F_{sing}), \tag{2.1}$$

in which $sE_{reg} - F_{reg} = \text{diag}(\ sI - W, \ sN - I\ )$, with nilpotent N, is a regular pencil and $sE_{sing} - F_{sing}$ is a singular pencil of the special form

$$sE_{sing} - F_{sing} = \text{diag}(\ sK_{\epsilon_1} - L_{\epsilon_1}, \ldots, sK_{\epsilon_\rho} - L_{\epsilon_\rho}, \ sK_{\gamma_1}^T - L_{\gamma_1}^T, \ldots, sK_{\gamma_\sigma}^T - L_{\gamma_\sigma}^T\ ), \tag{2.2}$$

$$L_\kappa = \begin{bmatrix} 0 & 1 & & \\ & \ddots & \ddots & \\ & & 0 & 1 \end{bmatrix} \in \mathbb{R}^{\kappa \times (\kappa+1)}, \quad K_\kappa = \begin{bmatrix} 1 & 0 & & \\ & \ddots & \ddots & \\ & & 1 & 0 \end{bmatrix} \in \mathbb{R}^{\kappa \times (\kappa+1)}, \tag{2.3}$$

with nonnegative integers

$$\epsilon_1 \geq \cdots \geq \epsilon_\rho \geq 0, \quad 0 \leq \gamma_1 \leq \cdots \leq \gamma_\sigma,$$

which are called *right and left Kronecker indices* , also *column minimal indices and row minimal indices*. Here, we share the convention to allow blocks of sizes 0×1 and 1×0 and declare the 1×1 blocks $\text{diag}(sK_0 - L_0, sK_0^T - L_0^T)$ to stand for the 1×1 blocks $s0 - 0$.[4]

There are different index notions concerning the index of the general matrix pencil (E, F). In [15, 16] the *index of the pencil* (E, F) is defined as the index of the regular part, that is,

$$\mu = \text{ind}(E, F) = \text{ind}(E_{reg}, F_{reg}) = \text{ind}(N), \quad \text{ind}(E_{sing}, F_{sing}) = 0. \tag{2.4}$$

It is argued for this notion in [16] that *the index of the matrix pencil is the maximum length of [. . .] a chain of differentiators. The blocks of the singular part correspond to undetermined and overdetermined ODEs, respectively.*

In contrast, [2, Definition 3.2] involves the maximal left Kronecker index γ_ρ into a further index notion stating

$$\mu_{BR} = \text{ind}_{BR}(E, F) = \max\{\text{ind}(E_{reg}, F_{reg}), \max_{\kappa=1,\ldots,\rho} \gamma_\kappa\}, \ \text{ind}_{BR}(E_{sing}, F_{sing}) = \max_{\kappa=1,\ldots,\rho} \gamma_\kappa.$$

---

[4]We refer to [3, 16] for details and further references. Here we note only that all involved matrices have solely real entries. In contrast, the Kronecker normal form is not necessarily real.

For regular pencils one has $\mu_{BR} = \mu$. As observed in [2, Page 19], one has $\mu_{BR} \leq 1$ if and only if Condition (A) is given. By Examples (1.1)–(1.3) we already know this index notion to be unsuitable for the characterization of normal solvability of the operator T.

On the other hand, the *tractability index* introduced in [10, Chapter 10] for possibly nonregular DAEs, is fully consistent with the notion from [16]. The tractability index μ_{trac} of the matrix pencil (E, F) is defined by means of *an admissible matrix sequence*,

$$G_0 = E, \quad G_1 = G_0 - FQ_0, \quad G_i = G_{i-1} - F\Pi_{i-2}Q_{i-1}, \quad i = 2, \ldots, r+2,$$

in which $r = \text{rank } E$, the matrix $Q_i \in \mathbb{R}^{m \times m}$ represents an *admissible projector* of $\mathbb{R}^m$ onto $\ker G_i$, and further $\Pi_0 = I - Q_0$, $\Pi_i = \Pi_{i-1}(I - Q_i)$. One possibility for Q_0 is the orthoprojector $I - E^+E$. Regarding that $\text{im } E + F \ker E = \text{im } E - F \ker E = \text{im } G_1$ we can express Condition (A) as $\text{im } F \subseteq \text{im } G_1$.

By construction, cf. [10, Chapter 10], it holds that $G_{r+1} = G_{r+2}$ and $\ker \Pi_r = \ker \Pi_{r+1}$, and further

$$\text{im } G_0 \subseteq \text{im } G_1 \subseteq \cdots \subseteq \text{im } G_{r+1} = \text{im } G_{r+2} \subseteq \text{im}[E \ F], \tag{2.5}$$

$$\ker \Pi_0 \subseteq \ker \Pi_1 \subseteq \cdots \subseteq \ker \Pi_r = \ker \Pi_{r+1} \subseteq \mathbb{R}^m. \tag{2.6}$$

The *tractability index of the matrix pencil* (E, F)[5] is defined to be $\mu_{\text{trac}} = \kappa$, where $\kappa \leq r + 1$ is the smallest integer indicating the maximal possible range in the sequence (2.5) such that $\text{im } G_\kappa = \text{im } G_{r+1}$, that is,

$$\text{im } G_{\kappa-1} \subset \text{im } G_\kappa = \cdots = \text{im } G_{r+1} = \text{im } G_{r+2} \subseteq \text{im}[E \ F],$$

if $\kappa \geq 1$, and, as the case may be, with $\kappa = 0$,

$$\text{im } G_0 = \text{im } G_1 = \cdots = \text{im } G_{r+1} = \text{im } G_{r+2} \subseteq \text{im}[E \ F].$$

Aside from that, there is an *additional index* in this context, which we designate by $\mu_{\text{ad}} = \nu$, where $\nu \leq r$ is the smallest integer such that $\ker \Pi_\nu = \ker \Pi_r$ in the sequence (2.6). Later on it will become clear that the subscript *ad* actually stands for *adjoint-differentiation*.

For regular matrix pencils, the tractability index equals the Kronecker index, that is, $\mu_{\text{trac}}(E_{reg}, F_{reg}) = \text{ind}(E_{reg}, F_{reg}) = \mu$, e.g., [10, Chapter 1]. Then, one has $m = k$ and the matrix G_μ is nonsingular, $\text{im } G_\mu = \text{im}[E, \ F] = \mathbb{R}^m$.

We return to DA operators T associated with possibly singular pencils (E, F), and assign the different indices of the pencil to the operator. We consider some examples of singular pencils and the corresponding DA operators.

---

[5]This is a special case of Definition 3.1 below.

Example 2.1 Consider the operator T generated by the singular matrix pencil (E, F)

$$E = \frac{1}{2} \begin{bmatrix} 1 & -1 \\ -1 & 1 \end{bmatrix}, \quad F = \frac{1}{2} \begin{bmatrix} -1 & -1 \\ 1 & 1 \end{bmatrix}.$$

It is shown to be closed and normally solvable in [17] by a quite involved reasoning via singular perturbations. Obviously, it holds that $r = 1$ here. An admissible matrix sequence reads

$$G_0 = E, \quad Q_0 = \frac{1}{2} \begin{bmatrix} 1 & 1 \\ 1 & 1 \end{bmatrix}, \quad \Pi_0 = E, \quad G_1 = \begin{bmatrix} 1 & 0 \\ -1 & 0 \end{bmatrix}, \quad Q_1 = \begin{bmatrix} 0 & 0 \\ -1 & 1 \end{bmatrix}, \quad \Pi_1 = 0,$$

$$G_1 = G_2 = G_3,$$

such that $\mu_{\text{trac}} = 0$ and $\mu_{\text{ad}} = 1$. Observe that here $\text{im}[E \ F] \subset \mathbb{R}^2$ is merely a one-dimensional subspace. $\qquad\qquad\square$

Example 2.2 We provide an admissible matrix sequence for the operator T in Example 1.1. We begin with

$$G_0 = E = \begin{bmatrix} 1 & 0 & 0 \\ 0 & 1 & 0 \end{bmatrix}, \quad F = \begin{bmatrix} 0 & 1 & 0 \\ 0 & 0 & 1 \end{bmatrix}, \quad Q_0 = \begin{bmatrix} 0 & 0 & 0 \\ 0 & 0 & 0 \\ 0 & 0 & 1 \end{bmatrix}, \quad \Pi_0 = \begin{bmatrix} 1 & 0 & 0 \\ 0 & 1 & 0 \\ 0 & 0 & 0 \end{bmatrix},$$

and derive

$$G_1 = \begin{bmatrix} 1 & 0 & 0 \\ 0 & 1 & -1 \end{bmatrix}, \quad Q_1 = \begin{bmatrix} 0 & 0 & 0 \\ 0 & 1 & 0 \\ 0 & 1 & 0 \end{bmatrix}, \quad \Pi_1 = \begin{bmatrix} 1 & 0 & 0 \\ 0 & 0 & 0 \\ 0 & 0 & 0 \end{bmatrix},$$

$$G_2 = \begin{bmatrix} 1 & -1 & 0 \\ 0 & 1 & -1 \end{bmatrix}, \quad Q_2 = \begin{bmatrix} 1 & 0 & 0 \\ 1 & 0 & 0 \\ 1 & 0 & 0 \end{bmatrix}, \quad \Pi_2 = 0, \ G_3 = G_2, Q_3 = Q_2, \Pi_3 = \Pi_2, G_4 = G_3.$$

All matrices G_i have rank 2, so that $\mu_{\text{trac}} = 0$ results. Here, obviously, regarding that E has already full row-rank would allow immediately to conclude $\mu_{\text{trac}} = 0$ without recurring to the sequence. We observe further $\mu_{\text{ad}} = 2$. Regarding that the operator S in Example 1.3 is the adjoint of T and looking at $\text{im } S$ given there, we know that, although $\text{im } S$ is closed, the second derivative of a component of p is involved. We emphasize that $\mu_{\text{ad}} = 2$ here. $\qquad\qquad\square$

Example 2.3 We provide an admissible matrix sequence for the operator S in Example 1.3 which is the adjoint to the operator T in Example 1.1. We begin with

$$
G_0 = -E^T = - \begin{bmatrix} 1 & 0 \\ 0 & 1 \\ 0 & 0 \end{bmatrix}, \quad F^T = \begin{bmatrix} 0 & 0 \\ 1 & 0 \\ 0 & 1 \end{bmatrix}, \quad Q_0 = 0, \quad \Pi_0 = I,
$$

and obtain

$$
G_0 = G_1 = G_2 = G_3 = G_4, \quad \Pi_0 = \Pi_1 = \Pi_2 = \Pi_3,
$$

and therefore $\mu_{\text{trac}} = 0$, and $\mu_{\text{ad}} = 0$ result. Here, obviously, regarding that E^T has already full column-rank allows immediately to conclude $\mu_{\text{trac}} = 0$ and also $\mu_{\text{ad}} = 0$. The adjoint operator T to S, see Example 1.1, is surjective so that no derivatives of q are involved. In this context we like to point out that we have $\mu_{\text{ad}} = 0$. □

Lemma 2.1 *If $\kappa \geq 0$ and the matrices $L_\kappa, K_\kappa \in \mathbb{R}^{\kappa \times (\kappa+1)}$ are given by (2.3), then the following holds:*

(1) *The flat singular pencil $s K_\kappa - L_\kappa$ has the indices*[6]

$$
\mu = 0, \quad \mu_{\text{trac}} = 0, \quad \mu_{\text{ad}} = \kappa, \quad \mu_{BR} = 0, \quad \mu_{\text{strangeness}} = 0.
$$

(2) *The steep singular pencils $s K_\kappa^T - L_\kappa^T$ and $-s K_\kappa^T - L_\kappa^T$ have the indices*

$$
\mu = 0, \quad \mu_{\text{trac}} = 0, \quad \mu_{\text{ad}} = 0, \quad \mu_{BR} = \kappa, \quad \mu_{\text{strangeness}} = \kappa.
$$

(3) *The operators T and $S = T^*$ associated to the pencils $s K_\kappa - L_\kappa$ and $-s K_\kappa^T - L_\kappa^T$ are normally solvable. T is surjective and S is injective. $\ker T$ and $(\text{im } S)^\perp$ are infinite-dimensional. The pencils inducing T and S share the indices $\mu = \mu_{\text{trac}}$.*

Proof The statements can be verified in a straightforward manner as it is done above for the special case $\kappa = 2$, see Examples 1.1, 2.2, 1.2, 1.3, 2.3.

Proposition 2.2 *For each matrix pencil $sE - F$, $E, F \in \mathbb{R}^{k \times m}$, it holds that $\mu = \mu_{\text{trac}}$*

Proof The tractability index and its constituent parts are invariant with respect to transformations, so that we may turn to the Quasi-Kronecker form (2.1). For regular matrix pencils, μ is the Kronecker index, and the identity $\mu = \mu_{\text{trac}}$ is known, e.g., [10, Chapter 1]. By Lemma 2.1, for each singular part one has $\mu = \mu_{\text{trac}} = 0$. Constructing an admissible matrix sequence blockwise according to structure of the Quasi-Kronecker form will complete the proof. □

---

[6]For the strangeness index we refer to [9].

We now reformulate statements given in [13, 14] in terms of the matrix pencil and its Quasi-Kronecker form by means of matrix sequences.

Theorem 2.3 *Let the DA operator* T : $\mathrm{dom}\,T \subset L^2((a, b), \mathbb{R}^m) \to L^2((a, b), \mathbb{R}^k)$ *be associated to the matrix pencil* $sE - F$, $E, F \in \mathbb{R}^{k \times m}$,
$\mathrm{dom}\,T = \{x \in L^2((a, b), \mathbb{R}^m) | Ex \in H^1((a, b), \mathbb{R}^m), (Ex)(a) = 0\}$. *Let* μ *be the index of the matrix pencil,* $r = \mathrm{rank}\,E$ *and* $G_0, G_1, \ldots, G_{r+2}$ *be an admissible matrix sequence. Then the following statements hold true:*

(1) T *and its adjoint* $S = T^*$ *share their index* μ.
(2) T *is normally solvable, if and only if* $\mathrm{im}\,G_1$ *is maximal in* (2.5), *that is,* $\mu \leq 1$
(3) T *is surjective, if and only if* $\mathrm{im}\,G_1 = \mathbb{R}^k$.
(4) T *is injective, if and only if* $\ker G_\mu = \{0\}$.
(5) T *is regular, if and only if* $\mathrm{im}\,G_\mu = \mathbb{R}^k$, $\ker G_\mu = \{0\}$, $m = k$.
(6) T *is Fredholm, if and only if* T *is regular with* $\mu \leq 1$.
(7) $\mathrm{im}\,T$ *is dense, if and only if* $\mathrm{im}\,G_\mu = \mathbb{R}^k$.

Proof It suffices to verify the statements for a pencil in Quasi-Kronecker form. Namely, the transformation

$$\mathcal{L}E\mathcal{K} = \tilde{E}, \quad \mathcal{L}F\mathcal{K} = \tilde{F},$$

is associated with [10, Section 2.3]

$$\mathcal{L}G_i\mathcal{K} = \tilde{G}_i, \quad \mathcal{K}^{-1}Q_i\mathcal{K} = \tilde{Q}_i, \quad \mathcal{K}^{-1}\Pi_i\mathcal{K} = \tilde{\Pi}_i, \quad , i \geq 0.$$

For the structured pencil in Quasi-Kronecker form, the matrix sequence can be formulated to meet the same structure.

(1) This statement is well-known for regular pencils, by definition also for arbitrary pencils.
(2) Since the singular part has index zero, see Lemma 2.1, the question reduces to the regular part. For the regular part the statement is well-known.
(3) T is surjective if and only if each of its structural parts is surjective. This is the case, if and only if the regular part has index $\mu \leq 1$, equivalently, $G_{reg,1}$ is nonsingular, and additionally, the singular part has a full-row-rank matrix G_{sing}, which congruously excludes steep and zero blocks.
(4) By definition, the matrix $G_{reg,\mu}$ is nonsingular and the corresponding T_{reg} is injective. Concerning the singular part, there must be a full-column-rank matrix $G_{sing,0}$ to exclude the flat and zero blocks. It holds that $G_{sing,0} = \ldots = G_{sing,\mu}$ owing to the steep blocks so that $\mathrm{diag}(G_{reg,\mu}, G_{sing,0})$ is injective.
(5) A matrix pencil is regular if and only if the admissible matrix sequences feature a nonsingular matrix $G_{reg,\mu}$. Singular matrix pencils are characterized by admissible matrix sequences of singular matrices [10, Theorems 1.31 and 13.4].

(6) Regularity excludes an infinite-dimensional nullspace and an infinite codimension. The condition $\mu \leq 1$ is necessary and sufficient for normal solvability according to statement (2).

(7) im T_{reg} is dense. For densely solvability, the steep and zero singular blocks must be excluded. We have im $G_{sing,0} = \ldots = $ im $G_{sing,\mu}$ for the flat blocks, they have full row-rank so that also diag($G_{reg,\mu}, G_{sing,\mu}$) is surjective. □

3 DA Operators with Time-Varying Coefficients

3.1 Preliminaries

We study in this section DA operators being closures and adjoints of the DA operator $\overset{\circ}{T} : \operatorname{dom} \overset{\circ}{T} \subset L^2((a,b), \mathbb{R}^m) \to L^2((a,b), \mathbb{R}^k)$, given by

$$\overset{\circ}{T}x = Ex' - Fx, \quad x \in \operatorname{dom} \overset{\circ}{T} = \{w \in H^1((a,b), \mathbb{R}^m) | (Ew)(a) = 0\},$$

with at least continuous coefficient functions $E, F : [a,b] \to \mathbb{R}^{k \times m}$. The leading coefficient function E has constant rank $r > 0$ and its nullspace ker E is a C^1-subspace in $\mathbb{R}^m$. Such an operator is associated with the possibly nonregular standard form DAE

$$E(t)x'(t) - F(t)x(t) = q(t), \quad t \in [a,b].$$

In this section we apply several routine notations and tools used in the projector based analysis of DAEs. We refer to the appendix for a short roundup and to [10, 12] for more details.

The basic tool of the projector based analysis consists in the construction of *admissible matrix function sequences* $G_0, \ldots, G_{r+2} : [a,b] \to \mathbb{R}^{k \times m}$, emanating from the coefficients E, F, with $G_0 = E$. By construction, the inclusions

$$\operatorname{im} G_0 \subseteq \operatorname{im} G_1 \subseteq \ldots \subseteq \operatorname{im} G_{r+1} = \operatorname{im} G_{r+2} \tag{3.1}$$

are valid pointwise. There are several special projector functions incorporated in an admissible matrix function sequence, among them admissible projectors Q_i onto ker G_i and $\Pi_i = \Pi_{i-1}(I - Q_i)$, $\Pi_0 = (I - Q_0)$, yielding the further inclusions

$$\operatorname{ker} \Pi_0 \subseteq \operatorname{ker} \Pi_1 \subseteq \ldots \subseteq \operatorname{ker} \Pi_r = \operatorname{ker} \Pi_{r+1}. \tag{3.2}$$

Each of the time-varying subspaces in (3.1) and (3.2) has constant dimension, which is ensured by several rank conditions. Denote $r_i = \operatorname{rank} G_i(t), t \in [a,b]$.

Remark 3.1 We emphasize that the admissible matrix function sequence constitutes an immediate generalization of the admissible matrix sequence applied in Sect. 2. In particular, (3.1) and (3.2) are consistent with (2.5) and (2.6), respectively. Now the time-dependencies are incorporated into the matrix function sequence. For instance, we may express

$$G_1 = E - F Q_0 + E Q_0'.$$

The subspaces involved in (3.1) and (3.2) are proved to be invariant with respect to special possible choices within the construction procedure and also with respect to the factorization of $E = AD$ in Proposition 3.1 below.

In Sect. 2 different index notions have been discussed. We have seen that solely the tractability index μ_{trac} coincides in the constant coefficient case with μ defined by (2.4) to be the Kronecker index of the regular part. This enables us to use the simpler symbol μ for the tractability index.

Definition 3.1 Let the coefficient function pair (E, F) have an admissible matrix function sequence $G_0, G_1, \ldots, G_{r+2}, r = \text{rank } E, \rho = r + 1$.

The *tractability index* of (E, F) is defined to be $\mu = \kappa$, where $\kappa \leq \rho$ is the smallest integer indicating the maximal possible range in the sequence (3.1), that is $\text{im } G_\kappa = \text{im } G_\rho$. The integers $0 < r_0 \leq r_1 \leq \cdots \leq r_\rho = r_{\rho+1}$ with $r_i = \text{rank } G_i$ are called characteristic values of (E, F).

The *additional index* is defined to be $\mu_{\text{ad}} = \nu$, where $\nu \leq r$ is the smallest integer indicating the maximal possible nullspace in the sequence (3.2), such that $\ker \Pi_\nu = \ker \Pi_r$.

The pair (E, F) is *regular*, if $m = k$ and $r_\mu = m$ and otherwise *nonregular*.

In regular cases and $\mu \geq 1$, it holds that $\mu_{\text{ad}} = \mu - 1$, which is why, so far, no extra notion μ_{ad} has been used in the context of the projector based analysis of regular DAEs.[7]

Definition 3.2 The tractability index μ, the characteristic values, and the additional index μ_{ad} of the DA operator $\overset{\circ}{T}$ and its closure T are defined as the corresponding quantities of their coefficient pair (E, F).

Remark 3.2 We mention that, except for the case $m = k$, G_1 being nonsingular, the so-called *local pencils* $(E(\bar{t}), F(\bar{t}))$, with frozen $\bar{t} \in [a, b]$, are improper for the characterization of time-varying pairs (E, F). This well-known fact will be underlined below by Examples 3.1 and 3.3 which are traditional textbook-examples picked up from the monographs [4, 6, 9]. In particular, Example 3.1 shows a nonregular pair (E, F) with $\mu = 0$ and local pencils being regular with Kronecker index two. In contrast, Example 3.3 shows a regular pair (E, F) with $\mu = 2$ and singular local pencils.

---

[7]Moreover, supposed the coefficients E and F are sufficiently smooth so that both, the regular tractability index and the regular strangeness index are well defined, then it holds $\mu - 1 = \mu_{\text{strangeness}}$, [10, Section 2.10]. Consequently, $\mu_{\text{ad}} = \mu_{\text{strangeness}}$ for regular pairs (E, F) and associated DAEs. In contrast, the situation is completely different in the nonregular case as Lemma 2.1 confirms.

3.2   Closed DA Operators and Adjoint Pairs

We start with the DA operator $\mathring{T} : \operatorname{dom} \mathring{T} \subset L^2((a, b), \mathbb{R}^m) \to L^2((a, b), \mathbb{R}^k)$,

$$\mathring{T}x = Ex' - Fx, \quad x \in \operatorname{dom} \mathring{T} = \{w \in H^1((a, b), \mathbb{R}^m) | (Ew)(a) = 0\}.$$

$\mathring{T}$ is unbounded and nonclosed, but densely defined. $\mathring{T}$ is closable since it has a closed extension provided by Proposition 3.1 below. We look for its minimal closed extension, the *closure*.

Proposition 3.1 *Let $E, F : [a, b] \to \mathbb{R}^{k \times m}$ be continuous, $\ker E$ be a C^1-subspace, and E has constant rank $r > 0$. Let $E = AD$, with continuous $A : [a, b] \to \mathbb{R}^{k \times n}$ and continuously differentiable $D : [a, b] \to \mathbb{R}^{n \times m}$, be any proper factorization of E (cf. Sect. 4.3), furthermore $B := -(F + AD')$.*
   Then the operator $T : L^2((a, b), \mathbb{R}^m) \to L^2((a, b), \mathbb{R}^k)$ given by

$$Tx = A(Dx)' + Bx, \quad x \in \operatorname{dom} T,$$

$$\operatorname{dom} T = \{x \in L^2((a, b), \mathbb{R}^m) | Dx \in H^1((a, b), \mathbb{R}^n), (Dx)(a) = 0\},$$

is densely defined and represents the closure of $\mathring{T}$.

Proof Let R be the continuously differentiable border projector according to (A.2) and let $A(t)^-$ denote the pointwise generalized inverse such that $A(t)^- A(t) = R(t)$. Regarding $D(a) = A(a)^- E(a)$ the inclusion $\operatorname{dom} \mathring{T} \subset \operatorname{dom} T$ is evident, thus T is an extension of $\mathring{T}$ and densely defined. We show that T is closed.
   Consider a sequence $\{x_i\} \subset \operatorname{dom} T$, $x_* \in L^2((a, b), \mathbb{R}^m)$, and $y_* \in L^2((a, b), \mathbb{R}^k)$, such that $x_i \xrightarrow{L^2} x_*, Tx_i \xrightarrow{L^2} y_*$.
   From $Tx_i = A(Dx_i)' + Bx_i$ we derive $(I - AA^-)(Tx_i - Bx_i) = 0$ yielding $(I - AA^-)(y_* - Bx_*) = 0$ on the one hand, and, on the other hand,

$$(Dx_i)' = A^- Tx_i + R' Dx_i - A^- Bx_i \xrightarrow{L^2} A^- y_* + R' Dx_* - A^- Bx_* =: v_*.$$

Owing to $Dx_i \xrightarrow{L^2} Dx_*, (Dx_i)' \xrightarrow{L^2} v_*$, it follows that $Dx_* \in H^1((a, b), \mathbb{R}^n)$ and $(Dx_*)' = v_*$. Since now $Dx_i \xrightarrow{H^1} Dx_*$ and $(Dx_i)(a) = 0$ it results that $(Dx_*)(a) = 0$, and hence $x_* \in \operatorname{dom} T$.
   Considering the relations $(I - AA^-)(y_* - Bx_*) = 0$ and $(Dx_*)' = v_* = A^- y_* + R' Dx_* - A^- Bx_*$ we obtain

$$A(Dx_*)' = AA^- y_* - AA^- Bx_* = y_* - Bx_*,$$

which means $y_* = Tx_*$ and proves the closedness of T.

Finally, we check if T is actually the closure of $\mathring{T}$. We have to show that for each arbitrary $x_* \in \operatorname{dom} T$ there is a sequence $\{x_i\} \subset \operatorname{dom} \mathring{T}$ such that $x_i \xrightarrow{L^2} x_*$, $\mathring{T} x_i \xrightarrow{L^2} T x_*$.

Denote by D^+ the pointwise Moore-Penrose inverse of D. Let $x_* \in \operatorname{dom} T$. We introduce $u_* = D x_* \in H^1((a,b), \mathbb{R}^n)$, $w_* = (I - D^+ D)x_* \in L^2((a,b), \mathbb{R}^m)$. Since $H^1((a,b), \mathbb{R}^m)$ is dense in $L^2((a,b), \mathbb{R}^m)$ there is a sequence $\{w_i\} \subset H^1((a,b), \mathbb{R}^m)$ such that $w_i \xrightarrow{L^2} w_*$.

Set $x_i := D^+ u_* + (I - D^+ D)w_i$ so that $x_i \in H^1((a,b), \mathbb{R}^m)$ and $E(a)x_i(a) = E(a)D(a)^+ u_*(a) = 0$, thus $x_i \in \operatorname{dom} \mathring{T}$. Moreover, we have

$$x_i \xrightarrow{L^2} D^+ u_* + (I - D^+ D)w_* = D^+ D x_* + (I - D^+ D)x_* = x_*,$$

$$\mathring{T} x_i = A(D x_*)' + B D^+ D x_* + B(I - D^+ D)w_i$$

$$\xrightarrow{L^2} A(D x_*)' + B D^+ D x_* + B(I - D^+ D)w_* = T x_*,$$

which completes the proof. □

Proposition 3.2 Let $E, F : [a,b] \to \mathbb{R}^{k \times m}$ be continuous, $\ker E$ be a C^1-subspace, and E has constant rank $r > 0$. Let $E = AD$, with continuous $A : [a,b] \to \mathbb{R}^{k \times n}$ and continuously differentiable $D : [a,b] \to \mathbb{R}^{n \times m}$, be any proper factorization of E, further $B := -(F + AD')$.

Then the operator $S : L^2((a,b), \mathbb{R}^k) \to L^2((a,b), \mathbb{R}^m)$ given by

$$Sy = -D^T(A^T y)' + B^T y, \quad y \in \operatorname{dom} S,$$

$$\operatorname{dom} S = \{y \in L^2((a,b), \mathbb{R}^k) | A^T y \in H^1((a,b), \mathbb{R}^n), (A^T y)(b) = 0\},$$

is densely defined, closed, and represents the adjoint of the operator T from Proposition 3.1.

Proof The coefficient A^T is continuous and $\operatorname{im} A^T = (\ker A)^\perp = (\ker R)^\perp$ is a C^1-subspace in $\mathbb{R}^n$, and hence, S is densely defined owing to Lemma 4.1. Its closedness can be verified analogously to Proposition 4.2 below. We compute for each $x \in \operatorname{dom} T$ and each $y \in \operatorname{dom} S$ (let $\langle \cdot, \cdot \rangle$ denote the Euclidean inner product in $\mathbb{R}^l$ for $l = k, m, n$)

$$(Tx, y) = \int_a^b \langle A(t)(Dx)'(t) + B(t)x(t), y(t) \rangle \mathrm{dt}$$

$$= \int_a^b \{\langle (Dx)'(t), A(t)^T y(t) \rangle + \langle x(t), B(t)^T y(t) \rangle\} \mathrm{dt}$$

$$= \int_a^b \{-\langle (Dx)(t), (A^T y)'(t)\rangle + \langle x(t), B(t)^T y(t)\rangle\}\mathrm{dt}$$

$$= \int_a^b \langle x(t), -D(t)^T (A^T y)'(t) + B(t)^T y(t))\}\mathrm{dt} = (x, Sy),$$

and hence, $S \subset T^*$. The equality can be established by following the lines of proof for simple differential operators in [8, Chapter III, Examples 2.7 and 5.31]. □

If, in addition to the assumptions in Propositions 3.1 and 3.2, the coefficient E itself is continuously differentiable, then we can choose a proper factorization $E = AD$ with both A and D being continuously differentiable. Then regarding that $E = AD$, $D = A^- E$ and $E^T = D^T A^T$, $A^T = D^{-T} E^T$ we find the further representations

$$\mathrm{dom}\, T = \{x \in L^2((a, b), \mathbb{R}^m)|Ex \in H^1((a, b), \mathbb{R}^m), (Ex)(a) = 0\},$$

$$\mathrm{dom}\, S = \{y \in L^2((a, b), \mathbb{R}^k)|E^T y \in H^1((a, b), \mathbb{R}^k), (E^T y)(b) = 0\},$$

which are in line with the constant coefficient case, and

$$Tx = A(Dx)' + Bx = (ADx)' - A'Dx - (F + AD')x$$
$$= (Ex)' - (F + E')x, \quad x \in \mathrm{dom}\, T, \tag{3.3}$$

$$Sy = -D^T (A^T y)' + B^T y = -(D^T A^T y)' + D'^T A^T y - (F + D'^T A^T)y$$
$$= -(E^T y)' - F^T y, \quad y \in \mathrm{dom}\, S. \tag{3.4}$$

Observe that, in contrast to the representations of T and S via proper factorizations of E, the formulas (3.3) and (3.4) display no symmetry, cf. also Sect. 4.1 in this context.

Furthermore, if E is continuously differentable, then S represents the closure of the additional DA operator $\mathring{S} : L^2([a, b], \mathbb{R}^k) \to L^2([a, b], \mathbb{R}^m)$,

$$\mathring{S}y = -E^T y' - (F + E')^T y, \quad y \in \mathrm{dom}\, \mathring{S},$$

$$\mathrm{dom}\, \mathring{S} = \{y \in H^1([a, b], \mathbb{R}^k)|(E^T y)(b) = 0\}.$$

The densely defined operators $\mathring{T}$ and $\mathring{S}$ are adjoint to each other, since

$$(\mathring{T}x, y) = (x, \mathring{S}y), \quad x \in \mathrm{dom}\, \mathring{T}, \ y \in \mathrm{dom}\, \mathring{S}.$$

Also $\mathring{T}$ and S from Proposition 3.2 are obviously adjoint to each other. S is the unique maximal operator adjoint to $\mathring{T}$, that means, the adjoint operator of $\mathring{T}$, and $\mathring{S}$ is a restriction of S.

Remark 3.3 The representation of the closed DA operators T and their adjoints via properly factorized leading matrix coefficients is closely related to the concept of DAEs with properly involved derivatives which has one origin in the desire for a certain symmetry of the formulation of adjoint pairs of DAEs. In this context, the notion of *factorization-adjoint DAEs* is introduced in [11, Definition 1].

Proposition 3.3 *If the DA operator T is regular with index μ and characteristic values $0 < r_0 \leq \cdots \leq r_{\mu-1} = r_\mu = m$, then its adjoint $S = T^*$ is likewise so.*

Proof This statement is an immediate consequence of [11, Theorem 3] concerning the common structure of factorization-adjoint pairs of DAEs. $\square$

We further elucidate the matter by examples. We pick up two textbook-examples discussed, e.g., in the monographs [4, 6, 9] and consider the associated DA operators and their adjoints.

Example 3.1 ([6, page 91],[4, page 23],[9, page 56]) The local matrix pencils of the pair

$$E(t) = \begin{bmatrix} -t & t^2 \\ -1 & t \end{bmatrix}, \quad F(t) = -I, \quad t \in [a, b],$$

are everywhere regular, $\det(sE(t) - F(t)) \equiv 1$. The homogeneous DAE has an infinite-dimensional solution space and the DAE is no longer solvable for all smooth inhomogeneities. Using the factorization

$$E(t) = \begin{bmatrix} -t \\ -1 \end{bmatrix} \begin{bmatrix} 1 & -t \end{bmatrix} =: A(t)D(t), \quad B(t) := -F(t) - A(t)D'(t) = \begin{bmatrix} 1 & -t \\ 0 & 0 \end{bmatrix}$$

we turn to the closed operator

$$Tx = A(Dx)' + Bx, \quad (Tx)(t) = \begin{bmatrix} -t(x_1(t) - tx_2(t))' + x_1(t) - tx_2(t) \\ -(x_1(t) - tx_2(t))' \end{bmatrix},$$

$$\operatorname{dom} T = \{x \in L^2([a, b], \mathbb{R}^2) | Dx \in H^1([a, b], \mathbb{R}), (Dx)(a) = 0\}.$$

The DAE has strangeness $\mu_{\text{strangeness}} = 1$, see [9, page 70].[8] On the other hand, its tractability index is $\mu = 0$, since the admissible matrix function sequences are

---

[8]As already mentioned in [10, Sections 2.10 and 10.2], though, for regular DAEs, it holds that $\mu_{\text{strangeness}} = \mu - 1$, if $\mu \geq 1$, the strangeness and the tractability index are quite different for nonregular DAEs.

stationary beginning with G_0, e.g.,

$$G_0 = E = AD, \quad Q_0(t) = \begin{bmatrix} 0 & t \\ 0 & 1 \end{bmatrix}, \quad BQ_0 = 0, \quad G_0 = G_1 = G_2 = G_3.$$

The characteristic values are $r_0 = r_1 = r_2 = r_3 = 1$, further $\mu_{ad} = 0$. The nullspace of T reads

$$\ker T = \{x \in \operatorname{dom} T \,|\, x_1(t) = t x_2(t), \ t \in [a, b]\}.$$

Regarding that $q = Tx$ implies $Dx = q_1 + t(Dx)' = q_1 - t q_2, q_2 = -(q_1 - t q_2)'$, we find

$$\operatorname{im} T = \{q \in L^2([a, b], \mathbb{R}^2) \,|\, q_1 - t q_2 \in H^1([a, b], \mathbb{R}), (q_1 - t q_2)(a) = 0, q_2 = -(q_1 - t q_2)'\}$$

which is a closed subspace in $L^2([a, b], \mathbb{R}^2)$. Therefore, T is normally solvable, and this is consistent with the expectation that $\mu = 0$ implies a closed range. □

Example 3.2 The local matrix pencil of the pair

$$E(t) = \begin{bmatrix} t & 1 \\ -t^2 & -t \end{bmatrix}, \quad F(t) = \begin{bmatrix} -2 & 0 \\ 2t & 0 \end{bmatrix}, \quad t \in [a, b],$$

is everywhere singular, $\det(s E(t) - F(t)) \equiv 0$, but, as in Example 3.1, the homogeneous DAE has an infinite-dimensional solution space and the DAE is no longer solvable for all smooth inhomogeneities. Using the factorization

$$E(t) = \begin{bmatrix} 1 \\ -t \end{bmatrix} \begin{bmatrix} t & 1 \end{bmatrix} =: A(t)D(t), \quad B(t) := -F(t) - A(t)D'(t) = \begin{bmatrix} 1 & 0 \\ -t & 0 \end{bmatrix}$$

we turn to the closed operator

$$Sy = A(Dy)' + By, \quad (Sy)(t) = \begin{bmatrix} (t y_1(t) + y_2(t))' + y_1(t) \\ -t(t y_1(t) + y_2(t))' - t y_1(t) \end{bmatrix},$$

$$\operatorname{dom} S = \{y \in L^2([a, b], \mathbb{R}^2) \,|\, Dy \in H^1([a, b], \mathbb{R}), (Dy)(b) = 0\}.$$

The DAE has tractability index is $\mu = 0$, which is documented by the admissible matrix function sequence

$$G_0 = E = AD, \quad Q_0(t) = \begin{bmatrix} 1 & 0 \\ -t & 0 \end{bmatrix}, \quad \Pi_0(t) = \begin{bmatrix} 0 & 0 \\ t & 1 \end{bmatrix}, \quad G_1(t) = \begin{bmatrix} t+1 & 1 \\ -t(t+1) & -t \end{bmatrix},$$

$$Q_1(t) = \begin{bmatrix} -t & -1 \\ t(t+1) & t+1 \end{bmatrix}, \quad \Pi_1 = 0, \quad G_1 = G_2 = G_3.$$

The characteristic values are $r_0 = r_1 = r_2 = r_3 = 1$, and $\mu_{ad} = 1$. The nullspace and range of S are

$$\ker S = \{y \in \text{dom } S| \ y_1 = -(Dy)'\}, \quad \text{im } S = \{p \in L^2([a, b], \mathbb{R}^2)| \ tp_1 + p_2 = 0\},$$

which are closed subspaces in $L^2([a, b], \mathbb{R}^2)$. Therefore, S is normally solvable, and this is consistent with the expectation that $\mu = 0$ implies a closed range. Regarding also the operator T from Example 3.1, the relations

$$\text{im } T = (\ker S)^{\perp}, \quad \text{im } S = (\ker T)^{\perp},$$

can be easily checked. Taking a closer look at the operators, we find that S is the adjoint to T, $S = T^*$. This property gives rise to the expectation that also adjoint pairs of DA operators associated with nonregular DAEs share their characteristic values and tractability index.

Observe that T in Example 3.1 features $\mu_{ad} = 0$ and $q \in \text{im } T$ is involved together with a first derivative, whereas the operator S in the present example shows $\mu_{ad} = 1$ and $p \in \text{im } S$ is involved with no derivative. This property further substantiates the idea mentioned above, that μ_{ad} indicates derivatives involved in the range of the adjoint operator.   □

Example 3.3 ([6, page 91],[4, page. 23],[9, page 56]) The local matrix pencils of the pair

$$E(t) = \begin{bmatrix} 0 & 0 \\ 1 & -t \end{bmatrix}, \quad F(t) = \begin{bmatrix} -1 & t \\ 0 & 0 \end{bmatrix},$$

are singular, $\det(sE(t) - F(t)) \equiv 0$. The homogeneous DAE has the trivial solution only and the DAE is solvable for all sufficiently smooth inhomogeneities. The DAE has strangeness index $\mu_{strangeness} = 1$, [9, page 70]. Using the factorization

$$E(t) = \begin{bmatrix} 0 \\ 1 \end{bmatrix} \begin{bmatrix} 1 & -t \end{bmatrix} =: A(t)D(t), \quad B(t) := -F(t) - A(t)D'(t) = \begin{bmatrix} 1 & -t \\ 0 & 1 \end{bmatrix}$$

we turn to the closed operator

$$Tx = A(Dx)' + Bx, \quad (Tx)(t) = \begin{bmatrix} x_1(t) - tx_2(t) \\ (x_1(t) - tx_2(t))' + x_2(t) \end{bmatrix},$$

$$\text{dom } T = \{x \in L^2([a, b], \mathbb{R}^2)| Dx \in H^1([a, b], \mathbb{R}), (Dx)(a) = 0\}.$$

The DAE is regular with tractability index $\mu = 2$, which is recognizable by the admissible matrix function sequence

$$G_0 = E = AD, \quad Q_0(t) = \begin{bmatrix} 0 & t \\ 0 & 1 \end{bmatrix}, \quad \Pi_0(t) = \begin{bmatrix} 1 & -t \\ 0 & 0 \end{bmatrix},$$

$$G_1(t) = \begin{bmatrix} 0 & 0 \\ 1 & 1-t \end{bmatrix}, \quad Q_1(t) = \begin{bmatrix} 1-t & -t(1-t) \\ -1 & t \end{bmatrix}, \quad \Pi_1 = 0,$$

$$G_2(t) = \begin{bmatrix} 1 & -t \\ 1 & 1-t \end{bmatrix},$$

such that $r_0 = r_1 = 1$, $r_2 = 2$, and $\mu_{ad} = 1$. T is injective and its range is

$$\text{im } T = \{q \in L^2([a, b], \mathbb{R}^2) | q_1 \in H^1([a, b], \mathbb{R}), q_1(a) = 0\},$$

which is a nonclosed subspace in $L^2([a, b], \mathbb{R}^2)$. Therefore, T is densely solvable and fails to be normally solvable. □

Example 3.4 The local matrix pencils of the pair

$$E(t) = \begin{bmatrix} 0 & -1 \\ 0 & t \end{bmatrix}, \quad F(t) = \begin{bmatrix} -1 & 0 \\ t & -1 \end{bmatrix},$$

are singular, $\det(s E(t) - F(t)) \equiv 0$. The homogeneous DAE has the trivial solution only and the DAE is solvable for all sufficiently smooth inhomogeneities. Using the factorization

$$E(t) = \begin{bmatrix} -1 \\ t \end{bmatrix} \begin{bmatrix} 0 & 1 \end{bmatrix} =: A(t)D(t), \quad B(t) := -F(t) - A(t)D'(t) = \begin{bmatrix} 1 & 0 \\ -t & 1 \end{bmatrix}$$

we turn to the closed operator

$$Sy = A(Dy)' + By, \quad (Sy)(t) = \begin{bmatrix} -y_2'(t) + y_1(t) \\ t y_2'(t) - t y_1(t) + y_2(t) \end{bmatrix},$$

$$\text{dom } S = \{y \in L^2([a, b], \mathbb{R}^2) | Dy \in H^1([a, b], \mathbb{R}), (Dy)(b) = 0\}.$$

The DAE is regular with tractability index $\mu = 2$, which is recognizable by the admissible matrix function sequence

$$G_0 = E = AD, \quad Q_0(t) = \begin{bmatrix} 1 & 0 \\ 0 & 0 \end{bmatrix}, \quad \Pi_0(t) = \begin{bmatrix} 0 & 0 \\ 0 & 1 \end{bmatrix},$$

$$G_1(t) = \begin{bmatrix} 1 & -1 \\ -t & t \end{bmatrix}, \quad Q_1(t) = \begin{bmatrix} 0 & 1 \\ 0 & 1 \end{bmatrix}, \quad \Pi_1 = 0,$$

$$G_2(t) = \begin{bmatrix} 1 & -1 \\ -t & t+1 \end{bmatrix},$$

such that $r_0 = r_1 = 1$ and $r_2 = 2$. S is injective and has the range

$$\operatorname{im} S = \{p \in L^2([a,b], \mathbb{R}^2) | tp_1 + p_2 \in H^1([a,b], \mathbb{R}), (tp_1 + p_2)(a) = 0\},$$

which is a nonclosed dense subspace in $L^2([a,b], \mathbb{R}^2)$. Therefore, S is densely solvable and fails to be normally solvable. Note that S is the adjoint operator of T from Example 3.3. We observe that T and $S = T^*$ share their tractability index $\mu = 2$ and the characteristic values $r_0 = r_1 = 2, r_2 = 2$ as well, further we have the additional index $\mu_{\text{ad}} = 1$, the differentiation index $\mu_D = 2$ and the strangeness index $\mu_{\text{strangeness}} = 1$.   □

3.3   Normal Solvability and Beyond

We continue to investigate the closed DA operators $T : \operatorname{dom} T \subset L^2((a,b), \mathbb{R}^m) \to L^2((a,b), \mathbb{R}^k)$ associated with the DAE

$$E(t)x'(t) - F(t)x(t) = q(t), \quad t \in [a,b],$$

with time-varying coefficients, as described by Propositions 3.1. We suppose that an admissible matrix function sequence is given and we are looking for criteria of normal solvability. Recall Condition (A), that is, $\operatorname{im} F \subseteq \operatorname{im} G_1$, to be a sufficient condition of normal solvability, e.g.,[12, Theorem 3.2]. At the same time, this condition ensures tractability index $\mu \le 1$. Moreover, by [12, Theorem 3.4], the condition $\ker G_1 \subseteq \ker E$ is a sufficient condition of normal solvability, too. This condition indicates index $\mu = 1$. In what follows, we generalize these results further, in particular [12, Proposition 3.7], and verify the conjecture stating that each closed DA operator T which has tractability index $\mu \le 1$ is normally solvable.

Theorem 3.4 *Let $E, F : [a,b] \to \mathbb{R}^{k \times m}$ be sufficiently smooth, at least continuous matrix valued functions, $\ker E$ be a C^1-subspaces in $\mathbb{R}^m$, $r = \operatorname{rank} E$, $\rho = r + 1$. If the pair (E, F) has tractability index $\mu \le 1$, then the following holds:*

(1) *The associated closed DA operator T is normally solvable and its range can be represented in terms of an admissible matrix function sequence by*

$$\operatorname{im} T = \{q \in L^2((a,b), \mathbb{R}^k) | W_1 q = -W_1 F D^- U \int_a (U^{-1} D \Pi_{\kappa-1} G_\kappa^-(I - W_1)q)(s) ds \},$$

in which $\kappa = \mu_{\mathrm{ad}} + 1$, $W_1 = I - G_1 G_1^+$ *is the orthoprojector function along* $\mathrm{im}\, G_1$, $U \in C^1([a, b], \mathbb{R}^{n \times n})$ *is given by*

$$U' - (D\Pi_\kappa D^-)'U + D\Pi_{\kappa-1}G_\kappa^- B_\kappa D^- U = 0, \quad U(a) = I, \tag{3.5}$$

and G_κ^- *is the pointwise generalized inverse of* G_κ *determined by*

$$G_\kappa^- G_\kappa G_\kappa^- = G_\kappa^-, \ G_\kappa G_\kappa^- G_\kappa = G_\kappa, \ G_\kappa^- G_\kappa = I - Q_\kappa, \ G_\kappa G_\kappa^- = I - W_1.$$

(2) *If Condition* (A) *is valid, that is,* $\mathrm{im}\, F(t) \subseteq \mathrm{im}\, G_1(t)$, $t \in [a, b]$, *then*

$$\mathrm{im}\, T = \{ q \in L^2((a, b), \mathbb{R}^k) |\ W_1 q = 0 \}$$

(3) *If* $\mathrm{im}\, G_1(t) \equiv \mathbb{R}^k$, *then* T *is surjective.*

Proof (1) We choose a proper factorization such that we can make use of the representation

$$\mathrm{dom}\, T = \{ x \in L^2((a, b), \mathbb{R}^m) | Dx \in H^1((a, b), \mathbb{R}^n), (Dx)(a) = 0 \},$$

$$Tx = A(Dx)' + Bx, \quad x \in \mathrm{dom}\, T,$$

set $\rho = r + 1$, and form a corresponding matrix function sequence G_i, $i = 0, \ldots, \rho + 1$. The operator T has index $\mu \leq 1$ and characteristic values $r_0 \leq r_1 = \cdots = r_\rho = r_{\rho+1}$. For $i \geq 1$ we introduce the orthoprojector functions $W_i = W_1$ and the pointwise generalized inverses G_i^- such that

$$G_i^- G_i G_i^- = G_i^-, \ G_i G_i^- G_i = G_i, \ G_i^- G_i = I - Q_i, \ G_i G_i^- = I - W_i.$$

Observe that, in particular, $G_\kappa^- = G_\kappa^- G_\kappa G_\kappa^- = G_\kappa^-(I - W_1)$ and $W_1 G_\kappa = 0$. By definition we have also $\ker \Pi_{\kappa-1} = \ker \Pi_\kappa$, thus $\Pi_{\kappa-1} = \Pi_{\kappa-1}\Pi_\kappa$.

Next, we consider an arbitrary $\hat{x} \in \mathrm{dom}\, T$ and $\hat{q} = T\hat{x}$. Owing to [10, Proposition 10.3] we can represent

$$\hat{q} = G_\kappa D^-(D\Pi_\kappa \hat{x})' + B_\kappa \hat{x} \tag{3.6}$$

$$+ G_\kappa \sum_{l=0}^{\kappa-1} \underbrace{\{ Q_l \hat{x} - (I - \Pi_l)Q_{l+1}D^-(D\Pi_l Q_{l+1}\hat{x})' + \mathscr{V}_l D\Pi_l \hat{x} + \mathscr{U}_l(D\Pi_l \hat{x})' \}}_{\in \ker \Pi_\kappa},$$

in which

$$\mathscr{U}_l = -(I - \Pi_l)\{ Q_l + Q_{l+1}(I - \Pi_l)Q_{l+1}P_l \}\Pi_l D^-,$$

$$\mathscr{V}_l = (I - \Pi_l)\{ (P_l + Q_{l+1}Q_l)D^-(D\Pi_l D^-)' - Q_{l+1}D^-(D\Pi_{l+1}D^-)' \}D\Pi_l D^-.$$

Regarding the properties $W_1 B_\kappa = W_1 B \Pi_{\kappa-1} = -W_1 F \Pi_{\kappa-1}$ and $B_\kappa = B_\kappa \Pi_{\kappa-1} = B_\kappa \Pi_{\kappa-1} \Pi_\kappa = B_\kappa \Pi_\kappa$, we obtain

$$W_1 \hat{q} = W_1 B_\kappa \hat{x} = W_1 B_\kappa \Pi_\kappa \hat{x} = W_1 B \Pi_\kappa \hat{x} = W_1 B D^- D \Pi_\kappa \hat{x} = -W_1 F D^- D \Pi_\kappa \hat{x}. \tag{3.7}$$

Additionally, multiplication of (3.6) by $D \Pi_{\kappa-1} G_\kappa^-$ and regarding $\Pi_{\kappa-1} G_\kappa^- G_\kappa = \Pi_\kappa$ leads to

$$D \Pi_\kappa D^- (D \Pi_\kappa \hat{x})' + D \Pi_{\kappa-1} G_\kappa^- B_\kappa \hat{x} = D \Pi_{\kappa-1} G_\kappa^- \hat{q},$$

thus

$$(D \Pi_\kappa \hat{x})' - (D \Pi_\kappa D^-)' D \Pi_\kappa \hat{x} + D \Pi_{\kappa-1} G_\kappa^- B_\kappa D^- D \Pi_\kappa \hat{x} = D \Pi_{\kappa-1} G_\kappa^- \hat{q},$$

Since $(D\hat{x})(a) = 0$ implies $(D \Pi_\kappa \hat{x})(a) = (D \Pi_\kappa D^-)(a)(D\hat{x})(a) = 0$, the function $\hat{u} := D \Pi_\kappa \hat{x}$ satisfies the initial value problem

$$u' - (D \Pi_\kappa D^-)'u + D \Pi_{\kappa-1} G_\kappa^- B_\kappa D^- u = D \Pi_{\kappa-1} G_\kappa^- \hat{q}, \quad u(a) = 0. \tag{3.8}$$

Note that this IVP is uniquely solvable and its solution features the property $u = D \Pi_\kappa D^- u$. We apply the fundamental solution matrix U given by (3.5) to represent

$$\hat{u} = U \int_a (U^{-1} D \Pi_{\kappa-1} G_\kappa^- \hat{q})(s) \mathrm{d}s.$$

Inserting this expression into (3.7) yields

$$W_1 \hat{q} = -W_1 F D^- D \Pi_\kappa \hat{x} = -W_1 F D^- U \int_a (U^{-1} D \Pi_{\kappa-1} G_\kappa^- \hat{q})(s) \mathrm{d}s$$

$$= -W_1 F D^- U \int_a (U^{-1} D \Pi_{\kappa-1} G_\kappa^- (I - W_1) \hat{q})(s) \mathrm{d}s. \tag{3.9}$$

It results that $\hat{q}$ belongs to the set $\mathfrak{A}$,

$$\mathfrak{A} = \{q \in L^2((a,b), \mathbb{R}^k) \mid W_1 q = -W_1 F D^- U \int_a (U^{-1} D \Pi_{\kappa-1} G_\kappa^- (I - W_1) q)(s) \mathrm{d}s\},$$

and hence $\mathrm{im}\, T \subseteq \mathfrak{A}$.

Next we consider an arbitrary $\hat{q} \in \mathfrak{A}$ and look for an $\hat{x} \in \mathrm{dom}\, T$ such that $T\hat{x} = \hat{q}$ and eventually $\mathfrak{A} = \mathrm{im}\, T$.

Given $\hat{q} \in \mathfrak{A}$, there is a unique solution $\tilde{u} \in H^1((a,b), \mathbb{R}^n)$ of the IVP

$$u' - R'u + DG_1^- BD^- u = DG_1^- \hat{q}, \quad u(a) = 0.$$

It holds that $\tilde{u} = R\tilde{u}$. Next we introduce $\tilde{x} = (I - Q_0 G_1^- B)D^- \tilde{u} + Q_0 G_1^- \hat{q}$ which belongs to $\operatorname{dom} T$, since $D\tilde{x} = R\tilde{u} = \tilde{u} \in H^1((a, b), \mathbb{R}^n)$, $(D\tilde{x})(a) = \tilde{u}(a) = 0$. Introducing also

$$\tilde{q} = T\tilde{x} = G_1(D^-(D\tilde{x})' + Q_0 \tilde{x}) + BD^- D\tilde{x},$$

we derive

$$\tilde{q} = G_1(D^- \tilde{u}' + Q_0 \tilde{x}) + BD^- \tilde{u}$$

$$= G_1(D^-\{R'\tilde{u} - DG_1^- BD^- \tilde{u} + DG_1^- \hat{q}\} - Q_0 G_1^- BD^- \tilde{u} + Q_0 G_1^- \hat{q}) + BD^- \tilde{u}$$

$$= G_1(-P_0 G_1^- BD^- \tilde{u} + P_0 G_1^- \hat{q} - Q_0 G_1^- BD^- \tilde{u} + Q_0 G_1^- \hat{q}) + BD^- \tilde{u}$$

$$= G_1(-G_1^- BD^- \tilde{u} + G_1^- \hat{q}) + BD^- \tilde{u}$$

$$= G_1(-G_1^- BD^- \tilde{u} + G_1^- \hat{q}) + G_1 G_1^- BD^- \tilde{u} + W_1 BD^- \tilde{u}$$

$$= G_1 G_1^- \hat{q} + W_1 BD^- \tilde{u}.$$

It follows that

$$G_1 G_1^- \tilde{q} = G_1 G_1^- \hat{q}, \tag{3.10}$$

$$W_1 \tilde{q} = W_1 BD^- D\tilde{x} = W_1 B\Pi_\kappa D^- D\tilde{x} = W_1 BD^- D\Pi_\kappa \tilde{x} = -W_1 FD^- D\Pi_\kappa \tilde{x}. \tag{3.11}$$

Owing to [10, Proposition 10.3], now applied to $\tilde{q} = T\tilde{x}$, we obtain the representations

$$D\Pi_\kappa \tilde{x} = U \int_a (U^{-1} D\Pi_{\kappa-1} G_\kappa^- \tilde{q})(s)\mathrm{ds},$$

$$W_1 \tilde{q} = -W_1 FD^- U \int_a (U^{-1} D\Pi_{\kappa-1} G_\kappa^- (I - W_1)\tilde{q})(s)\mathrm{ds}.$$

Since (3.10) corresponds to $(I - W_1)\tilde{q} = (I - W_1)\hat{q}$, we arrive at

$$W_1 \tilde{q} = -W_1 FD^- U \int_a (U^{-1} D\Pi_{\kappa-1} G_\kappa^- (I - W_1)\hat{q})(s)\mathrm{ds} = W_1 \hat{q},$$

and hence, $\hat{q} = \tilde{q} \in \operatorname{im} T$. This proves $\mathfrak{A} = \operatorname{im} T$.

It remains to show that $\mathfrak{A}$ is closed in $L^2((a, b), \mathbb{R}^k)$. Consider a sequence $\{q_l\} \subset \mathfrak{A}$ and a $q_* \in L^2((a, b), \mathbb{R}^k)$ so that $q_l \xrightarrow{L^2} q_*$. We have

$$W_1 q_l = \underbrace{-W_1 FD^- U \int_a (U^{-1} D\Pi_{\kappa-1} G_\kappa^- (I - W_1)q_l)(s)\mathrm{ds}}_{=:u_l},$$

with $u_l \in H^1((a,b), \mathbb{R}^n)$, $u_l(a) = 0$, and

$$u_l' - (D\Pi_\kappa D^-)' u_l + D\Pi_{\kappa-1} G_\kappa^- B_\kappa D^- u_l = D\Pi_{\kappa-1} G_\kappa^- q_l.$$

Set

$$u_* = U \int_a (U^{-1} D\Pi_{\kappa-1} G_\kappa^- (I - W_1) q_*)(s) ds$$

such that $u_* \in H^1((a,b), \mathbb{R}^n)$, $u_*(a) = 0$, and $u_l \xrightarrow{H^1} u_*$. From

$$0 = W_1 q_l + W_1 F D^- u_l \xrightarrow{L^2} W_1 q_* + W_1 F D^- u_*$$

it follows that $W_1 q_* = -W_1 F D^- u_*$, thus $q_* \in \mathfrak{A}$, which completes the proof of Statement (1). Statement (2) is a simple consequence of (1), since Condition (A) is equivalent to $W_1 F = 0$. Statement (3) is then evident.   □

Note that the condition $W_1 F(I - Q_0) = 0$ is applied in [12, Theorem 3.2], which is equivalent to $W_1 F = 0$, and also to Condition (A). In [12, Theorem 3.4] the condition $\ker(E - FQ_0 + EQ_0') \subseteq \ker E$ is assumed, eqivalently, $\ker G_1 \subseteq \ker E$. In both cases, the DA operator has evidently index $\mu \leq 1$. In contrast, [12, Proposition 3.7] uses the condition $W_1 F Q_1 = 0$ together with the somewhat nontransparent condition [12, (3.31) on page 198]. These conditions can now be verified under the assumptions of Theorem 3.4.

Theorem 3.5 : *Let $E, F : [a,b] \to \mathbb{R}^{k \times m}$ be sufficiently smooth, at least continuous matrix valued functions, $\ker E$ be a C^1-subspaces in $\mathbb{R}^m$, $r = \operatorname{rank} E$, $\rho = r + 1$. If the pair (E, F) has tractability index $\mu \leq 1$, then the following holds:*

(1) *The associated closed DA operator T is normally solvable and its nullspace can be represented in terms of an admissible matrix function sequence by*

$$\ker T = \{x \in \operatorname{dom} T \mid x = (I - Q_0 G_1^- B) D^- V \int_a (V^{-1} D\omega)(s) ds + Q_0 \omega,$$

$$\omega \in L^2((a,b), \mathbb{R}^k), \ G_1 \omega = 0\},$$

in which $V \in C^1([a,b], \mathbb{R}^{n \times n})$ is given by

$$V' - R'U + DG_1^- B D^- V = 0, \quad V(a) = I, \tag{3.12}$$

and G_1^- is the pointwise generalized inverse of G_1 determined by

$$G_1^- G_1 G_1^- = G_1^-, \ G_1 G_1^- G_1 = G_1, \ G_1^- G_1 = I - Q_1, \ G_1 G_1^- = I - W_1.$$

(2) *If* $\ker G_1 = \{0\}$, *then T is injective.*

Proof (1) T is normally solvable owing to Theorem 3.4. Consider $x \in \ker T$. $Tx=0$ yields $G_1\{D^-(Dx)' + Q_0x\} + BD^-Dx = 0$, thus

$$G_1\{D^-(Dx)' + Q_0x + G_1^-BD^-Dx\} = 0, \qquad (3.13)$$

$$W_1BD^-Dx = 0. \qquad (3.14)$$

We derive from (3.13) that

$$D^-(Dx)' + Q_0x + G_1^-BD^-Dx = \omega \in L^2((a,b), \mathbb{R}^k), \quad G_1\omega = 0,$$

and further

$$(Dx)' - R'Dx + DG_1^-BD^-Dx = D\omega, \quad Q_0x + Q_0G_1^-BD^-Dx = Q_0\omega,$$

leading to the representation

$$x = D^-Dx + Q_0x = (I - Q_0G_1^-B)D^-V\int_a (V^{-1}D\omega)(s)\mathrm{d}s + Q_0\omega.$$

On the other hand, for each arbitrary $\omega \in L^2((a,b), \mathbb{R}^k)$, with $G_1\omega = 0$, the IVP

$$v' - R' + DG_1^-BD^-v = D\omega, \quad v(a) = 0$$

has a unique solution $v \in H^1((a,b), \mathbb{R}^n)$ and $v = Rv$. Set $x = (I - Q_0G_1^-B)D^-v + Q_0\omega$ such that $Dx = DD^-v = Rv = v \in H^1((a,b), \mathbb{R}^n)$ and $(Dx)(a) = 0$, thus $x \in \mathrm{dom}\, T$. Next we compute

$$Tx = G_1\{D^-v' + Q_0x + G_1^-BD^-v\} + W_1BD^-v = G_1\omega + W_1BD^-v$$

$$= W_1BD^-v = \tilde{q}.$$

It follows that $\tilde{q}$ belongs to $\mathrm{im}\, T$. Owing to the representation of $\mathrm{im}\, T$ in Theorem 3.4, and regarding that $(I - W_1)\tilde{q} = 0$ we arrive at $\tilde{q} = 0$, and hence $Tx = 0$. This completes the proof of the first statement. The second statement is then a direct consequence. □

Corollary 3.6 *Under the assumptions of Theorems 3.4 and 3.5, T is a Fredholm operator if T is regular and $\mu \le 1$.*

Remark 3.4 We have shown that the index condition $\mu \le 1$ is sufficient for normal solvability. This condition is supposably also necessary. This is shown for time-varying regular pairs (E, F) and for arbitrary constant pairs (E, F). A proof of the general case would be very technical and voluminous.

Remark 3.5 We conjecture that the statements of Theorem 2.3 are valid in the same way also for DA operators with time-varying coefficients.

4 Generalizations, Modifications, and Further Comments

4.1 *Continuous A and D, with Continuously Differentiable Border Projector Function*

It has been proposed in [1] to compose the leading term of a DAE at the very beginning by means of a well-matched pair of continuous matrix functions A and D featuring C^1-subspaces ker A and im D which satisfy the transversality condition (A.1). This pursues and generalizes the approach of [6] and means, instead of applying the standard form $Ex' - Fx = q$ one should start at once from a DAE with properly stated leading term, $A(Dx)' + Bx = q$, see also [10, 11]. We quote from [1, Page 785]: *the new form brings more symmetry, transparency and beauty into the theory.*

The results concerning the DA operators T and S of the present paper can be immediately modified to be valid for given forms with properly stated leading terms owing to the following lemma and proposition.

Lemma 4.1 *Let $D : [a, b] \to \mathbb{R}^{n \times m}$ be continuous and let* im D *be a C^1-subspace in $\mathbb{R}^n$. Then each of the sets*

$$\mathcal{M} = \{x \in L^2((a, b), \mathbb{R}^m)|Dx \in H^1((a, b), \mathbb{R}^n)\},$$

$$\mathcal{M}_a = \{x \in L^2((a, b), \mathbb{R}^m)|Dx \in H^1((a, b), \mathbb{R}^n), (Dx)(a) = 0\},$$

$$\mathcal{M}_b = \{x \in L^2((a, b), \mathbb{R}^m)|Dx \in H^1((a, b), \mathbb{R}^n), (Dx)(b) = 0\},$$

$$\mathcal{M}_{a,b} = \{x \in L^2((a, b), \mathbb{R}^m)|Dx \in H^1((a, b), \mathbb{R}^n), (Dx)(a) = 0, (Dx)(b) = 0\},$$

is dense in $L^2((a, b), \mathbb{R}^m)$.

Proof Denote by $\tilde{R} : [a, b] \to \mathbb{R}^{n \times n}$ the orthoprojector function such that $\tilde{R}(t)^2 = \tilde{R}(t) = \tilde{R}(t)^T$ and im $\tilde{R}(t) =$ im $D(t)$, $t \in [a, b]$. Then $\tilde{R}$ is continuously differentiable and inherits the constant rank r from D. We may represent $\tilde{R} = DD^+$, where D^+ is the pointwise Moore–Penrose inverse of D. Note that $D^+ : [a, b] \to \mathbb{R}^{m \times n}$ itself is continuous and the projector function $Q = I - D^+D : [a, b] \to \mathbb{R}^{m \times m}$ onto the nullspace of D is also continuous. Denote $P = I - Q$.

For any $x_* \in L^2((a, b), \mathbb{R}^m)$, we have $Qx_* \in L^2((a, b), \mathbb{R}^m)$, $Px_* \in L^2((a, b), \mathbb{R}^m)$, $Dx_* \in L^2((a, b), \mathbb{R}^n)$, and the decomposition $x_* = Px_* + Qx_* = D^+Dx_* + Qx_*$ as well. Since $H_0^1 := \{u \in H^1((a, b), \mathbb{R}^n)|u(a) = 0, u(b) = 0\}$ is dense in $L^2((a, b), \mathbb{R}^n)$, there is a sequence $\{u_l\} \subset H_0^1$, $u_l \xrightarrow{L^2} Dx_*$. Now

$x_l := D^+ u_l + Q x_*$ belongs to $\mathcal{M}$, $\mathcal{M}_a$, $\mathcal{M}_b$ and $\mathcal{M}_{a,b}$, since $D x_l = D D^+ u_l = \tilde{R} u_l \in H^1((a,b), \mathbb{R}^n)$. Finally, $x_l \xrightarrow{L^2} D^+ D x_* + Q x_* = x_*$, which completes the proof. □

Proposition 4.2 *Let* $A : [a,b] \to \mathbb{R}^{k \times n}$, $D : [a,b] \to \mathbb{R}^{n \times m}$, $B : [a,b] \to \mathbb{R}^{k \times m}$ *be continuous, let* $\ker A$ *and* $\operatorname{im} D$ *be* C^1-*subspaces in* $\mathbb{R}^n$, *and let the transversality condition* (A.1) *be valid. Then the DA operator* $T : L^2((a,b), \mathbb{R}^m) \to L^2((a,b), \mathbb{R}^k)$ *given by*

$$T x = A(Dx)' + Bx, \quad x \in \operatorname{dom} T,$$

$$\operatorname{dom} T = \{x \in L^2((a,b), \mathbb{R}^m) \,|\, Dx \in H^1((a,b), \mathbb{R}^n), (Dx)(a) = 0\},$$

is densely defined and closed.

Proof T is densely defined owing to Lemma 4.1. We show that T is closed. The following reasoning follows closely the lines of the proof of Proposition 3.1.

Let R be the continuously differentiable border projector according to (A.1) and let A^- denote a continuous pointwise generalized inverse such that $A(t)^- A(t) = R(t)$, $t \in [a,b]$. Consider a sequence $\{x_i\} \subset \operatorname{dom} T$, $x_* \in L^2((a,b), \mathbb{R}^m)$, and $y_* \in L^2((a,b), \mathbb{R}^k)$, such that $x_i \xrightarrow{L^2} x_*$, $T x_i \xrightarrow{L^2} y_*$.

From $T x_i = A(D x_i)' + B x_i$ we derive $(I - A A^-)(T x_i - B x_i) = 0$ yielding $(I - A A^-)(y_* - B x_*) = 0$ on the one hand, and, on the other hand,

$$(D x_i)' = A^- T x_i + R' D x_i - A^- B x_i \xrightarrow{L^2} A^- y_* + R' D x_* - A^- B x_* =: v_*.$$

Owing to $D x_i \xrightarrow{L^2} D x_*$, $(D x_i)' \xrightarrow{L^2} v_*$, it follows that $D x_* \in H^1((a,b), \mathbb{R}^n)$ and $(D x_*)' = v_*$. Since now $D x_i \xrightarrow{H^1} D x_*$ and $(D x_i)(a) = 0$ it results that $(D x_*)(a) = 0$, and hence $x_* \in \operatorname{dom} T$.

Considering the relations $(I - A A^-)(y_* - B x_*) = 0$ and $(D x_*)' = v_* = A^- y_* + R' D x_* - A^- B x_*$ we obtain

$$A(D x_*)' = A A^- y_* - A A^- B x_* = y_* - B x_*,$$

which means $y_* = T x_*$ and proves the closedness of T. □

4.2 Different Assignment of Boundary Conditions

One might be interested in the operator $\tilde{T} : L^2((a,b), \mathbb{R}^m) \to L^2([a,b], \mathbb{R}^k)$,

$$\tilde{T} x = A(Dx)' + Bx, \quad x \in \operatorname{dom} \tilde{T},$$

$$\operatorname{dom} \tilde{T} = \{x \in L^2((a,b), \mathbb{R}^m) \,|\, Dx \in H^1((a,b), \mathbb{R}^n)\},$$

which is an extension of the previous DA operator T. $\tilde{T}$ is densely defined and closed owing to Lemma 4.1 and Proposition 3.1. Here, we are led to the adjoint $\tilde{S} : L^2((a,b), \mathbb{R}^k) \to L^2((a,b), \mathbb{R}^m)$,

$$\tilde{S}x = -A^T(D^Ty)' + B^Ty, \quad y \in \operatorname{dom}\tilde{S},$$

$$\operatorname{dom}\tilde{S} = \{y \in L^2((a,b), \mathbb{R}^k) | A^Ty \in H^1((a,b), \mathbb{R}^n), (A^Ty)(a) = 0, (A^Ty)(b) = 0\},$$

which is a restriction of the previous S and, by Lemma 4.1 and Proposition 3.1, a closed densely defined DA operator.

We assign to $\tilde{T}$ the same tractability index as to T. If T has regular index $\mu \leq 1$, (that is, $k = m, r_\mu = m$), then T is bijective, thus Fredholm. Then $\tilde{T}$ inherits the surjectivity, but not the injectivity and ker $\tilde{T}$ has then dimension $d = r_0 = r$.

If T has regular index $\mu > 1$, then it is injective and densely solvable, and $\tilde{T}$ is also densely solvable and ker $\tilde{T}$ has finite dimension $d = m - \sum_{i=0}^{\mu-1}(m - r_i) < r$.

4.3 Integrable Coefficients

In an earlier paper [7], integrable coefficients have been considered in the case $k = m$. More precisely, in the notation of the present paper, the assumptions

$$E \in W^{1,\infty}((a,b), \mathbb{R}^{m \times m}), \quad F \in L^\infty((a,b), \mathbb{R}^{m \times m}),$$
$$Q \in W^{1,\infty}((a,b), \mathbb{R}^{m \times m}) \tag{4.1}$$

for the pointwise projector $Q(t)$ onto the nullspace ker $E(t)$ almost everywhere are used. Then, E is factorized into $E = AD$ with $E = A$ and $D = P = I - Q$. This becomes a proper factorization for integrable coefficients.

Under these conditions, the operator $\mathring{T} : \operatorname{dom}\mathring{T} \subset L^2((a,b), \mathbb{R}^m) \to L^2((a,b), \mathbb{R}^m)$ given by

$$\mathring{T}x = Ex' - Fx, \quad \operatorname{dom}\mathring{T} = H^1((a,b), \mathbb{R}^m)$$

is well-defined. In [7] it is shown that $\mathring{T}$ is closable, and for the closure $T = \overline{\mathring{T}}$ it holds

$$Tx = E(Px)' - (F + EP')x, \quad \operatorname{dom}T = \{x \in L^2((a,b), \mathbb{R}^m) | Px \in H^1((a,b), \mathbb{R}^m)\}.$$

It turns out that $H_P^1(a,b) := \operatorname{dom}T$ is a Hilbert space with the scalar product

$$(x, \bar{x})_{H_P^1} = (x, \bar{x})_{L^2} + ((Px)', (P\bar{x})')_{L^2}, \quad x, \bar{x} \in \operatorname{dom}T.$$

Appropriate boundary conditions are considered as being (finitely many) continuous linear functionals $\{l_1, \ldots, l_s\}$ on $H_P^1(a, b)$ thus determining a closed subspace $V \subseteq H_P^1(a, b)$ by $V = \{x \in H_P^1(a, b) | l_i(x) = 0, i = 1, \ldots, s\}$. The main result of [7] can be summarized as follows.

Proposition 4.3 *Let* (4.1) *be fulfilled. Moreover, let* $R \in W^{1,\infty}((a, b), \mathbb{R}^{m \times m})$ *be such that* $R(t)$ *is a pointwise projector onto* $\operatorname{im} A(t)$ *almost everywhere.*[9] *Set* $S = I - R$ *and* $B = -(F + AP')$. *If* $H := A + SBQ$ *is bijective for almost every* $t \in (a, b)$ *and* $H^{-1} \in L^{\infty}((a, b), \mathbb{R}^{m \times m})$, *then* $T|_V$ *is normally solvable and* $\dim \ker T|_V < \infty$ *for every closed subspace* $V \subseteq H_P^1(a, b)$.

Remark 4.1

(1) $A(t) + S(t)B(t)Q(t)$ is bijective if and only if the matrix pencil $(A(t), B(t))$ is regular and has index 1.
(2) Let the assumptions of Proposition 4.3 hold. If V is determined by finitely many boundary conditions, then $T|_V$ is Fredholm.

Appendix

Proper Factorization and Properly Stated DAEs

We say that N *is a* C^1- *subspace in* $\mathbb{R}^n$, if $N(t) \subseteq \mathbb{R}^n$ is a time-varying subspace, $t \in [a, b]$, and the projector-valued function $Q : [a, b] \to \mathbb{R}^{n \times n}$, with $Q(t) = Q(t)^2 = Q(t)^T$, $\operatorname{im} Q(t) = N(t)$, $t \in [a, b]$, is continuously differentiable. Note that any C^1- subspace in $\mathbb{R}^n$ has constant dimension.

Each continuous matrix function $E : [a, b] \to \mathbb{R}^{k \times m}$ with constant rank r and a nullspace which is a C^1-subspace in $\mathbb{R}^m$ can be factorized into $E = AD$ so that A is continuous, D is continuously differentiable, $\ker A$ and $\operatorname{im} D$ are a C^1-subspace in $\mathbb{R}^n$, and

$$E(t) = A(t)D(t), \quad A(t) \in \mathbb{R}^{k \times n}, \quad D(t) \in \mathbb{R}^{n \times m},$$

$$\ker A(t) \oplus \operatorname{im} D(t) = \mathbb{R}^n, \quad t \in [a, b]. \tag{A.1}$$

A possible choice is $n = m$, $A = E$, $D = E^+E$. If E itself is continuously differentiable, then also the factor A can be chosen to be continuously differentiable, for instance $A = EE^+$, $D = E$. Owing to the condition (A.1) this factorization is called *proper factorization*. Note that then the function $R : [a, b] \to \mathbb{R}^{n \times n}$, projecting pointwise onto $\operatorname{im} D$ along $\ker A$ is also continuously differentiable and

---

[9]For example, $R(t) = A(t)A(t)^+$.

one has

$$\text{im } E = \text{im } A, \ \ker E = \ker D, \quad A = AR, \ D = RD. \tag{A.2}$$

R is then called *border-projector function*.

Using any proper factorization of the leading coefficient E, the standard form DAE

$$Ex' - Fx = q$$

can be rewritten with $B = -(F + AD')$ as *DAE with properly stated leading term or DAE with properly involved derivative*,

$$A(Dx)' + Bx = q.$$

Admissible Matrix Function Sequences

Given are at least continuous matrix functions $E, F : [a, b] \to \mathbb{R}^{k \times m}$, E has a C^1-nullspace and constant rank r. We use a proper factorization $E = AD$, $A : [a, b] \to \mathbb{R}^{k \times n}$, $D : [a, b] \to \mathbb{R}^{n \times m}$, and $B = -(F + AD')$. $R : [a, b] :\to \mathbb{R}^{n \times n}$ denotes the continuously differentiable projector-valued function such that $\text{im } D = \text{im } R$ and $\ker A = \ker R$.

Let $Q_0 : [a, b] \to \mathbb{R}^{n \times n}$ denote any continuously differentiable projector-valued function such that $\text{im } Q_0 = \ker D$, for instance, $Q_0 = I - D^+ D$ with the pointwise Moore-Penrose inverse D^+. Set $P_0 = I - Q_0$ and let D^- denote the pointwise generalized inverse of D determined by

$$D^- D D^- = D^-, \quad D D^- D = D, \quad D D^- = R, \quad D^- D = P_0.$$

Set $G_0 = AD$, $B_0 = B$, $\Pi_0 = P_0$. For a given level $\kappa \in \mathbb{N}$, the sequence $G_0, \ldots, G_\kappa$ is called an *admissible matrix function sequence* associated to the pair (E, F) and triple (A, D, B), respectively, e.g.,[10, Definition 2.6], if it is built by the rule

$$G_i = G_{i-1} + B_{i-1} Q_{i-1},$$

$$\qquad B_i = B_{i-1} P_{i-1} - G_i D^- (D \Pi_i D^-)' D \Pi_{i-1},$$

$$\qquad N_i = \ker G_i, \quad \widehat{N}_i := (N_0 + \cdots + N_{i-1}) \cap N_i, \quad N_0 + \cdots + N_{i-1} =: \widehat{N}_i \oplus X_i,$$

$$\qquad \text{choose } Q_i \text{ such that } Q_i = Q_i^2, \ \text{im } Q_i = N_i, \ X_i \subseteq \ker Q_i,$$

$$\qquad P_i = I - Q_i, \ \Pi_i = \Pi_{i-1} P_i,$$

$$i = 1, \ldots, \kappa,$$

and, additionally,

(a) G_i has constant rank r_i, $i = 0, \ldots, \kappa$,
(b) $\widehat{N_i}$ has constant dimension u_i, $i = 1, \ldots, \kappa$,
(c) Π_i is continuous and $D\Pi_i D^-$ is continuously differentiable, $i = 0, \ldots, \kappa$.

The admissible matrix functions G_i are continuous. The construction is supported by constant-rank conditions.

We mention that, for time-invariant E and F, the matter simplifies to

$$G_i = G_0 + B_0(Q_0 + \ldots + \Pi_{i-1}Q_i) = G_0 + B_0(I - \Pi_i) = E - F(I - \Pi_i).$$

Set $\rho = r + 1$. If $G_0, \ldots, G_{\rho+1}$ is an admissible matrix function sequence, then

$$\operatorname{im} G_0 \subseteq \operatorname{im} G_1 \subseteq \cdots \subseteq \operatorname{im} G_\rho = \operatorname{im} G_{\rho+1},$$

and

$$\ker \Pi_0 \subseteq \ker \Pi_1 \subseteq \cdots \subseteq \ker \Pi_{\rho-1} = \ker \Pi_\rho,$$

that is, the related subspace sequences become stationary at least at level $\rho = r + 1$ and ρ, respectively, [10, Section 10.2]. Note that the there are actually regular DAEs featuring $\operatorname{im} G_{\rho-1} \subset \operatorname{im} G_\rho = \operatorname{im} G_{\rho+1}$, see [10, Example 2.11].

A series of useful properties is incorporated into admissible matrix function sequences, e.g., [10, Propositions 2.5 and 2.7]. In particular, the products Π_i, $\Pi_{i-1}Q_i$, $D\Pi_i D^-$ are projectors, too, and

$$\ker \Pi_i = N_0 + \cdots + N_i,$$

$$B_{i+1} = B_{i+1}\Pi_i.$$

The subspaces $\operatorname{im} G_i$, $\ker \Pi_i = N_0 + \cdots + N_i$, and the numbers r_i, u_i are independent of the special choice of the projector functions Q_j, [10, Theorem 2.8], and also invariant under so-called refactorization $AD = \bar{A}\bar{D}$, [10, Theorem 2.21]. Moreover, the numbers r_i, u_i persist under transformations, which allow to call them *characteristic values* of the given pair (E, F) and triple (A, D, B), respectively.

Finally we quote further useful tools to deal with admissible matrix function sequences. Choose continuous projector-valued functions $W_i : [a, b] \to \mathbb{R}^{k \times k}$ such that $\ker W_i = \operatorname{im} G_i$ and then pointwise generalized inverses G_i^- of G_i determined by

$$G_i^- G_i G_i^- = G_i^-, \quad G_i G_i^- G_i = G_i, \quad I - G_i G_i^- = W_i, \quad G_i^- G_i = P_i, \quad i \geq 0.$$

Owing to [10, Proposition 2.5] one has then

$$W_i B_i = W_i B,$$
$$\operatorname{im} G_i = \operatorname{im}(G_{i-1} + W_{i-1} B Q_{i-1}) = \operatorname{im} G_{i-1} \oplus \operatorname{im} W_{i-1} B Q_{i-1}.$$

References

1. Balla, K., März, R.: A unified approach to linear differential algebraic equations and their adjoints. J. Analy. Appl. **21**(3), 783–802 (2002)
2. Berger, T., Reis, T.: Controllability of linear differential-algebraic systems – A survey. In: Ilchmann, A., Reis, T. (eds.) Surveys in Differential-Algebraic Equations I. Differential-Algebraic Equations Forum, pp. 1–61. Springer, Heidelberg (2013)
3. Berger, T., Trenn, S.: The Quasi-Kronecker form for matrix pencils. SIAM J. Matrix Anal. Appl. **33**(2), 336–368 (2012)
4. Brenan, K.E., Campbell, S.L., Petzold, L.R.: Numerical Solution of Initial-Value Problems in Differential-Algebraic Equations. Elsevier Science Publishing, Amsterdam (1989)
5. Chistyakov, V.: Algebro-differential operators with a finite-dimensional kernel. Nauka, Novosibirsk (1996). (In Russian)
6. Griepentrog, E., März, R.: Differential-Algebraic Equations and Their Numerical Treatment. Teubner Texte zur Mathematik, vol. 88. BSB Teubner, Leipzig (1986)
7. Hanke, M.: Linear differential-algebraic equations in spaces of integrable functions. J. Diff. Eq. **79**(1), 14–30 (1989)
8. Kato, T.: Perturbation Theory for Linear Operators. Classics in Mathematics. Springer, Berlin (1995). Reprint of the 1980 Edition
9. Kunkel, P., Mehrmann, V.: Differential-Algebraic Equations. Analysis and Numerical Solution. Textbooks in Mathematics. European Mathematical Society, Zürich (2006)
10. Lamour, R., März, R., Tischendorf, C.: Differential-Algebraic Equations: A Projector Based Analysis. Ilchmann, A., Reis, T. (eds.) Differential-Algebraic Equations Forum. Springer, Berlin (2013).
11. Linh, V.H., März, R.: Adjoint pairs of differential-algebraic equations and their Lyapunov exponents. J Dyn. Diff. Eq. **29**(2), 655–684 (2017)
12. März, R.: Surveys in differential-algebraic equations II. In: Ilchmann, A., Reis, T. (eds.) Differential-Algebraic Equations from a Functional-Analytic Viewpoint: A Survey. Differential-Algebraic Equations Forum, pp. 163–285. Springer, Heidelberg (2015)
13. Puche, M., Reis, T., Schwenninger, F.L.: Constant-coefficient differential-algebraic operators and the kronecker form. Linear Algebra Appl. **552**, 29–41 (2018)
14. Puche, M., Reis, T., Schwenninger, F.L.: Differential-algebraic operators and the Kronecker form. Lecture at the Workshop "Descriptor 2019" in Paderborn, Germany, March 17–20, 2019 (2019)
15. Röbenack, K.: Beitrag zur Analyse von Deskriptorsystemen. Berichte aus der Steuerungs- und Regelungstechnik. Shaker Verlag Aachen (1999)
16. Röbenack, K., Reinschke, K.J.: On generalized inverses of singular pencils. Int. J. Appl. Math. Comput. Sci. **21**(1), 161–172 (2011)
17. Zhuk, S.M.: Closedness and normal solvability of an operator generated by a degenerate linear differential equation with variable coefficients. Nonlinear Oscil. **10**(4), 469–486 (2007)

Part II
Numerical Methods and Model Order Reduction

Inter/Extrapolation-Based Multirate Schemes: A Dynamic-Iteration Perspective

Andreas Bartel and Michael Günther

Abstract Multirate behavior of ordinary differential equations (ODEs) and differential-algebraic equations (DAEs) is characterized by widely separated time constants in different components of the solution or different additive terms of the right-hand side. Here, classical multirate schemes are dedicated solvers, which apply (e.g.) micro and macro steps to resolve fast and slow changes in a transient simulation accordingly. The use of extrapolation and interpolation procedures is a genuine way for coupling the different parts, which are defined on different time grids.

This paper contains for the first time, to the best knowledge of the authors, a complete convergence theory for inter/extrapolation-based multirate schemes for both ODEs and DAEs of index one, which are based on the fully-decoupled approach, the slowest-first and the fastest-first approach. The convergence theory is based on linking these schemes to multirate dynamic iteration schemes, i.e., dynamic iteration schemes without further iterations. This link defines naturally stability conditions for the DAE case.

Keywords ODEs · DAEs · Multirate schemes · Convergence theory

Mathematics Subject Classification (2010) 65L80, 65L99, 65M12

The authors are indebted to the EU project ROMSOC (EID).

A. Bartel (✉) · M. Günther
Fakultät für Mathematik und Naturwissenschaften, IMACM – Institute of Mathematical Modelling, Analysis and Computational Mathematics, Bergische Universität Wuppertal, Wuppertal, Germany
e-mail: bartel@uni-wuppertal.de; guenther@uni-wuppertal.de

T. Reis et al. (eds.), *Progress in Differential-Algebraic Equations II*,
Differential-Algebraic Equations Forum,
https://doi.org/10.1007/978-3-030-53905-4_3

1 Introduction

In practice, technical applications are often modeled as coupled systems of ordinary differential equations (ODEs) or differential algebraic equations (DAEs). Furthermore, it is a very common aspect of technical applications that the transient behavior is characterized by different time constants. At a given instance of time, certain parts of a dynamical system are slowly evolving, while others have a fast dynamics in the direct comparison. Here, this is referred to as *multirate behavior*. To name but a few applications: multibody systems [1, 10], electric circuits [11, 17], climate models [21] and, of course, multiphysical systems, e.g. field/circuit coupling [20]. Now, to have an efficient numerical treatment of systems with multirate behavior, special integration schemes are developed, so-called multirate schemes. To the best knowledge of the authors, the multirate history goes back to Rice [22] in 1960, where step sizes for time integration are adapted to the activity level of subsystems. Many works followed, and we give only a partial list here: based on BDF-methods [13], based on ROW methods [14], based on extrapolation methods [12] partitioned RK and compound step [16], mixed multirate with ROW [4], based on a refinement strategy [23], for conservation laws [8], compound-fast [24], infinitesimal step [25], implicit-explicit [9], based on GARK-methods [15].

The fundamental idea of a multirate scheme is the following: an efficient algorithm should (if there are no stability issues) sample a certain component/subsystem according to the activity level. The more active a component is, the shorter are the time scales and the higher the sampling rate should be chosen to achieve a given level of accuracy. In other words, there is not a global time step, but a local one, which should reflect the inherent time scale of an unknown or some subsystem. For simplicity, we work here with only two time scales. That is, we allow for a fast subsystem (of higher dynamics), which employs a small step of size h (*micro step*) and a slow subsystem, which employs a larger step size H (*macro step*). Furthermore, we assume for simplicity the relation $H = mh$ with $m \in \mathbb{N}$. In fact, the main feature of a certain multirate scheme is to define the coupling variables in an appropriate way. Here we focus on inter- and extrapolation strategies for coupling both subsystems, since we aim at highlighting the connection to dynamic iteration schemes.

The work is structured as follows: In Sect. 2, the formulation of multirate initial value problems is given on the basis of ordinary differential equations (ODEs). Furthermore, various known versions of extra- and interpolation coupling are explained. Following this, the consistency of multirate one-step methods are discussed for ODEs (Sect. 3). Then, in Sect. 4, the ODE results are generalized to the DAE case. Conclusions complete the presentation.

2 Notation for Coupled Systems and Multirate Extra/Interpolation

We start from an initial value problem (IVP) based on a model of ordinary differential equations (ODEs):

$$\dot{w} = h(t, w), \qquad w(t_0) = w_0, \qquad t \in (t_0, t_{\text{end}}], \qquad (2.1)$$

where h is continuous and Lipschitz continuous in w, $w_0 \in \mathbb{R}^n$ is given. Moreover, let h or w, resp., be comprised of some slower changing parts (in time domain), whereas the remaining parts are faster changing. This is referred to as multirate behavior. Now, there are two equivalent ways of partitioning:

(a) The *component-wise partitioning* splits the unknown into slow $y_S(t) \in \mathbb{R}^m$ and fast components $y_F(t) \in \mathbb{R}^{n-m}$, such that $w^\top = (y_S^\top, y_F^\top)$ and

$$\begin{aligned} \dot{y}_S &= f_S(t, y_S, y_F), \ y_S(t_0) = y_{S,0}, \\ \dot{y}_F &= f_F(t, y_S, y_F), \ y_F(t_0) = y_{F,0}, \end{aligned} \qquad (2.2)$$

with corresponding splitting of the right-hand side.

(b) The *right-hand side partitioning* is an additive splitting of h into slow and fast summands:

$$\dot{w} = h_s(t, w) + h_f(t, w), \qquad w(t_0) = w_0, \qquad (2.3)$$

such that $w = w_s + w_f$ with $\dot{w}_s = h_s(t, w_s + w_f)$ and $\dot{w}_f = h_f(t, w_s + w_f)$. Of course, the initial data needs to be split in a suitable way. If the dynamics are solely determined by h_s and h_f, the splitting is arbitrary to some extent.

Since both ways of partitioning are equivalent, i.e., a component-wise partitioning can be written as a right-hand side partitioning and vice-versa [15], we choose for the work at hand the formulation (2.2), without loss of generality. Moreover, the partitioning (2.2) can be generalized to the case of differential algebraic equations (DAEs) with certain index-1 assumptions. This DAE setting is treated in Sect. 4.

In this work, we study multirate methods, which belong to the framework of one-step-methods (and multi-step schemes, too, see Remark 3.3 below) and which are based on extrapolation and interpolation for the coupling variables. To describe these methods, let us assume that the computation of the coupled system (2.2) has reached time $t = \bar{t}$ with

$$\begin{aligned} \dot{y}_S &= f_S(t, y_S, y_F), \qquad y_S(\bar{t}) = y_{S,\bar{t}}, \\ \dot{y}_F &= f_F(t, y_S, y_F), \qquad y_F(\bar{t}) = y_{F,\bar{t}}. \end{aligned} \qquad (2.4)$$

Now, the multirate integration of the whole coupled system is defined for one macro step, i.e., on $[\bar{t}, \bar{t} + H] \subseteq [t_0, t_{\text{end}}]$. It comprises a single step of macro step size H for the subsystem y_S and $m \in \mathbb{N}$ steps of (micro step) size h for y_F. To this end, the respective coupling variables need to be evaluated. Here, our presentation is restricted to extrapolation and interpolation for the coupling variables, although there are several other techniques such as, just to name a few, compound-step [16], Multirate GARK [15] or extrapolation based [12] schemes. Depending on the sequence of computation of the unknowns y_S and y_F, one distinguishes the following three versions of extra-/and interpolation techniques:

(i) *fully-decoupled approach [7]:* fast and slow variables are integrated in parallel using in both cases extrapolated waveforms based on information from the initial data of the current macro step at $\bar{t}$;

(ii) *slowest-first approach [13]:* in a first step, the slow variables are integrated, using an extrapolated waveform of y_F based on information available at $\bar{t}$ for evaluating the coupling variable y_F in the current macro step. In a second step, m micro steps are performed to integrate the fast variables y_F from $\bar{t}$ to $\bar{t} + H$, using an interpolated waveform of y_S based on information from the current macro step size $[\bar{t}, \bar{t} + H]$ for evaluating the coupling variable y_F.

(iii) *fastest-first approach [13]:* in a first step, m micro steps are performed to integrate the fast variables, using an extrapolated waveform of y_S based on information available at $\bar{t}$ for evaluating the coupling variable y_S in the current macro step. In a second step, one macro step is performed to integrate the slow variables y_S from $\bar{t}$ to $\bar{t} + H$, using an interpolated waveform of y_F based on information from the current macro step size $[\bar{t}, \bar{t} + H]$ for evaluating the coupling variable y_F.

Remark 2.1 The restriction that the extrapolation can only be based on the information at $\bar{t}$ can be relaxed to the data of the preceding macro step $[\bar{t} - H, \bar{t}]$. In fact, one can encode such an information e.g. as a spline model, which is also updated and transported from macro step to macro step.

3   The ODE Case

The details presented in this section are based on a result first presented in [7]. Starting from this result, we use the underlying strategy to extend it to our case of the three multirate versions named in the previous section. Basically, for ODE systems, all variants of extrapolation/interpolation-based multirate schemes have convergence order p (in the final asymptotic phase) provided that it holds:

(i) the basic integration scheme (i.e., the scheme for both the slow and the fast subsystems with given coupling data) has order p and

(ii) the extrapolation/interpolation schemes are of approximation order $p - 1$.

For the fully decoupled approach, this is a consequence of the following result, which is a generalization of Theorem 1 in [3] for constant extrapolation:

Theorem 3.1 (Consistency of Fully-Decoupled Multirate Schemes) *Given the coupled ODE-IVP (2.2), where f_S and f_F are Lipschitz w.r.t. the sought solution. Furthermore, we apply two basic integration schemes of order p: one for y_S with macro step size H, a second for y_F with fixed multirate factor $m(\in \mathbb{N})$ steps of size h. If these integration schemes are combined with two extrapolation procedures for the coupling variables of order $p-1$, the resulting fully decoupled multirate scheme has order p.*

Proof We consider the case that we have computed the IVP system (2.2) until time $\bar{t}$ with initial data $y_S(\bar{t}) = y_{S,\bar{t}}$, $y_F(\bar{t}) = y_{F,\bar{t}}$, i.e., we have the setting given in system (2.4). Moreover, the unique solution of (2.4) is referred to as

$$(y_S(t; y_{S,\bar{t}}, y_{F,\bar{t}})^\mathsf{T}, \quad y_F(t; y_{S,\bar{t}}, y_{F,\bar{t}})^\mathsf{T}) \quad \text{or} \quad (y_S(t)^\mathsf{T}, \quad y_F(t)^\mathsf{T}) \text{ as short-hand.}$$

Next, we provide extrapolated, known quantities $\tilde{y}_S$ and $\tilde{y}_F$ for the coupling variables of order $p - 1$: (for constants respective L_S, $L_F > 0$)

$$\begin{aligned}
y_S(t) - \tilde{y}_S(t) &= L_S \cdot H^p + \mathcal{O}(H^{p+1}) & \text{for any } t \in [\bar{t}, \bar{t} + H], & \quad \text{and} \\
y_F(t) - \tilde{y}_F(t) &= L_F \cdot H^p + \mathcal{O}(H^{p+1}) & \text{for any } t \in [\bar{t}, \bar{t} + H].
\end{aligned}$$
(3.1)

Replacing the coupling variables in (2.4) by $\tilde{y}_S$ and $\tilde{y}_F$, we obtain the following modified system

$$\begin{aligned}
\dot{y}_S &= f_S(t, y_S, \tilde{y}_F) =: \tilde{f}_S(t, y_S), & y_S(\bar{t}) &= y_{S,\bar{t}}, \\
\dot{y}_F &= f_F(t, \tilde{y}_S, y_F) =: \tilde{f}_F(t, y_F), & y_F(\bar{t}) &= y_{F,\bar{t}},
\end{aligned}$$
(3.2)

which is fully decoupled (for $t \in [\bar{t}, \bar{t} + H]$). Its unique solution is referred to as

$$(\hat{y}_S(t; y_{S,\bar{t}}, y_{F,\bar{t}})^\mathsf{T}, \quad \hat{y}_F(t; y_{S,\bar{t}}, y_{F,\bar{t}})^\mathsf{T}).$$

Now, we apply the two basic integration schemes of order p in multirate fashion to the decoupled model (3.2) and we refer to the numerical solution at $t^* = \bar{t} + H$ as

$$(y_{S,H}(t^*), y_{F,H}(t^*))^\mathsf{T}.$$

Then, the distance between multirate and exact solution can be estimated as follows:

$$\begin{pmatrix} \|y_{S,H}(t^*) - y_S(t^*)\| \\ \|y_{F,H}(t^*) - y_F(t^*)\| \end{pmatrix} \leq \begin{pmatrix} \|y_{S,H}(t^*) - \hat{y}_S(t^*)\| \\ \|y_{F,H}(t^*) - \hat{y}_F(t^*)\| \end{pmatrix} + \begin{pmatrix} \|\hat{y}_S(t^*) - y_S(t^*)\| \\ \|\hat{y}_F(t^*) - y_F(t^*)\| \end{pmatrix}.$$
(3.3)

The fully decoupled multirate scheme gives for the first term on the right-hand side:

$$\begin{pmatrix} \|y_{S,\,H}(t^*) - \widehat{y}_S(t^*)\| \\ \|y_{F,\,H}(t^*) - \widehat{y}_F(t^*)\| \end{pmatrix} \leq \begin{pmatrix} c_S H^{p+1} + \mathcal{O}(H^{p+2}) \\ c_F H^{p+1} + \mathcal{O}(H^{p+2}) \end{pmatrix} \tag{3.4}$$

employing constants c_S, $c_F > 0$ (for leading errors). Using Lipschitz continuity of f_S, f_F for the second summand on the right-hand side of (3.3), we find

$$\begin{pmatrix} \|\widehat{y}_S(t^*) - y_S(t^*)\| \\ \|\widehat{y}_F(t^*) - y_F(t^*)\| \end{pmatrix} \leq \int_{\bar{t}}^{t^*} \begin{pmatrix} \|f_S(\tau, \widehat{y}_S(\tau), \widetilde{y}_F(\tau)) - f_S(\tau, y_S(\tau), y_F(\tau))\| \\ \|f_F(\tau, \widetilde{y}_S(\tau), \widehat{y}_F(\tau)) - f_F(\tau, y_S(\tau), y_F(\tau))\| \end{pmatrix} d\tau$$

$$\leq \int_{\bar{t}}^{t^*} \begin{pmatrix} L_{S,S}\|\widehat{y}_S(\tau) - y_S(\tau)\| + L_{S,F}\|\widetilde{y}_F(\tau) - y_F(\tau)\| \\ L_{F,S}\|\widetilde{y}_S(\tau) - y_S(\tau)\| + L_{F,F}\|\widehat{y}_F(\tau) - y_F(\tau)\| \end{pmatrix} d\tau \tag{3.5}$$

with respective Lipschitz constants $L_{i,j}$ (for system i and dependent variables j). We remark that this estimate is decoupled. Inserting the extrapolation estimates (3.1), we deduce further

$$\begin{pmatrix} \|\widehat{y}_S(t^*) - y_S(t^*)\| \\ \\ \|\widehat{y}_F(t^*) - y_F(t^*)\| \end{pmatrix} \leq \begin{pmatrix} L_{S,F} \cdot L_F \cdot H^{p+1} + L_{S,S}\int_{\bar{t}}^{t^*} \|\widehat{y}_S(\tau) - y_S(\tau)\| d\tau + \mathcal{O}(H^{p+2}) \\ \\ L_{F,S} \cdot L_S \cdot H^{p+1} + L_{F,F}\int_{\bar{t}}^{t^*} \|\widehat{y}_F(\tau) - y_F(\tau)\| d\tau + \mathcal{O}(H^{p+2}) \end{pmatrix}.$$

Via Gronwall's lemma, we deduce:

$$\begin{pmatrix} \|\widehat{y}_S(t^*) - y_S(t^*)\| \\ \|\widehat{y}_F(t^*) - y_F(t^*)\| \end{pmatrix} \leq \begin{pmatrix} L_{S,F}L_F\, e^{L_{S,S}(t^*-\bar{t})}\, H^{p+1} + \mathcal{O}(H^{p+2}) \\ L_{F,S}L_S\, e^{L_{F,F}(t^*-\bar{t})}\, H^{p+1} + \mathcal{O}(H^{p+2}) \end{pmatrix}. \tag{3.6}$$

In combination with the integration estimate (3.4), the error (3.3) of the fully-decoupled multirate scheme has consistency order p on the macro scale level, which is the claim.   □

The proof can be slightly adapted to verify the convergence result for both remaining variants as well:

Corollary 3.2 (Consistency of Slowest-First Multirate Schemes) *The convergence result of Theorem 3.1 remains valid if the fully-decoupled approach is replaced by the slowest-first approach, i.e., the coupling variables y_S (during the integration of y_F) are evaluated using interpolation of the already computed slow data in the current macro step.*

Proof We just give the changes of the above proof. For the slowest-first variant, the modified equation on the current macro step $[\bar{t}, \bar{t} + H]$ reads

$$
\begin{aligned}
\dot{y}_S &= f_S(t, y_S, \widetilde{y}_F) =: \widetilde{f}_S(t, y_S), & y_S(\bar{t}) &= y_{S,\bar{t}}, \\
\dot{y}_F &= f_F(t, y_S^{\text{int}}, y_F) =: \widetilde{f}_F(t, y_F), & y_F(\bar{t}) &= y_{F,\bar{t}}
\end{aligned}
\tag{3.7}
$$

with extrapolated values $\widetilde{y}_F$ as in the fully-decoupled approach and interpolated values y_S^{int} of order $p - 1$ based on the numerical approximations $y_{S,H}(t_k)$ with $t_k \in [\bar{t}, \bar{t} + H]$ such that it holds:

$$
\widehat{y}_S(t) - y_S^{\text{int}}(t) = \widetilde{L}_S \cdot H^p + \mathcal{O}(H^{p+1}) \quad \text{for any } t \in [\bar{t}, \bar{t} + H].
\tag{3.8}
$$

Again, the hat-notation is employed for the exact solution of system (3.7). The computation of the slow part still employs extrapolated coupling variables. This decouples the slow part from the fast part as before and hence the error estimates of y_S are unchanged. In fact, we can use the estimates (3.4$_1$) and (3.6$_1$): for any time $\tau \in (\bar{t}, \bar{t} + H]$.

Now, for the fast part, the corresponding estimate to (3.5$_2$) reads (with using $y_S^{\text{int}}(t) - y_S(t) = y_S^{\text{int}}(t) - \widehat{y}_S(t) + \widehat{y}_S(t) - y_S(t)$)

$$
\|\widehat{y}_F(t^*) - y_F(t^*)\| \leq \int_{\bar{t}}^{t^*} L_{F,S}\left(\|\widehat{y}_S(\tau) - y_S^{\text{int}}(\tau)\| + \|\widehat{y}_S(\tau) - y_S(\tau)\|\right) \\
+ L_{F,F}\|\widehat{y}_F(\tau) - y_F(\tau)\|\, d\tau.
$$

Using (3.6$_1$) (with τ instead of t^*) and using (3.8), we find

$$
\|\widehat{y}_F(t^*) - y_F(t^*)\|
$$

$$
\leq \int_{\bar{t}}^{t^*} \left(L_{F,S}\widetilde{L}_S H^p + \mathcal{O}(H^{p+1}) + L_{F,S}L_{S,F}L_F\, e^{L_{S,S}(\tau-\bar{t})} H^{p+1} + \mathcal{O}(H^{p+2}) \right.
$$

$$
\left. + L_{F,F}\|\widehat{y}_F(\tau) - y_F(\tau)\| \right) d\tau
$$

$$
\leq L_{F,S}\widetilde{L}_S H^{p+1} + L_{F,F}\int_{\bar{t}}^{t^*} \|\widehat{y}_F(\tau) - y_F(\tau)\| d\tau + \mathcal{O}(H^{p+2}).
$$

Now, the application of Gronwall's lemma leads to

$$
\|\widehat{y}_F(t^*) - y_F(t^*)\| \leq L_{F,S}\widetilde{L}_S e^{L_{F,F}H} H^{p+1} + \mathcal{O}(H^{p+2}).
$$

Finally, we need to form the total error in the fast components, the equivalent to (3.3$_2$). Since the numerical scheme for the fast component is of order p, we can still employ (3.4$_2$), and we get the estimate

$$\|y_{F,H}(t^*) - y_F(t^*)\| \leq \left(c_F + L_{F,s}\widetilde{L}_s e^{L_{F,F}H}\right) H^{p+1} + \mathcal{O}(H^{p+2}). \quad (3.9)$$

$\square$

Remark 3.1 If one uses interpolation schemes of order p instead of $p-1$, one has to replace the term $\widetilde{L}_S H^p$ by $\widetilde{L}_S H^{p+1}$, which yields the estimate

$$\|y_{F,H}(t^*) - y_F(t^*)\| \leq c_F H^{p+1} + \mathcal{O}(H^{p+2}), \quad (3.10)$$

that is, the extra-/interpolation error is dominated by the error of the numerical integration scheme.

Corollary 3.3 (Consistency of Fastest-First Multirate Schemes) *The convergence result of Theorem 3.1 remains valid if the fully-decoupled approach is replaced by the fastest-first one, i.e., the coupling variables y_F (during the integration of y_S) are evaluated using interpolation instead of extrapolation.*

Proof For the fastest-first variant, the modified equation (3.2) reads on $[\bar{t}, \bar{t} + H]$

$$\begin{aligned} \dot{y}_S &= f_S(t, y_S, y_F^{\text{int}}) =: \widetilde{f}_S(t, y_S), & y_S(\bar{t}) &= y_{S,\bar{t}}, \\ \dot{y}_F &= f_F(t, \widetilde{y}_S, y_F) =: \widetilde{f}_F(t, y_F), & y_F(\bar{t}) &= y_{F,\bar{t}}, \end{aligned} \quad (3.11)$$

with extrapolated values $\widetilde{y}_S$ as in the fully-decoupled approach and interpolated values y_F^{int} of order $p-1$ based on the numerical approximations $y_{F,H}(t_k)$ with $t_k \in [\bar{t}, \bar{t} + H]$:

$$\widehat{y}_F(t) - y_F^{\text{int}}(t) = \widetilde{L}_F \cdot H^p + \mathcal{O}(H^{p+1}) \quad \text{for any } t \in [\bar{t}, \bar{t} + H]. \quad (3.12)$$

The second equation of (3.11) for y_F is unchanged with respect to Theorem 3.1, since the extrapolation of y_S is still used. Hence, we still have all respective estimates for the fast part, in particular (3.4$_2$) and (3.6$_2$). For the slow part, the corresponding estimate to (3.5$_1$) now reads (with using $y_F^{\text{int}}(t) - y_F(t) = y_F^{\text{int}}(t) - \widehat{y}_F(t) + \widehat{y}_F(t) - y_F(t)$)

$$\|\widehat{y}_S(t^*) - y_S(t^*)\| \leq \int_{\bar{t}}^{t^*} \left(L_{S,F}\left(\|\widehat{y}_F(\tau) - y_F^{\text{int}}(\tau)\| + \|\widehat{y}_F(\tau) - y_F(\tau)\|\right)\right.$$

$$\left. + L_{S,S}\|\widehat{y}_S(\tau) - y_S(\tau)\|\right) d\tau.$$

Using 3.6_2 (with τ replaced by $t^\star$) and using (3.12), we find

$$\|\widehat{y}_S(t^\star)-y_S(t^\star)\| \le \int_{\bar{t}}^{t^\star} \left(L_{S,F}\widetilde{L}_F H^p + \mathcal{O}(H^{p+1}) + L_{S,F}L_{F,S}L_S\, e^{L_{F,F}(\tau-\bar{t})}\, H^{p+1} \right.$$

$$\left. + \mathcal{O}(H^{p+2}) + L_{S,S}\|\widehat{y}_S(\tau) - y_S(\tau)\| \right) d\tau$$

$$\le L_{S,F}\widetilde{L}_F H^{p+1} + \int_{\bar{t}}^{t^\star} L_{S,S}\|\widehat{y}_F(\tau) - y_F(\tau)\|d\tau + \mathcal{O}(H^{p+2}).$$

Applying now Gronwall's lemma leads to

$$\|\widehat{y}_S(t^\star) - y_S(t^\star)\| \le L_{S,F}\widetilde{L}_F e^{L_{S,S}H}\, H^{p+1} + \mathcal{O}(H^{p+2}).$$

Finally, we use both the above deduced error and the numerical error (3.4_1) in the general error sum (3.3_1) and we find for the slow part

$$\|y_{S,\,H}(t^\star) - y_S(t^\star)\| \le \left(c_S + L_{S,F}\widetilde{L}_F e^{L_{S,S}H} \right) H^{p+1} + \mathcal{O}(H^{p+2}). \quad (3.13)$$

i.e., the numerical integration error is dominated by the extrapolation/interpolation error.   □

Remark 3.2 If one uses interpolation schemes of order p instead of $p - 1$, one has to replace the term $\widetilde{L}_F H^p$ by $\widetilde{L}_F H^{p+1}$, which yields the estimate

$$\|y_{S,\,H}(t^\star) - y_S(t^\star)\| \le c_S H^{p+1} + \mathcal{O}(H^{p+2}), \quad (3.14)$$

that is, the extra-/interpolation error is dominated by the error of the numerical integration scheme.

Remark 3.3 For the basic integration schemes employed in Theorem 3.1, Corollaries 3.2 and 3.3 we can use either

(a) one-step integration schemes, or
(b) multistep schemes, where both schemes are 0-stable.

Remark 3.4 (Schemes) Extrapolation of order 0 and 1 can be easily obtained from the initial data at $t = \bar{t}$ and a derivative information, which is provided by the ODE. This allows directly the construction of multirate methods of order 2.

Remark 3.5 Notice that for a working multirate scheme, we still have to specify the extrapolation/interpolation formulas. In fact, arbitrary high orders of the extra-/interpolation are only possible if information of previous time steps is used. Generally, this may turn a one-step scheme into a multi-step scheme, and raise questions concerning stability. However, if the extrapolation is computed sequentially in a spline-oriented fashion (see Remark 2.1), the modified functions $\widetilde{f}_S$ and $\widetilde{f}_F$ are the same for all time intervals inside $[t_0, t_{end}]$, and the extrapolation/interpolation based multirate scheme can still be considered as a one-step scheme applied to the modified ODE equations.

4 The DAE Case

The component-wise partitioning (2.2) (as well as the right-hand side partitioning (2.3)) can be generalized to the case of differential algebraic equations (DAEs). Let us assume that the slow and the fast subsystem can be written as semi-explicit system of index-1, each for given corresponding coupling terms as time functions. This reads:

$$\dot{y}_S = f_S(t, y_S, y_F, z_S, z_F), \quad y_S(t_0) = y_{S,0}, \quad \dot{y}_F = f_F(t, y_S, y_F, z_S, z_F), \quad y_F(t_0) = y_{F,0},$$

$$0 = g_S(t, y_S, y_F, z_S, z_F), \qquad\qquad 0 = g_F(t, y_S, y_F, z_S, z_F). \tag{4.1}$$

Moreover, the overall system is assumed to be index-1 as well. All index-1 conditions lead to the assumption that the following Jacobians

$$\frac{\partial g_S}{\partial z_S}, \quad \frac{\partial g_F}{\partial z_F} \quad \text{and} \quad \begin{pmatrix} \frac{\partial g_S}{\partial z_S} & \frac{\partial g_S}{\partial z_F} \\ \frac{\partial g_F}{\partial z_S} & \frac{\partial g_F}{\partial z_F} \end{pmatrix} \quad \text{are regular} \tag{4.2}$$

in a neighborhood of the solution. For later use, we introduce Lipschitz constants with respect to the algebraic variables:

$$\|g_S(t, y_S, y_F, z_S, z_F) - g_S(t, y_S, y_F, \widehat{z}_S, \widehat{z}_F)\| \le L_S^{g_S}\|z_S - \widehat{z}_S\| + L_F^{g_S}\|z_F - \widehat{z}_F\| \tag{4.3}$$

and analogously $L_S^{g_F}$, $L_F^{g_F}$ and $L_\rho^{f_\lambda}$ with $\lambda, \rho \in \{F, S\}$. Furthermore, for the Lipschitz constants with respect to the differential variables, we use the symbol M_λ^j (with $j \in \{f_S, f_F\}$), e.g.,

$$\|f_S(t, y_S, y_F, z_S, z_F) - f_S(t, \widehat{y}_S, \widehat{y}_F, z_S, z_F)\| \le M_S^{f_S}\|y_S - \widehat{y}_S\| + M_F^{f_S}\|y_F - \widehat{y}_F\|. \tag{4.4}$$

To analyze inter-/extrapolation based multirate schemes for these general index-1 DAEs, we consider dynamic iteration schemes with old, known iterates $y_\lambda^{(i)}$, $z_\lambda^{(i)}$

and to be computed, new iterates $y_\lambda^{(i+1)}$, $z_\lambda^{(i+1)}$ defined by the following dynamic system

$$\dot{y}_S^{(i+1)} = F_S(t, y_S^{(i+1)}, y_F^{(i+1)}, z_S^{(i+1)}, z_F^{(i+1)}, y_S^{(i)}, y_F^{(i)}, z_S^{(i)}, z_F^{(i)}),$$

$$0 = G_S(t, y_S^{(i+1)}, y_F^{(i+1)}, z_S^{(i+1)}, z_F^{(i+1)}, y_S^{(i)}, y_F^{(i)}, z_S^{(i)}, z_F^{(i)}),$$

$$\tag{4.5}$$

$$\dot{y}_F^{(i+1)} = F_F(t, y_S^{(i+1)}, y_F^{(i+1)}, z_S^{(i+1)}, z_F^{(i+1)}, y_S^{(i)}, y_F^{(i)}, z_S^{(i)}, z_F^{(i)}),$$

$$0 = G_F(t, y_S^{(i+1)}, y_F^{(i+1)}, z_S^{(i+1)}, z_F^{(i+1)}, y_S^{(i)}, y_F^{(i)}, z_S^{(i)}, z_F^{(i)})$$

based on splitting functions F_S, G_S, F_F and G_F. To have a simpler notation, we introduce the abbreviations

$$x := (y_S, y_F, z_S, z_F). \quad x_S := (y_S, z_S), \quad x_F := (y_F, z_F).$$

The above splitting functions have to be consistent, this reads,

$$F_\lambda(t, x, x) = f_\lambda(t, x), \qquad G_\lambda(t, x, x) = g_\lambda(t, x), \quad \text{for } \lambda \in \{F, S\}.$$

For the different multirate approaches, we have the following splitting functions:

(i) Fully-decoupled approach:

$$F_S(t, x^{(i+1)}, x^{(i)}) = f_S(t, x_S^{(i+1)}, x_F^{(i)}), \qquad F_F(t, x^{(i+1)}, x^{(i)}) = f_F(t, x_S^{(i)}, x_F^{(i+1)}),$$

$$G_S(t, x^{(i+1)}, x^{(i)}) = g_S(t, x_S^{(i+1)}, x_F^{(i)}), \qquad G_F(t, x^{(i+1)}, x^{(i)}) = g_F(t, x_S^{(i)}, x_F^{(i+1)}).$$

(ii) Slowest-first approach:

$$F_S(t, x^{(i+1)}, x^{(i)}) = f_S(t, x_S^{(i+1)}, x_F^{(i)}), \qquad F_F(t, x^{(i+1)}, x^{(i)}) = f_F(t, x_S^{(i+1)}, x_F^{(i+1)}),$$

$$G_S(t, x^{(i+1)}, x^{(i)}) = g_S(t, x_S^{(i+1)}, x_F^{(i)}), \qquad G_F(t, x^{(i+1)}, x^{(i)}) = g_F(t, x_S^{(i+1)}, x_F^{(i+1)}).$$

(iii) Fastest-first approach:

$$F_S(t, x^{(i+1)}, x^{(i)}) = f_S(t, x_S^{(i+1)}, x_F^{(i+1)}), \qquad F_F(t, x^{(i+1)}, x^{(i)}) = f_F(t, x_S^{(i)}, x_F^{(i+1)}),$$

$$G_S(t, x^{(i+1)}, x^{(i)}) = g_S(t, x_S^{(i+1)}, x_F^{(i+1)}), \qquad G_F(t, x^{(i+1)}, x^{(i)}) = g_F(t, x_S^{(i)}, x_F^{(i+1)}).$$

It has been shown that convergence of a dynamic iteration scheme for DAEs can no longer be guaranteed by choosing a window step size H small enough, see e.g. [2, 19]. An additional contractivity condition has to hold to guarantee convergence of the dynamic iteration scheme with respect to the number of iterations for fixed window step size H. We have to distinguish the following two aspects for contraction:

(a) *Convergence within one window* $[\bar{t}, \bar{t} + H]$: In this case, it is sufficient to have (see [6] as a generalization of [19]):

$$\max_{\bar{t} \leq \tau \leq \bar{t}+H} \left\| \left(\begin{matrix} \frac{\partial G_S}{\partial z_S^{(i+1)}} & \frac{\partial G_S}{\partial z_F^{(i+1)}} \\ \frac{\partial G_F}{\partial z_S^{(i+1)}} & \frac{\partial G_F}{\partial z_F^{(i+1)}} \end{matrix} \right)^{-1} \cdot \left(\begin{matrix} \frac{\partial G_S}{\partial z_S^{(i)}} & \frac{\partial G_S}{\partial z_F^{(i)}} \\ \frac{\partial G_F}{\partial z_S^{(i)}} & \frac{\partial G_F}{\partial z_F^{(i)}} \end{matrix} \right) \Big|_{(\tau,\, x(\tau),\, x(\tau))} \right\| \leq \alpha < 1$$

using the L^∞-norm and evaluation at the analytic solution x. The quantity $\alpha \in \mathbb{R}^+$ is referred to as contraction number. For the type of norm employed on the above left-hand side, we use later the following short-hand

$$\left\| \left(\frac{\partial G_\rho}{\partial x_\lambda^{(i+1)}} \right)^{-1} \frac{\partial G_\lambda}{\partial x_\tau^{(i)}} \right\| := \max_{\bar{t} \leq \tau \leq \bar{t}+H} \left\| \left(\frac{\partial G_\rho}{\partial x_\lambda^{(i+1)}} \right)^{-1} \cdot \frac{\partial G_\lambda}{\partial x_\tau^{(i)}} \Big|_{(\tau,\, x(\tau),\, x(\tau))} \right\|$$

(for $\rho, \lambda, \tau \in \{F, S\}, x \in \{y, z\}$).

(b) *Stable error propagation from window to window:* Let us assume that k iterations are performed on the current time window. Then a sufficient condition for a stable error propagation from window to window is given by [2]

$$L_\Phi \alpha^k < 1$$

with Lipschitz constant L_Φ for the extrapolation operator. Note that for $k = 1$ a stable error propagation implies convergence within one window, as $L_\Phi \alpha < 1$ implies $\alpha < 1$ for $L_\Phi \geq 1$.

Remark 4.1

(i) Notice that for the stable error propagation in b) it might be necessary that more than one iteration is performed, although the error reduction (i.e., $\alpha < 1$) holds.
(ii) If one employs a dynamic iteration with only one iteration (one solve of the DAEs), then a multirate scheme is obtained. These schemes are referred to as *multirate co-simulation*, see [5].

As we did for the ODE case, interpolation/extrapolation based multirate schemes of convergence order p for coupled index-1 DAEs can now be obtained by replacing the exact solution of the DAE system with splitting functions

(i) by a numerical integration of convergence order p,
(ii) with stopping after the first iteration (i.e., $k = 1$), plus

(iii) employing extrapolation/interpolation schemes of order $p - 1$ and

(iv) having satisfied the contractivity condition $L_\Phi \alpha < 1$.

For the different coupling strategies, this condition reads

(i) fully-decoupled approach:

$$
L_\Phi \max_{\bar{\imath} \leq \tau \leq \bar{\imath}+H} \left\| \begin{pmatrix} \dfrac{\partial G_S}{\partial z_S^{(i+1)}} & 0 \\[2mm] 0 & \dfrac{\partial G_F}{\partial z_F^{(i+1)}} \end{pmatrix}^{-1} \cdot \begin{pmatrix} 0 & \dfrac{\partial G_S}{\partial z_F^{(i)}} \\[2mm] \dfrac{\partial G_F}{\partial z_S^{(i)}} & 0 \end{pmatrix} \right\| < 1
$$

$$
\Leftrightarrow \max_{\bar{\imath} \leq \tau \leq \bar{\imath}+H} \left\| \begin{pmatrix} \left(\dfrac{\partial G_S}{\partial z_S^{(i+1)}}\right)^{-1} \dfrac{\partial G_S}{\partial z_F^{(i)}} & 0 \\[4mm] 0 & \left(\dfrac{\partial G_F}{\partial z_F^{(i+1)}}\right)^{-1} \dfrac{\partial G_F}{\partial z_S^{(i)}} \end{pmatrix} \right\| < \frac{1}{L_\Phi}.
$$

Sufficient conditions for this are

$$
\left\| \left(\frac{\partial G_S}{\partial z_S^{(i+1)}}\right)^{-1} \frac{\partial G_S}{\partial z_F^{(i)}} \right\| < \frac{1}{L_\Phi} \quad \text{and} \quad \left\| \left(\frac{\partial G_F}{\partial z_F^{(i+1)}}\right)^{-1} \frac{\partial G_F}{\partial z_S^{(i)}} \right\| < \frac{1}{L_\Phi}.
$$

Introducing the ratios of Lipschitz-constants:

$$
\alpha_S := \frac{L_F^{g_S}}{L_S^{g_S}}, \qquad \alpha_F := \frac{L_S^{g_F}}{L_F^{g_F}}
$$

for g_S and g_F (see (4.3)), the last conditions can be reformulated as:

$$
\alpha_S < \frac{1}{L_\Phi} \quad \text{and} \quad \alpha_F < \frac{1}{L_\Phi}. \tag{4.6}
$$

(ii) slowest-first approach:

$$
\max_{\bar{\imath} \leq \tau \leq \bar{\imath}+H} \left\| \begin{pmatrix} \dfrac{\partial G_S}{\partial z_S^{(i+1)}} & 0 \\[2mm] \dfrac{\partial G_F}{\partial z_S^{(i+1)}} & \dfrac{\partial G_F}{\partial z_F^{(i+1)}} \end{pmatrix}^{-1} \cdot \begin{pmatrix} 0 & \dfrac{\partial G_S}{\partial z_F^{(i)}} \\[2mm] 0 & 0 \end{pmatrix} \right\| < 1
$$

$$
\Leftrightarrow \max_{\bar{\imath} \leq \tau \leq \bar{\imath}+H} \left\| \begin{pmatrix} 0 & \left(\dfrac{\partial G_S}{\partial z_S^{(i+1)}}\right)^{-1} \dfrac{\partial G_S}{\partial z_F^{(i)}} \\[4mm] 0 & \left(\dfrac{\partial G_F}{\partial z_F^{(i+1)}}\right)^{-1} \dfrac{\partial G_F}{\partial z_S^{(i+1)}} \left(\dfrac{\partial G_S}{\partial z_S^{(i+1)}}\right)^{-1} \dfrac{\partial G_S}{\partial z_F^{(i)}} \end{pmatrix} \right\| < \frac{1}{L_\Phi}.
$$

For this, sufficient conditions are

$$\left\| \left(\frac{\partial G_S}{\partial z_S^{(i+1)}} \right)^{-1} \frac{\partial G_S}{\partial z_F^{(i)}} \right\| < \frac{1}{L_\Phi} \quad \text{and} \quad \left\| \left(\frac{\partial G_F}{\partial z_F^{(i+1)}} \right)^{-1} \frac{\partial G_F}{\partial z_S^{(i+1)}} \left(\frac{\partial G_S}{\partial z_S^{(i+1)}} \right)^{-1} \frac{\partial G_S}{\partial z_F^{(i)}} \right\| < \frac{1}{L_\Phi}.$$

Formulated with ratios of Lipschitz-constants, we have

$$\alpha_S < \frac{1}{L_\Phi} \quad \text{and} \quad \alpha_F \alpha_S < \frac{1}{L_\Phi}, \tag{4.7}$$

which is equivalent to

$$\alpha_S < \frac{1}{L_\Phi} \quad \text{and} \quad \alpha_F < 1. \tag{4.8}$$

(iii) fastest-first approach: we obtain analogously to (ii)

$$\left\| \begin{pmatrix} \left(\frac{\partial G_S}{\partial z_S^{(i+1)}} \right)^{-1} \frac{\partial G_S}{\partial z_F^{(i+1)}} \left(\frac{\partial G_F}{\partial z_F^{(i+1)}} \right)^{-1} \frac{\partial G_F}{\partial z_S^{(i)}} & 0 \\ \left(\frac{\partial G_F}{\partial z_F^{(i+1)}} \right)^{-1} \frac{\partial G_F}{\partial z_S^{(i)}} & 0 \end{pmatrix} \right\| < \frac{1}{L_\Phi}.$$

For this, sufficient conditions for this are

$$\left\| \left(\frac{\partial G_F}{\partial z_F^{(i+1)}} \right)^{-1} \frac{\partial G_F}{\partial z_S^{(i)}} \right\| < \frac{1}{L_\Phi} \quad \text{and} \quad \left\| \left(\frac{\partial G_S}{\partial z_S^{(i+1)}} \right)^{-1} \frac{\partial G_S}{\partial z_F^{(i+1)}} \left(\frac{\partial G_F}{\partial z_F^{(i+1)}} \right)^{-1} \frac{\partial G_F}{\partial z_S^{(i)}} \right\| < \frac{1}{L_\Phi}.$$

In ratios of Lipschitz-constants, this reads

$$\alpha_F < \frac{1}{L_\Phi} \quad \text{and} \quad \alpha_S \alpha_F < \frac{1}{L_\Phi}, \tag{4.9}$$

which is equivalent to

$$\alpha_F < \frac{1}{L_\Phi} \quad \text{and} \quad \alpha_S < 1. \tag{4.10}$$

In all cases, convergence is given for problems that are coupled weakly enough, i.e., the respective above estimates for $L_\Phi \alpha < 1$ hold. If not, additional iteration of the multirate scheme will be necessary. This will, in fact, destroy the multirate benefit.

Remark 4.2 One shall notice that the stability criteria are relaxed if the multirate scheme is not fully decoupled: a larger fast ratio α_F is allowed in the case of slowest-first approach, and a larger slow ratio α_S is in the case of fastest-first approach.

These stability conditions, together with appropriate numerical time integration, are sufficient to obtain convergent multirate schemes based on dynamic iteration schemes, as shown in

Theorem 4.1 *Given the split DAE problem* (4.1) *with the index-1 conditions for the overall system and the subsystems* (4.2). *The above variants of multirate methods based on dynamic iteration (with sufficiently small window size H) are convergent on the macro step level of order p if*

(a) the respective basic integration schemes are of order p,
(b) the applied inter-/extrapolation procedure are of order $p - 1$, and
(c) the respective stability restriction

$$\text{(i) fully-decoupled: (4.6),}\quad \text{(ii) slowest-first: (4.8),}\quad \text{(iii) fastest-first: (4.10),}$$

are satisfied. The latter conditions guarantee stability.

Remark 4.3 This theorem combines the stability and convergence results of [2, 6] for dynamic iteration in the case of only one sweep $k = 1$ with multirate time integration for different coupling strategies: Jacobi iteration (fully decoupled) and Gauss-Seidel iteration (slowest-first and fastest-first strategy, depending on the order of the subsystems).

Proof (Sketch) We first inspect the time integration within one window $[t_n, t_{n+1}]$ $(n = 0, \ldots, N)$ in the case, where we solve the time interval of interest $[t_0, T]$ with $N = (T - t_0)/H$ windows. The overall error within this window is the difference between the exact solution of the split system (4.1) and the approximation given by the numerical time integration of the dynamic iteration system (4.5) (superscript 'dyn.it,h'):

$$\begin{pmatrix} y_S(t_{n+1}) - y_S^{\text{dyn.it},h}(t_{n+1}) \\ y_F(t_{n+1}) - y_F^{\text{dyn.it},h}(t_{n+1}) \\ z_S(t_{n+1}) - z_S^{\text{dyn.it},h}(t_{n+1}) \\ z_F(t_{n+1}) - z_F^{\text{dyn.it},h}(t_{n+1}) \end{pmatrix}$$

$$= \begin{pmatrix} y_S(t_{n+1}) - y_S^{\text{dyn.it}}(t_{n+1}) \\ y_F(t_{n+1}) - y_F^{\text{dyn.it}}(t_{n+1}) \\ z_S(t_{n+1}) - z_S^{\text{dyn.it}}(t_{n+1}) \\ z_F(t_{n+1}) - z_F^{\text{dyn.it}}(t_{n+1}) \end{pmatrix} + \begin{pmatrix} y_S^{\text{dyn.it}}(t_{n+1}) - y_S^{\text{dyn.it},h}(t_{n+1}) \\ y_F^{\text{dyn.it}}(t_{n+1}) - y_F^{\text{dyn.it},h}(t_{n+1}) \\ z_S^{\text{dyn.it}}(t_{n+1}) - z_S^{\text{dyn.it},h}(t_{n+1}) \\ z_F^{\text{dyn.it}}(t_{n+1}) - z_F^{\text{dyn.it},h}(t_{n+1}) \end{pmatrix},$$

which can be split into two contributions employing the exact solution of dynamic iteration system (4.5) (superscript 'dyn.it'), i.e., into the splitting error of the dynamic iteration without time discretization errors and the error due to discrete time stepping. Consequently, the overall error can be estimated by the sum of both differences.

We discuss now both error contributions.

(a) For the special coupling structure investigated in [2] the splitting error (not including time discretization) has been shown to be of order $\mathcal{O}(H^p)$ provided that the stability restrictions apply and extra-/interpolation procedures of order $p - 1$ are used. This result can be generalized straightforward to the general case of system (4.1).

(b) For the multirate variant, the dynamic iteration consists of only one sweep, i.e., system (4.5) defines a non-autonomous DAE system of index one. Employing an order p scheme with step size h as assumed, we obtain a global error of size $\mathcal{O}(h^p)$, which is bounded by $\mathcal{O}(H^p)$.

Consequently for the whole time interval of interest $[t_0, T]$, the error recursion from window to window is the nearly the same as the one based on exact time integration: only the coefficient of the iteration matrix are perturbed by an additional term of order $\mathcal{O}(H^p)$. Thus the convergence analysis of the dynamic iteration scheme yields convergence order p provided that $\alpha + \mathcal{O}(H^p)$ is bounded by one above, which is always feasible by using H small enough.   □

Remark 4.4 In the special case of DAE-ODE coupling, G_S and G_F do not depend on old iterates of the algebraic variables; hence $\alpha = 0$, and convergence can always be guaranteed for H small enough. For the case, where the fast system is an ODE, and implicit Euler approaches are used, explicit conditions for convergence are given in [18] and read in our notation:

$$ H < \frac{1}{M_S^{fs} + L_S^{fs} M_S^{gs}}, \quad h < \frac{1}{M_S^{fF} + L_S^{fF} M_F^{gs}} $$

We note that these conditions are quite strong assumptions in the case of stiff equations.

Remark 4.5 (Schemes) Compared with the ODE case, the first order extrapolation needs Jacobian information for the G-parts. In fact, this is needed for an implicit integration scheme anyways.

5   Conclusion and Outlook

The presented work contains a full convergence theory for the quite straightforward approach of inter/extrapolation-based multirate schemes for both the ODE and index-1 DAE case. We linked our theory to the concept of multirate dynamic iteration schemes. Thereby, sufficient conditions for the convergence of the dynamic iteration of DAE are transferred to the multirate setting.

As these conditions can be restrictive for stiff differential equations [18], one-sided Lipschitz-conditions might yield more realistic results. This will be investigated in future work.

References

1. Arnold, M.: Multi-rate time integration for large scale multibody system models. In: Eberhard, P. (ed.) IUTAM Symposium on Multiscale Problems in Multibody System Contacts, pp. 1–10. Springer, Berlin (2007)
2. Arnold, M., Günther, M.: Preconditioned dynamic iteration for coupled differential-algebraic systems. BIT Numer. Math. **41**(1), 1–25 (2001)
3. Arnold, M.: Modular time integration of coupled problems in system dynamics. In: Günther, M., Schilders, W.H.A. (eds.) Novel Mathematics Inspired by Industrial Applications. Springer, Berlin (2020)
4. Bartel, A.: Multirate ROW methods of mixed type for circuit simulation. In: van Rienen, U., Günther, M., Hecht, D. (eds.) Scientific Computing in Electrical Engineering. Lecture Notes in Computational Science and Engineering, vol. 18, pp. 241–249. Springer, Berlin (2001)
5. Bartel, A.: Partial Differential-Algebraic Models in Chip Design–Thermal and Semiconductor Problems, VDI-Verlag, Düsseldorf (2004)
6. Bartel, A., Brunk, M., Günther, M., Schöps, S.: Dynamic iteration for coupled problems of electric circuits and distributed devices. SIAM J. Sci. Comput. **35**(2), B315–B335 (2013)
7. Bartel, A., Günther, M.: Multirate Schemes — An Answer of Numerical Analysis to a Demand from Applications. IMACM Preprint, No. 2019-12, University of Wuppertal
8. Constantinescu, E.M., Sandu, A.: Multirate timestepping methods for hyperbolic conservation laws. J. Sci. Comput. **33**(3), 239–278 (2007)
9. Constantinescu, E.M., Sandu, A.: Extrapolated implicit-explicit time stepping. SIAM J. Sci. Comput. **31**, 4452–4477 (2010)
10. Eich-Soellner, E., Führer, C: Numerical Methods in Multibody Dynamics. Teubner, Stuttgart (1998)
11. El Guennouni, A., Verhoeven, A., ter Maten, E.J.W., Beelen T.G.J.: Aspects of multirate time integration methods in circuit simulation problems. In: Di Bucchianico, A., Mattheij, R.M.M., Peletier, M.A. (eds.) Progress in Industrial Mathematics at ECMI 2004. Springer, Berlin (2006)
12. Engstler, C., Lubich, C.: Multirate extrapolation methods for differential equations with different time scales. Computing **58**(2), 173–185 (1997)
13. Gear, C.W., Wells, D.: Multirate linear multistep methods. BIT Numer. Math. **24**, 484–502 (1984)
14. Günther M., Rentrop P.: Multirate ROW methods and latency of electric circuits. Appl. Num. Math. **13**, 83–102 (1992)
15. Günther M., Sandu, A.: Multirate generalized additive Runge Kutta methods. Numer. Math. **133**, 497–524 (2016)
16. Günther, M., Kværnø, A. Rentrop, P.: Multirate partitioned Runge-Kutta methods. BIT Numer. Math. **41**(3), 504–514 (2001)
17. Günther, M., Feldmann, U., ter Maten, J.: Modelling and discretization of circuit problems. In: Handbook of Numerical Analysis, vol. 13, pp. 523–659. Elsevier, Amsterdam (2005)
18. Hachtel, Ch., Bartel, A., Günther, M., Sandu, A.: Multirate implicit euler schemes for a class of differential-algebraic equations of index-1. J. Comput. Appl. Math. (2019). https://doi.org/10.1016/j.cam.2019.112499
19. Jackiewicz, Z., Kwapisz, M.: Convergence of waveform relaxation methods for differential-algebraic systems. SIAM J. Numer. Anal. **33**, 2303–2317 (1996)

20. Schöps, S., De Gersem, H., Bartel, A.: A Co-simulation framework for multirate time-integration of field/circuit coupled problems. IEEE Trans. Magn. **46**(8), 3233–3236 (2010)
21. Stocker, Th.: Introduction to Climate Modelling. Springer, Heidelberg (2011)
22. Rice, J.R.: Split Runge-Kutta method for simultaneous equations. J. Res. Nat. Bur. Standar. **64B**, 151–170 (1960)
23. Savcenco, V., Hundsdorfer. W., Verwer, J.G.: A multirate time stepping strategy for stiff ordinary differential equations. BIT Numer. Math.**47**, 137–155, 579–584 (2007)
24. Verhoeven, A., Tasić, B., Beelen, T.G.J., ter Maten, E.J.W., Mattheij, R.M.M.: BDF compound-fast multirate transient analysis with adaptive stepsize control. J. Numer. Analy. Industr. Appl. Math. **3**(3–4), 275–297 (2008)
25. Wensch, J., Knoth, O., Galant, A.: Multirate infinitesimal step methods for atmospheric flow simulation. BIT Numer. Math. **49**(2), 449–473 (2009)

Least-Squares Collocation for Higher-Index DAEs: Global Approach and Attempts Toward a Time-Stepping Version

Michael Hanke and Roswitha März

Abstract Overdetermined polynomial least-squares collocation for two-point boundary value problems for higher index differential-algebraic equations shows excellent convergence properties while at the same time being only slightly more expensive than the widely used collocation method for ordinary differential equations by piecewise polynomials. In the present paper, basic properties of this method when applied to initial value problems by a windowing technique are proven. Some examples are provided in order to show the potential of time-stepping approach.

Keywords Differential-algebraic equation · Higher index · Initial-value problem · Essentially ill-posed problem · Least-squares problem · Polynomial collocation

Mathematics Subject Classification (2010) 65L80, 65L08

1 Introduction

In a number of recent papers [7–10] convergence results for an overdetermined polynomial least-squares collocation   for two-point boundary value problems for higher index differential-algebraic equations (DAEs) have been established. This method is comparable in computational efficiency with the widely used collocation

M. Hanke (✉)
School of Engineering Sciences, Department of Mathematics, KTH Royal Institute
of Technology, Stockholm, Sweden
e-mail: hanke@nada.kth.se

R. März
Institute of Mathematics, Humboldt-University of Berlin, Berlin, Germany
e-mail: maerz@math.hu-berlin.de

© The Editor(s) (if applicable) and The Author(s), under exclusive licence   91
to Springer Nature Switzerland AG 2020
T. Reis et al. (eds.), *Progress in Differential-Algebraic Equations II*,
Differential-Algebraic Equations Forum,
https://doi.org/10.1007/978-3-030-53905-4_4

method for ordinary differential equations using piecewise polynomials. For initial value problems (IVPs), a considerable increase in numerical efficiency of the overdetermined polynomial least-squares collocation method is expected if one can construct time-stepping or windowing techniques. Below, we consider some key issues in this respect. Our ultimate goal is that overdetermined collocation is used on succeeding individual time-windows, though we emphasize that the present note deals with the very first attempts in this context only.

We are interested in general initial-value problems (IVPs),

$$f((Dx)'(t), x(t), t) = 0, \quad t \in [a, b], \quad G_a x(a) = r. \tag{1.1}$$

$x : [a, b] \to \mathbb{R}^m$ is the unknown vector-valued function defined on the finite interval $[a, b] \subset \mathbb{R}$. We assume an explicit partitioning of the unknowns into differentiated and nondifferentiated (also called algebraic) components by selecting

$$D \in \mathbb{R}^{k \times m}, \quad D = [I_k \ 0]$$

with the identity matrix $I_k \in \mathbb{R}^{k \times k}$. The function $f : \mathbb{R}^k \times \mathbb{R}^m \times \mathbb{R} \to \mathbb{R}^m$ is assumed to be sufficiently smooth, at least continuous and with continuous partial derivatives with respect to the first and second arguments.

The initial values deserve some special attention. For a solution to exist they must be consistent. We will ensure this by requiring special properties on the matrix G_a. It is reasonable to assume that at most the differentiated components $x_1, \ldots, x_k$ are fixed by initial conditions, which leads to the requirement

$$G_a \in \mathbb{R}^{l \times m}, \quad \ker G_a \supseteq \ker D,$$

such that $G_a x(a) = G_a D^+ Dx(a)$. Moreover, we will assume that the initial conditions are independent of each other, that is rank $G_a = l$, where l denotes the actual dynamical degree of freedom. Later on, more detailed requirements, depending on the DAE will be posed.

Let the interval $[a, b]$ be decomposed into L subintervals,

$$a = w_0 < w_1 < \cdots < w_L = b,$$

with lengths $H_\lambda = w_\lambda - w_{\lambda-1}$, $\lambda = 1, \ldots, L$. First, for $\lambda = 1$, we provide an approximating segment $\tilde{x}^{[1]} : [w_0, w_1] \to \mathbb{R}^m$ by applying overdetermined collocation to the IVP

$$f((D\tilde{x}^{[1]})'(t), \tilde{x}^{[1]}(t), t) = 0, \quad t \in [w_0, w_1], \quad G_a \tilde{x}^{[1]}(a) = r. \tag{1.2}$$

For $\lambda > 1$, having already the segment $\tilde{x}^{[\lambda-1]} : [w_{\lambda-2}, w_{\lambda-1}] \to \mathbb{R}^m$, we intend to determine the next segment $\tilde{x}^{[\lambda]} : [w_{\lambda-1}, w_\lambda] \to \mathbb{R}^m$ by solving the DAE

$$f((D\tilde{x}^{[\lambda]})'(t), \tilde{x}^{[\lambda]}(t), t) = 0, \quad t \in [w_{\lambda-1}, w_\lambda]. \tag{1.3}$$

In order to obtain an appropriate approximation to the solution of (1.1), we need to compensate the now unavailable initial conditions by certain transfer conditions using $\tilde{x}^{[\lambda-1]}$. Below we investigate two different approaches, namely,

$$G(w_{\lambda-1})\tilde{x}^{[\lambda]}(w_{\lambda-1}) = G(w_{\lambda-1})\tilde{x}^{[\lambda-1]}(w_{\lambda-1}), \tag{1.4}$$

with a suitably prescribed matrix function $G : [a, b] \to \mathbb{R}^{l \times m}$, and

$$D\tilde{x}^{[\lambda]}(w_{\lambda-1}) = D\tilde{x}^{[\lambda-1]}(w_{\lambda-1}). \tag{1.5}$$

The construction of appropriate transfer conditions is crucial for the success of the method.[1]

In the present note we merely deal with the linear version of the IVP,

$$A(t)(Dx)'(t) + B(t)x(t) - q(t) = 0, \quad t \in [a, b], \tag{1.6}$$

$$G_a x(a) = r, \tag{1.7}$$

in which the right-hand side $q : [a, b] \to \mathbb{R}^m$ and the matrix coefficients $A : [a, b] \to \mathbb{R}^{m \times k}$ and $B : [a, b] \to \mathbb{R}^{m \times m}$ are assumed to be sufficiently smooth, however at least continuous, thus uniformly bounded.

As it is well-known,[2] conventional time-stepping methods such as the BDF in the famous DAE solver DASSL work well only when applied to index-1 DAEs and special form index-2 DAEs. The so far available time-stepping solvers for more general higher-index DAEs are definitely bound to the construction and evaluation of so-called derivative array systems,[3] e.g., [3, 4, 12, 16, 17], which accounts for a serious limitation in view of applications. The recently discussed ansatz of overdetermined least-squares collocation [7–10] fully avoids the use of derivative arrays and no reduction procedures are incorporated, which is highly beneficial. However, this is a global ansatz over the entire interval, not a time-stepping method and large ill-conditioned discrete systems may arise. For this reason, eventually, a time-stepping version would be much more advantageous. Recall that we come up with very first related ideas here.

The paper is organized as follows: We describe the global overdetermined collocation procedure in Sect. 2 and collect there the relevant convergence results. In Sect. 3 we derive basic error estimates for overdetermined collocation on arbitrary individual subintervals corresponding to both procedures (1.2)–(1.3) and (1.4). A corresponding result for the approach (1.2)–(1.3) and (1.5) is provided in Sect. 4. We study the simpler time-stepping version with uniform window-size H and the

---

[1] It should be noted that also an appropriate continuous functional of $\tilde{x}^{[\lambda-1]}$ can be considered as a suitable candidate for defining a transfer condition.

[2] We refer to [1, 6] for an early discussion and to [2, 13] for a topical one.

[3] Also called prolongation. The necessary differentiations have to be provided analytically or via automatic differentiation.

same uniform stepsize on all subintervals in Sect. 5. Convergence of the method using the transfer condition (1.4) is shown in Sect. 5.1. However, our estimates in Sect. 4 are not sufficient to show convergence for the case (1.3), (1.5). Therefore, an investigation of a very special system in Sect. 5.3 provides some hints on what could be expected in that case. In order to demonstrate the behavior of the method, we provide a more complex example in Sect. 6 using both approaches, (1.4) as well as (1.5).

2 Global Overdetermined Collocation

2.1 The Global Procedure

Let us consider first the case of global overdetermined collocation, that is $L = 1$ and $H = b - a$. Let, for a given $n \in \mathbb{N}$, a grid π on the interval $[a, b]$ be defined:

$$\pi : \quad a = t_0 < \cdots < t_n = b,$$

where $t_j = a + jh$ and $h = (b - a)/n$.[4]

In order to be able to introduce collocation conditions we will need a space of piecewise continuous functions. Let $C_\pi([a, b], \mathbb{R}^m)$ denote the space of all functions $x : [a, b] \to \mathbb{R}^m$ which are continuous on each subinterval (t_{j-1}, t_j) and feature continuous extensions onto $[t_{j-1}, t_j]$, $j = 1, \ldots, n$. Furthermore, let $\mathscr{P}_N$ denote the set of polynomials of degree less than or equal to N, $N \geq 1$. We define the ansatz space

$$X_\pi = \{p \in C_\pi([a, b], \mathbb{R}^m) | Dp \in C([a, b], \mathbb{R}^k),$$

$$p_\kappa|_{(t_{j-1}, t_j)} \in \mathscr{P}_N, \kappa = 1, \ldots, k,$$

$$p_\kappa|_{(t_{j-1}, t_j)} \in \mathscr{P}_{N-1}, \kappa = k + 1, \ldots, m,$$

$$j = 1, \ldots, n\}.$$

Let now M points τ_i be given such that $0 < \tau_1 < \cdots < \tau_M < 1$. The set of collocation points is given by

$$S_{\pi, M} = \{t_{ji} = t_{j-1} + \tau_i h | \ j = 1, \ldots, n, \ i = 1, \ldots, M\}. \tag{2.1}$$

---

[4]A generalization to quasi-uniform grids is easily possible.

Using this set $S_{\pi,M}$, an interpolation operator $R_{\pi,M} : C_\pi([a,b], \mathbb{R}^m) \to C_\pi([a,b], \mathbb{R}^m)$ is given by assigning, to each $w \in C_\pi([a,b], \mathbb{R}^m)$, the piecewise polynomial $R_{\pi,M} w$ with

$$R_{\pi,M} w|_{(t_{j-1}, t_j)} \in \mathscr{P}_{M-1}, \quad j = 1, \dots, n, \quad R_{\pi,M} w(t) = w(t), \ t \in S_{\pi,M}.$$

The functional

$$\Phi_{\pi,M}(x) = \| R_{\pi,M}(f((Dx)'(\cdot), x(\cdot), \cdot)) \|_{L^2}^2 + |G_a x(a) - r|^2, \quad x \in X_\pi,$$

can be represented as (cf. [10, Subsection 2.3], also [8, 9])

$$\Phi_{\pi,M}(x) = W^T \mathscr{L} W + |G_a x(a) - r|^2, \quad x \in X_\pi,$$

with the vector $W \in \mathbb{R}^{mMn}$,

$$W = \begin{bmatrix} W_1 \\ \vdots \\ W_n \end{bmatrix} \in \mathbb{R}^{mMn}, \quad W_j = \left(\frac{h}{M}\right)^{1/2} \begin{bmatrix} f((Dx)'(t_{j1}), x(t_{j1}), t_{j1}) \\ \vdots \\ f((Dx)'(t_{jM}), x(t_{jM}), t_{jM}) \end{bmatrix} \in \mathbb{R}^{mM},$$

with the matrix $\mathscr{L}$ being positive definite, symmetric and independent[5] of h. Moreover there are constants $\kappa_l, \kappa_u > 0$ such that

$$\kappa_l |V|^2 \leq V^T \mathscr{L} V \leq \kappa_u |V|^2, \quad V \in \mathbb{R}^{mMn}. \tag{2.2}$$

If the DAE in (1.1) is regular with index one, $l = k$, and $M = N$, then there is an element $\tilde{x}_\pi \in X_\pi$ such that $\Phi_{\pi,M}(\tilde{x}_\pi) = 0$, which corresponds to the classical collocation method resulting in a system of $nMm + l$ equations for $nNm + k = nMm + l$ unknowns. Though classical collocation works well for regular index-1 DAEs (e.g., [14]), it is known to be useless for higher-index DAEs.

Reasonably, one applies l initial conditions in compliance with the dynamical degree of freedom of the DAE. In the case of higher-index DAEs, the dynamical degree of freedom is always less than k. For $0 \leq l \leq k$ and $M \geq N+1$, necessarily an overdetermined collocation system results since $nMm + l > nNm + k$. *Overdetermined least-squares collocation* consists of choosing $M \geq N + 1$ and then determining an element $\tilde{x}_\pi \in X_\pi$ which minimizes the functional $\Phi_{\pi,M}$, i.e.,

$$\tilde{x}_\pi \in \operatorname{argmin}\{\Phi_{\pi,M}(x) | x \in X_\pi\}.$$

This runs astonishingly well [9, 10], see also Sect. 6.

---

[5]The entries of $\mathscr{L}$ are fully determined by the corresponding M Lagrangian basis polynomials, thus, by M and $\tau_1, \dots, \tau_M$.

2.2   Convergence Results for the Global Overdetermined Collocation Applied to Linear IVPs

We now specify results obtained for boundary value problems in [8–10] for a customized application to IVPs. Even though we always assume a sufficiently smooth classical solution $x_* : [a, b] \to \mathbb{R}^m$ of the IVP (1.6), (1.7) to exist, for the following, an operator setting in Hilbert spaces will be convenient. The spaces to be used are:

$$L^2 = L^2((a, b), \mathbb{R}^m\}, \quad H_D^1 = \{x \in L^2 | Dx \in H^1((a, b), \mathbb{R}^k)\}, \quad Y = L^2 \times \mathbb{R}^l.$$

The operator $T : H_D^1 \to L^2$ given by

$$(Tx)(t) = A(t)(Dx)'(t) + B(t)x(t), \quad a.e.\, t \in (a, b), \quad x \in H_D^1,$$

is bounded. Since, for $x \in H_D^1$, the values $Dx(a)$ and thus $G_a x(a) = G_a D^+ Dx(a)$ are well-defined, the composed operator $\mathscr{T} : X \to Y$ given by

$$\mathscr{T}x = \begin{bmatrix} Tx \\ G_a x(a) \end{bmatrix}, \quad x \in H_D^1,$$

is well-defined and also bounded.

Let $U_\pi : H_D^1 \to H_D^1$ denote the orthogonal projector of the Hilbert space H_D^1 onto X_π.

For a more concise notation later on, we introduce the composed interpolation operator $\mathscr{R}_{\pi, M} : C_\pi([a, b], \mathbb{R}^m) \times \mathbb{R}^l \to Y$,

$$\mathscr{R}_{\pi, M} \begin{bmatrix} w \\ r \end{bmatrix} = \begin{bmatrix} R_{\pi, M} & 0 \\ 0 & I \end{bmatrix} \begin{bmatrix} w \\ r \end{bmatrix}.$$

With these settings, overdetermined least-squares collocation reduces to the minimization of

$$\Phi_{\pi, M}(x) = \|R_{\pi, M}(Tx - q)\|_{L^2}^2 + |G_a x(a) - r|^2 = \|\mathscr{R}_{\pi, M}(\mathscr{T}x - y)\|_Y^2, \quad x \in X_\pi,$$

that is, to find

$$\tilde{x}_\pi \in \mathrm{argmin}\{\Phi_{\pi, M}(x) | x \in X_\pi\}.$$

Later on, we will provide conditions which ensure that $\ker \mathscr{R}_{\pi, M} \mathscr{T} U_\pi = X_\pi^\perp$ such that $\tilde{x}_\pi$ is uniquely defined. Therefore,

$$\tilde{x}_\pi = (\mathscr{R}_{\pi, M} \mathscr{T} U_\pi)^+ \mathscr{R}_{\pi, M}\, y.$$

We consider also the related functional

$$\Phi(x) = \|Tx - q\|_{L^2}^2 + |G_a x(a) - r|^2 = \|\mathscr{T}x - y\|_Y^2, \quad x \in H_D^1,$$

and the corresponding method for approximating the solution x_* by determining

$$x_\pi \in \operatorname{argmin}\{\Phi(x)|x \in X_\pi\}.$$

As before, the conditions assumed below will guarantee that the minimizer x_π is unique such that

$$x_\pi = (\mathscr{T}U_\pi)^+ y.$$

Below, the operator $\mathscr{T}$ is ensured to be injective. Since $\mathscr{T}$ is associated with a higher-index DAE, the inverse $\mathscr{T}^{-1}$ is unbounded and the IVP is essentially ill-posed in the sense of Tikhonov. Following ideas to treat ill-posed problems, e.g., [11], the proofs in [8–10] are based on estimates of the type

$$\|x_\pi - x_*\|_{H_D^1} \leq \frac{\beta_\pi}{\gamma_\pi} + \alpha_\pi,$$

$$\|\tilde{x}_\pi - x_*\|_{H_D^1} \leq \frac{\tilde{\beta}_\pi}{\tilde{\gamma}_\pi} + \alpha_\pi,$$

in which

$$\alpha_\pi = \|(I - U_\pi)x_*\|_{H_D^1},$$

$$\beta_\pi = \|\mathscr{T}(I - U_\pi)x_*\|_Y,$$

$$\tilde{\beta}_\pi = \|\mathscr{R}_{\pi,M}\mathscr{T}(I - U_\pi)x_*\|_Y,$$

$$\gamma_\pi = \inf_{p \in X_\pi, p \neq 0} \frac{\|\mathscr{T}p\|_Y}{\|p\|_{H_D^1}} = \inf_{p \in X_\pi, p \neq 0} \left(\frac{\|Tp\|_{L^2}^2 + |G_a p(a)|^2}{\|p\|_{H_D^1}} \right)^{1/2},$$

$$\tilde{\gamma}_\pi = \inf_{p \in X_\pi, p \neq 0} \frac{\|\mathscr{R}_{\pi,M}\mathscr{T}p\|_Y}{\|p\|_{H_D^1}} = \inf_{p \in X_\pi, p \neq 0} \left(\frac{\|\mathscr{R}_{\pi,M}Tp\|_{L^2}^2 + |G_a p(a)|^2}{\|p\|_{H_D^1}} \right)^{1/2}.$$

The most challenging task in this context is to provide suitable positive lower bounds of the *instability thresholds* γ_π and $\tilde{\gamma}_\pi$, [8–10] and, what is the same, upper bounds for the Moore-Penrose inverses

$$\|(\mathscr{T}U_\pi)^+\| = \frac{1}{\gamma_\pi}, \quad \|(\mathscr{R}_{\pi,M}\mathscr{T}U_\pi)^+\| = \frac{1}{\tilde{\gamma}_\pi}.$$

It should be noted that $\mathscr{T}$ and $\mathscr{R}_{\pi,M}\mathscr{T}$ are of very different nature: While $\mathscr{T}$ is bounded, $\mathscr{R}_{\pi,M}\mathscr{T}$ is unbounded owing to the fact that $R_{\pi,M}$ is an unbounded operator in L^2, see [8].

We now briefly summarize the relevant estimations resulting from [8, 9] for IVPs. For details we refer to [8, 9].

The general assumptions with respect to the DAE and the initial conditions are:[6]

1. The operator T is fine with tractability index $\mu \geq 2$ and characteristic values $0 < r_0 \leq \cdots \leq r_{\mu-1} < r_\mu = m$.
2. The initial conditions are accurately stated such that $l = m - \sum_{i=0}^{\mu-1}(m - r_i)$ and $G_a = G_a \Pi_{can}(a)$, with the canonical projector Π_{can}. This implies im $\mathscr{T} =$ im $T \times \mathbb{R}^l$, see [14, Theorem 2.1].
3. The coefficients A, B, the right-hand side $q \in$ im T, and the solution x_* are sufficiently smooth.

Result (a), see [9]: Assume $M \geq N+1$. Then there are positive constants c_α, c_β, c_γ and c such that, for all sufficiently small stepsizes $h > 0$,

$$\gamma_\pi \geq c_\gamma h^{\mu-1}, \quad \alpha_\pi \leq c_\alpha h^N, \quad \beta_\pi \leq c_\beta h^N,$$

and eventually

$$\|x_\pi - x_*\|_{H_D^1} \leq c\, h^{N-\mu+1}.$$

Result (b), see [8]: Assume $M \geq N+\mu$. Then there are positive constants $c_\alpha, \tilde{c}_\beta$, $\tilde{c}_\gamma$, and $\tilde{c}$ such that, for all sufficiently small stepsizes $h > 0$,

$$\tilde{\gamma}_\pi \geq \tilde{c}_\gamma h^{\mu-1}, \quad \alpha_\pi \leq c_\alpha h^N, \quad \tilde{\beta}_\pi \leq \tilde{c}_\beta h^N,$$

and eventually

$$\|\tilde{x}_\pi - x_*\|_{H_D^1} \leq \tilde{c}\, h^{N-\mu+1}.$$

By [8], one can do with $\tilde{c}_\gamma = c_\gamma/2$. We refer to [9, 10] for a series of tests which confirm these estimations or perform even better. Recall that so far, IVPs for higher-index DAEs are integrated by techniques which evaluate derivative arrays, e.g., [5]. Comparing with those methods even the global overdetermined collocation method features beneficial properties. However, a time-stepping version could be much more advantageous.

---

[6]The following results are also valid for index-1 DAEs. However, we do not recommend this approach for $\mu = 1$ since standard collocation methods work well, see [14].

3 Overdetermined Collocation on an Arbitrary Subinterval $[\bar{t}, \bar{t} + H] \subset [a, b]$

3.1 Preliminaries

We continue to consider the IVP (1.6), (1.7) as described above, but instead of the global approach immediately capturing the entire interval $[a, b]$ we now aim at stepping forward by means of consecutive time-windows applying overdetermined least-squares collocation on each window. As special cases, we have in mind the two windowing procedures outlined by (1.2), (1.3), and (1.4), and by (1.2), (1.3), and (1.5). At the outset we ask how overdetermined collocation works on an arbitrary subinterval,

$$[\bar{t}, \bar{t} + H] \subseteq [a, b].$$

It will become important to relate global quantities (valid for overdetermined least-squares collocation on $[a, b]$) to their local counterparts (appropriate on subintervals of length H). We introduce the function spaces related to this subinterval,

$$L^2_{sub} = L^2((\bar{t}, \bar{t} + H), \mathbb{R}^m\}, \quad H^1_{sub} = H^1((\bar{t}, \bar{t} + H), \mathbb{R}^k),$$

$$H^1_{D,sub} = \{x \in L^2_{sub} | Dx \in H^1_{sub}\}, \quad Y_{sub} = L^2_{sub} \times \mathbb{R}^l, \quad \hat{Y}_{sub} = L^2_{sub} \times \mathbb{R}^k,$$

equipped with natural norms, in particular,

$$\|x\|_{H^1_{D,sub}} = (\|x\|^2_{L^2_{sub}} + \|(Dx)'\|^2_{L^2_{sub}})^{1/2}, \quad x \in H^1_{D,sub}.$$

Note that we indicate quantities associated to the subinterval by the extra subscript *sub* only if necessary and otherwise misunderstandings could arise.

Now we assume that the grid π is related to the subinterval only,

$$\pi : \quad \bar{t} = t_0 < \cdots < t_n = \bar{t} + H,$$

where $t_j = \bar{t} + jh$ and $h = H/n$. The ansatz space reads now

$$X_\pi = \{p \in C_\pi([\bar{t}, \bar{t} + H], \mathbb{R}^m)| \ Dp \in C([\bar{t}, \bar{t} + H], \mathbb{R}^k),$$

$$p_\kappa|_{(t_{j-1},t_j)} \in \mathscr{P}_N, \ \kappa = 1, \ldots, k, \ p_\kappa|_{(t_{j-1},t_j)} \in \mathscr{P}_{N-1}, \ \kappa = k+1, \ldots, n,$$

$$j = 1, \ldots, n\}.$$

With $0 < \tau_1 < \cdots < \tau_M < 1$, the set of collocation points

$$S_{\pi,M} = \{t_{ji} = t_{j-1} + \tau_i h| \ j = 1, \ldots, n, \ i = 1, \ldots, M\} \tag{3.1}$$

belongs to the subinterval $[\bar{t}, \bar{t} + H]$. Correspondingly, the interpolation operator $R_{\pi,M}$ acts on $C_\pi([\bar{t}, \bar{t}+H], \mathbb{R}^m)$. We introduce the operator $T_{sub} : H^1_{D,sub} \to L^2_{sub}$,

$$(T_{sub}x)(t) = A(t)(Dx)'(t) + B(t)x(t), \ a.e. \ t \in (\bar{t}, \bar{t} + H), \ x \in H^1_{D,sub},$$

and the composed operators $\mathscr{T}_{sub} : H^1_{D,sub} \to Y_{sub}$ and $\hat{\mathscr{T}}_{sub} : H^1_{D,sub} \to \hat{Y}_{sub}$,

$$\mathscr{T}_{sub}x = \begin{bmatrix} T_{sub}x \\ G(\bar{t})x(\bar{t}) \end{bmatrix}, \quad \hat{\mathscr{T}}_{sub}x = \begin{bmatrix} T_{sub}x \\ Dx(\bar{t}) \end{bmatrix}, \quad x \in H^1_{D,sub}.$$

Occasionally, we also use the operators $T_{IC,sub} : H^1_{D,sub} \to \mathbb{R}^l$ and $T_{ICD,sub} : H^1_{D,sub} \to \mathbb{R}^k$ given by

$$T_{IC,sub}x = G(\bar{t})x(\bar{t}), \quad T_{ICD,sub}x = Dx(\bar{t}), \quad x \in H^1_{D,sub},$$

which are associated with the initial condition posed at $\bar{t}$. Here, aiming for injective composed operators, we suppose a function $G : [a, b] \to \mathbb{R}^l$ such that

$$\ker G(t) = \ker \Pi_{can}(t), \ \operatorname{im} G(t) = \mathbb{R}^l, \ |G(t)| \le c_G, \ t \in [a, b]. \tag{3.2}$$

Since T_{sub} inherits the tractability index, the characteristic values of T, and also the canonical projector (restricted to the subinterval, see [13, Section 2.6]), the local initial condition at $\bar{t}$, $G(\bar{t})x(\bar{t}) = r$, is accurately stated. Then $\operatorname{im} \mathscr{T}_{sub} = \operatorname{im} T_{sub} \times \mathbb{R}^l$ and $\ker \mathscr{T}_{sub} = \{0\}$, so that the overdetermined least-squares collocation on $[\bar{t}, \bar{t} + H]$ works analogously to the global one described in Sect. 2.

The composed interpolation operators $\mathscr{R}_{\pi,M}$ and $\hat{\mathscr{R}}_{\pi,M}$ act now on $C_\pi([\bar{t}, \bar{t} + H], \mathbb{R}^m) \times \mathbb{R}^l$ and $C_\pi([\bar{t}, \bar{t} + H], \mathbb{R}^m) \times \mathbb{R}^k$,

$$\mathscr{R}_{\pi,M}\begin{bmatrix} w \\ r \end{bmatrix} = \begin{bmatrix} R_{\pi,M} & 0 \\ 0 & I_l \end{bmatrix}\begin{bmatrix} w \\ r \end{bmatrix}, \quad \hat{\mathscr{R}}_{\pi,M}\begin{bmatrix} w \\ \hat{r} \end{bmatrix} = \begin{bmatrix} R_{\pi,M} & 0 \\ 0 & I_k \end{bmatrix}\begin{bmatrix} w \\ \hat{r} \end{bmatrix}.$$

Let $U_{\pi,sub} : H^1_{D,sub} \to H^1_{D,sub}$ be the orthogonal projector of $H^1_{D,sub}$ onto $X_\pi \subset H^1_{D,sub}$.

Accordingly, we define $\alpha_{\pi,sub}$ and, furthermore, $\beta_{\pi,sub}, \gamma_{\pi,sub}, \tilde{\beta}_{\pi,sub}, \tilde{\gamma}_{\pi,sub}$, associated with the operator $\mathscr{T}_{sub}$ and, similarly, $\hat{\beta}_{\pi,sub}, \hat{\gamma}_{\pi,sub}, \hat{\tilde{\beta}}_{\pi,sub}, \hat{\tilde{\gamma}}_{\pi,sub}$ associated with $\hat{\mathscr{T}}_{sub}$.

The following lemma provides conditions for the existence of a function $G : [a, b] \to \mathbb{R}$ having the properties (3.2). The latter is a necessary prerequisite for the transition condition (1.4).

Lemma 3.1 *Let the operator T be fine with tractability index $\mu \geq 2$, characteristic values $0 < r_0 \leq \cdots \leq r_{\mu-1} < r_\mu = m$, $l = m - \sum_{i=0}^{\mu-1}(m - r_i)$, and the canonical projector function Π_{can}.*
Then there are continuously differentiable functions $G : [a, b] \to \mathbb{R}^{l \times m}$ and $K : [a, b] \to \mathbb{R}^{k \times k}$ such that

$$\operatorname{im} G(t) = \mathbb{R}^l, \quad \ker G(t) = \ker \Pi_{can}(t), \quad [I_l \ 0]K(t)D = G(t), \quad t \in [a, b],$$

$K(t)$ remains nonsingular on $[a, b]$, and, with $\kappa = (\max_{a \leq t \leq b}|K(t)|)^{-1}$,

$$|Dz| = |K(t)^{-1}K(t)Dz| \geq \kappa|K(t)Dz| \geq \kappa|G(t)z|, \quad z \in \mathbb{R}^k, \ t \in [a, b].$$

Proof We choose an admissible matrix function sequence with admissible projector functions $Q_0, \ldots, Q_{\mu-1}$, see [13, Section 2.2]. Denote $P_i = I - Q_i$, $\Pi_i = P_0 \cdots P_i$. Then, $\Pi_{\mu-1}$ and $D\Pi_{\mu-1}D^+$ are also projector functions, both with constant rank l. Since $D\Pi_{\mu-1}D^+$ is continuously differentiable, we find a continuously differentiable matrix function $\Gamma_{dyn} : [a, b] \to \mathbb{R}^{l \times k}$ so that

$$\operatorname{im} \Gamma_{dyn}(t) = \mathbb{R}^l, \quad \ker \Gamma_{dyn}(t) = \ker(D\Pi_{\mu-1}D^+)(t), \quad t \in [a, b].$$

Furthermore, there is a pointwise reflexive generalized inverse $\Gamma_{dyn}^- : [a, b] \to \mathbb{R}^{k \times l}$, also continuously differentiable, such that $\Gamma_{dyn}\Gamma_{dyn}^- = I$ and $\Gamma_{dyn}^-\Gamma_{dyn} = D\Pi_{\mu-1}D^+$. Similarly, we find constant-rank continuously differentiable matrix functions $\Gamma_{nil,i} : [a, b] \to \mathbb{R}^{(m-r_i) \times k}$ and pointwise generalized inverses $\Gamma_{nil,i}^- : [a, b] \to \mathbb{R}^{k \times (m-r_i)}$ such that

$$\Gamma_{nil,i}\Gamma_{nil,i}^- = I, \quad \Gamma_{nil,i}^-\Gamma_{nil,i} = D\Pi_{i-1}Q_iD^+, \quad i = 1, \ldots, \mu - 1.$$

The resulting $k \times k$ matrix function

$$K = \begin{bmatrix} \Gamma_{dyn} \\ \Gamma_{nil,1} \\ \vdots \\ \Gamma_{nil,\mu-1} \end{bmatrix} = \begin{bmatrix} \Gamma_{dyn} \\ \Gamma_{nil} \end{bmatrix}$$

remains nonsingular on $[a, b]$ owing to the decomposition $I_k = DD^+ = D\Pi_0 Q_1 D^+ + \cdots + D\Pi_{\mu-2}Q_{\mu-1}D^+ + D\Pi_{\mu-1}D^+$.

Set $G = \Gamma_{dyn}D = [I_l \ 0]KD$. This implies $\ker G(t) = \ker \Pi_{\mu-1}$. Taking into account the fact that $\ker \Pi_{\mu-1} = \ker \Pi_{can}$, see [13, Theorem 2.8], one has actually $\ker G(t) = \ker \Pi_{can}$.

Finally, we derive for $z \in \mathbb{R}^k$, $t \in [a, b]$,

$$|Dz|^2 = |K(t)^{-1} K(t) Dz|^2 \geq \kappa^2 |K(t) Dz|^2 = \kappa^2 (|G(t)z|^2 + |\Gamma_{nil}(t) Dz|^2)$$
$$\geq \kappa^2 |G(t)z|^2,$$

which completes the proof. □

Lemma 3.2 *For $\bar{t} \in [a, b]$, $0 < H \leq b - \bar{t}$, and*

$$C_H = \left(\max \left(\frac{2}{H}, 2H \right) \right)^{1/2}$$

it holds that

$$|Dx(t)| \leq C_H \|Dx\|_{H^1_{sub}} \leq C_H \|x\|_{H^1_{D,sub}}, \quad t \in [\bar{t}, \bar{t} + H], \quad x \in H^1_{D,sub}.$$

Proof By definition, $x \in H^1_{D,sub}$ implies $u = Dx \in H^1_{sub}$. Since H^1_{sub} is continuously embedded in C_{sub}, it follows that

$$u(t) = u(s) + \int_s^t u'(\tau) d\tau, \quad t, s \in [\bar{t}, \bar{t} + H],$$

which gives

$$|u(t)|^2 \leq 2|u(s)|^2 + 2 \left(\int_s^t |u'(\tau)| d\tau \right)^2 \leq 2|u(s)|^2 + 2H \int_{\bar{t}}^{\bar{t}+H} |u'(\tau)|^2 d\tau.$$

Integrating this inequality with respect to s leads to

$$H|u(t)|^2 \leq 2 \int_{\bar{t}}^{\bar{t}+H} |u(s)|^2 ds + 2H^2 \int_{\bar{t}}^{\bar{t}+H} |u'(\tau)|^2 d\tau.$$

Finally, with C_H as defined in the assertion, it holds that

$$\|u\|_{C_{sub}}^2 \leq C_H^2 \|u\|_{H^1_{sub}}^2 \leq C_H^2 \|x\|_{H^1_{D,sub}}^2$$

and the assertion follows. □

Lemma 3.3 *Let the function G fulfilling* (3.2) *with the bound c_G be given, and denote* $c_T = (2 \max\{\|A\|_\infty^2, \|B\|_\infty^2\})^{1/2}$.

(1) *Then, for each subinterval, the inequalities*

$$\|T_{sub}x\|_{L^2_{sub}} \le c_T \|x\|_{H^1_{D,sub}}, \quad x \in H^1_{D,sub},$$

$$|T_{IC,sub}x| \le c_G C_H \|x\|_{H^1_{D,sub}}, \quad |T_{ICD,sub}x| \le C_H \|x\|_{H^1_{D,sub}}, \quad x \in H^1_{D,sub},$$

$$(3.3)$$

are valid.

(2) *If $M \ge N+1$ and A, B are of class C^M, then there are constants C_{AB1}, C_{AB2}, both independent of the size H of the subinterval, such that*

$$\|R_{\pi,M}T_{sub}U_\pi x\|_{L^2_{sub}} \le C_{AB1}\|x\|_{H^1_{D,sub}}, \quad x \in H^1_{D,sub},$$

$$\|R_{\pi,M}T_{sub}U_\pi x - T_{sub}U_\pi x\|_{L^2_{sub}} \le C_{AB1}h^{M-N-1/2}\|x\|_{H^1_{D,sub}}, \quad x \in H^1_{D,sub}.$$

Proof

(1) Regarding that A, B are given on $[a,b]$, by straightforward computation we obtain

$$\|T_{sub}x\|_{L^2_{sub}}^2 \le 2 \max\{\|A\|_{\infty,sub}^2, \|B\|_{\infty,sub}^2\}\|x\|_{H^1_{D,sub}}^2 \le c_T \|x\|_{H^1_{D,sub}}^2.$$

Applying Lemma 3.2 we find the inequalities (3.3).

(2) These inequalities can be verified analogously to the first two items of [8, Proposition 4.2].   □

We are now prepared to estimate the values $\alpha_{\pi,sub}$, $\beta_{\pi,sub}$, $\tilde{\beta}_{\pi,sub}$, $\hat{\beta}_{\pi,sub}$, and $\hat{\tilde{\beta}}_{\pi,sub}$.

Theorem 3.4 *Let the operator T described in Sect. 2 be fine with tractability index $\mu \ge 2$ and characteristic values $0 < r_0 \le \cdots \le r_{\mu-1} < r_\mu = m$, $l = m - \sum_{i=0}^{\mu-1}(m-r_i)$. Let the coefficients A, B, as well as the solution x_* of the IVP* (1.6), (1.7) *be sufficiently smooth. Let the function G with* (3.2) *be given and $[\bar{t}, \bar{t}+H] \subset [a,b]$.*

Then there are positive constants $\alpha_{\pi,sub}$, C_β, $\tilde{C}_\beta$, $\hat{C}_\beta$, $\hat{\tilde{C}}_\beta$ such that

$$\alpha_{\pi,sub} \le C_\alpha H^{1/2}h^N,$$

$$\beta_{\pi,sub} \le C_\beta h^N, \quad \tilde{\beta}_{\pi,sub} \le \tilde{C}_\beta h^N,$$

$$\hat{\beta}_{\pi,sub} \le \hat{C}_\beta h^N, \quad \hat{\tilde{\beta}}_{\pi,sub} \le \hat{\tilde{C}}_\beta h^N.$$

uniformly for all individual subintervals $[\bar{t}, \bar{t}+H]$ and all sufficient fine grids X_π.

Proof First we choose N nodes $0 < \tau_{*,1} < \cdots < \tau_{*,N} < 1$ and construct the interpolating function $p_{*,int} \in X_\pi$ so that

$$Dp_{*,int}(\bar{t}) = Dx_*(\bar{t}), \quad p_{*,int}(t_j + \tau_{*,i}h) = x_*(t_j + \tau_{*,i}h), \quad i = 1, \ldots, N, \ j = 1, \ldots, n,$$

yielding

$$\|x_* - p_{*,int}\|_{\infty,sub} + \|(Dx_*)' - (Dp_{*,int})'\|_{\infty,sub} \leq C_* h^N,$$

with a uniform constant C_* for all subintervals. C_* is determined by x_* and its derivatives given on $[a, b]$. Now we have also

$$\|x_* - p_{*,int}\|_{H^1_{D,sub}} \leq C_* \sqrt{2H} h^N,$$

and therefore, with $C_\alpha = C_* \sqrt{2}$,

$$\alpha_{\pi,sub} = \|(I - U_{\pi,sub})x_*\|_{H^1_{D,sub}} = \|(I - U_{\pi,sub})(x_* - p_{*,int})\|_{H^1_{D,sub}} \leq C_\alpha \sqrt{H} h^N.$$

Set $C_D = \sqrt{2} \max\{1, b - a\}C_\alpha$ such that $C_H \sqrt{H} C_\alpha \leq C_D$ for all H. Using Lemma 3.2 we derive

$$|D((I - U_{\pi,sub})x_*)(\bar{t})| \leq C_H \alpha_{\pi,sub} \leq C_D h^N.$$

We derive further

$$\beta^2_{\pi,sub} = \|\mathscr{T}_{sub}(I - U_{\pi,sub})x_*\|^2_{Y_{sub}}$$
$$= \|T_{sub}(I - U_{\pi,sub})x_*\|^2_{L^2_{sub}} + |G(\bar{t})D^+ D((I - U_{\pi,sub})x_*)(\bar{t})|^2$$
$$\leq \|T_{sub}\|^2 \alpha^2_{\pi,sub} + c_G^2 C_D^2 h^{2N} \leq (c_T^2 C_\alpha^2 (b - a) + c_G^2 C_D^2)h^{2N} = C_\beta^2 h^{2N},$$

$$\hat{\beta}^2_{\pi,sub} = \|\hat{\mathscr{T}}_{sub}(I - U_{\pi,sub})x_*\|^2_{Y_{sub}}$$
$$= \|T_{sub}(I - U_{\pi,sub})x_*\|^2_{L^2_{sub}} + |D((I - U_{\pi,sub})x_*)(\bar{t})|^2$$
$$\leq \|T_{sub}\|^2 \alpha^2_{\pi,sub} + c_G^2 C_D^2 h^{2N} \leq (c_T^2 C_\alpha^2 (b - a) + C_D^2)h^{2N} = \hat{C}_\beta^2 h^{2N}.$$

Following [8, Section 2.3], we investigate also $w_* = T_{sub}(x_* - p_{*,int}) \in C_\pi([\bar{t}, \bar{t} + H], \mathbb{R}^m)$ and use the estimate (cf. [8, Section 2.3])

$$H^{-1/2}\|R_{\pi,M}w_*\|_{L^2,sub} \leq \|R_{\pi,M}w_*\|_{\infty,sub} \leq C_L \|w_*\|_{\infty,sub} \leq \max\{\|A\|_\infty, \|B\|_\infty\}C_L h^N.$$

Here, C_L denotes a constant that depends only on the choice of the interpolation nodes $\tau_{*,1}, \ldots, \tau_{*,N}$. Then we derive

$$\|R_{\pi,M}T_{sub}(I - U_{\pi,sub})x_*\|_{L^2,sub} \leq \|R_{\pi,M}T_{sub}(I - U_{\pi,sub})(x_* - p_{*,int})\|_{L^2,sub}$$

$$\leq \|R_{\pi,M}T_{sub}(x_* - p_{*,int})\|_{L^2,sub}$$

$$+ \|R_{\pi,M}T_{sub}U_{\pi,sub}(x_* - p_{*,int})\|_{L^2,sub}$$

$$\leq \|R_{\pi,M}w_*\|_{L^2,sub} + C_{AB1}\|x_* - p_{*,int}\|_{H^1_{D,sub}}$$

$$\leq C_{RT}\sqrt{H}h^N,$$

where $C_{RT} = C_L \max\{\|A\|_\infty, \|B\|_\infty\} + \sqrt{2}C_*C_{AB1}$. Therefore,

$$\tilde{\beta}^2_{\pi,sub} = \|\mathscr{R}_{\pi,m}\mathscr{T}_{sub}(I - U_{\pi,sub})x_*\|^2_{\hat{Y}_{sub}}$$

$$= \|R_{\pi,M}T_{sub}(I - U_{\pi,sub})x_*\|^2_{L^2_{sub}} + |G(\bar{t})D^+D((I - U_{\pi,sub})x_*)(\bar{t})|^2$$

$$\leq C^2_{RT}Hh^{2N} + c^2_G C^2_D h^{2N} \leq C^2_{RT}(b-a)h^{2N} + c^2_G C^2_D h^{2N} = \tilde{C}^2_\beta h^{2N},$$

$$\hat{\tilde{\beta}}^2_{\pi,sub} = \|\mathscr{R}_{\pi,m}\hat{\mathscr{T}}_{sub}(I - U_{\pi,sub})x_*\|^2_{\hat{Y}_{sub}}$$

$$= \|R_{\pi,M}T_{sub}(I - U_{\pi,sub})x_*\|^2_{L^2_{sub}} + |D((I - U_{\pi,sub})x_*)(\bar{t})|^2$$

$$\leq C^2_{RT}Hh^{2N} + C^2_D h^{2N} \leq C^2_{RT}(b-a)h^{2N} + C^2_D h^{2N} = \hat{\tilde{C}}^2_\beta h^{2N}. \qquad \square$$

3.2 Overdetermined Collocation on $[\bar{t}, \bar{t} + H] \subset [a, b]$, with Accurately Stated Initial Condition at $\bar{t}$

We ask if there are positive constants c_γ and $\tilde{c}_\gamma$ serving as lower bounds for all the individual constants characterizing the instability thresholds associated to each arbitrary subinterval $[\bar{t}, \bar{t} + H] \subset [a, b]$.

Theorem 3.5 *Let the operator T described in Sect. 2 be fine with tractability index $\mu \geq 2$ and characteristic values $0 < r_0 \leq \cdots \leq r_{\mu-1} < r_\mu = m$, $l = m - \sum_{i=0}^{\mu-1}(m - r_i)$. Let the coefficients A, B, the right-hand side $q \in \operatorname{im} T$, as well as the solution x_* of the IVP (1.6), (1.7) be sufficiently smooth. Let q_{sub} denote the restriction of q onto the subinterval $[\bar{t}, \bar{t} + H] \subset [a, b]$.*

Let a function G with (3.2) be given.

(1) *Then, for each arbitrary $r \in \mathbb{R}^l$, there is exactly one solution $x_{[r]}$ of the equation*
$$\mathscr{T}_{sub} x = (q_{sub}, r) \text{ and}$$

$$\|x_{[r]} - x_*\|_{H^1_{D,sub}} \leq c_{sub} |r - G(\bar{t})x_*(\bar{t})|.$$

$x_{[r]}$ *coincides on the subinterval with x_*, if and only if $r = G(\bar{t})x_*(\bar{t})$. Furthermore, there is a bound C_p such that $c_{sub} \leq C_p$ is valid for all subintervals.*

(2) *If $M \geq N + 1$, there is a constant $C_\gamma > 0$ such that,*

$$\gamma_{\pi,sub} \geq C_\gamma h^{\mu-1}, \quad \|(\mathscr{T}_{sub} U_{\pi,sub})^+\|_{Y_{sub} \to H^1_{D,sub}} = \frac{1}{\gamma_{\pi,sub}} \leq \frac{1}{C_\gamma h^{\mu-1}}$$

uniformly for all subintervals and sufficiently small stepsizes $h > 0$.

(3) *If $M \geq N + \mu$, there is a positive constant $\tilde{C}_\gamma = \frac{C_\gamma}{2}$ such that*

$$\|(\mathscr{R}_{\pi,M} \mathscr{T}_{sub} U_{\pi,sub})^+\|_{Y_{sub} \to H^1_{D,sub}} = \frac{1}{\tilde{\gamma}_{\pi,sub}} \leq \frac{1}{\tilde{C}_\gamma h^{\mu-1}}$$

uniformly for all subintervals and sufficiently small stepsizes $h > 0$.

Proof

(1) This is a consequence of Proposition A.1 in the Appendix.
(2) The constant C_γ can be obtained by a careful inspection and adequate modification of the proof of [9, Theorem 4.1] on the basis of Proposition A.1 below instead of [9, Proposition 4.3]. Similarly to [9, Lemma 4.4], we provide the inequality

$$\|q\|^2_{Z_{sub}} \leq \|q\|^2_\pi := \|q\|^2_{L^2_{sub}} + \sum_{i=1}^{\mu-1} \sum_{s=0}^{\mu-i} d_{i,s} \|(D\mathscr{L}_{\mu-i} q)^{(s)}\|^2_{L^2_{sub}}, \quad q \in Z_\pi,$$

with $Z_\pi = \{q \in L^2_{sub} | D\mathscr{L}_{\mu-i} q \in C^{\mu-i}_\pi([\bar{t}, \bar{t} + H], \mathbb{R}^k), i = 1, \ldots, \mu - 1\} \subset T_{sub} X_\pi$, with coefficients $d_{i,s}$ being independent of the subinterval.
(3) This statement proves by a slight modification of [8, Proposition 4.2]. □

Theorem 3.5 allows to apply homogeneous error estimations on all subintervals. Note that the involved constants C_α etc. may depend on N and M. For providing the function G with (3.2), the canonical nullspace $N_{can} = \ker \Pi_{can}$ must be available, not necessarily the canonical projector itself. Owing to [13, Theorem 2.8], it holds that $N_{can} = \ker \Pi_{\mu-1}$ for any admissible matrix function sequence, which makes N_{can} easier accessible. Nevertheless, though the function G is very useful in theory it is hardly available in practice.

For problems with dynamical degree $l = 0$ the canonical projector Π_{can} vanishes identically, that is, the initial condition is absent, and T_{sub} itself is injective. This happens, for example, for Jordan systems, see also Sect. 5.3. In those cases, with no initial conditions and no transfer the window-wise forward stepping works well.

Let $\tilde{x}_{\pi,old}$ be already computed as approximation of the solution x_* on an certain *old* subinterval of length H_{old} straight preceding the current one $[\bar{t}, \bar{t} + H]$. Motivated by Theorems 3.4 and 3.5 assume

$$\|\tilde{x}_{\pi,old} - x_*\|_{H^1_{sub,old}} \leq Ch_{old}^{N-\mu+1}$$

for sufficiently small stepsize h_{old}. Applying Lemma 3.2 we obtain

$$|D\tilde{x}_{\pi,old}(\bar{t}) - Dx_*(\bar{t})| \leq C_{H_{old}} Ch_{old}^{N-\mu+1}.$$

Next we apply overdetermined least-squares collocation on the current subinterval $[\bar{t}, \bar{t} + H]$. We use the transfer condition $r = G(\bar{t})\tilde{x}_{\pi,old}(\bar{t})$ to state the initial condition for the current subinterval. The overdetermined collocation generates the new segment $\tilde{x}_\pi$,

$$\tilde{x}_\pi = \text{argmin}\{\|R_{\pi,M}(T_{sub}x - q)\|^2_{H^1_{D,sub}} + |G(\bar{t})x(\bar{t}) - G(\bar{t})\tilde{x}_{\pi,old}(\bar{t})|^2 | x \in X_\pi\},$$

which is actually an approximation of $x_{[r]}$ being neighboring to x_*, such that

$$\|\tilde{x}_\pi - x_{[r]}\|_{H^1_{D,sub}} \leq \tilde{c}h^{N-\mu+1}.$$

Owing to Theorem 3.5 we have also

$$\|x_{[r]} - x_*\|_{H^1_{D,sub}} \leq c_{sub}|r - G(\bar{t})x_*(\bar{t})| = c_{sub}|G(\bar{t})\tilde{x}_{\pi,old}(\bar{t}) - G(\bar{t})x_*(\bar{t})|$$

$$\leq c_{sub}c_G C_{H_{old}} Ch_{old}^{N-\mu+1}.$$

If $h = h_{old}$, it follows that

$$\|\tilde{x}_\pi - x_*\|_{H^1_{D,sub}} \leq C_{sub}h^{N-\mu+1}$$

with $C_{sub} = c_{sub}c_G C_{H_{old}} C + \tilde{c}$. This is the background which ensures the windowing procedure (1.2), (1.3), (1.4) to work.

4   Overdetermined Collocation on a Subinterval $[\bar{t}, \bar{t} + H] \subset [a, b]$, with Initial Conditions Related to $Dx(\bar{t})$

Here we proceed as in the previous section, but now we use the initial condition $Dx(\bar{t}) = \hat{r}$ instead of $G(\bar{t})x(\bar{t}) = r$, to avoid the use of the function G. Obviously, this formulation is easier to use in practice since D is given. However, in contrast to the situation in Theorem 3.5, the equation $\hat{\mathscr{T}}_{sub}x = (q_{sub}, \hat{r})$ is no longer solvable for arbitrary $\hat{r} \in \mathbb{R}^k$. For solvability, $\hat{r}$ must be consistent.

Theorem 4.1 *Let the operator T described in Sect. 2.2 be fine with tractability index $\mu \geq 2$ and characteristic values $0 < r_0 \leq \cdots \leq r_{\mu-1} < r_\mu = m$, $l = m - \sum_{i=0}^{\mu-1}(m - r_i)$. Let the coefficients A, B, the right-hand side $q \in \text{im}\,T$, as well as the solution x_* of the IVP (1.6),(1.7) be sufficiently smooth. Then the following holds:*

(1) *$\hat{\mathscr{T}}_{sub}$ is injective.*
(2) *If $M \geq N + 1$, there is a constant $\hat{C}_\gamma$ uniformly for all possible subintervals and sufficiently small stepsizes $h > 0$ such that*

$$\hat{\gamma}_{\pi,sub} \geq \hat{C}_\gamma h^{\mu-1}.$$

and hence

$$\|(\hat{\mathscr{T}}_{sub}U_{\pi,sub})^+\|_{\hat{Y}_{sub}} = \frac{1}{\hat{\gamma}_{\pi,sub}} \leq \frac{1}{\hat{C}_\gamma h^{\mu-1}}.$$

(3) *If $M \geq N + \mu$, there is a constants $\tilde{C}_\gamma > 0$ uniformly for all possible subintervals and sufficiently small stepsizes $h > 0$, such that*

$$\|(\hat{\mathscr{R}}_{\pi,M}\hat{\mathscr{T}}_{sub}U_{\pi,sub})^+\|_{\hat{Y}_{sub}} = \frac{1}{\tilde{\hat{\gamma}}_{\pi,sub}} \leq \frac{1}{\tilde{C}_\gamma h^{\mu-1}}.$$

Proof The assertions are straightforward consequences of Theorem 3.5 and Lemma 3.1.
$\hat{\mathscr{T}}x = 0$ means $Tx = 0$ and $Dx(\bar{t}) = 0$, thus also $G(\bar{t})x(\bar{t}) = [I_l\ 0]K(\bar{t})Dx(\bar{t}) = 0$, finally $\mathscr{T}x = 0$. Since $\mathscr{T}$ is injective it follows that $x = 0$. For $p \in X_\pi$,

$$\|\hat{\mathscr{T}}_{sub}p\|^2_{\hat{Y}_{sub}} = \|T_{sub}p\|^2_{L^2_{sub}} + |Dp(\bar{t})|^2 \geq \|T_{sub}p\|^2_{L^2_{sub}} + \kappa^2|G(\bar{t})p(\bar{t})|^2$$

$$\geq \min\{1, \kappa^2\}\|\mathscr{T}_{sub}p\|^2_{Y_{sub}} \geq \min\{1, \kappa^2\}\left(C_\gamma h^{\mu-1}\|p\|_{H^1_{D,sub}}\right)^2,$$

and

$$\|\hat{\mathscr{R}}_{\pi,M}\hat{\mathscr{T}}_{sub}p\|^2_{\hat{Y}_{sub}} = \|R_{\pi,M}T_{sub}p\|^2_{L^2_{sub}} + |Dp(\bar{t})|^2 \geq \|R_{\pi,M}T_{sub}p\|^2_{L^2_{sub}} + \kappa^2|G(\bar{t})p(\bar{t})|^2$$

$$\geq \min\{1,\kappa^2\}\|\mathscr{R}_{\pi,M}\mathscr{T}_{sub}p\|^2_{\hat{Y}_{sub}} \geq \min\{1,\kappa^2\}\left(\tilde{C}_\gamma h^{\mu-1}\|p\|_{H^1_{D,sub}}\right)^2. \quad \square$$

In contrast to the situation in Sect. 3.2 the equation $\hat{\mathscr{T}}_{sub}x = (q_{sub}, \hat{r})$ is no longer solvable for all $\hat{r} \in \mathbb{R}^k$. Recall that q_{sub} is the restriction of $q = Tx_*$ so that $q_{sub} \in \text{im } T_{sub}$. Denote

$$\hat{y} = \begin{bmatrix} q_{sub} \\ Dx_*(\bar{t}) \end{bmatrix}, \quad \hat{y}^{[\delta]} = \begin{bmatrix} q_{sub} \\ \hat{r} \end{bmatrix}, \quad \delta = \|\hat{y} - \hat{y}^{[\delta]}\| = |Dx_*(\bar{t}) - \hat{r}|,$$

and, following [11], we take $\hat{y}^{[\delta]}$ as noisy data and compute

$$\tilde{\hat{x}}^{[\delta]}_\pi = \text{argmin}\{\|\hat{\mathscr{R}}_{\pi,M}(\hat{\mathscr{T}}_{sub}x - y^{[\delta]})\|^2_{L^2_{sub}\times\mathbb{R}^k}|x \in X_\pi\}$$

$$= \text{argmin}\{\|R_{\pi,M}(T_{sub}x - q_{sub})\|^2_{L^2_{sub}} + |Dx(\bar{t}) - \hat{r}|^2|x \in X_\pi\}$$

and similarly,

$$\hat{x}^{[\delta]}_\pi = \text{argmin}\{\|\hat{\mathscr{T}}_{sub}x - y^{[\delta]}\|^2_{L^2_{sub}\times\mathbb{R}^k}|x \in X_\pi\}$$

$$= \text{argmin}\{\|T_{sub}x - q_{sub}\|^2_{L^2_{sub}} + |Dx(\bar{t}) - \hat{r}|^2|x \in X_\pi\}.$$

Applying the error representation [11, Equation (2.9)] we arrive at

$$\tilde{\hat{x}}^{[\delta]}_\pi - x_* = (\hat{\mathscr{R}}_{\pi,M}\hat{\mathscr{T}}U_\pi)^+(\hat{y}^{[\delta]} - \hat{y})$$

$$+ (\hat{\mathscr{R}}_{\pi,M}\hat{\mathscr{T}}U_\pi)^+\hat{\mathscr{R}}_{\pi,M}\hat{\mathscr{T}}_{sub}(I - U_\pi)x_* - (I - U_\pi)x_*$$

and, correspondingly,

$$\hat{x}^{[\delta]}_\pi - x_* = (\hat{\mathscr{T}}U_\pi)^+(\hat{y}^{[\delta]} - \hat{y}) + (\hat{\mathscr{T}}U_\pi)^+\hat{\mathscr{T}}_{sub}(I - U_\pi)x_* - (I - U_\pi)x_*.$$

Thus,

$$\|\hat{x}^{[\delta]}_\pi - x_*\|_{H^1_{D,sub}} \leq \frac{1}{\hat{C}_\gamma h^{\mu-1}}\{\|\hat{y}^{[\delta]} - \hat{y}\| + \hat{\beta}_{\pi,sub}\} + \alpha_\pi = \frac{1}{\hat{C}_\gamma h^{\mu-1}}\{\delta + \hat{\beta}_{\pi,sub}\} + \alpha_\pi,$$

$$\|\tilde{\hat{x}}^{[\delta]}_\pi - x_*\|_{H^1_{D,sub}} \leq \frac{1}{\tilde{C}_\gamma h^{\mu-1}}\{\|\hat{y}^{[\delta]} - \hat{y}\| + \tilde{\hat{\beta}}_{\pi,sub}\} + \alpha_\pi = \frac{1}{\tilde{C}_\gamma h^{\mu-1}}\{\delta + \tilde{\hat{\beta}}_{\pi,sub}\} + \alpha_\pi.$$

All these estimations can be put together in order to arrive at a recursive error estimation for the application of (1.3), (1.5). Unfortunately, this estimate is not sufficient for proving convergence of the windowing technique in contrast to the approach using accurately stated initial conditions of Sect. 3.2!

5   Time-Stepping with $b - a = LH$ and $H = nh$

We set now $H = (b - a)/L$, $w_\lambda = a + \lambda H$, $\lambda = 0, \ldots, L$, and $h = H/n$, and study the somehow uniform time-stepping procedures.

5.1   Time-Stepping with Accurate Transfer Conditions

In the time-stepping approach corresponding to (1.3)–(1.4), the transfer conditions are given so that G is chosen according to (3.2). Let $\tilde{x}^{[\lambda]}$ be the approximation provided by the overdetermined least-squares collocation for the subinterval $[a + (\lambda - 1)H, a + \lambda H]$ corresponding to the initial and transfer conditions

$$G_a \tilde{x}_\pi^{[1]}(a) = r,$$

$$G(w_\lambda)\tilde{x}_\pi^{[\lambda]}(a + (\lambda - 1)H) = G(w_\lambda)\tilde{x}_\pi^{\lambda-1}(a + (\lambda - 1)H), \quad \lambda > 1.$$

Then we obtain from Theorem 3.5 and Lemma 3.2, for $\lambda = 1$,

$$\|\tilde{x}_\pi^{[1]} - x_*\|_{H^1_{D,sub}} \le \tilde{C}h^{N-\mu+1} =: d_1.$$

For $\lambda > 1$, let $r = G_\lambda \tilde{x}_\pi^{[\lambda-1]}(a + (\lambda - 1)H)$. Then it holds

$$\begin{aligned}
\|\tilde{x}_\pi^{[\lambda]} - x_*\|_{H^1_{D,sub}} &\le \|\tilde{x}_\pi^{[\lambda]} - x_{[r]}\|_{H^1_{D,sub}} + \|x_{[r]} - x_*\|_{H^1_{D,sub}} \\
&\le \tilde{C}h^{N-\mu+1} + C_p|r - G_\lambda x_*(a + (\lambda - 1)H)| \\
&\le \tilde{C}h^{N-\mu+1} + C_p c_G C_H \|\tilde{x}_\pi^{[\lambda-1]} - x_*\|_{H^1_{D,sub}} \\
&\le \bar{C}(h^{N-\mu+1} + C_H \|\tilde{x}_\pi^{[\lambda-1]} - x_*\|_{H^1_{D,sub}}) =: d_\lambda
\end{aligned}$$

where $\bar{C} = \max\{C_p c_G, \tilde{C}\}$. Hence,

$$d_1 \le \bar{C}h^{N-\mu+1}, \quad d_\lambda \le \bar{C}(C_H d_{\lambda-1} + h^{N-\mu+1}).$$

A solution of this recursion provides us with

$$d_\lambda \leq \sum_{\iota=0}^{\lambda-1} \bar{C}(\bar{C}C_H)^\iota h^{N-\mu+1} = \bar{C} \frac{1 - (\bar{C}C_H)^\lambda}{1 - \bar{C}C_H} h^{N-\mu+1}.$$

A similar estimation can be derived for the least-squares approximations using the operator $(\mathscr{T}_{sub}U_{\pi,sub})^+$.

Example 5.1 The index-2 DAE with $k = 2$, $m = 3$, $l = 1$,

$$\begin{bmatrix} 1 & 0 \\ 0 & 1 \\ 0 & 0 \end{bmatrix} \left(\begin{bmatrix} 1 & 0 & 0 \\ 0 & 1 & 0 \end{bmatrix} x \right)'(t) + \begin{bmatrix} \theta & -1 & -1 \\ \eta t(1 - \eta t) - \eta & \theta & -\eta t \\ 1 - \eta t & 1 & 0 \end{bmatrix} x(t) = q(t), \qquad (5.1)$$

is taken from [10, Example 1.1]. One has $N_{can}(t) = \{z \in \mathbb{R}^3 | \ \eta t z_1 - z_2 = 0\}$ so that

$$G(t) = \begin{bmatrix} \eta t & -1 & 0 \end{bmatrix}$$

will do. We consider the DAE on the interval $(0,1)$. The right-hand side q is chosen in such a way that

$$x_1(t) = e^{-t} \sin t,$$

$$x_2(t) = e^{-2t} \sin t,$$

$$x_3(t) = e^{-t} \cos t$$

is a solution. This solution becomes unique if an appropriate initial condition is added. With $G_a = G(0)$, the initial condition becomes

$$G_a x(0) = G_a \begin{bmatrix} 0 & 0 & 1 \end{bmatrix}^T = 0.$$

In the following experiments, $\eta = -25$ and $\theta = -1$ have been chosen. This allows for a comparison with the experiments in [10].

This problem is solved on equidistant grids using, for each polynomial degree N, $M = N + 1$ Gaussian collocation points scaled to $(0, 1)$. The tables show the errors of the approximate solutions in $H_D^1(0, 1)$. The columns labeled order contain an estimation k_{est} of the order

$$k_{est} = \log(\|x_\pi - x_*\|_{H_D^1(0,1)} / \|x_{\pi'} - x_*\|_{H_D^1(0,1)}) / \log 2.$$

Here, π' is obtained from π by stepsize halving. It should be noted that the norm is taken for the complete interval $(0, 1)$ even in the windowing approach. In order

Table 1 Errors and estimation of the convergence order for (5.1) and $\bar{t} = 0$, $H = 1$ using $M = N + 1$

	N = 1		N = 2		N = 3		N = 4		N = 5	
n	Error	Order	Error	Order	Error	Order	Error	Order	Error	Order
10	1.21e+0		1.65e−1		2.84e−3		7.55e−6		2.82e−7	
20	1.12e+0	0.1	3.74e−2	2.1	5.04e−4	2.5	9.66e−7	3.0	1.51e−8	4.2
40	1.29e−0	−0.2	1.55e−2	1.3	9.59e−5	2.4	1.25e−7	2.9	7.74e−10	4.3
80	1.16e−0	0.2	6.65e−3	1.2	1.83e−5	2.4	1.31e−8	3.3	1.32e−10	2.6
160	9.80e−1	0.2	3.21e−3	1.0	3.05e−6	2.6	1.31e−9	3.3	1.75e−10	−0.4
320	8.63e−1	0.2	1.60e−3	1.0	4.94e−7	2.6	2.00e−10	2.7	3.62e−10	−1.1

Table 2 Errors and estimation of the convergence order for (5.1) and $n = 1$ using $H = 1/L$

	N = 1		N = 2		N = 3		N = 4		N = 5	
L	Error	Order	Error	Order	Error	Order	Error	Order	Error	Order
10	3.76e+0		2.19e−1		2.82e−3		9.34e−6		2.84e−7	
20	2.67e+0	0.5	7.62e−2	1.5	5.06e−4	2.5	1.29e−6	2.9	1.53e−8	4.2
40	1.77e+0	0.6	3.30e−2	1.2	9.72e−5	2.4	1.92e−7	2.7	7.90e−10	4.3
80	1.62e+0	0.1	1.39e−2	1.2	1.89e−5	2.4	2.38e−8	3.0	4.67e−11	4.1
160	1.65e+0	−0.0	5.06e−3	1.5	3.20e−6	2.6	2.26e−9	3.4	1.13e−10	−1.3
320	1.66e+0	−0.0	1.91e−3	1.4	5.26e−7	2.6	2.21e−10	3.4	1.46e−10	−0.4

to enable a comparison, we provide the results for solving the problem without windowing in Table 1. This corresponds to $\bar{t} = 0$ and $H = 1$.

In the next experiment, the time-stepping approach using accurately stated transfer conditions has been tested with $n = 1$. The results are shown in Table 2. □

A more complex example is presented in Sect. 6.

5.2 Time-Stepping with Transfer Conditions Based on D

In our experiments in fact, the situation is much better than indicated by the estimates in Sect. 4. The latter are not sufficient to show convergence of the present time-stepping approach when the transfer conditions are based on D, see (1.5).

Example 5.2 (Continuation of Example 5.1) We apply the time-stepping procedure under the same conditions as in Example 5.1, however, this time the transfer conditions are chosen as

$$\tilde{x}_i^{[\lambda]}(\bar{t}) = \tilde{x}_i^{[\lambda-1]}(\bar{t}), \quad i = 1, 2.$$

The results are presented in Table 3. The errors are slightly worse than those of Table 2 where accurately stated transfer conditions are used. However, the observed orders of convergence are similar, at least for $N \geq 2 = \mu - 1$. The values for $n = 2$ and $n = 3$ have also been checked. The orders are identical to those of Table 3

Table 3 Errors and estimation of the convergence order for (5.1) and $n = 1$ using $H = 1/L$

	$N = 1$		$N = 2$		$N = 3$		$N = 4$		$N = 5$	
L	Error	Order	Error	Order	Error	Order	Error	Order	Error	Order
10	1.80e+0		1.46e−1		3.27e−3		9.85e−6		3.16e−7	
20	2.36e+0	−0.4	4.65e−2	1.6	5.84e−4	2.5	1.35e−6	2.9	1.71e−8	4.2
40	2.77e+1	−3.5	1.66e−2	1.5	1.09e−4	2.4	1.75e−7	2.9	8.78e−10	4.3
80	5.07e+2	−4.2	6.64e−3	1.3	2.03e−5	2.4	1.76e−8	3.3	6.65e−11	3.7
160	1.11e+3	−1.1	3.19e−3	1.1	3.51e−6	2.5	1.60e−9	3.5	1.50e−10	−1.2
320	7.46e+2	0.6	1.59e−3	1.0	6.44e−7	2.4	1.85e−10	3.1	3.07e−10	−1.0

even if the errors are smaller due to the smaller stepsize h. For $N = 1$, divergent approximations are obtained. However, this is beyond the scope of our theoretical results even in the case of accurate transfer conditions. □

5.3 Studying the Damping of Inconsistent Transition Values

The results of the previous sections show that the windowing method converges if the transfer conditions used refer to the dynamic components, only. The latter are, in general, not easily available unless a detailed analysis of the DAE is available. However, so far we do not know any conditions for convergence if the practically accessible values of the differentiated components Dx are used in the transfer conditions.[7] Example 5.2 indicates, however, that the use of (1.5) may be possible. In order to gain some more insight into what could be expected in the setting of Sect. 5.2, we will consider a simple special case in this section.

The model problem in question here is a simple system featuring only one Jordan block,

$$J(Dx)' + x = 0,$$

$$Dx(\bar{t}) = r.$$

Here, $J \in \mathbb{R}^{\mu \times (\mu-1)}$, $D \in \mathbb{R}^{(\mu-1) \times \mu}$ where

$$J = \begin{bmatrix} 0 \\ 1 & 0 \\ & \ddots & \ddots \\ & & 1 \end{bmatrix}, \quad D = \begin{bmatrix} 1 & 0 \\ & \ddots & \ddots \\ & & 1 & 0 \end{bmatrix}.$$

---

[7] In the index-1 case, Dx describes just the dynamic components such that convergence is assured for using all differentiated components. However, for index-1 DAEs, much more efficient collocation methods are available.

This system has index μ and no dynamic components, $l = 0$. The system is solvable for $r = 0$, only, leading to the unique solution $x_*(t) \equiv 0$. When trying to solve the system using the proposed windowing technique, the only information transferred from the subinterval $[\bar{t}, \bar{t}+H]$ to the next one is the value of the approximate solution x_π at the end of the interval, $Dx_\pi(\bar{t}+H)$. The latter is an approximation to the exact solution $Dx_*(\bar{t} + H) \equiv 0$ that cannot be guaranteed to be consistent with the DAE. Therefore, we ask the question of how $Dx_\pi(\bar{t} + H)$ depends on r.

Let

$$x_{[r],\pi} = \text{argmin}\{\|\hat{\mathscr{T}}_{sub}x\|_{L^2_{sub} \times \mathbb{R}^k}^2 \mid x \in X_\pi\}$$

$$= \text{argmin}\{\|T_{sub}x\|_{L^2_{sub}}^2 + |Dx(\bar{t}) - r|_{\mathbb{R}^k}^2 \mid x \in X_\pi\}$$

where $Tx = J(Dx)' + x$. Obviously, $Dx_{[r],\pi}(\bar{t} + H)$ depends linearly on r. There exists a matrix $S = S(N, H, n)$ such that $Dx_{[r],\pi}(\bar{t} + H) = Sr$ which we will denote as the transfer matrix. For convergence of the method, it is necessary that the spectral radius $\rho(S)$ of the transfer matrix is bounded by 1.

The analytical computation of S is rather tedious. After some lengthy calculations, we found that, for $\mu = 2$, it holds, with $\eta = (N + 1)^{-1}$,

$$\rho(S(N, H, n) = \eta^n \left| \frac{2}{\left(-1 + \sqrt{1 - \eta^2}\right)^n + \left(-1 - \sqrt{1 - \eta^2}\right)^n} \right|$$

$$\approx \eta^n 2^{1-n}.$$

In particular, $\rho(S)$ is independent of H and n can be chosen arbitrarily. Moreover, the damping of the inconsistent value r is the better the larger n is. This result can be compared to the experiments in Example 5.2 (an index-2 problem) where we cannot identify any influence of an inaccuracy due to inconsistent transfer conditions.

For larger values of μ, we determined $\rho(S)$ by numerical means. Results are shown in Tables 4, 5 and 6. We observe that, for an index $\mu > 2$, n must be chosen

Table 4 Spectral radius of the transfer matrix $S(N, H, n)$ for $n = 1$ and $H = 0.1$ (left panel) and $H = 0.01$ (right panel). The column headings show the index μ

N	2	3	4	5	N	2	3	4	5
2	3.3e−1	2.1e+0	1.3e+0	1.1e+0	2	3.3e−1	2.1e+0	1.1e+0	1.0e+0
3	2.5e−1	1.8e+0	5.9e+0	2.9e+0	3	2.5e−1	1.8e+0	5.9e+0	1.5e+0
4	2.0e−1	1.5e+0	7.1e+0	1.4e+1	4	2.0e−1	1.5e+0	7.1e+0	1.4e+1
5	1.7e−1	1.3e+0	7.0e+0	2.3e+1	5	1.7e−1	1.3e+0	7.0e+0	2.3e+1
6	1.5e−1	1.1e+0	6.5e+0	2.7e+1	6	1.5e−1	1.1e+0	6.6e+0	2.8e+1
7	1.2e−1	9.7e−1	6.1e+0	2.9e+1	7	1.2e−1	9.7e−1	6.1e+0	2.9e+1
8	1.1e−1	8.7e−1	5.6e+0	2.9e+1	8	1.1e−1	8.7e−1	5.6e+0	2.9e+1

Table 5 Spectral radius of the transfer matrix $S(N, H, n)$ for $n = 2$ and $H = 0.1$ (left panel) and $H = 0.01$ (right panel). The column headings show the index μ

N	2	3	4	5	N	2	3	4	5
2	5.9e−2	1.4e+0	1.5e+0	1.2e+0	2	5.9e−2	1.4e+0	1.2e+0	1.0e+0
3	3.2e−2	6.4e−1	8.0e+0	9.9e+0	3	3.2e−2	6.4e−1	8.2e+0	2.5e+0
4	2.0e−2	3.7e−1	5.0e+0	2.0e+1	4	2.0e−2	3.7e−1	5.0e+0	3.6e+2
5	1.4e−2	2.5e−1	3.1e+0	3.0e+1	5	1.4e−2	2.5e−1	3.1e+0	3.2e+2
6	1.0e−2	1.8e−1	2.1e+0	2.2e+1	6	1.0e−2	1.8e−1	2.1e+0	8.1e−1
7	7.9e−3	1.3e−1	1.5e+0	1.6e+1	7	7.9e−3	1.3e−1	1.5e+0	2.1e+0
8	6.2e−3	1.0e−1	1.2e+0	1.2e+1	8	6.2e−3	1.0e−1	1.2e+0	1.2e+1

Table 6 Spectral radius of the transfer matrix $S(N, H, n)$ for $n = 3$ and $H = 0.1$ (left panel) and $H = 0.01$ (right panel). The column headings show the index μ

N	2	3	4	5	N	2	3	4	5
2	1.0e−2	6.8e−1	1.8e+0	1.4e+0	2	1.0e−2	6.8e−1	1.3e+0	1.0e+0
3	4.1e−3	1.8e−2	6.1e+0	2.5e+0	3	4.1e−3	1.8e−1	6.3e+0	5.5e+0
4	2.1e−3	8.1e−2	2.1e+0	1.7e+1	4	2.1e−3	8.1e−2	2.1e+0	4.2e+1
5	1.2e−3	4.3e−2	9.2e−1	1.8e+1	5	1.2e−3	4.3e−2	9.2e−1	7.8e−1
6	7.4e−4	2.6e−2	5.1e−1	8.5e+0	6	7.4e−4	2.6e−2	5.1e−1	7.5e−1
7	4.9e−4	1.7e−2	3.1e−1	4.8e+0	7	4.9e−4	1.7e−2	3.1e−1	2.8e−1
8	3.5e−4	1.2e−2	2.1e−1	3.0e+0	8	3.5e−4	1.2e−2	2.1e−1	2.3e−1

larger than 1 in order to ensure $\rho(S) < 1$. Moreover, $\rho(S)$ depends on H only marginally for the investigated cases.

Details of the derivations are collected in the appendix.

6 A More Complex Example

In order to show the merits of the windowing technique, we will continue to use the example considered in [9]. This example is the linearized version of a test example from [5]. We consider an initial value problem for the DAE

$$A(Dx)'(t) + B(t)x(t) = y(t), \quad t \in [0, 5] \tag{6.1}$$

with

$$A = \begin{bmatrix} 1 & 0 & 0 & 0 & 0 & 0 \\ 0 & 1 & 0 & 0 & 0 & 0 \\ 0 & 0 & 1 & 0 & 0 & 0 \\ 0 & 0 & 0 & 1 & 0 & 0 \\ 0 & 0 & 0 & 0 & 1 & 0 \\ 0 & 0 & 0 & 0 & 0 & 1 \\ 0 & 0 & 0 & 0 & 0 & 0 \end{bmatrix}, D = \begin{bmatrix} 1 & 0 & 0 & 0 & 0 & 0 & 0 \\ 0 & 1 & 0 & 0 & 0 & 0 & 0 \\ 0 & 0 & 1 & 0 & 0 & 0 & 0 \\ 0 & 0 & 0 & 1 & 0 & 0 & 0 \\ 0 & 0 & 0 & 0 & 1 & 0 & 0 \\ 0 & 0 & 0 & 0 & 0 & 1 & 0 \end{bmatrix},$$

the smooth matrix coefficient

$$B(t) = \begin{bmatrix} 0 & 0 & 0 & -1 & 0 & 0 & 0 \\ 0 & 0 & 0 & 0 & -1 & 0 & 0 \\ 0 & 0 & 0 & 0 & 0 & -1 & 0 \\ 0 & 0 & \sin t & 0 & 1 & -\cos t & -2\rho\cos^2 t \\ 0 & 0 & -\cos t & -1 & 0 & -\sin t & -2\rho\sin t\cos t \\ 0 & 0 & 1 & 0 & 0 & 0 & 2\rho\sin t \\ 2\rho\cos^2 t & 2\rho\sin t\cos t & -2\rho\sin t & 0 & 0 & 0 & 0 \end{bmatrix}, \quad \rho = 5.$$

This DAE is obtained if the test example from [5] is linearized in the solution $x_*(t)$ considered there.[8] It has tractability index $\mu = 3$ and dynamical degree of freedom $l = 4$. In order to use the windowing technique with accurately stated initial conditions, we will need a function $G : [0, 5] \to \mathbb{R}^{4 \times 7}$ fulfilling the assumptions of Theorem 3.5. The nullspace of the projector Π_2 has the representation

$$\ker \Pi_2 = \ker \begin{bmatrix} I - \Omega & 0 & 0 \\ \Omega'\Omega & I - \Omega & 0 \\ 0 & 0 & 0 \end{bmatrix}, \quad \Omega = b(t)b(t)^T, \quad b(t) = \begin{bmatrix} -\cos^2 t \\ -\cos t \sin t \\ \sin t \end{bmatrix}.$$

Based on this representation, we can use

$$G(t) = \begin{bmatrix} \sin t & -\cos t & 0 & 0 & 0 & 0 & 0 \\ 0 & 1 & \cos t & 0 & 0 & 0 & 0 \\ -\cos^3 t & -\sin t \cos^2 t & \sin t \cos t & \sin t & -\cos t & 0 & 0 \\ -(\sin t \cos t)^2 & -\sin^3 t \cos t & \sin^3 t & 0 & 1 & \cos t & 0 \end{bmatrix}. \quad (6.2)$$

In the following numerical experiments we choose the exact solution

$$\begin{aligned} x_1 &= \sin t, & x_4 &= \cos t, \\ x_2 &= \cos t, & x_5 &= -\sin t, \\ x_3 &= 2\cos^2 t, & x_6 &= -2\sin 2t, \\ x_7 &= -\rho^{-1}\sin t, \end{aligned}$$

---

[8]Compare also [9, Sections 6.3 and 6.4].

Table 7 Errors and estimation of the convergence order for (6.1) and $\bar{t} = 0$, $H = 5$ using $M = N + 3$

	$N = 1$		$N = 2$		$N = 3$		$N = 4$		$N = 5$	
n	Error	Order	Error	Order	Error	Order	Error	Order	Error	Order
10	2.64e+0		5.24e−1		6.29e−2		6.33e−3		5.73e−4	
20	1.54e+0	0.8	1.99e−1	1.4	1.77e−2	1.8	9.39e−4	2.8	6.12e−5	3.2
40	8.79e−1	0.8	9.36e−2	1.1	6.44e−3	1.5	1.66e−4	2.5	7.31e−6	3.1
80	4.69e−1	0.9	4.63e−2	1.0	2.84e−3	1.2	3.42e−5	2.3	9.02e−7	3.0
160	3.00e−1	0.6	2.33e−2	1.0	1.37e−3	1.1	7.69e−6	2.2	1.12e−7	3.0
320	2.30e−1	0.4	1.18e−2	1.0	6.75e−4	1.0	1.82e−6	2.1	1.40e−8	3.0

which is also the one used in [9]. Setting $G_a = G(0)$, this provides us with the initial condition[9]

$$G_a x(0) = \begin{bmatrix} -1 \\ 3 \\ 0 \\ 0 \end{bmatrix}.$$

The problem is solved on equidistant grids using, for each polynomial degree N, $M = N + 3$ Gaussian collocation points scaled to $(0, 1)$. This number of collocation points has been chosen such that the assumptions of Theorem 3.5(3) are fulfilled. The tables show the errors of the approximate solutions in $H_D^1(0, 5)$. Similarly as in previous examples, the columns labeled order contain an estimation k_{est} of the order

$$k_{est} = \log(\|x_\pi - x_*\|_{H_D^1(0,5)} / \|x_{\pi'} - x_*\|_{H_D^1(0,5)}) / \log 2.$$

Here, π' is obtained from π by stepsize halving.

In order to enable a comparison, we provide the results for solving the problem without windowing in Table 7. This corresponds to $\bar{t} = 0$ and $H = 5$. Note that the results are almost identical to those obtained in [9] using a slightly different formulation of the initial condition and a different number of collocation points.

In Tables 8, 9 and 10 the results using the windowing technique with transfer conditions (1.5) for different numbers of subdivisions n of the individual windows $[\bar{t}, \bar{t} + H]$ are shown. Since the transfer condition is based on all of the differentiated components Dx, they are expected to be inconsistent away from the initial point $t = 0$. For $n = 1$ and $N \leq 3$, the method delivers exponentially divergent approximations.

---

[9]This initial condition is slightly different from the one used in [9]. However, both conditions are equivalent.

Table 8 Errors and estimation of the convergence order for (6.1) and $n = 1$, $H = 5/L$ using $M = N + 3$

	$N = 4$		$N = 5$	
L	Error	Order	Error	Order
10	1.21e−2		7.18e−4	
20	2.28e−3	2.4	7.65e−5	3.2
40	5.16e−4	2.1	9.36e−6	3.0
80	1.25e−4	2.0	1.18e−6	3.0
160	3.10e−5	2.0	1.48e−7	3.0
320	7.74e−6	2.0	1.93e−8	2.9

Table 9 Errors and estimation of the convergence order for (6.1) and $n = 2$, $H = 5/L$ using $M = N + 3$

	$N = 1$		$N = 2$		$N = 3$		$N = 4$		$N = 5$	
L	Error	Order	Error	Order	Error	Order	Error	Order	Error	Order
10	2.30e+0		2.66e−1		2.99e−2		1.99e−3		7.64e−5	
20	1.64e+0	0.5	2.98e−1	−0.2	1.25e−2	1.3	4.89e−4	2.0	9.24e−6	3.0
40	1.49e+0	0.1	2.41e+1	−6.3	5.99e−3	1.1	1.22e−4	2.0	1.16e−6	3.0
80	1.45e+0	0.0	4.16e+5	−14.1	3.03e−3	1.0	3.06e−5	2.0	1.46e−7	3.0
160	1.44e+0	0.0	1.15e+14	−28.0	1.54e−3	1.0	7.65e−6	2.0	1.84e−8	3.0
320	1.44e+0	0.0	1.48e+31	−56.8	7.77e−4	1.0	1.91e−6	2.0	1.09e−8	0.8

Table 10 Errors and estimation of the convergence order for (6.1) and $n = 3$, $H = 5/L$ using $M = N + 3$

	$N = 1$		$N = 2$		$N = 3$		$N = 4$		$N = 5$	
L	Error	Order	Error	Order	Error	Order	Error	Order	Error	Order
10	1.74e+0		1.38e−1		1.38e−2		7.31e−4		2.05e−5	
20	1.64e+0	0.0	6.92e−2	1.0	6.20e−3	1.2	1.83e−4	2.0	2.53e−6	3.0
40	1.66e+0	0.0	3.95e−2	0.8	3.07e−3	1.0	4.61e−5	2.0	3.18e−7	3.0
80	1.67e+0	0.0	2.75e−2	0.5	1.55e−3	1.0	1.15e−5	2.0	3.99e−8	3.0
160	1.68e+0	0.0	2.35e−2	0.2	7.81e−4	1.0	2.89e−6	2.0	6.43e−9	2.6
320	1.68e+0	0.0	2.23e−2	0.1	3.93e−4	1.0	7.22e−7	2.0	2.41e−8	−1.9

Finally, we consider the case of using accurately stated initial conditions as transfer conditions. So they correspond to choosing $G(\bar{t})$ according to (6.2). The results are collected in Table 11. The latter can be compared to the behavior of the global method as shown in Table 7. The results are rather close to each other.

Table 11 Errors and estimation of the convergence order for (6.1) and accurately posed transfer conditions with $n = 1$, $H = 5/L$ and $M = N + 3$

L	$N = 1$		$N = 2$		$N = 3$		$N = 4$		$N = 5$	
	Error	Order	Error	Order	Error	Order	Error	Order	Error	Order
10	5.32e+0		5.12e−1		8.46e−2		1.20e−2		1.03e−3	
20	2.56e+0	1.1	2.67e−1	0.9	2.64e−2	1.7	2.47e−3	2.3	8.85e−5	3.5
40	2.20e+0	0.2	2.03e−1	0.4	1.09e−2	1.3	5.85e−4	2.1	9.51e−6	3.2
80	2.17e+0	0.0	1.88e−1	0.1	5.14e−3	1.1	1.44e−4	2.0	1.14e−6	3.1
160	2.17e+0	0.0	1.84e−1	0.0	2.53e−3	1.0	3.59e−5	2.0	1.40e−7	3.0
320	2.17e+0	0.0	1.83e−1	0.0	1.26e−3	1.0	8.97e−6	2.0	1.76e−8	3.0

7 Conclusions

We continued the investigation of overdetermined least-squares collocation using piecewise polynomial ansatz functions. This method is known to efficiently produce accurate numerical approximations of solutions for two-point boundary value problems for higher-index DAEs including IVPs as a special case. Since a further increase in computational efficiency is expected if modified for a customized application to IVPs, we considered time-stepping techniques for IVPs in this paper. It turned out that the success of such techniques depends strongly on the transfer conditions used. In the case that the intrinsic structure is available, meaning in particular that the dynamic solution components are known, the time-stepping method has convergence properties similar to the boundary value approach. However, if only the information about the differentiated components of the DAE is used, so far our estimates do not secure convergence of the time-stepping approach. Investigations of a model problem indicate that even in this case convergence can be obtained provided that the method parameters are chosen appropriately.

The overdetermined least-squares collocation method shows impressive convergence results in our experiments. On one hand, the accuracy is impressive, on the other hand, the computational efficiency is comparable to widely used collocation methods for ordinary differential equations. Opposed to that, there are severe difficulties to theoretically justify these methods. The underlying reason is the ill-posedness of higher-index DAEs. To the best of our knowledge, available convergence results are rather sparse and important questions of practical relevance for constructing efficient algorithms are completely open, e.g., *a-posteriori* error estimations, the choice of grids, polynomial orders, collocation points etc. However, the results so far are encouraging.

A   Proof of Theorem 3.5

The Proposition A.1 below plays its role when verifying the statements of Theorem 3.5. We collect the necessary ingredients of the projector based DAE analysis to prove Proposition A.1. We refer to [13, 15] for more details. Let the DAE (1.6) be fine with tractability index $\mu \geq 2$ and characteristic values

$$0 < r_0 \leq \cdots \leq r_{\mu-1} < r_\mu = m, \quad l = m - \sum_{i=0}^{\mu-1} (m - r_i). \tag{A.1}$$

Recall that this property is determined by the given coefficients $A : [a, b] \to \mathbb{R}^{m \times k}$, $D = [I \ 0] \in \mathbb{R}^{k \times m}$, and $B : [a, b] \to \mathbb{R}^{m \times m}$. A and B are sufficiently smooth, at least continuous. Then there are an admissible sequence of matrix valued functions starting from $G_0 := AD$ and ending up with a nonsingular G_μ, see [13, Definition 2.6], as well as associated projector valued functions

$$P_0 := D^+ D \quad \text{and} \quad P_1, \ldots, P_{\mu-1} \in C([a, b], \mathbb{R}^{m \times m})$$

which provide a fine decoupling of the DAE. We have then the further projector valued functions

$$Q_i = I - P_i, \ i = 0, \ldots, \mu - 1,$$

$$\Pi_0 := P_0, \ \Pi_i := \Pi_{i-1} P_i \in C([a, b], \mathbb{R}^{m \times m}), \ i = 1, \ldots, \mu - 1,$$

$$D\Pi_i D^+ \in C^1([a, b], \mathbb{R}^{k \times k}), \ i = 1, \ldots, \mu - 1.$$

By means of the projector functions we decompose the unknown x and decouple the DAE itself into their characteristic parts, see [13, Section 2.4].

The component $u = D\Pi_{\mu-1} x = D\Pi_{\mu-1} D^+ Dx$ satisfies the explicit regular ODE residing in $\mathbb{R}^k$,

$$u' - (D\Pi_{\mu-1} D^+)' u + D\Pi_{\mu-1} G_\mu^{-1} B\Pi_{\mu-1} D^+ u = D\Pi_{\mu-1} G_\mu^{-1} q. \tag{A.2}$$

The components $v_i = \Pi_{i-1} Q_i x = \Pi_{i-1} Q_i D^+ D x$, $i = 1, \ldots, \mu - 1$, satisfy the triangular subsystem involving several differentiations,

$$
\begin{bmatrix}
0 & \mathscr{N}_{12} & \cdots & \mathscr{N}_{1,\mu-1} \\
 & 0 & \ddots & \vdots \\
 & & \ddots & \mathscr{N}_{\mu-2,\mu-1} \\
 & & & 0
\end{bmatrix}
\begin{bmatrix}
(Dv_1)' \\
\vdots \\
\\
(Dv_{\mu-1})'
\end{bmatrix}
\tag{A.3}
$$

$$
+
\begin{bmatrix}
I & \mathscr{M}_{12} & \cdots & \mathscr{M}_{1,\mu-1} \\
 & I & \ddots & \vdots \\
 & & \ddots & \mathscr{M}_{\mu-2,\mu-1} \\
 & & & I
\end{bmatrix}
\begin{bmatrix}
v_1 \\
\vdots \\
\\
v_{\mu-1}
\end{bmatrix}
=
\begin{bmatrix}
\mathscr{L}_1 \\
\vdots \\
\\
\mathscr{L}_{\mu-1}
\end{bmatrix}
q.
$$

The coefficients $\mathscr{N}_{i,j}$, $\mathscr{M}_{ij}$, and $\mathscr{L}_i$ are subsequently given. Finally, one has for $v_0 = Q_0 x$ the representation

$$
v_0 = \mathscr{L}_0 y - \mathscr{H}_0 D^+ u - \sum_{j=1}^{\mu-1} \mathscr{M}_{0j} v_j - \sum_{j=1}^{\mu-1} \mathscr{N}_{0j} (Dv_j)'. \tag{A.4}
$$

The subspace $\operatorname{im} D\Pi_{\mu-1}$ is an invariant subspace for the ODE (A.2). The components $v_0, v_1, \ldots, v_{\mu-1}$ remain within their subspaces $\operatorname{im} Q_0$, $\operatorname{im} \Pi_{\mu-2} Q_1, \ldots$, $\operatorname{im} \Pi_0 Q_{\mu-1}$, respectively. The structural decoupling is associated with the decomposition

$$
x = D^+ u + v_0 + v_1 + \cdots + v_{\mu-1}.
$$

All coefficients in (A.2)–(A.4) are continuous on $[a, b]$ and explicitly given in terms of the used admissible matrix function sequence as

$$
\mathscr{N}_{01} := -Q_0 Q_1 D^+
$$

$$
\mathscr{N}_{0j} := -Q_0 P_1 \cdots P_{j-1} Q_j D^+, \qquad j = 2, \ldots, \mu - 1,
$$

$$
\mathscr{N}_{i,i+1} := -\Pi_{i-1} Q_i Q_{i+1} D^+, \qquad i = 1, \ldots, \mu - 2,
$$

$$
\mathscr{N}_{ij} := -\Pi_{i-1} Q_i P_{i+1} \cdots P_{j-1} Q_j D^+, \qquad j = i+2, \ldots, \mu-1, \ i = 1, \ldots, \mu-2,
$$

$$
\mathscr{M}_{0j} := Q_0 P_1 \cdots P_{\mu-1} \mathscr{M}_j D\Pi_{j-1} Q_j, \qquad j = 1, \ldots, \mu-1,
$$

$$
\mathscr{M}_{ij} := \Pi_{i-1} Q_i P_{i+1} \cdots P_{\mu-1} \mathscr{M}_j D\Pi_{j-1} Q_j, \ j = i+1, \ldots, \mu-1, \ i = 1, \ldots, \mu-2,
$$

$$
\mathscr{L}_0 := Q_0 P_1 \cdots P_{\mu-1} G_\mu^{-1},
$$

$$
\mathscr{L}_i := \Pi_{i-1} Q_i P_{i+1} \cdots P_{\mu-1} G_\mu^{-1}, \qquad i = 1, \ldots, \mu-2,
$$

$$
\mathscr{L}_{\mu-1} := \Pi_{\mu-2} Q_{\mu-1} G_\mu^{-1},
$$

$$
\mathscr{H}_0 := Q_0 P_1 \cdots P_{\mu-1} \mathscr{K} \Pi_{\mu-1},
$$

in which

$$\mathscr{K} := (I - \Pi_{\mu-1})G_{\mu}^{-1}B_{\mu-1}\Pi_{\mu-1} + \sum_{\lambda=1}^{\mu-1}(I - \Pi_{\lambda-1})(P_{\lambda} - Q_{\lambda})(D\Pi_{\lambda}D^{+})'D\Pi_{\mu-1},$$

$$\mathscr{M}_j := \sum_{k=0}^{j-1}(I - \Pi_k)\{P_kD^{+}(D\Pi_kD^{+})' - Q_{k+1}D^{+}(D\Pi_{k+1}D^{+})'\}D\Pi_{j-1}Q_lD^{+},$$

$$j = 1, \ldots, \mu - 1.$$

Consider an arbitrary subinterval $[\bar{t}, \bar{t} + H] \subseteq [a, b]$ and use the function spaces

$$L_{sub}^2 = L^2((\bar{t}, \bar{t} + H), \mathbb{R}^m), \quad H_{sub}^1 = H^1((\bar{t}, \bar{t} + H), \mathbb{R}^k), \quad H_{D,sub}^1 = \{x \in L_{sub}^2 | Dx \in H_{sub}^1\},$$

equipped with their natural norms. Additionally, we introduce the function space (cf., [9, 15])

$$Z_{sub} := \{q \in L_{sub}^2 : v_{\mu-1} := \mathscr{L}_{\mu-1}q, \quad Dv_{\mu-1} \in H_{sub}^1,$$

$$v_{\mu-j} := \mathscr{L}_{\mu-j}q - \sum_{i=1}^{j-1}\mathscr{N}_{\mu-j,\mu-j+i}(Dv_{\mu-j+i})' - \sum_{i=1}^{j-1}\mathscr{M}_{\mu-j,\mu-j+i}v_{\mu-j+i},$$

$$Dv_{\mu-j} \in H_{sub}^1, \quad \text{for} \quad j = 2, \ldots, \mu - 1\}$$

and its norm

$$\|q\|_{Z_{sub}} := \left(\|q\|_{L_{sub}^2}^2 + \sum_{i=1}^{\mu-1}\|(Dv_i)'\|_{L_{sub}^2}^2\right)^{1/2}, \quad q \in Z_{sub}.$$

The latter function space is very special, it strongly depends on the decoupling coefficients which in turn are determined by the given data A, D, B.

We also assume a function $G : [a, b] \to \mathbb{R}^l$ with $G(t) = G(t)D^{+}D$ for all $t \in [a, b]$ to be given, and introduce the operator related to the subinterval $T_{sub} : H_{D,sub}^1 \to L_{sub}^2$ and the composed operator $\mathscr{T}_{sub} : H_{D,sub}^1 \to L_{sub}^2 \times \mathbb{R}^l$, by

$$T_{sub}x = A(Dx)' + Bx, \quad \mathscr{T}_{sub}x = \begin{bmatrix} T_{sub}x \\ G(\bar{t})x(\bar{t}) \end{bmatrix}, \quad x \in H_{D,sub}^1.$$

Here, trivially, the restrictions of A and B to the subinterval are meant. The operators T_{sub} and $\mathcal{T}_{sub}$ are well-defined and bounded. Regarding

$$\|T_{sub}x\|^2_{L^2_{sub}} = \int_{\bar{t}}^{\bar{t}+H} |A(t)(Dx)'(t) + B(t)x(t)|^2 \mathrm{dt}$$

$$\leq 2\max\{\max_{t\in[\bar{t},\bar{t}+H]} |A(t)|^2, \max_{t\in[\bar{t},\bar{t}+H]} |B(t)|^2\}\|x\|^2_{H^1_{D,sub}}$$

$$\leq 2\max\{\max_{t\in[a,b]} |A(t)|^2, \max_{t\in[a,b]} |B(t)|^2\}\|x\|^2_{H^1_{D,sub}}$$

we see that there is an upper bound on the operator norm of T_{sub} uniformly for all subintervals. Similarly, supposing G to be bounded on $[a, b]$, there is a uniform upper bound for the norm of $\mathcal{T}_{sub}$, too.

Proposition A.1 *Let the DAE be fine on $[a, b]$ with characteristic values (A.1) and index $\mu \geq 2$.*

Let the function $G : [a, b] \to \mathbb{R}^l$ be such that

$$\ker G(t) = \ker \Pi_{\mu-1}(t), \quad |G(t)| \leq c_G, \quad |G(t)^-| \leq c_{G^-}, \quad t \in [a, b],$$

in which c_G and c_{G^-} denote constants and $G(t)^-$ is a reflexive generalized inverse of $G(t)$. Then it holds:

(1) $\operatorname{im} T_{sub} = Z_{sub}$, $\operatorname{im} \mathcal{T}_{sub} = Z_{sub} \times \mathbb{R}^l$, $\ker \mathcal{T}_{sub} = \{0\}$.
(2) *The function space Z_{sub} equipped with the norm $\|\cdot\|_{Z_{sub}}$ is complete.*
(3) *There is a constant c_Z, uniformly for all subintervals $[\bar{t}, \bar{t} + H] \subseteq [a, b]$, such that*

$$\|x\|_{H^1_{D,sub}} \leq c_Z \, (\|q\|^2_{Z_{sub}} + |r|^2)^{1/2} \text{ for all } q \in Z_{sub}, r \in \mathbb{R}^l, x = \mathcal{T}^{-1}_{sub}(q,r).$$

Note that such a functions G exists always. For instance, applying Lemma 3.1 one can set $G(t) = [I_l \ 0]K(t)D$ and supplement it by $G(t)^- = D^+ K(t)^{-1}[I_l \ 0]^+$.

Proof

(1) The first assertions can be verified by means of the above decoupling formulas, which are given on $[a, b]$, and which are valid in the same way on each arbitrary subinterval, too. In particular, examining the equation $\mathcal{T}_{sub}x = 0$, we know from (A.3) that $q \in L^2_{sub}$, $q = 0$ implies $v_j = 0$ on the subinterval successively for $j = \mu - 1, \ldots, 1$. On the other hand, $G(\bar{t})x(\bar{t}) = 0$ leads to $u(\bar{t}) = D\Pi_{\mu-1}(\bar{t})x(\bar{t}) = D\Pi_{\mu-1}(\bar{t})G(\bar{t})^- G(\bar{t})x(\bar{t}) = 0$. Since $u \in H^1_{sub}$ solves the homogeneous ODE (A.2) on the subinterval, u vanishes there identically. Finally, from (A.4) it follows that $v_0 = 0$, and hence, $x = 0$.
(2) Let $q_n \in Z_{sub}$ be a fundamental sequence with respect to the $\|\cdot\|_{Z_{sub}}$-norm, and $v_{n,i} \in H^1_{D,sub}$, $i = 1, \ldots, \mu - 1$, correspondingly defined by (A.3), further $w_{n,i} = (Dv_{n,i})'$, $i = 1, \ldots, \mu - 1$. Then there exists an elements $q_* \in L^2_{sub}$

such that $q_n \overset{L^2}{\longrightarrow} q_*$ and there are further elements $w_{*,i} \in L^2((\bar{t}, \bar{t} + H), \mathbb{R}^k)$ so that $w_{n,i} \overset{L^2}{\longrightarrow} w_{*,i}, i = 1, \dots, \mu - 1$. The first line of the associated relations (A.3) leads to $v_{n,\mu-1} = \mathscr{L}_{\mu-1} q_n \overset{L^2}{\longrightarrow} \mathscr{L}_{\mu-1} q_* =: v_{*,\mu-1}, Dv_{n,\mu-1} = D\mathscr{L}_{\mu-1} q_n \overset{L^2}{\longrightarrow} Dv_{*,\mu-1}$, thus $Dv_{*,\mu-1} \in H^1_{sub}, (Dv_{*,\mu-1})' = w_{*,\mu-1}$. The next lines of (A.3) successively for $j = 2, \dots, \mu - 1$ provide

$$v_{n,\mu-j} = \mathscr{L}_{\mu-j} q_n - \sum_{i=1}^{j-1} \mathscr{N}_{\mu-j,\mu-j+i} (Dv_{n,\mu-j+i})' - \sum_{i=1}^{j-1} \mathscr{M}_{\mu-j,\mu-j+i} v_{n,\mu-j+i}$$

$$\overset{L^2}{\longrightarrow} \mathscr{L}_{\mu-j} q_* - \sum_{i=1}^{j-1} \mathscr{N}_{\mu-j,\mu-j+i} (Dv_{*,\mu-j+i})' - \sum_{i=1}^{j-1} \mathscr{M}_{\mu-j,\mu-j+i} v_{*,\mu-j+i} =: v_{*,\mu-j},$$

$$Dv_{*,\mu-j} \in H^1_{sub}, \quad (Dv_{*,\mu-j})' = w_{*,\mu-j},$$

and eventually we arrive at $q_* \in Z_{sub}$.

(3) The operator T_{sub} is bounded also with respect to the new image space Z_{sub} equipped with the norm $\|\cdot\|_{Z_{sub}}$. Namely, for each $x \in H^1_{D,sub}$ owing to the decoupling it holds that

$$Dv_i = D\Pi_{i-1} Q_i x = D\Pi_{i-1} Q_i D^+ Dx,$$

$$(Dv_i)' = (D\Pi_{i-1} Q_i D^+)' Dx + D\Pi_{i-1} Q_i D^+ (Dx)', \quad i = 1, \dots, \mu - 1.$$

This leads to $\|T_{sub} x\|_{Z_{sub}} \leq c^Z_{T_{sub}} \|x\|_{H^1_{D,sub}}$, with a uniform constant $c^Z_{T_{sub}}$ for all subintervals. In the new setting, the associated operator $\mathscr{T}_{sub} : H^1_{D,sub} \rightarrow Z_{sub} \times \mathbb{R}^l$ is a homeomorphism, and hence, its inverse is bounded. It remains to verify the existence of a uniform upper bound c_Z of the norm of $\mathscr{T}^{-1}_{sub}$. □

Let an arbitrary pair $(q, r) \in Z_{sub} \times \mathbb{R}^l$ be given and the solution $x \in H^1_{D,sub}$ of $\mathscr{T}_{sub} x = (q, r)$, i.e., $T_{sub} x = q, G(\bar{t}) x(\bar{t}) = r$. We apply again the decomposition of the solution $x = D^+ u + v_0 + v_1 + \dots + v_{\mu-1}$ and the decoupling (A.2), (A.3), (A.4). Owing to the properties of the function G it holds that $u(\bar{t}) = D\Pi_{\mu-1}(\bar{t}) x(\bar{t}) = D\Pi_{\mu-1}(\bar{t}) G(\bar{t})^- G(\bar{t}) x(\bar{t}) = D\Pi_{\mu-1}(\bar{t}) G(\bar{t})^- r$ and thus

$$|u(\bar{t})| \leq k_1 |r|,$$

with a constant k_1 being independent of the subinterval. Below, all the further constants k_i are also uniform ones for all subintervals.

Let $U(t, \bar{t})$ denote the fundamental solution matrix normalized at $\bar{t}$ of the ODE (A.2). U is defined on the original interval $[a, b]$, there continuously differentiable and nonsingular. $U(t, \bar{t})$ and $U(t, \bar{t})^{-1} = U(\bar{t}, t)$ are uniformly bounded on [a, b]. Turning back to the subinterval we apply the standard solution representation

$$u(t) = U(t, \bar{t})u(\bar{t}) + \int_{\bar{t}}^{t} U(t, s)D\Pi_{\mu-1}(s)G_{\mu}^{-1}(s)q(s)ds$$

$$= U(t, \bar{t})D\Pi_{\mu-1}(\bar{t})G(\bar{t})^{-}r + \int_{\bar{t}}^{t} U(t, s)D\Pi_{\mu-1}(s)G_{\mu}^{-1}(s)q(s)ds, \quad t \in [\bar{t}, \bar{t} + H].$$

Taking into account that the involved coefficients are defined on [a, b] and continuous there we may derive an inequality

$$\|u\|_{H_{sub}^1}^2 \leq k_2|r|^2 + k_3\|q\|_{sub}^2.$$

Next we rearrange system (A.3) to

$$\begin{bmatrix} v_1 \\ \vdots \\ v_{\mu-1} \end{bmatrix} = \mathfrak{M}^{-1} \begin{bmatrix} \mathscr{L}_1 \\ \vdots \\ \mathscr{L}_{\mu-1} \end{bmatrix} q - \mathfrak{M}^{-1} \begin{bmatrix} 0 & \mathscr{N}_{12} & \cdots & \mathscr{N}_{1,\mu-1} \\ & 0 & \ddots & \vdots \\ & & \ddots & \mathscr{N}_{\mu-2,\mu-1} \\ & & & 0 \end{bmatrix} \begin{bmatrix} (Dv_1)' \\ \vdots \\ (Dv_{\mu-1})' \end{bmatrix}$$

,

in which the inverse of the matrix function

$$\mathfrak{M} = \begin{bmatrix} I & \mathscr{M}_{12} & \cdots & \mathscr{M}_{1,\mu-1} \\ & I & \ddots & \vdots \\ & & \ddots & \mathscr{M}_{\mu-2,\mu-1} \\ & & & I \end{bmatrix}$$

is again continuous on [a, b] and upper triangular. This allows to derive the inequalities

$$\|v_j\|_{L_{sub}^2}^2 \leq k_4\|q\|_{L_{sub}^2}^2 + k_5 \sum_{i=1}^{\mu-1} \|(Dv_i)'\|_{L_{sub}^2}^2, \quad j = 1, \ldots, \mu-1.$$

Considering also (A.4) we obtain

$$\|x\|_{L_{sub}^2}^2 \leq k_6\|q\|_{L_{sub}^2}^2 + k_7 \sum_{i=1}^{\mu-1} \|(Dv_i)'\|_{L_{sub}^2}^2 + k_8|r|^2.$$

Since $(Dx)' = u' + \sum_{i=1}^{\mu-1}(Dv_i)'$ we have further

$$\|x\|^2_{H^1_{D,sub}} \le k_9 \left\{ \|q\|^2_{L^2_{sub}} + \sum_{i=1}^{\mu-1}\|(Dv_i)'\|^2_{L^2_{sub}} + |r|^2 \right\} = k_9(\|q\|^2_{Z_{sub}} + |r|^2).$$

B   On the Derivation of the Transfer Matrix $S(N, H, n)$

Consider an interval $(0, h)$. For the representation of polynomials we will use the Legendre polynomials P_k [18]. They have the properties

1. $\int_{-1}^{1} P_k(t)P_l(t)dt = \frac{2}{2k+1}\delta_{kl}, k, l = 0, 1, \ldots.$
2. $P_k(1) = 1, P_k(-1) = (-1)^k, k = 0, 1, \ldots.$
3. $P'_{k+1} - P'_{k-1} = (2k+1)P_k, k = 1, 2, \ldots.$

Let

$$p_k(t) = a_k P_k(1 - \frac{2}{h}t), \quad a_k = \left(\frac{2k+1}{h}\right)^{1/2}.$$

Then it holds

$$\int_0^h p_k(t)p_l(t)dt = \delta_{kl}, \quad p_k(0) = a_k, \quad p_k(h) = (-1)^k a_k.$$

From the representation for the derivatives, we obtain

$$\frac{h}{2}(c_k p'_{k-1} - d_k p'_{k+1}) = (2k+1)p_k$$

where

$$c_k = \frac{a_k}{a_{k-1}} = \left(\frac{2k+1}{2k-1}\right)^{1/2}, \quad d_k = \frac{a_k}{a_{k+1}} = \left(\frac{2k+1}{2k+3}\right)^{1/2}.$$

Since $p_0(t) \equiv a_0$ and $p'_1(t) = -2a_1/h$, we have the representation

$$\frac{h}{2}\bar{\Gamma}\begin{bmatrix} p'_1 \\ \vdots \\ p'_N \end{bmatrix} = D_- \begin{bmatrix} p_0 \\ \vdots \\ p_{N-1} \end{bmatrix},$$

with

$$
\bar{\Gamma} = \begin{bmatrix} -d_0 & & & & \\ 0 & -d_1 & & & \\ c_2 & 0 & -d_2 & & \\ & \ddots & \ddots & \ddots & \\ & & c_{N-1} & 0 & -d_{N-1} \end{bmatrix}, \quad D_- = \begin{bmatrix} 1 & & & \\ & 3 & & \\ & & \ddots & \\ & & & 2N-1 \end{bmatrix}.
$$

This provides

$$
\begin{bmatrix} p'_0 \\ \vdots \\ p'_N \end{bmatrix} = \frac{2}{h} \Gamma \begin{bmatrix} p_0 \\ \vdots \\ p_N \end{bmatrix}, \quad \Gamma = \begin{bmatrix} 0 & 0 \\ \bar{\Gamma}^{-1} D_- & 0 \end{bmatrix}.
$$

A representation of Γ being more suitable for the subsequent derivations can be obtained by observing that

$$
\bar{\Gamma} = D_-^{1/2} \begin{bmatrix} -1 & & & & \\ 0 & -1 & & & \\ 1 & 0 & -1 & & \\ & \ddots & \ddots & \ddots & \\ & & 1 & 0 & -1 \end{bmatrix} D_+^{-1/2}, \quad D_+ = \begin{bmatrix} 3 & & & \\ & 5 & & \\ & & \ddots & \\ & & & 2N+1 \end{bmatrix}.
$$

Let Z denote the tridiagonal matrix in this decomposition. Then it holds

$$
(Z^{-1})_{ij} = \begin{cases} -1, & i \geq j, i - j \text{ even}, \\ 0, & \text{else}. \end{cases}
$$

Hence,

$$
\Gamma = \begin{bmatrix} 0 & 0 \\ D_+^{1/2} Z^{-1} D_-^{1/2} & 0 \end{bmatrix} = D^{1/2} Y D^{1/2}
$$

where

$$
D = \begin{bmatrix} 1 & & & \\ & 3 & & \\ & & \ddots & \\ & & & 2N+1 \end{bmatrix}, \quad Y = - \begin{bmatrix} 0 & & & & & & \\ 1 & 0 & & & & & \\ 0 & 1 & 0 & & & & \\ 1 & 0 & 1 & 0 & & & \\ 0 & 1 & 0 & 1 & 0 & & \\ 1 & 0 & 1 & 0 & 1 & 0 & \\ 0 & 1 & 0 & 1 & 0 & 1 & 0 \\ \vdots & \ddots & \ddots & \ddots & \ddots & \ddots & \ddots \\ & \cdots & 0 & 1 & 0 & 1 & 0 & 1 & 0 \end{bmatrix}.
$$

Assume now

$$
x_i = \sum_{k=0}^{N} \alpha_{ik} p_k, \quad \mathbf{a} = (a_0, \ldots, a_N)^T.
$$

Then

$$
x_i(0) = \sum_{n=0}^{N} \alpha_{in} p_n(0) = \sum_{n=0}^{N} \alpha_{in} a_n = \mathbf{a}^T \alpha_i \quad i = 1, \ldots, \mu - 1.
$$

We collect the coefficients $\alpha = (\alpha_1, \ldots, \alpha_{\mu-1})^T$ and set

$$
A = \begin{bmatrix} I & & & \\ \frac{2}{h}\Gamma^T & I & & \\ & \ddots & \ddots & \\ & & \frac{2}{h}\Gamma^T & I \end{bmatrix}, \quad C = \begin{bmatrix} \mathbf{a}^T & & \\ & \ddots & \\ & & \mathbf{a}^T \end{bmatrix}.
$$

Let now $H > 0$ be fixed and $h = H/n$ for a given positive integer n. The functional to be minimized is

$$
\Phi_{sub}(x) = \frac{1}{2}\|x_1\|^2_{L^2_{sub}} + \sum_{i=2}^{\mu} \frac{1}{2}\|x'_{i-1} - x_i\|^2_{L^2_{sub}}, \quad x = (x_1, \ldots, x_\mu)^T
$$

on X_π under the condition $x_i = r_i$, $i = 1, \ldots, \mu - 1$. The term for $i = \mu$ in this sum can be omitted since, for given $x_{\mu-1} \in \mathscr{P}_N$, $x_\mu \in \mathscr{P}_{N-1}$ can always be set to $x'_{\mu-1}$ such that the last term amounts to 0.

For a shorthand notation, define $x_i^\nu = x_i|_{((\nu-1)h,\nu h)}$. Assuming the representation

$$x_i^\nu = \sum_{k=0}^N \alpha_{ik}^\nu p_k^\nu$$

on $((\nu-1)h, \nu h)$ with p_k^ν being the polynomials p_k transformed onto $((\nu-1)h, \nu h)$, we obtain

$$\Phi_{sub}(x) = \sum_{\nu=1}^n \left(\frac{1}{2}|\alpha_1^\nu|^2 + \frac{1}{2}\sum_{i=2}^{\mu-1} \left| \frac{2}{h}\Gamma^T \alpha_{i-1}^\nu + \alpha_i^\nu \right|^2 \right)$$

where $\alpha_i^\nu = (\alpha_{i0}^\nu, \ldots, \alpha_{iN}^\nu)^T$. Furthermore,

$$x_i^{\nu-1}(\nu h) = \sum_{k=0}^N \alpha_{ik}^{\nu-1} p_k(h) = \sum_{k=0}^N \alpha_{ik}^{\nu-1} a_k(-1)^k$$

for $i = 1, \ldots, \mu - 1$. Define $\mathbf{b} = (a_0, \ldots, (-1)^N a_N)^T$.

All these equations can be conveniently written down in a matrix fashion. The initial condition becomes

$$C\alpha^1 = r$$

while the transfer conditions read

$$B\alpha^{\nu-1} = C\alpha^\nu, \quad \nu = 2, \ldots, n$$

with

$$B = \begin{bmatrix} \mathbf{b}^T & & \\ & \ddots & \\ & & \mathbf{b}^T \end{bmatrix}.$$

Let $\boldsymbol{\alpha} = (\alpha^1, \ldots, \alpha^n)^T$ and

$$\mathscr{A} = \begin{bmatrix} A & & & \\ & A & & \\ & & \ddots & \\ & & & A \end{bmatrix}, \quad \mathscr{C} = \begin{bmatrix} C & & & \\ -B & C & & \\ & \ddots & \ddots & \\ & & -B & C \end{bmatrix}.$$

Note that $\mathscr{A}$ is nonsingular since A is so. Similarly, $\mathscr{C}$ has full row rank since C has the same property.

Finally, we obtain

$$\Phi_{sub}(x) = \Phi_{sub}(\boldsymbol{\alpha}) = \frac{1}{2}|\mathscr{A}\boldsymbol{\alpha}|^2 \to \min \text{ such that } \mathscr{C}\boldsymbol{\alpha} = (r, 0, \dots, 0)^T.$$

The transfer matrix is then given by

$$S(N, H, n)r = B\alpha^n(r) \text{ for all } r \in \mathbb{R}^{\mu-1}.$$

In the case $\mu = 2$, a simple analytical solution is feasible.

B.1   The Case $\mu = 2$

In order to simplify the notation, the index i will be omitted. The transfer matrix reduces to a scalar

$$\rho_n = \left| \frac{x^n(H)}{r} \right|.$$

The Lagrange functional belonging to the present optimization problem reads

$$\varphi(\alpha, \lambda) = \sum_{\nu=1}^{n} \sum_{k=0}^{N} (\alpha_k^1)^2 + \lambda_1 \left(\sum_{k=0}^{N} a_k \alpha_k^1 - r \right)$$

$$+ \sum_{\nu=2}^{n} \lambda_\nu \left(\sum_{k=0}^{N} a_k \alpha_k^\nu - \sum_{k=0}^{N} (-1)^k a_k \alpha_k^{\nu-1} \right).$$

In the following, we will use the notations

$$a = \sum_{k=0}^{N} a_k^2 = \frac{1}{h}(N+1)^2, \quad b = \sum_{k=0}^{N} (-1)^k a_k^2 = \frac{(-1)^N}{h}(N+1), \quad c = \left| \frac{b}{a} \right|.$$

The derivatives of the Lagrange functional are

$$\frac{\partial \varphi}{\partial \lambda_1} = \sum_{k=0}^{N} a_k \alpha_k^1 - r,$$

$$\frac{\partial \varphi}{\partial \lambda_\nu} = \sum_{k=0}^{N} a_k \alpha_k^\nu - \sum_{k=0}^{N} a_k (-1)^k \alpha_k^{\nu-1}, \quad \nu = 2, \ldots, n$$

$$\frac{\partial \varphi}{\partial \alpha_k^n} = \alpha_k^n + \lambda_n a_k,$$

$$\frac{\partial \varphi}{\partial \alpha_k^\nu} = \alpha_n^\nu + \lambda_\nu a_k - \lambda_{\nu+1} a_k (-1)^k, \quad \nu = 1, \ldots, n-1.$$

Hence, for $\nu = 1$,

$$r = \sum_{k=0}^{N} a_k \alpha_k^1 = \sum_{k=0}^{N} a_k \left(\lambda_2 a_k (-1)^k - \lambda_1 a_k \right)$$

$$= b\lambda_2 - a\lambda_1.$$

Similarly, for $\nu = n$,

$$0 = \sum_{k=0}^{N} a_k \alpha_k^n - \sum_{k=0}^{N} a_k (-1)^k \alpha_k^{\nu-1}$$

$$= -\sum_{k=0}^{N} a_k^2 \lambda_n + \sum_{k=0}^{N} a_k (-1)^k \left(\lambda_{n-1} a_k - \lambda_n a_k (-1)^k \right)$$

$$= b\lambda_{n-1} - 2a\lambda_n.$$

And finally, for $1 < \nu < n$,

$$0 = \sum_{k=0}^{N} a_k \alpha_k^\nu - \sum_{k=0}^{N} a_k (-1)^k \alpha_k^{\nu-1}$$

$$= \sum_{k=0}^{N} a_k \left(\lambda_{\nu+1} a_k (-1)^k - \lambda_\nu a_k \right) - \sum_{k=0}^{N} a_k (-1)^k \left(\lambda_\nu a_k (-1)^k - \lambda_{\nu-1} a_k \right)$$

$$= b\lambda_{\nu+1} - 2a\lambda_\nu + b\lambda_{\nu-1}.$$

This provides us with the linear system of equations

$$
\begin{bmatrix}
-a & b & & & \\
b & -2a & b & & \\
 & \ddots & \ddots & \ddots & \\
 & & b & -2a & b \\
 & & & b & -2a
\end{bmatrix}
\begin{bmatrix}
\lambda_1 \\
\lambda_2 \\
\vdots \\
\lambda_{n-1} \\
\lambda_n
\end{bmatrix}
=
\begin{bmatrix}
r \\
0 \\
\vdots \\
0 \\
0
\end{bmatrix}.
$$

Since $x^n(H) = \sum_{k=0}^{N} a_k (-1)^k \alpha_k^n = -\sum_{k=0}^{N} a_k (-1)^k \lambda_n a_k = -b\lambda_n$ it is sufficient to compute the last component λ_n of the solution to this system. Let A_n denote the system matrix and $\tilde{A}_n$ the matrix obtained from A_n by replacing the last column of A_n by the right-hand side. According to Cramer's rule it holds

$$
\lambda_n = \frac{\det A_n}{\det \tilde{A}_n}.
$$

Let $u_n = \det A_n$ and $v_n = \det \tilde{A}_n$. Then we obtain the recursion

$$
v_1 = r,
$$
$$
v_\nu = -b v_{\nu-1}.
$$

Its solution is given by

$$
v_\nu = (-b)^{\nu-1} r.
$$

Analogously, we have

$$
u_1 = -a,
$$
$$
u_2 = 2a^2 - b^2,
$$
$$
u_\nu = -2a u_{\nu-1} - b^2 u_{\nu-2}.
$$

This recursion is a simple difference equation with the general solution

$$
u_\nu = c_1 z_1^\nu + c_2 z_2^\nu,
$$

where

$$
z_{1,2} = a\left(-1 \pm \sqrt{1 - c^2}\right).
$$

Application of the initial condition leads to $c_1 = c_2 = 1/2$. Inserting these expressions we obtain

$$\rho_n = \left| \frac{x^n(H)}{r} \right|$$

$$= \left| \frac{-b\lambda_n}{r} \right|$$

$$= \left| \frac{2b^n}{a^n \left[\left(-1 + \sqrt{1 - c^2} \right)^n + \left(-1 - \sqrt{1 - c^2} \right)^n \right]} \right|$$

$$= c^n \left| \frac{2}{\left(-1 + \sqrt{1 - c^2} \right)^n + \left(-1 - \sqrt{1 - c^2} \right)^n} \right|.$$

From the definition of c we obtain $c = (N + 1)^{-1}$. Hence, $\sqrt{1 - c^2} \approx 1$ such that

$$\rho_L \approx c^L 2^{1-L}.$$

B.2 An Approach for $\mu > 2$

In the case $\mu > 2$, the steps taken in the case $\mu = 2$ can be repeated. The Lagrangian system for the constraint optimization problem reads

$$\begin{bmatrix} \mathscr{A}^T \mathscr{A} & \mathscr{C}^T \\ \mathscr{C} & 0 \end{bmatrix} \begin{bmatrix} \alpha \\ \lambda \end{bmatrix} = \begin{bmatrix} 0 \\ \mathbf{r} \end{bmatrix}$$

where

$$\mathbf{r} = (r, 0, \dots, 0)^T.$$

The computation steps are then

(i) $\alpha = -(\mathscr{A}^T \mathscr{A})^{-1} \mathscr{C}^T \lambda$
(ii) $\lambda = -[\mathscr{C}(\mathscr{A}^T \mathscr{A})^{-1} \mathscr{C}^T]^{-1} \mathbf{r}$
(iii) $\alpha = (\mathscr{A}^T \mathscr{A})^{-1} \mathscr{C}^T [\mathscr{C}(\mathscr{A}^T \mathscr{A})^{-1} \mathscr{C}^T]^{-1} \mathbf{r}$
(iv) $x(H) = B\alpha = B(\mathscr{A}^T \mathscr{A})^{-1} \mathscr{C}^T [\mathscr{C}(\mathscr{A}^T \mathscr{A})^{-1} \mathscr{C}^T]^{-1} \mathbf{r}.$

In the end, this yields

$$S(N, H, n) = B(\mathscr{A}^T \mathscr{A})^{-1} \mathscr{C}^T [\mathscr{C}(\mathscr{A}^T \mathscr{A})^{-1} \mathscr{C}^T]^{-1}.$$

This representation can easily be evaluated using symbolic computations. It should be mentioned that most terms in $S(N, H, n)$ lead to simple rational expressions in N. However, the results presented in Sect. 5.3 have been computed numerically.

References

1. Brenan, K.E., Campbell, S.L., Petzold, L.R.: Numerical Solution of Initial-Value Problems in Differential-Algebraic Equations. North-Holland, Elsevier Science Publishing, Amsterdam (1989)
2. Burger, M., Gerdts, M.: In: Ilchmann, A., Reis, T. (eds.) Surveys in Differential-Algebraic Equations IV, chap. A Survey on Numerical Methods for the Simulation of Initial Value Problems with sDAEs, pp. 221–300. Differential-Algebraic Equations Forum. Springer, Heidelberg (2017)
3. Campbell, S.L.: The numerical solution of higher index linear time varying singular systems of differential equations. SIAM J. Sci. Stat. Comput. **6**(2), 334–348 (1985)
4. Campbell, S.L.: A computational method for general nonlinear higher index singular systems of differential equations. IMACS Trans. Sci. Comput. **1**(2), 555–560 (1989)
5. Campbell, S.L., Moore, E.: Constraint preserving integrators for general nonlinear higher index DAEs. Num. Math. **69**, 383–399 (1995)
6. Griepentrog, E., März, R.: Differential-Algebraic Equations and Their Numerical Treatment. Teubner Texte zur Mathematik 88. BSB Teubner Leipzig (1986)
7. Hanke, M., März, R.: Convergence analysis of least-squares collocation metods for nonlinear higher-index differential-algebraic equations. J. Comput. Appl. Math. (2019). doi:10.1016/j.cam.2019.112514
8. Hanke, M., März, R.: A reliable direct numerical treatment of differential-algebraic equations by overdetermined collocation: An operator approach. J. Comput. Appl. Math. (2019). doi:10.1016/j.cam.2019.112510
9. Hanke, M., März, R., Tischendorf, C.: Least-squares collocation for higher-index linear differential-algebaic equations: Estimating the stability threshold. Math. Comput. **88**(318), 1647–1683 (2019). https://doi.org/10.1090/mcom/3393
10. Hanke, M., März, R., Tischendorf, C., Weinmüller, E., Wurm, S.: Least-squares collocation for linear higher-index differential-algebraic equations. J. Comput. Appl. Math. **317**, 403–431 (2017). http://dx.doi.org/10.1016/j.cam.2016.12.017
11. Kaltenbacher, B., Offtermatt, J.: A convergence analysis of regularization by discretization in preimage space. Math. Comput. **81**(280), 2049–2069 (2012)
12. Kunkel, P., Mehrmann, V.: Differential-Algebraic Equations. Analysis and Numerical Solution. Textbooks in Mathematics. European Mathematical Society, Zürich (2006)
13. Lamour, R., März, R., Tischendorf, C.: In: Ilchmann, A., Reis, T. (eds.) Differential-Algebraic Equations: A Projector Based Analysis. Differential-Algebraic Equations Forum. Springer, Berlin, Heidelberg, New York, Dordrecht, London (2013)
14. Lamour, R., März, R., Weinmüller, E.: In: Ilchmann, A., Reis, T. (eds.) Surveys in Differential-Algebraic Equations III, chap. Boundary-Value Problems for Differential-Algebraic Equations: A Survey, pp. 177–309. Differential-Algebraic Equations Forum. Springer, Heidelberg (2015)
15. März, R.: In: Ilchmann, A., Reis, T. (eds.) Surveys in Differential-Algebraic Equations II, chap. Differential-Algebraic Equations from a Functional-Analytic Viewpoint: A Survey, pp. 163–285. Differential-Algebraic Equations Forum. Springer, Heidelberg (2015)

16. Pryce, J.D.: Solving high-index DAEs by Taylor series. Numer. Algorithms **19**(1–4), 195–211 (1998)
17. Schwarz, D.E., Lamour, R.: A new approach for computing consistent initial values and Taylor coefficients for DAEs using projector-based constrained optimization. Numer. Algorithms **78**(2), 355–377 (2018)
18. Suetin, P.K.: Classical Orthogonal Polynomials (in Russian), 2nd edn. Nauka, Moskva (1979)

Exponential Integrators for Semi-linear Parabolic Problems with Linear Constraints

Robert Altmann and Christoph Zimmer

Abstract This paper is devoted to the construction of exponential integrators of first and second order for the time discretization of constrained parabolic systems. For this extend, we combine well-known exponential integrators for unconstrained systems with the solution of certain saddle point problems in order to meet the constraints throughout the integration process. The result is a novel class of semi-explicit time integration schemes. We prove the expected convergence rates and illustrate the performance on two numerical examples including a parabolic equation with nonlinear dynamic boundary conditions.

Keywords PDAE · Exponential integrator · Parabolic equations · Time discretization

Mathematics Subject Classification (2010) 65M12, 65J15, 65L80

1 Introduction

Exponential integrators provide a powerful tool for the time integration of stiff ordinary differential equations as well as parabolic partial differential equations (PDE), cf. [8, 19, 24]. Such integrators are based on the possibility to solve the linear part – which is responsible for the stiffness of the system – in an exact manner. As a result, large time steps are possible which makes the method well-suited for time stepping, especially for parabolic systems where CFL conditions may be

R. Altmann (✉)
Department of Mathematics, University of Augsburg, Augsburg, Germany
e-mail: robert.altmann@math.uni-augsburg.de

C. Zimmer
Institute of Mathematics MA 4-5, Technical University Berlin, Berlin, Germany
e-mail: zimmer@math.tu-berlin.de

© The Editor(s) (if applicable) and The Author(s), under exclusive licence   137
to Springer Nature Switzerland AG 2020
T. Reis et al. (eds.), *Progress in Differential-Algebraic Equations II*,
Differential-Algebraic Equations Forum,
https://doi.org/10.1007/978-3-030-53905-4_5

very restrictive. For semi-linear ODEs and parabolic PDEs exponential integrators are well-studied in the literature. This includes explicit and implicit exponential Runge-Kutta methods [11, 17, 18], exponential Runge-Kutta methods of high order [26], exponential Rosenbrock-type methods [21], and multistep exponential integrators [10].

In this paper, we construct and analyze exponential integrators for parabolic PDEs which underlie an additional (linear) constraint. This means that we aim to approximate the solution to

$$\dot{u}(t) + \mathscr{A}u(t) = f(t, u)$$

which at the same time satisfies a constraint of the form $\mathscr{B}u(t) = g(t)$. Such systems can be considered as differential-algebraic equations (DAEs) in Banach spaces, also called partial differential-algebraic equations (PDAEs), cf. [1, 13, 25]. PDAEs of parabolic type include the transient Stokes problem (where $\mathscr{B}$ equals the divergence operator) as well as problems with nontrivial boundary conditions (with $\mathscr{B}$ being the trace operator). On the other hand, PDAEs of hyperbolic type appear in the modeling of gas and water networks [5, 14, 22] and in elastic multibody modeling [33].

Besides for the special application of the incompressible Navier-Stokes equations [23, 28], exponential integrators have not been considered for PDAEs so far. In the finite-dimensional case, however, exponential integrators have been analyzed for DAEs of (differential) index 1 [20]. We emphasize that the parabolic PDAEs within this paper generalize index-2 DAEs in the sense that a standard spatial discretization by finite elements leads to DAEs of index 2. Known time stepping methods for the here considered parabolic PDAEs include splitting methods [3], algebraically stable Runge-Kutta methods [4], and discontinuous Galerkin methods [37].

In the first part of the paper we discuss the existence and uniqueness of solutions for semi-linear PDAEs of parabolic type with linear constraints. Afterwards, we propose two exponential integrators of first and second order for such systems. The construction of this novel class of time integration schemes benefits from the interplay of well-known time integration schemes for unconstrained systems and stationary saddle point problems in order to meet the constraints. Since the latter is done in an implicit manner, the combination with explicit schemes for the dynamical part of the system leads to so-called *semi-explicit* time integration schemes. As exponential integrators are based on the exact evaluation of semigroups, we need to extend this to the constrained case. The proper equivalent is the solution of a homogeneous but transient saddle point problem, which is a linear PDAE.

The resulting exponential Euler scheme requires the solution of three stationary and a single transient saddle point problem in each time step. All these systems are linear, require in total only one evaluation of the nonlinear function, and do not call for another linearization step. Further, the transient system is homogeneous such that it can be solved without an additional regularization (or index reduction in the finite-dimensional case). The corresponding second-order scheme requires the solution of additional saddle point problems. Nevertheless, all these systems are linear and easy to solve. In a similar manner – but under additional regularity assumptions – one may translate more general exponential Runge-Kutta schemes to the constrained case. Here, however, we restrict ourselves to schemes of first and second order.

The paper is organized as follows. In Sect. 2 we recall the most important properties of exponential integrators for parabolic problems in the unconstrained case. Further, we introduce the here considered parabolic PDAEs, summarize all needed assumptions, and analyze the existence of solutions. The exponential Euler method is then subject of Sect. 3. Here we discuss two approaches to tackle the occurrence of constraints and prove first-order convergence. An exponential integrator of second order is then introduced and analyzed in Sect. 4. Depending on the nonlinearity, this scheme converges with order 2 or reduced order $3/2$. Comments on the efficient computation and numerical experiments for semi-linear parabolic systems illustrating the obtained convergence results are presented in Sect. 5.

2 Preliminaries

In this preliminary section we recall basic properties of exponential integrators when applied to PDEs of parabolic type. For this (and the later analysis) we consider the well-known φ functions. Further, we introduce the precise setting for the here considered parabolic systems with constraints and discuss their solvability.

2.1 Exponential Integrators for Parabolic Problems

As exponential integrators are based on the exact solution of linear homogeneous problems, we consider the recursively defined φ-functions, see, e.g. [35, Ch. 11.1],

$$\varphi_0(z) := e^z, \qquad \varphi_{k+1}(z) := \frac{\varphi_k(z) - \varphi_k(0)}{z}. \tag{2.1}$$

For $z = 0$ the values are given by $\varphi_k(0) = 1/k!$. The importance of the φ-functions comes from the fact that they can be equivalently written as integrals of certain exponentials. More precisely, we have for $k \geq 1$ that

$$\varphi_k(z) = \int_0^1 e^{(1-s)z} \frac{s^{k-1}}{(k-1)!} \, ds. \tag{2.2}$$

We will consider these functions in combination with differential operators. For a bounded and invertible operator $\mathscr{A}: X \to X$, where $e^{t\mathscr{A}} := \exp(t\mathscr{A})$ is well-defined, we can directly use the formula in (2.1) using the notion $\frac{1}{\mathscr{A}} = \mathscr{A}^{-1}$. More generally, we can apply (2.2) to define φ_k with non-invertible or unbounded differential operators as arguments. Moreover, the exact solution of a linear abstract

ODE with a bounded operator and a polynomial right-hand side can be expressed in terms of the φ-functions. More precisely, the solution of

$$\dot{u}(t) + \mathscr{A}u(t) = \sum_{k=1}^{n} \frac{f_k}{(k-1)!} t^{k-1} \in X \tag{2.3}$$

with initial condition $u(0) = u_0$ and coefficients $f_k \in X$ is given by

$$u(t) = \varphi_0(-t\mathscr{A})\, u_0 + \sum_{k=1}^{n} \varphi_k(-t\mathscr{A})\, f_k\, t^k. \tag{2.4}$$

If $-\mathscr{A}\colon D(\mathscr{A}) \subset X \to X$ is an unbounded differential operator which generates a strongly continuous semigroup, then we obtain the following major property for the corresponding φ-functions.

Theorem 2.1 (cf. [19, Lem. 2.4]) *Assume that the linear operator $-\mathscr{A}$ is the infinitesimal generator of a strongly continuous semigroup $e^{-t\mathscr{A}}$. Then, the operators $\varphi_k(-\tau\mathscr{A})$ are linear and bounded in X*

With the interpretation of the exponential as the corresponding semigroup, the solution formula for bounded operators (2.4) directly translates to linear parabolic PDEs of the form (2.3) with an unbounded differential operator $\mathscr{A}$, cf. [19].

The construction of exponential integrators for $\dot{u}(t) + \mathscr{A}u(t) = f(t, u)$ is now based on the replacement of the nonlinearity f by a polynomial and (2.4). Considering the interpolation polynomial of degree 0, i.e., evaluating the nonlinearity only in the starting value of u, we obtain the *exponential Euler scheme*. The corresponding scheme for constrained systems is discussed in Sect. 3 and a second-order scheme in Sect. 4.

2.2 Parabolic Problems with Constraints

In this subsection, we introduce the constrained parabolic systems of interest and gather assumptions on the involved operators. Throughout this paper we consider semi-explicit and semi-linear systems meaning that the constraints are linear and that the nonlinearity only appears in the low-order terms of the dynamic equation. Thus, we consider the following parabolic PDAE: find $u\colon [0, T] \to \mathscr{V}$ and $\lambda\colon [0, T] \to \mathscr{Q}$ such that

$$\dot{u}(t) + \mathscr{A}u(t) + \mathscr{B}^*\lambda(t) = f(t, u) \quad \text{in } \mathscr{V}^*, \tag{2.5a}$$

$$\mathscr{B}u(t) = g(t) \quad \text{in } \mathscr{Q}^*. \tag{2.5b}$$

Therein, $\mathscr{V}$ and $\mathscr{Q}$ denote Hilbert spaces with respective duals $\mathscr{V}^*$ and $\mathscr{Q}^*$. The space $\mathscr{V}$ is part of a Gelfand triple $\mathscr{V}, \mathscr{H}, \mathscr{V}^*$, cf. [38, Ch. 23.4]. This means that $\mathscr{V}$ is continuously (and densely) embedded in the pivot space $\mathscr{H}$ which implies $\mathscr{H}^* \hookrightarrow \mathscr{V}^*$, i.e., the continuous embedding of the corresponding dual spaces. In this setting, the Hilbert space $\mathscr{H}$ is the natural space for the initial data. Note, however, that the initial condition may underlie a consistency condition due to the constraint (2.5b), cf. [13]. For the here considered analysis we assume slightly more regularity, namely $u(0) = u_0 \in \mathscr{V}$, and consistency of the form $\mathscr{B}u_0 = g(0)$.

The assumptions on the operators $\mathscr{A} \in \mathscr{L}(\mathscr{V}, \mathscr{V}^*)$ and $\mathscr{B} \in \mathscr{L}(\mathscr{V}, \mathscr{Q}^*)$ are summarized in the following.

Assumption 2.1 (Constraint Operator $\mathscr{B}$) *The operator* $\mathscr{B}: \mathscr{V} \to \mathscr{Q}^*$ *is linear, continuous, and satisfies an inf-sup condition, i.e., there exists a constant* $\beta > 0$ *such that*

$$\inf_{q \in \mathscr{Q} \setminus \{0\}} \sup_{v \in \mathscr{V} \setminus \{0\}} \frac{\langle \mathscr{B}v, q \rangle}{\|v\|_{\mathscr{V}} \|q\|_{\mathscr{Q}}} \geq \beta.$$

Assumption 2.2 (Differential Operator $\mathscr{A}$) *The operator* $\mathscr{A}: \mathscr{V} \to \mathscr{V}^*$ *is linear, continuous, and has the form* $\mathscr{A} = \mathscr{A}_1 + \mathscr{A}_2$ *with* $\mathscr{A}_1 \in \mathscr{L}(\mathscr{V}, \mathscr{V}^*)$ *being self-adjoint and* $\mathscr{A}_2 \in \mathscr{L}(\mathscr{V}, \mathscr{H})$. *Further, we assume that* $\mathscr{A}$ *is elliptic on* $\mathscr{V}_{ker} := \ker \mathscr{B}$, *i.e., on the kernel of the constraint operator.*

Without loss of generality, we may assume under Assumption 2.2 that $\mathscr{A}_1$ is elliptic on $\mathscr{V}_{ker}$. This can be seen as follows: With $\mu_{\mathscr{A}}$ denoting the ellipticity constant of $\mathscr{A}$ and $c_{\mathscr{A}_2}$ the continuity constant of $\mathscr{A}_2$, we set

$$\mathscr{A}_1 \leftarrow \mathscr{A}_1 + \frac{c_{\mathscr{A}_2}^2}{2\mu_{\mathscr{A}}} \, \mathrm{id}_{\mathscr{H}} \quad \text{and} \quad \mathscr{A}_2 \leftarrow \mathscr{A}_2 - \frac{c_{\mathscr{A}_2}^2}{2\mu_{\mathscr{A}}} \, \mathrm{id}_{\mathscr{H}}.$$

This then implies

$$\langle \mathscr{A}_1 v_{ker}, v_{ker} \rangle \geq \mu_{\mathscr{A}} \|v_{ker}\|_{\mathscr{V}}^2 - c_{\mathscr{A}_2} \|v_{ker}\|_{\mathscr{V}} \|v_{ker}\|_{\mathscr{H}} + \frac{c_{\mathscr{A}_2}^2}{2\mu_{\mathscr{A}}} \|v_{ker}\|_{\mathscr{H}}^2 \geq \frac{\mu_{\mathscr{A}}}{2} \|v_{ker}\|_{\mathscr{V}}^2$$

for all $v_{ker} \in \mathscr{V}_{ker}$. Hence, we assume throughout this paper that, given Assumption 2.2, $\mathscr{A}_1$ is elliptic on $\mathscr{V}_{ker}$. As a result, $\mathscr{A}_1$ induces a norm which is equivalent to the $\mathscr{V}$-norm on $\mathscr{V}_{ker}$, i.e.,

$$\mu \|v_{ker}\|_{\mathscr{V}}^2 \leq \|v_{ker}\|_{\mathscr{A}_1}^2 \leq C \|v_{ker}\|_{\mathscr{V}}^2. \tag{2.6}$$

Remark 2.1 The results of this paper can be extended to the case where $\mathscr{A}$ only satisfies a Gårding inequality on $\mathscr{V}_{ker}$. In this case, we add to $\mathscr{A}$ the term $\kappa \, \mathrm{id}_{\mathscr{H}}$ such that $\mathscr{A} + \kappa \, \mathrm{id}_{\mathscr{H}}$ is elliptic on $\mathscr{V}_{ker}$ and add it accordingly to the nonlinearity f.

Assumption 2.1 implies that $\mathscr{B}$ is onto such that there exists a right-inverse denoted by $\mathscr{B}^- : \mathscr{Q}^* \to \mathscr{V}$. This in turn motivates the decomposition

$$\mathscr{V} = \mathscr{V}_{\ker} \oplus \mathscr{V}_c \qquad \text{with} \qquad \mathscr{V}_{\ker} = \ker \mathscr{B}, \quad \mathscr{V}_c = \operatorname{im} \mathscr{B}^-.$$

We emphasize that the choice of the right-inverse (and respectively $\mathscr{V}_c$) is, in general, not unique and allows a certain freedom in the modeling process. Within this paper, we define the complementary space $\mathscr{V}_c$ as in [4] in terms of the annihilator of $\mathscr{V}_{\ker}$, i.e.,

$$\mathscr{V}_c := \{v \in \mathscr{V} \mid \mathscr{A}v \in \mathscr{V}_{\ker}^0\} = \{v \in \mathscr{V} \mid \langle \mathscr{A}v, w \rangle = 0 \text{ for all } w \in \mathscr{V}_{\ker}\}.$$

The analysis of constrained systems such as (2.5) is heavily based on the mentioned decomposition of $\mathscr{V}$. Furthermore, we need the restriction of the differential operator to the kernel of $\mathscr{B}$, i.e.,

$$\mathscr{A}_{\ker} := \mathscr{A}|_{\mathscr{V}_{\ker}} : \mathscr{V}_{\ker} \to \mathscr{V}_{\ker}^* := (\mathscr{V}_{\ker})^*.$$

Note that we use here the fact that functionals in $\mathscr{V}^*$ define functionals in $\mathscr{V}_{\ker}^*$ simply through the restriction to $\mathscr{V}_{\ker}$. The closure of $\mathscr{V}_{\ker}$ in the $\mathscr{H}$-norm is denoted by $\mathscr{H}_{\ker} := \overline{\mathscr{V}_{\ker}}^{\mathscr{H}}$. Assumption 2.2 now states that $\mathscr{A}_{\ker}$ is an elliptic operator. This in turn implies that $-\mathscr{A}_{\ker}$ generates an analytic semigroup on $\mathscr{H}_{\ker}$, see [30, Ch. 7, Th. 2.7].

Finally, we need assumptions on the nonlinearity f. Here, we require certain smoothness properties such as local Lipschitz continuity in the second component. The precise assumptions will be given in the respective theorems.

Example 2.1 The (weak) formulation of semi-linear parabolic equations with dynamical (or Wentzell) boundary conditions [34] fit into the given framework. For this, the system needs to be formulated as a coupled system which leads to the PDAE structure (2.5), cf. [2]. We emphasize that also the boundary condition may include nonlinear reaction terms. We will consider this example in the numerical experiments of Sect. 5.

2.3 Existence of Solutions

In this section we discuss the existence of solutions to (2.5), where we use the notion of Sobolev-Bochner spaces $L^2(0, T; X)$ and $H^1(0, T; X)$ for a Banach space X, cf. [38, Ch. 23]. For the case that f is independent of u, the existence of solutions is well-studied, see [1, 13, 36]. We recall the corresponding result in the special of $\mathscr{A}$ being self-adjoint, which is needed in later proofs.

Lemma 2.2 *Let $\mathscr{A} \in \mathscr{L}(\mathscr{V}, \mathscr{V}^*)$ be self-adjoint and elliptic on $\mathscr{V}_{\mathrm{ker}}$ and let $\mathscr{B}$ satisfy Assumption 2.1. Further, assume $f \in L^2(0, T; \mathscr{H})$, $g \in H^1(0, T; \mathscr{Q}^*)$, and $u_0 \in \mathscr{V}$ with $\mathscr{B}u_0 = g(0)$. Then, the PDAE (2.5) with right-hand sides f, g – independent of u – and initial value u_0 has a unique solution*

$$u \in C([0, T]; \mathscr{V}) \cap H^1(0, T; \mathscr{H}), \qquad \lambda \in L^2(0, T; \mathscr{Q})$$

with $u(0) = u_0$. The solution depends continuously on the data and satisfies

$$\|u(t) - \mathscr{B}^- g(t)\|_{\mathscr{A}}^2 \leq \|u_0 - \mathscr{B}^- g(0)\|_{\mathscr{A}}^2 + \int_0^t \|f(s) - \mathscr{B}^- \dot{g}(s)\|_{\mathscr{H}}^2 \, ds. \qquad (2.7)$$

Proof A sketch of the proof can be found in [36, Lem. 21.1]. For more details we refer to [40, Ch. 3.1.2.2]. □

In order to transfer the results of Lemma 2.2 to the semi-linear PDAE (2.5) we need to reinterpret the nonlinearity $f: [0, T] \times \mathscr{V} \to \mathscr{H}$ as a function which maps an abstract measurable function $u: [0, T] \to \mathscr{V}$ to $f(\cdot, u(\cdot)): [0, T] \to \mathscr{H}$. For this, we assume the classical Carathéodory condition, see [15, Rem. 1], i.e.,

i.) $v \mapsto f(t, v)$ is a continuous function for almost all $t \in [0, T]$,
ii.) $t \mapsto f(t, v)$ is a measurable function for all $v \in \mathscr{V}$.

Furthermore, we need a boundedness condition such that the Nemytskii map induced by f maps $C([0, T]; \mathscr{V})$ to $L^2(0, T; \mathscr{H})$. We will assume in the following that there exists a function $k \in L^2(0, T)$ such that

$$\|f(t, v)\|_{\mathscr{H}} \leq k(t)(1 + \|v\|_{\mathscr{V}}) \qquad (2.8)$$

for all $v \in \mathscr{V}$ and almost all $t \in [0, T]$. We emphasize that condition (2.8) is sufficient but not necessary for f to induce a Nemytskii map, cf. [15, Th. 1(ii)]. We will use this condition to prove the existence and uniqueness of a global solution to (2.5).

Theorem 2.3 *Assume that $\mathscr{A}$ and $\mathscr{B}$ satisfy Assumptions 2.1 and 2.2. Further, let $g \in H^1(0, T; \mathscr{Q}^*)$ and suppose that $f: [0, T] \times \mathscr{V} \to \mathscr{H}$ satisfies the Carathéodory conditions as well as the uniform bound (2.8). Assume that for every $v \in \mathscr{V}$ there exists an open ball $B_r(v) \subseteq \mathscr{V}$ with radius $r = r(v) > 0$ and a constant $L = L(v) \geq 0$, such that for almost every $t \in [0, T]$ it holds that*

$$\|f(t, v_1) - f(t, v_2)\|_{\mathscr{H}} \leq L \|v_1 - v_2\|_{\mathscr{V}} \qquad (2.9)$$

for all $v_1, v_2 \in B_r(v)$. Then, for a consistent initial value $u_0 \in \mathscr{V}$, i.e., $\mathscr{B}u_0 = g(0)$, the semi-linear PDAE (2.5) has a unique solution

$$u \in C([0, T]; \mathscr{V}) \cap H^1(0, T; \mathscr{H}), \qquad \lambda \in L^2(0, T; \mathscr{Q})$$

with $u(0) = u_0$.

Proof Without loss of generality, we assume that $\mathscr{A} = \mathscr{A}_1$. For this, we redefine $f(t, v) \leftarrow f(t, v) - \mathscr{A}_2 v$, leading to an update of the involved constants $L \leftarrow L + c_{\mathscr{A}_2}$ and $k \leftarrow k + c_{\mathscr{A}_2}$ but leaving the radius r unchanged.

To prove the statement we follow the steps of [30, Ch. 6.3]. Let $t' \in (0, T]$ be arbitrary but fixed. With (2.8) we notice that the Nemyskii map induced by f maps $C([0, t']; \mathscr{V})$ into $L^2(0, t'; \mathscr{H})$, cf. [15, Th. 1]. Therefore, the solution map $S_{t'} \colon C([0, t']; \mathscr{V}) \to C([0, t']; \mathscr{V})$, which maps $y \in C([0, t']; \mathscr{V})$ to the solution of

$$\dot{u}(t) + \mathscr{A}u(t) + \mathscr{B}^*\lambda(t) = f(t, y(t)) \qquad \text{in } \mathscr{V}^*, \tag{2.10a}$$

$$\mathscr{B}u(t) \qquad\qquad = g(t) \qquad \text{in } \mathscr{Q}^* \tag{2.10b}$$

with initial value u_0, is well-defined, cf. Lemma 2.2. To find a solution to (2.5) we have to look for a unique fixed point of $S_{t'}$ and show that t' can be extended to T.

Let $\tilde{u} \in C([0, T]; \mathscr{V})$ be the solution of the PDAE (2.5) for $f \equiv 0$ and initial value u_0. With $r = r(u_0)$ and $L = L(u_0)$ we now choose $t_1 \in (0, T]$ such that

$$\|\tilde{u}(t) - u_0\|_{\mathscr{V}} \le \frac{r}{2}, \tag{a}$$

$$\int_0^t |k|^2 \, ds \le \frac{\mu r^2}{4 \, (1 + r + \|u_0\|_{\mathscr{V}})^2}, \tag{b}$$

$$L^2 t_1 < \mu, \tag{c}$$

$$\int_0^t \frac{3}{\mu} |k|^2 (1 + \|\tilde{u}\|_{\mathscr{V}}^2) \, ds \le \frac{r^2}{4} \cdot \exp\left(-\frac{3}{\mu} \int_0^t |k|^2 \, ds\right) \tag{d}$$

for all $t \in [0, t_1]$. This is well-defined, since $\tilde{u} - u_0$ and the integrals in (b) and (d) are continuous functions in t, which vanish for $t = 0$. We define

$$D := \left\{ y \in C([0, t_1]; \mathscr{V}) \mid \|y - \tilde{u}\|_{C([0,t_1], \mathscr{V})} \le r/2 \right\}$$

and consider $y_1, y_2 \in D$. By (a) we have $\|y_i - u_0\|_{C([0,t_1], \mathscr{V})} \le r$. Using that $\tilde{u}$ and $S_{t_1} y_i$ satisfy the constraint (2.10b), we obtain the estimate

$$\mu \, \|(S_{t_1} y_i - \tilde{u})(t)\|_{\mathscr{V}}^2 \overset{(2.7)}{\le} \int_0^t \|f(s, y_i(s))\|_{\mathscr{H}}^2 \, ds$$

$$\overset{(2.8)}{\le} \int_0^t |k(s)|^2 \left(1 + \|y_i(s) - u_0\|_{\mathscr{V}} + \|u_0\|_{\mathscr{V}}\right)^2 ds$$

$$\le \left(1 + r + \|u_0\|_{\mathscr{V}}\right)^2 \int_0^t |k(s)|^2 \, ds$$

which implies with (b) that S_{t_1} maps D into itself. Further, we have

$$\mu \, \|(S_{t_1} y_1 - S_{t_1} y_2)(t)\|_{\mathscr{V}}^2 \overset{(2.7)}{\leq} \int_0^t \|f(s, y_1(s)) - f(s, y_2(s))\|_{\mathscr{H}}^2 \, ds$$

$$\overset{(2.9)}{\leq} L^2 t_1 \|y_1 - y_2\|_{C(0,t_1;\mathscr{V})}^2$$

for all $t \leq t_1$, $i = 1, 2$. Together with the previous estimate and (c), this shows that S_{t_1} is a contraction on D. Hence, there exists a unique fixed point $u \in D \subset C([0, t_1]; \mathscr{V})$ of S_{t_1} by the Banach fixed point theorem [39, Th. 1.A]. On the other hand, for every fixed point $u^\star = S_{t_1} u^\star$ in $C([0, t_1]; \mathscr{V})$, we have the estimate

$$\mu \, \|(u^\star - \widetilde{u})(t)\|_{\mathscr{V}}^2 = \mu \, \|(S_{t_1} u^\star - \widetilde{u})(t)\|_{\mathscr{V}}^2$$

$$\leq \int_0^t |k(s)|^2 \big(1 + \|(u^\star - \widetilde{u})(s)\|_{\mathscr{V}} + \|\widetilde{u}(s)\|_{\mathscr{V}}\big)^2 \, ds.$$

Using $(a + b + c)^2 \leq 3\,(a^2 + b^2 + c^2)$ and Gronwall's inequality it follows that

$$\|(u^\star - \widetilde{u})(t)\|_{\mathscr{V}}^2 \leq \int_0^t \tfrac{3}{\mu} |k(s)|^2 \big(1 + \|\widetilde{u}(s)\|_{\mathscr{V}}^2\big) \, ds \cdot \exp\left(\tfrac{3}{\mu} \int_0^t |k(s)|^2 \, ds\right) \qquad (2.11)$$

for every $t \leq t_1$. Because of (d), this shows that $u^\star$ is an element of D and thus, $u^\star = u$.

By considering problem (2.5) iteratively from $[t_{i-1}, T]$, $t_0 := 0$, to $[t_i, T]$ with consistent initial value $u_0 = u(t_i)$, we can extend u uniquely on an interval $\mathbb{I}$ with $u \in C(\mathbb{I}; \mathscr{V})$ and $u = S_{t'} u$ for every $t' \in \mathbb{I}$. Note that either $\mathbb{I} = [0, T]$ or $\mathbb{I} = [0, T')$ with $T' \leq T$. The second case is only possible if $\|u(t)\|_{\mathscr{V}} \to \infty$ for $t \to T'$, otherwise we can extend u by starting at T'. But, since the estimate (2.11) also holds for $u = u^\star$ and $t < T'$, we have in limit that $\|u(T')\|_{\mathscr{V}} \leq \|u(T') - \widetilde{u}(T')\|_{\mathscr{V}} + \|\widetilde{u}(T')\|_{\mathscr{V}}$ is bounded. Therefore, $u = S_T u \in C([0, T]; \mathscr{V})$. Finally, the stated spaces for u and λ follow by Lemma 2.2 with right-hand side $f = f(\cdot, u(\cdot))$. $\square$

Remark 2.2 Under the given assumptions on f from Theorem 2.3, one finds a radius $r_u > 0$ and a Lipschitz constant $L_u \in [0, \infty)$, both based on the solution u, such that (2.9) holds for all $v_1, v_2 \in B_{r_u}(u(s))$ with $L = L_u$ and arbitrary $s \in [0, T]$. With these uniform constants one can show that the mapping of the data $u_0 \in \mathscr{V}$ and $g \in H^1(0, T; \mathscr{Q})$ with $\mathscr{B} u_0 = g(0)$ to the solution (u, λ) is continuous.

Remark 2.3 It is possible to weaken the assumption (2.8) of Theorem 2.3 for an arbitrary $p > 1$ to $\|f(t, v)\|_{\mathscr{H}} \leq k(t)(1 + \|v\|_{\mathscr{V}}^p)$. Under this assumption one can show the existence of a unique solution of (2.5), which may only exists locally.

Remark 2.4 The assumptions considered in [30, Ch. 6.3] are stronger then the one in Theorem 2.3. If these additional assumptions are satisfied, then the existence and uniqueness of a solution to (2.5) follows directly by Lemma 2.2, [30, Ch. 6, Th. 3.1 & 3.3], and the fact that every self-adjoint, elliptic operator $\mathscr{A} \in \mathscr{L}(\mathscr{V}, \mathscr{V}^*)$ has a unique invertible square root $\mathscr{A}^{1/2} \in \mathscr{L}(\mathscr{V}, \mathscr{H})$ with $\langle \mathscr{A} v_1, v_2 \rangle = (\mathscr{A}^{1/2} v_1, \mathscr{A}^{1/2} v_2)$ for all $v_1, v_2 \in \mathscr{V}$. This can be proven by interpreting $\mathscr{A}$ as an (unbounded) operator $\mathbf{A} \colon D(\mathbf{A}) \subset \mathscr{H} \to \mathscr{H}$ with domain $D(\mathbf{A}) := \mathscr{A}^{-1} \mathscr{H} \subset \mathscr{V} \hookrightarrow \mathscr{H}$ and the results of [6, Ch. 6, Th. 4 & Ch. 10, Th. 1] and [30, Th. 6.8].

2.4   A Solution Formula for the Linear Case

In the linear case, the solution of (2.5) can be expressed by the variation-of-constants formula (Duhamel's principle), cf. [13]. In the semi-linear case, we consider the term $f(t, u)$ as a right-hand side which leads to an implicit formula only. This, however, is still of value for the numerical analysis of time integration schemes.

The solution formula is based on the decomposition $u = u_{\mathrm{ker}} + u_{\mathrm{c}}$ with $u_{\mathrm{ker}} \colon [0, T] \to \mathscr{V}_{\mathrm{ker}}$ and $u_{\mathrm{c}} \colon [0, T] \to \mathscr{V}_{\mathrm{c}}$. The latter is fully determined by the constraint (2.5b), namely $u_{\mathrm{c}}(t) = \mathscr{B}^- g(t) \in \mathscr{V}_{\mathrm{c}}$. For u_{ker} we consider the restriction of (2.5a) to the test space $\mathscr{V}_{\mathrm{ker}}$. Since the Lagrange multiplier disappears in this case, we obtain

$$\dot{u}_{\mathrm{ker}} + \mathscr{A}_{\mathrm{ker}} u_{\mathrm{ker}} = \dot{u}_{\mathrm{ker}} + \mathscr{A} u_{\mathrm{ker}} = f(t, u_{\mathrm{ker}} + u_{\mathrm{c}}) - \dot{u}_{\mathrm{c}} \qquad \text{in } \mathscr{V}_{\mathrm{ker}}^*. \qquad (2.12)$$

Note that the right-hand side is well-defined as functional in $\mathscr{V}_{\mathrm{ker}}^*$ using the trivial restriction of $\mathscr{V}^*$ to $\mathscr{V}_{\mathrm{ker}}^*$. Further, the term $\mathscr{A} u_{\mathrm{c}}$ disappears under test functions in $\mathscr{V}_{\mathrm{ker}}$ due to the definition of $\mathscr{V}_{\mathrm{c}}$. If this orthogonality is not respected within the implementation, then this term needs to be reconsidered.

The solution to (2.12) can be obtained by an application of the variation-of-constants formula. Since the semigroup can only be applied to functions in $\mathscr{H}_{\mathrm{ker}}$, we introduce the operator

$$\iota_0 \colon \mathscr{H} \equiv \mathscr{H}^* \to \mathscr{H}_{\mathrm{ker}}^* \equiv \mathscr{H}_{\mathrm{ker}}.$$

This operator is again based on a simple restriction of test functions and leads to the solution formula

$$u(t) = u_{\mathrm{c}}(t) + u_{\mathrm{ker}}(t)$$

$$= \mathscr{B}^- g(t) + e^{-t \mathscr{A}_{\mathrm{ker}}} u_{\mathrm{ker}}(0) + \int_0^t e^{-(t-s) \mathscr{A}_{\mathrm{ker}}} \iota_0 \big[f(s, u_{\mathrm{ker}}(s) + u_{\mathrm{c}}(s)) - \dot{u}_{\mathrm{c}}(s) \big] \, \mathrm{d}s.$$

Assuming a partition of the time interval $[0, T]$ by $0 = t_0 < t_1 < \cdots < t_N = T$, we can write the solution formula in the form

$$u(t_{n+1}) - \mathscr{B}^- g_{n+1} \tag{2.13}$$

$$= e^{-(t_{n+1}-t_n)\mathscr{A}_{\mathrm{ker}}}\big[u(t_n) - \mathscr{B}^- g_n\big] + \int_{t_n}^{t_{n+1}} e^{-(t_{n+1}-s)\mathscr{A}_{\mathrm{ker}}} \iota_0\big[f(s, u(s)) - \dot u_{\mathrm{c}}(s)\big]\,\mathrm{d}s.$$

Note that we use here the abbreviation $g_n := g(t_n)$. In the following two sections we construct exponential integrators for constrained semi-linear systems of the form (2.5). Starting point is a first-order scheme based on the exponential Euler method applied to equation (2.12).

3   The Exponential Euler Scheme

The idea of exponential integrators is to approximate the integral term in (2.13) by an appropriate quadrature rule. Following the construction for PDEs [19], we consider in this section the function evaluation at the beginning of the interval. This then leads to the scheme

$$u_{n+1} - \mathscr{B}^- g_{n+1} = e^{-\tau\mathscr{A}_{\mathrm{ker}}}\big[u_n - \mathscr{B}^- g_n\big] + \int_0^{\tau} e^{-(\tau-s)\mathscr{A}_{\mathrm{ker}}} \iota_0\big[f(t_n, u_n) - \dot u_{\mathrm{c}}(t_n)\big]\,\mathrm{d}s$$

$$= \varphi_0(-\tau\mathscr{A}_{\mathrm{ker}})\big(u_n - \mathscr{B}^- g_n\big) + \tau\varphi_1(-\tau\mathscr{A}_{\mathrm{ker}})\big(\iota_0\big[f(t_n, u_n) - \mathscr{B}^- \dot g_n\big]\big). \tag{3.1}$$

As usual, u_n denotes the approximation of $u(t_n)$. Further, we restrict ourselves to a uniform partition of $[0, T]$ with step size τ for simplicity. Assuming that the resulting approximation satisfies the constraint in every step, we have $u_n - \mathscr{B}^- g_n \in \mathscr{V}_{\mathrm{ker}} \hookrightarrow \mathscr{H}_{\mathrm{ker}}$ such that the semigroup $e^{-\tau\mathscr{A}_{\mathrm{ker}}}$ is applicable. The derived formula (3.1) is beneficial for the numerical analysis but lacks the practical access which we tackle in the following.

3.1   The Practical Method

Since the evaluation of the φ-functions with the operator $\mathscr{A}_{\mathrm{ker}}$ is not straightforward, we reformulate the method by a number of saddle point problems. Furthermore, we need evaluations of $\mathscr{B}^-$ applied to the right-hand side g (or its time derivative). Also this is replaced by the solution of a saddle point problem.

Consider $x := \mathscr{B}^- g_n = \mathscr{B}^- g(t_n) \in \mathscr{V}_c \subseteq \mathscr{V}$. Then, x can be written as the solution of the stationary auxiliary problem

$$\mathscr{A}x + \mathscr{B}^* v = 0 \qquad \text{in } \mathscr{V}^*, \tag{3.2a}$$

$$\mathscr{B}x \qquad = g_n \qquad \text{in } \mathscr{Q}^*. \tag{3.2b}$$

Note that equation (3.2b) enforces the connection of x to the right-hand side g whereas the first equation of the system guarantees the desired $\mathscr{A}$-orthogonality. The Lagrange multiplier v is not of particular interest and simply serves as a dummy variable. The unique solvability of system (3.2) is discussed in the following lemma.

Lemma 3.1 *Let the operators $\mathscr{A}$ and $\mathscr{B}$ satisfy Assumptions 2.1 and 2.2. Then, for every $g_n \in \mathscr{Q}^*$ there exists a unique solution $(x, v) \in \mathscr{V}_c \times \mathscr{Q}$ to system (3.2).*

Proof Under the given assumptions on the operators $\mathscr{A}$ and $\mathscr{B}$ there exists a unique solution $(x, v) \in \mathscr{V} \times \mathscr{Q}$ to (3.2), even in the case with an inhomogeneity in the first equation, see [7, Ch. II, Prop. 1.3]. It remains to show that x is an element of $\mathscr{V}_c$. For this, note that x satisfies $\langle \mathscr{A}x, w \rangle = 0$ for all $w \in \mathscr{V}_{\ker}$, since the $\mathscr{B}^*$-term vanishes for these test functions. This, however, is exactly the definition of the complement space $\mathscr{V}_c$. □

Being able to compute $\mathscr{B}^- g_n$, we are now interested in the solution of problems involving the operator $\mathscr{A}_{\ker}$. This will be helpful for the reformulation of the exponential Euler method (3.1). We introduce the auxiliary variable $w_n \in \mathscr{V}_{\ker}$ as the solution of

$$\mathscr{A}_{\ker} w_n = f(t_n, u_n) - \dot{u}_c(t_n) = f(t_n, u_n) - \mathscr{B}^- \dot{g}_n \qquad \text{in } \mathscr{V}_{\ker}^*.$$

This is again equivalent to a stationary saddle point problem, namely

$$\mathscr{A}w_n + \mathscr{B}^* v_n = f(t_n, u_n) - \mathscr{B}^- \dot{g}_n \qquad \text{in } \mathscr{V}^*, \tag{3.3a}$$

$$\mathscr{B}w_n \qquad = 0 \qquad \text{in } \mathscr{Q}^*. \tag{3.3b}$$

As above, the Lagrange multiplier is only introduced for a proper formulation and not of particular interest in the following. The unique solvability of system (3.3) follows again by Lemma 3.1, since the right-hand side of the first equation is an element of $\mathscr{V}^*$. In order to rewrite (3.1), we further note that the recursion formula for φ_1 implies

$$\tau \varphi_1(-\tau \mathscr{A}_{\ker})h = -\big[\varphi_0(-\tau \mathscr{A}_{\ker}) - \mathrm{id}\big] \mathscr{A}_{\ker}^{-1} h$$

for all $h \in \mathscr{H}_{\ker}$. Recall that $\mathscr{A}_{\ker}$ is indeed invertible due to Assumption 2.2. Thus, the exponential Euler scheme can be rewritten as

$$u_{n+1} = \mathscr{B}^- g_{n+1} + \varphi_0(-\tau \mathscr{A}_{\ker})\big(u_n - \mathscr{B}^- g_n - w_n\big) + w_n.$$

Finally, we need a way to compute the action of $\varphi_0(-\tau \mathscr{A}_{\mathrm{ker}})$. For this, we consider the corresponding PDAE formulation. The resulting method then reads $u_{n+1} = \mathscr{B}^- g_{n+1} + z(t_{n+1}) + w_n$, where z is the solution of the *linear* homogeneous PDAE

$$\dot{z}(t) + \mathscr{A} z(t) + \mathscr{B}^* \mu(t) = 0 \qquad \text{in } \mathscr{V}^*, \tag{3.4a}$$

$$\mathscr{B} z(t) \qquad\qquad\quad = 0 \qquad \text{in } \mathscr{Q}^* \tag{3.4b}$$

with initial condition $z(t_n) = u_n - \mathscr{B}^- g_n - w_n$. Thus, the exponential Euler scheme given in (3.1) can be computed by a number of saddle point problems. We summarize the necessary steps in Algorithm 1.

Algorithm 1 Exponential Euler scheme

1: **Input**: step size τ, consistent initial data $u_0 \in \mathscr{V}$, right-hand sides f, g

2: **for** $n = 0$ **to** $N - 1$ **do**
3:     compute $\mathscr{B}^- g_n$, $\mathscr{B}^- g_{n+1}$, and $\mathscr{B}^- \dot{g}_n = \mathscr{B}^- \dot{g}(t_n)$ by (3.2)
4:     compute w_n by (3.3)
5:     compute z as solution of (3.4) on $[t_n, t_{n+1}]$ with initial data $u_n - \mathscr{B}^- g_n - w_n$
6:     set $u_{n+1} = \mathscr{B}^- g_{n+1} + z(t_{n+1}) + w_n$
7: **end for**

Remark 3.1 One step of the exponential Euler scheme consists of the solution of four (from the second step on only three) stationary and a single transient saddle point problem, including only one evaluation of the nonlinear function f. We emphasize that all these systems are linear such that no Newton iteration is necessary in the solution process. Furthermore, the time-dependent system is homogeneous such that it can be solved without the need of a regularization.

3.2 Convergence Analysis

In this section we analyze the convergence order of the exponential Euler method for constrained PDEs of parabolic type. For the unconstrained case it is well-known that the convergence order is one. Since our approach is based on the unconstrained PDE (2.12) of the dynamical part in $\mathscr{V}_{\mathrm{ker}}$, we expect the same order for the solution of Algorithm 1. For the associated proof we will assume that the approximation u_n lies within a strip of radius r around u, where f is locally Lipschitz continuous with constant $L > 0$. Note that by Remark 2.2 there exists such a uniform radius and local Lipschitz constant. Furthermore, a sufficiently small step size τ guarantees that u_n stays within this strip around u, since the solution z of (3.4) and $\mathscr{B}^- g$ are continuous.

Theorem 3.2 (Exponential Euler) *Consider the assumptions of Theorem 2.3 including Assumptions 2.1 and 2.2. Further, let the step size τ be sufficiently small such that the derived approximation u_n lies within a strip along u in which f is locally Lipschitz continuous with a uniform constant $L > 0$. For the right-hand side of the constraint we assume $g \in H^2(0, T; \mathcal{Q}^*)$. If the exact solution of (2.5) satisfies $\frac{d}{dt} f(\cdot, u(\cdot)) \in L^2(0, T; \mathcal{H})$, then the approximation u_n obtained by the exponential Euler scheme of Algorithm 1 satisfies*

$$\|u_n - u(t_n)\|_{\mathcal{V}}^2 \lesssim \tau^2 \int_0^{t_n} \|\tfrac{d}{dt} f(t, u(t))\|_{\mathcal{H}}^2 + \|\mathcal{B}^-\ddot{g}(t)\|_{\mathcal{H}}^2 \, dt.$$

Note that the involved constant only depends on t_n, L, and the operator $\mathcal{A}$.

Proof With w_n and z from (3.3) and (3.4), respectively, we define $U(t) := z(t) + w_n + \mathcal{B}^- g(t)$ for $t \in [t_n, t_{n+1}]$, $n = 0, \ldots, N-1$. This function satisfies

$$U(t_n) = z(t_n) + w_n + \mathcal{B}^- g_n = u_n \quad \text{and} \quad U(t_{n+1}) = z(t_{n+1}) + w_n + \mathcal{B}^- g_{n+1} = u_{n+1}.$$

Furthermore, since $\dot{U}(t) = \dot{z}(t) + \mathcal{B}^- \dot{g}(t)$, the function U solves the PDAE

$$\dot{U}(t) + \mathcal{A}U(t) + \mathcal{B}^*\Lambda(t) = f(t_n, u_n) + \mathcal{B}^-(\dot{g}(t) - \dot{g}_n) \quad \text{in } \mathcal{V}^*,$$
$$\mathcal{B}U(t) = g(t) \quad \text{in } \mathcal{Q}^*$$

on $[t_n, t_{n+1}]$, $n = 0, \ldots, N-1$ with initial value $U(t_0) = u_0$. To shorten notation we define $\triangle u := u - U$ and $\triangle\lambda := \lambda - \Lambda$, which satisfy

$$\tfrac{d}{dt}\triangle u + \mathcal{A}_1\triangle u + \mathcal{B}^*\triangle\lambda = f(\cdot, u(\cdot)) - f(t_n, u_n) - \mathcal{A}_2\triangle u - \mathcal{B}^-(\dot{g} - \dot{g}_n) \quad \text{in } \mathcal{V}^*,$$
$$\mathcal{B}\triangle u = 0 \quad \text{in } \mathcal{Q}^*$$

on each interval $[t_n, t_{n+1}]$ with initial value $\triangle u(t_0) = 0$ if $n = 0$ and $\triangle u(t_n) = u(t_n) - u_n$ otherwise. In the following, we derive estimates of $\triangle u$ on all subintervals. Starting with $n = 0$, we have by Lemma 2.2 that

$$\|\triangle u(t)\|_{\mathcal{A}_1}^2 \overset{(2.7)}{\leq} \int_0^t \|f(s, u(s)) - f(0, u_0) - \mathcal{A}_2\triangle u(s) - \mathcal{B}^-(\dot{g}(s) - \dot{g}_0)\|_{\mathcal{H}}^2 \, ds$$
$$\leq 2\int_0^t \left\|\int_0^s \tfrac{d}{d\eta} f(\eta, u(\eta)) \, d\eta - \mathcal{B}^-\ddot{g}(\eta) \, d\eta\right\|_{\mathcal{H}}^2 + \tfrac{c_{\mathcal{A}_2}^2}{\mu}\|\triangle u(s)\|_{\mathcal{A}_1}^2 \, ds.$$

By Gronwall's lemma and $t = t_1 = \tau$ we obtain with $c := 2c_{\mathscr{A}_2}^2 \mu^{-1}$ the bound

$$
\begin{aligned}
\|u(t_1) - u_1\|_{\mathscr{A}_1}^2 &\leq 2\, e^{c\tau} \int_0^\tau \left\| \int_0^s \frac{\mathrm{d}}{\mathrm{d}\eta} f(\eta, u(\eta)) - \mathscr{B}^- \ddot{g}(\eta)\, \mathrm{d}\eta \right\|_{\mathscr{H}}^2 \mathrm{d}s \\
&\leq 2\, e^{c\tau} \int_0^\tau s \int_0^s \left\| \frac{\mathrm{d}}{\mathrm{d}\eta} f(\eta, u(\eta)) - \mathscr{B}^- \ddot{g}(\eta) \right\|_{\mathscr{H}}^2 \mathrm{d}\eta\, \mathrm{d}s \\
&\leq 2\, e^{c\tau} \tau^2 \underbrace{\int_0^\tau \left\| \tfrac{\mathrm{d}}{\mathrm{d}s} f(s, u(s)) \right\|_{\mathscr{H}}^2 + \left\| \mathscr{B}^- \ddot{g}(s) \right\|_{\mathscr{H}}^2 \mathrm{d}s}_{=: \mathscr{I}(\frac{\mathrm{d}}{\mathrm{d}t} f, \ddot{g}, 0, t_1)} .
\end{aligned}
\tag{3.5}
$$

With the uniform Lipschitz constant L we have for $n \geq 1$ that

$$
\int_{t_n}^{t_{n+1}} \| f(s, u(s)) - f(t_n, u_n) \|_{\mathscr{H}}^2 \, \mathrm{d}s
$$

$$
\leq 2 \int_{t_n}^{t_{n+1}} \| f(t_n, u(t_n)) - f(t_n, u_n) \|_{\mathscr{H}}^2 + \| f(s, u(s)) - f(t_n, u(t_n)) \|_{\mathscr{H}}^2 \, \mathrm{d}s
$$

$$
\leq 2\tau \frac{L^2}{\mu} \| u(t_n) - u_n \|_{\mathscr{A}_1}^2 + 2 \int_{t_n}^{t_{n+1}} (s - t_n) \int_{t_n}^s \left\| \tfrac{\mathrm{d}}{\mathrm{d}\eta} f(\eta, u(\eta)) \right\|_{\mathscr{H}}^2 \mathrm{d}\eta\, \mathrm{d}s.
$$

With this, we obtain similarly as in (3.5) and with Young's inequality,

$$
\| u(t_{n+1}) - u_{n+1} \|_{\mathscr{A}_1}^2 \leq e^{c\tau} \left[\left(1 + 3\tau \frac{L^2}{\mu} \right) \| u(t_n) - u_n \|_{\mathscr{A}_1}^2 + 3\tau^2 \, \mathscr{I}\left(\tfrac{\mathrm{d}}{\mathrm{d}t} f, \ddot{g}, t_n, t_{n+1} \right) \right].
\tag{3.6}
$$

Therefore, with $(1 + x) \leq e^x$, estimate (3.5), and an iterative application of the estimate (3.6) we get

$$
\| u(t_{n+1}) - u_{n+1} \|_{\mathscr{A}_1}^2 \leq \tau^2\, 3 \sum_{k=0}^n \exp(c\tau)^{n+1-k} \left(1 + 3\tau \frac{L^2}{\mu} \right)^{n-k} \mathscr{I}\left(\tfrac{\mathrm{d}}{\mathrm{d}t} f, \ddot{g}, t_k, t_{k+1} \right)
$$

$$
\leq \tau^2\, 3 \exp(c\, t_{n+1}) \exp\left(3 \frac{L^2}{\mu} t_n \right) \mathscr{I}\left(\tfrac{\mathrm{d}}{\mathrm{d}t} f, \ddot{g}, 0, t_{n+1} \right)
$$

for all $n = 0, \ldots, N - 1$. The stated estimate finally follows by the equivalence of $\| \cdot \|_{\mathscr{V}}$ and $\| \cdot \|_{\mathscr{A}}$ on $\mathscr{V}_{\mathrm{ker}}$, see (2.6). $\qquad \square$

Remark 3.2 The assumption on the step size τ only depends on the nonlinearity f and not on the operator $\mathscr{A}$. Thus, this condition does not depend on the stiffness of the system and still allows large time steps.

Remark 3.3 In the case of a self-adjoint operator $\mathscr{A}$, i.e., $\mathscr{A}_2 = 0$, the convergence result can also be proven by the restriction to test functions in $\mathscr{V}_{\mathrm{ker}}$ and the application of corresponding results for the unconstrained case, namely [19, Th. 2.14]. This requires similar assumptions but with $\frac{d}{dt} f(\cdot, u(\cdot)) \in L^\infty(0, T; \mathcal{H})$.

We like to emphasize that this procedure is also applicable if $\mathcal{A}_2 \neq 0$ by moving $\mathcal{A}_2$ into the nonlinearity f. This, however, slightly changes the proposed scheme, since then only $\mathcal{A}_2 u_n$ enters the approximation instead of $\mathcal{A}_2 u(t)$. In practical applications this would also require an additional effort in order to find the symmetric part of the differential operator $\mathcal{A}$ which is still elliptic on $\mathcal{V}_{\mathrm{ker}}$.

3.3 An Alternative Approach

A second approach to construct an exponential Euler scheme which is applicable to constrained systems is to formally apply the method to the corresponding singularly perturbed PDE. This approach was also considered in [20] for DAEs of index 1. In the present case, we add a small term $\varepsilon \dot{\lambda}$ into the second equation of (2.5). Thus, we consider the system

$$\dot{u}(t) + \mathcal{A}u(t) + \mathcal{B}^*\lambda(t) = f(t, u) \qquad \text{in } \mathcal{V}^*, \tag{3.7a}$$

$$\varepsilon \dot{\lambda}(t) + \mathcal{B}u(t) \qquad\qquad = g(t) \qquad \text{in } \mathcal{Q}^*, \tag{3.7b}$$

which can be written in operator matrix form as

$$\begin{bmatrix} \dot{u} \\ \dot{\lambda} \end{bmatrix} = \begin{bmatrix} \mathrm{id} \\ & \frac{1}{\varepsilon}\,\mathrm{id} \end{bmatrix} \left\{ -\begin{bmatrix} \mathcal{A} & \mathcal{B}^* \\ \mathcal{B} \end{bmatrix} \begin{bmatrix} u \\ \lambda \end{bmatrix} + \begin{bmatrix} f(t, u) \\ g(t) \end{bmatrix} \right\}.$$

For this, an application of the exponential Euler method yields the scheme

$$\begin{bmatrix} u_{n+1} \\ \lambda_{n+1} \end{bmatrix} = \varphi_0\left(-\tau \begin{bmatrix} \mathcal{A} & \mathcal{B}^* \\ \frac{1}{\varepsilon}\mathcal{B} \end{bmatrix} \right) \begin{bmatrix} u_n \\ \lambda_n \end{bmatrix} + \tau\varphi_1\left(-\tau \begin{bmatrix} \mathcal{A} & \mathcal{B}^* \\ \frac{1}{\varepsilon}\mathcal{B} \end{bmatrix} \right) \begin{bmatrix} f(t_n, u_n) \\ \frac{1}{\varepsilon}g_n \end{bmatrix}.$$

We introduce the auxiliary variables $(\bar{w}_n, \bar{v}_n) \in \mathcal{V} \times \mathcal{Q}$ as the unique solution to the stationary saddle point problem

$$\mathcal{A}\bar{w}_n + \mathcal{B}^*\bar{v}_n = f(t_n, u_n) \qquad\qquad \text{in } \mathcal{V}^*,$$

$$\mathcal{B}\bar{w}_n \qquad\quad = \theta g_n + (1-\theta)g_{n+1} \qquad \text{in } \mathcal{Q}^*.$$

The included parameter $\theta \in [0, 1]$ controls the consistency as outlined below. Then, the exponential Euler method can be rewritten as

$$\begin{bmatrix} u_{n+1} \\ \lambda_{n+1} \end{bmatrix} = \varphi_0\left(-\tau \begin{bmatrix} \mathcal{A} & \mathcal{B}^* \\ \frac{1}{\varepsilon}\mathcal{B} \end{bmatrix} \right) \begin{bmatrix} u_n - \bar{w}_n \\ \lambda_n - \bar{\mu}_n \end{bmatrix} + \begin{bmatrix} \bar{w}_n \\ \bar{\mu}_n \end{bmatrix},$$

which allows an interpretation as the solution of a linear (homogeneous) PDE. Finally, we set $\varepsilon = 0$, which leads to the following time integration scheme: Given $\bar{w}_n$, solve on $[t_n, t_{n+1}]$ the linear system

$$\dot{z}(t) + \mathscr{A} z(t) + \mathscr{B}^* \mu(t) = 0 \qquad \text{in } \mathscr{V}^*,$$

$$\mathscr{B} z(t) \qquad\qquad = 0 \qquad \text{in } \mathscr{Q}^*$$

with initial condition $z(t_n) = u_n - \bar{w}_n$. The approximation of $u(t_{n+1})$ is then defined through $u_{n+1} := z(t_{n+1}) + \bar{w}_n$.

We emphasize that the initial value of z may be inconsistent. In this case, the initial value needs to be projected to $\mathscr{H}_{\text{ker}}$, cf. Sect. 2.4. If the previous iterate satisfies $\mathscr{B} u_n = g_n$, then the choice $\theta = 1$ yields $\mathscr{B} z(t_n) = 0$ and thus, consistency. This, however, does not imply $\mathscr{B} u_{n+1} = g_{n+1}$. On the other hand, $\theta = 0$ causes an inconsistency for z in the sense that $\mathscr{B} z(t_n) \neq 0$ but guarantees $\mathscr{B} u_{n+1} = g_{n+1}$. We now turn to an exponential integrator of higher order.

4 Exponential Integrators of Second Order

This section is devoted to the construction of an exponential integrator of order two for constrained parabolic systems of the form (2.5). In particular, we aim to transfer the method given in [35, Exp. 11.2.2], described by the *Butcher tableau*

$$\begin{array}{c|cc} 0 & & \\ 1 & \varphi_1 & \\ \hline & \varphi_1 - \varphi_2 & \varphi_2 \end{array} \qquad (4.1)$$

to the PDAE case. In the unconstrained case, i.e., for $\dot{v} + \mathscr{A}_{\text{ker}} v = \tilde{f}(t, v)$ in $\mathscr{V}_{\text{ker}}^*$, one step of this method is defined through

$$v_{n+1}^{\text{Eul}} := \varphi_0(-\tau \mathscr{A}_{\text{ker}}) v_n + \tau \varphi_1(-\tau \mathscr{A}_{\text{ker}}) \tilde{f}(t_n, v_n), \qquad (4.2a)$$

$$v_{n+1} := v_{n+1}^{\text{Eul}} + \tau \varphi_2(-\tau \mathscr{A}_{\text{ker}}) \big[\tilde{f}(t_{n+1}, v_{n+1}^{\text{Eul}}) - \tilde{f}(t_n, v_n) \big]. \qquad (4.2b)$$

Similarly as for the exponential Euler method, we will define a number of auxiliary problems in order to obtain an applicable method for parabolic systems with constraints.

4.1 The Practical Method

We translate the numerical scheme (4.2) to the constrained case. Let u_n denote the given approximation of $u(t_n)$. Then, the first step is to perform one step of the exponential Euler method, cf. Algorithm 1, leading to u_{n+1}^{Eul}. Second, we compute w_n' as the solution of the stationary problem

$$\mathscr{A} w_n' + \mathscr{B}^* v_n' = f(t_{n+1}, u_{n+1}^{\text{Eul}}) - \mathscr{B}^- \dot{g}_{n+1} - f(t_n, u_n) + \mathscr{B}^- \dot{g}_n \quad \text{in } \mathscr{V}^*, \quad (4.3a)$$

$$\mathscr{B} w_n' \qquad\qquad = 0 \qquad\qquad\qquad\qquad\qquad \text{in } \mathscr{Q}^* \quad (4.3b)$$

and w_n'' as the solution of

$$\mathscr{A} w_n'' + \mathscr{B}^* v_n'' = \tfrac{1}{\tau} w_n' \qquad \text{in } \mathscr{V}^*, \qquad\qquad (4.4a)$$

$$\mathscr{B} w_n'' \qquad = 0 \qquad \text{in } \mathscr{Q}^*. \qquad\qquad (4.4b)$$

Note that, due to the recursion formula (2.1), w_n' and w_n'' satisfy the identity

$$\tau \varphi_2(-\tau \mathscr{A}_{\text{ker}}) \iota_0 \big[f(t_{n+1}, u_{n+1}^{\text{Eul}}) - \mathscr{B}^- \dot{g}_{n+1} - f(t_n, u_n) + \mathscr{B}^- \dot{g}_n \big]$$
$$= -\varphi_1(-\tau \mathscr{A}_{\text{ker}}) w_n' + w_n'$$
$$= \varphi_0(-\tau \mathscr{A}_{\text{ker}}) w_n'' - w_n'' + w_n'.$$

It remains to compute $\varphi_0(-\tau \mathscr{A}_{\text{ker}}) w_n''$ and thus, to solve a linear dynamical system with starting value w_n''. More precisely, we consider the homogeneous system (3.4) on the time interval $[t_n, t_{n+1}]$ with initial value $z(t_n) = w_n''$. The solution at time t_{n+1} then defines the new approximation by

$$u_{n+1} := u_{n+1}^{\text{Eul}} + z(t_{n+1}) - w_n'' + w_n'.$$

Note that the consistency is already guaranteed by the exponential Euler step which yields $\mathscr{B} u_{n+1} = \mathscr{B} u_{n+1}^{\text{Eul}} = g_{n+1}$. The resulting exponential integrator is summarized in Algorithm 2.

Algorithm 2 A second-order exponential integrator

1: **Input**: step size τ, consistent initial data $u_0 \in \mathscr{V}$, right-hand sides f, g

2: **for** $n = 0$ **to** $N - 1$ **do**
3:     compute one step of the exponential Euler method for u_n leading to u_{n+1}^{Eul}
4:     compute $\mathscr{B}^- \dot{g}_n$ and $\mathscr{B}^- \dot{g}_{n+1}$ by (3.2)
5:     compute w_n' by (4.3)
6:     compute w_n'' by (4.4)
7:     compute z as solution of (3.4) on $[t_n, t_{n+1}]$ with initial condition $z(t_n) = w_n''$
8:     set $u_{n+1} = u_{n+1}^{\text{Eul}} + z(t_{n+1}) - w_n'' + w_n'$
9: **end for**

4.2 Convergence Analysis

In this subsection we aim to prove the second-order convergence of Algorithm 2 when applied to parabolic PDAEs of the form (2.5). For this, we examine two cases. First, we consider a nonlinearity with values in $\mathcal{V}$, i.e., we assume $f(\cdot, u(\cdot))\colon [0, T] \to \mathcal{V}$. Further, we assume $\mathcal{A}$ to be self-adjoint, meaning that $\mathcal{A}_2 = 0$. Note that this may be extended to general $\mathcal{A}$ as mentioned in Remark 3.3. In this case, the convergence analysis is based on the corresponding results for unconstrained systems. Second, we consider the more general case with nonlinearities $f\colon [0, T] \times \mathcal{V} \to \mathcal{H}$. Here, it can be observed that the convergence order drops to $3/2$. Note, however, that this already happens in the pure PDE case.

Theorem 4.1 (Second-Order Scheme) *In the setting of Sect. 2.2, including Assumptions 2.1 and 2.2, we assume that $\mathcal{A}$ is self-adjoint and that for the exact solution u the map $t \mapsto f(t, u(t))$ is two times differentiable with values in $\mathcal{V}$. Further we assume that the right-hand side g and u are sufficiently smooth, the latter with derivatives in $\mathcal{V}$. Then, the approximation obtained by Algorithm 2 is second-order accurate, i.e.,*

$$\| u_n - u(t_n) \|_{\mathcal{V}} \lesssim \tau^2.$$

Proof We reduce the procedure performed in Algorithm 2 to the unconstrained case. For this, assume that $u_n = u_{\mathrm{ker},n} + \mathcal{B}^- g_n \in \mathcal{V}$ is given with $u_{\mathrm{ker},n} \in \mathcal{V}_{\mathrm{ker}}$ and that u_{n+1}^{Eul} denotes the outcome of a single Euler step, cf. Algorithm 1. By $u_{\mathrm{ker},n+1}^{\mathrm{Eul}}$ we denote the outcome of a Euler step for the unconstrained system

$$\dot{u}_{\mathrm{ker}}(t) + \mathcal{A}_{\mathrm{ker}} u_{\mathrm{ker}}(t) = \tilde{f}(t, u_{\mathrm{ker}}(t)) \qquad \text{in } \mathcal{V}_{\mathrm{ker}}^*$$

with $\tilde{f}$ defined by $\tilde{f}(t, u_{\mathrm{ker}}) := \iota_0 \, [f(t, u_{\mathrm{ker}} + \mathcal{B}^- g(t)) - \mathcal{B}^- \dot{g}(t)]$ and initial data $u_{\mathrm{ker},n}$. For this, we know that $u_{n+1}^{\mathrm{Eul}} = u_{\mathrm{ker},n+1}^{\mathrm{Eul}} + \mathcal{B}^- g_{n+1}$. By the given assumptions, it follows from [17, Th. 4.3] that

$$u_{\mathrm{ker},n+1} := u_{\mathrm{ker},n+1}^{\mathrm{Eul}} + \tau \varphi_2(-\tau \mathcal{A}_{\mathrm{ker}}) \big[\tilde{f}(t_{n+1}, u_{\mathrm{ker},n+1}^{\mathrm{Eul}}) - \tilde{f}(t_n, u_{\mathrm{ker},n}) \big]$$

defines a second-order approximation of $u_{\mathrm{ker}}(t_{n+1})$. This in turn implies that

$$
\begin{aligned}
u_{n+1} &:= u_{\mathrm{ker},n+1} + \mathcal{B}^- g_{n+1} \\
&= u_{n+1}^{\mathrm{Eul}} + \tau \varphi_2(-\tau \mathcal{A}_{\mathrm{ker}}) \big[\tilde{f}(t_{n+1}, u_{\mathrm{ker},n+1}^{\mathrm{Eul}}) - \tilde{f}(t_n, u_{\mathrm{ker},n}) \big] \\
&= u_{n+1}^{\mathrm{Eul}} + \tau \varphi_2(-\tau \mathcal{A}_{\mathrm{ker}}) \, \iota_0 \big[f(t_{n+1}, u_{n+1}^{\mathrm{Eul}}) - \mathcal{B}^- \dot{g}_{n+1} - f(t_n, u_n) + \mathcal{B}^- \dot{g}_n \big]
\end{aligned}
$$

satisfies the error estimate

$$\|u_{n+1} - u(t_{n+1})\|_{\mathscr{V}} = \|u_{\ker,n+1} - u_{\ker}(t_{n+1})\|_{\mathscr{V}} \lesssim \tau^2.$$

It remains to show that u_{n+1} is indeed the outcome of Algorithm 2. Following the construction in Sect. 4.1, we conclude that

$$u_{n+1} = u_{n+1}^{\mathrm{Eul}} + \varphi_0(-\tau \mathscr{A}_{\ker})w_n'' - w_n'' + w_n'$$

with w_n' and w_n'' denoting the solutions of (4.3) and (4.4), respectively. Finally, note that $\varphi_0(-\tau \mathscr{A}_{\ker})w_n''$ is computed in line 7 of Algorithm 2.   □

Up to now we have assumed that $f(\cdot, u(\cdot))$ maps to $\mathscr{V}$, leading to the desired second-order convergence. In the following, we reconsider the more general case in which $f(\cdot, u(\cdot))$ only maps to $\mathscr{H}$. For PDEs it is well-known that the exponential integrator given by the Butcher tableau (4.1) has, in general, a reduced convergence order if $\frac{\mathrm{d}}{\mathrm{d}t} f(\cdot, u(\cdot)) \in L^{\infty}(0, T; \mathscr{H})$, cf. [17, Th. 4.3]. This carries over to the PDAE case.

Theorem 4.2 (Convergence Under Weaker Assumptions on f) *Consider the assumptions from Theorem 2.3 and let the step size τ be sufficiently small such that the discrete solution u_n lies in a strip along u, where f is locally Lipschitz continuous with a uniform constant $L > 0$. Further assume that $g \in H^3(0, T; \mathscr{Q}^*)$. If the exact solution of (2.5) satisfies $f(\cdot, u(\cdot)) \in H^2(0, T; \mathscr{H})$, then the approximation u_n obtained by Algorithm 2 satisfies the error bound*

$$\|u_n - u(t_n)\|_{\mathscr{V}}^2 \lesssim \tau^3 \int_0^{t_n} \|\tfrac{\mathrm{d}}{\mathrm{d}t} f(t, u(t))\|_{\mathscr{H}}^2 + \|\mathscr{B}^- \ddot{g}(t)\|_{\mathscr{H}}^2 \, \mathrm{d}t$$

$$+ \tau^4 \int_0^{t_n} \|\tfrac{\mathrm{d}^2}{\mathrm{d}t^2} f(t, u(t))\|_{\mathscr{H}}^2 + \|\mathscr{B}^- \dddot{g}(t)\|_{\mathscr{H}}^2 \, \mathrm{d}t.$$

Note that the involved constant only depends on t_n, L, and the operator $\mathscr{A}$.

Proof Let U^{Eul} be the function constructed in the proof of Theorem 3.2 which satisfies $U^{\mathrm{Eul}}(t_n) = u_n$ and $U^{\mathrm{Eul}}(t_{n+1}) = u_{n+1}^{\mathrm{Eul}}$ and set $U(t) := U^{\mathrm{Eul}}(t) + z(t) - w_n'' + \frac{t - t_n}{\tau} w_n'$. This function satisfies

$$U(t_n) = U^{\mathrm{Eul}}(t_n) = u_n, \qquad U(t_{n+1}) = U^{\mathrm{Eul}}(t_{n+1}) + z(t_{n+1}) - w_n'' + w_n' = u_{n+1}.$$

Note that the estimates (3.5) and (3.6) are still valid if one replaces u_{n+1} by $U^{\mathrm{Eul}}(t_{n+1})$ on the left-hand side of these estimates. As in the proof of Theorem 3.2, we can interpret U as the solution of a PDAE on $[t_n, t_{n+1}]$. The corresponding right-hand sides are then given by

$$f(t_n, u_n) + \tfrac{t - t_n}{\tau} \left(f(t_{n+1}, U^{\mathrm{Eul}}(t_{n+1})) - f(t_n, u_n) \right) + \mathscr{B}^- \left(\dot{g}(t) - \dot{g}_n - \tfrac{t - t_n}{\tau} (\dot{g}_{n+1} - \dot{g}_n) \right)$$

for the dynamic equation and $g(t)$ for the constraint. Then, by Young's inequality, Gronwall's lemma, and error bounds for the Taylor expansion we get

$$\|u(t_{n+1}) - u_{n+1}\|_{\mathscr{A}_1}^2 \le e^{c\tau} \left[\left(1 + 4\tau \frac{L^2}{\mu}\right) \|u(t_n) - u_n\|_{\mathscr{A}_1}^2 + 4\tau \frac{L^2}{\mu} \|(u - U^{\mathrm{Eul}})(t_{n+1})\|_{\mathscr{A}_1}^2 \right.$$

$$\left. + \tau^4 \int_{t_n}^{t_{n+1}} \frac{2}{15} \left\| \frac{\mathrm{d}^2}{\mathrm{d}t^2} f(t, u(t)) \right\|_{\mathscr{H}}^2 + \frac{2}{45} \|\mathscr{B}^- \dddot{g}(t)\|_{\mathscr{H}}^2 \, \mathrm{d}t \right]$$

with $c = 2c_{\mathscr{A}_2}^2 \mu^{-1}$. The stated error bound then follows by an iterative application of the previous estimate together with the estimates (3.5), (3.6) and the norm equivalence of $\| \cdot \|_{\mathscr{V}}$ and $\| \cdot \|_{\mathscr{A}_1}$.   □

We like to emphasize that the previous result is sharp in the sense that there exist examples leading to a convergence order of 1.5. The performance of the proposed scheme is presented in the numerical experiments of Sect. 5. We close this section with remarks on alternative second-order schemes.

4.3 A Class of Second-Order Schemes

The analyzed scheme (4.1) is a special case of a one-parameter family of exponential Runge-Kutta methods described by the tableau

$$\begin{array}{c|cc} 0 & & \\ c_2 & c_2 \varphi_{1,2} & \\ \hline & \varphi_1 - \frac{1}{c_2}\varphi_2 & \frac{1}{c_2}\varphi_2 \end{array}$$

with positive parameter $c_2 > 0$, cf. [19]. Therein, φ_1 stands for $\varphi_1(-\tau\mathscr{A}_{\mathrm{ker}})$, whereas $\varphi_{1,2}$ is defined by $\varphi_1(-c_2\tau\mathscr{A}_{\mathrm{ker}})$. Obviously, we regain (4.1) for $c_2 = 1$.

For $c_2 \ne 1$, the resulting scheme for constrained systems calls for two additional saddle point problems in order to compute $\mathscr{B}^- g(t_n + c_2\tau)$ and $\mathscr{B}^- \dot{g}(t_n + c_2\tau)$. This then leads to an exponential integrator summarized in Algorithm 3 with the abbreviations

$$g_{n,2} := g(t_n + c_2\tau), \qquad \dot{g}_{n,2} := \dot{g}(t_n + c_2\tau), \qquad t_{n,2} := t_n + c_2\tau.$$

We emphasize that all convergence results of Theorems 4.1 and 4.2 transfer to this family of second-order integrators. In a similar manner, Runge-Kutta schemes of higher order may be translated to the here considered constrained case.

Algorithm 3 A class of second-order exponential integrators

1: **Input**: step size τ, consistent initial data $u_0 \in \mathscr{V}$, right-hand sides f, g

2: **for** $n = 0$ **to** $N - 1$ **do**
3:     compute $\mathscr{B}^- g_n, \mathscr{B}^- g_{n,2}, \mathscr{B}^- g_{n+1}, \mathscr{B}^- \dot{g}_n, \mathscr{B}^- \dot{g}_{n,2}$, and $\mathscr{B}^- \dot{g}_{n+1}$ by (3.2)
4:     compute w_n by (3.3)
5:     solve (3.4) on $[t_n, t_{n,2}]$ with initial condition $z(t_n) = u_n - \mathscr{B}^- g_n - w_n$
6:     set $u_{n,2} = z(t_{n,2}) + w_n + \mathscr{B}^- g_{n,2}$
7:     compute w_n' by (4.3) with right-hand side

$$\frac{1}{c_2} \left(f(t_{n,2}, u_{n,2}) - f(t_n, u_n) - \mathscr{B}^- \dot{g}_{n,2} + \mathscr{B}^- \dot{g}_n \right)$$

8:     compute w_n'' by (4.4)
9:     solve (3.4) on $[t_n, t_{n+1}]$ with initial condition $z(t_n) = u_n - \mathscr{B}^- g_n - w_n + w_n''$
10:    set $u_{n+1} = z(t_{n+1}) + w_n + w_n' - w_n'' + \mathscr{B}^- g_{n+1}$.
11: **end for**

---

5   Numerical Examples

In this final section we illustrate the performance of the introduced time integration schemes for two numerical examples. The first example is a heat equation with nonlinear dynamic boundary conditions. In the second experiment, we consider the case of a non-symmetric differential operator for which the theory is not applicable.

Since exponential integrators for PDAEs are based on the exact solution of homogeneous systems of the form (3.4), we first discuss the efficient solution of such systems.

5.1   Efficient Solution of Homogeneous DAEs with Saddle Point Structure

This subsection is devoted to the approximation of $z(t)$, which is needed in line 5 of Algorithm 1 and in line 7 of Algorithm 2. Given a spatial discretization, e.g., by a finite element method, the PDAE (3.4) turns into a DAE of index 2, namely

$$M\dot{x}(t) + Ax(t) + B^T\lambda(t) = 0, \tag{5.1a}$$

$$Bx(t) \qquad\qquad = 0 \tag{5.1b}$$

with consistent initial value $x(0) = x_0$, $Bx_0 = 0$. The matrices satisfy $M, A \in \mathbb{R}^{n \times n}$ and $B \in \mathbb{R}^{m \times n}$ with $m \leq n$. Here, the mass matrix M is symmetric, positive definite and B has full rank. The goal is to find an efficient method to calculate the solution x at a specific time point $t \in [0, T]$.

Let us first recall the corresponding ODE case. There exist various methods to approximate the solution $x(t) = e^{-At}x_0$ of the linear ODE $\dot{x}(t) + Ax(t) = 0$ with initial condition $x(0) = x_0$, $A \in \mathbb{R}^{n \times n}$, for an overview see [27]. This includes Krylov subspace methods to approximate the action of the matrix exponential e^{-At} to a vector, see [12, 16, 32], but also methods based on an interpolation of $e^{-At}x_0$ by Newton polynomials [9]. The first approach is based on the fact that the solution $e^{-At}x_0 = \sum_{k=0}^{\infty} \frac{1}{k!}(-At)^k x_0$ is an element of the Krylov subspace

$$\mathcal{K}_n := \mathcal{K}_n(-A, x_0) := \text{span}\{x_0, -Ax_0, \ldots, (-A)^{n-1}x_0\}.$$

Now, we approximate $e^{-At}x_0$ by an element of $\mathcal{K}_r$ with r relatively small compared to n. For this, we generate an orthogonal basis of $\mathcal{K}_r$ using the Arnoldi algorithm with $v_1 = x_0/\|x_0\|$ as initial vector. This yields $-V_r^T A V_r = H_r$ with an isometric matrix $V_r \in \mathbb{R}^{n \times r}$ and an upper Hessenberg matrix $H_r \in \mathbb{R}^{r \times r}$. Since the columns of V_r are orthonormal and span $\mathcal{K}_r$, H_r is the orthogonal projection of $-A$ onto $\mathcal{K}_r$. Therefore, it is reasonable to use the approximation

$$e^{-At}x_0 \approx \|x_0\| \, V_r \, e^{H_r t} e_1$$

with unit basis vector $e_1 \in \mathbb{R}^r$, cf. [16]. We like to emphasize that the Arnoldi algorithm does not use the explicit representation of A but only its action onto a vector.

We return to the DAE case (5.1). By [13, Th. 2.2] there exists a matrix $X \in \mathbb{R}^{n \times n}$ such that the solution x of (5.1) with arbitrary consistent initial value $x_0 \in \ker B$ is given by $x(t) = e^{Xt}x_0$. Furthermore, there exists a function $\Lambda \in C^{\infty}([0, \infty); \mathbb{R}^{m \times n})$ with $\lambda(t) = \Lambda(t)x_0$. To calculate the action of X we note that by (5.1b) also $B\dot{x} = 0$ holds. We define $y := Xx_0$ and $\mu := \Lambda(0)x_0$. Then with equation (5.1a), $B\dot{x} = 0$, and $t \to 0^+$ we get

$$My + B^T \mu = -Ax_0, \tag{5.2a}$$

$$By = 0. \tag{5.2b}$$

Since the solution of (5.2) is unique, its solution y describes the action of X applied to x_0. As a result, we can approximate the solution of the DAE (5.1) in an efficient manner by using $x(t) = e^{Xt}x_0$, the saddle point problem (5.2), and Krylov subspace methods. For the numerical experiments we have adapted the code provided in [29].

Remark 5.1 Given an approximation $x_t \approx x(t)$, the solution μ of (5.2) with right-hand side $-Ax_t$ provides an approximation of the Lagrange multiplier $\lambda(t)$.

Remark 5.2 Since the saddle point problem (5.2) has to be solved several times in every time step, the numerical solution $\tilde{x}$ of (5.1) may not satisfy the constraint (5.1b) due to round-off errors. To prevent a drift-off, one can project $\tilde{x}$ onto the kernel of B – by solving an additional saddle point problem.

5.2   Nonlinear Dynamic Boundary Conditions

In this first example we revisit Example 2.1 and consider the linear heat equation
with nonlinear dynamic boundary conditions, cf. [34]. More precisely, we consider
the system

$$\dot{u} - \kappa \, \Delta u = 0 \qquad\qquad \text{in } \Omega := (0, 1)^2, \tag{5.3a}$$

$$\dot{u} + \partial_n u + \alpha \, u = f_\Gamma(t, u) \quad \text{on } \Gamma_{\text{dyn}} := (0, 1) \times \{0\} \tag{5.3b}$$

$$u = 0 \qquad\qquad \text{on } \Gamma_D := \partial\Omega \setminus \Gamma_{\text{dyn}} \tag{5.3c}$$

with $\alpha = 1$, $\kappa = 0.02$, and the nonlinearity $f_\Gamma(t, u)(x) = 3\cos(2\pi t) - \sin(2\pi x) - u^3(x)$. As initial condition we set $u(0) = u_0 = \sin(\pi x)\cos(5\pi y/2)$. Following [2],
we can write this in form of a PDAE, namely as

$$\begin{bmatrix} \dot{u} \\ \dot{p} \end{bmatrix} + \begin{bmatrix} \mathscr{K} & \\ & \alpha \end{bmatrix} \begin{bmatrix} u \\ p \end{bmatrix} + \mathscr{B}^* \lambda = \begin{bmatrix} 0 \\ f_\Gamma(t, p) \end{bmatrix} \qquad \text{in } \mathscr{V}^*, \tag{5.4a}$$

$$\mathscr{B} \begin{bmatrix} u \\ p \end{bmatrix} = 0 \qquad\qquad \text{in } \mathscr{Q}^* \tag{5.4b}$$

with spaces $\mathscr{V} = H^1_{\Gamma_D}(\Omega) \times H^{1/2}_{00}(\Gamma_{\text{dyn}})$, $\mathscr{H} = L^2(\Omega) \times L^2(\Gamma_{\text{dyn}})$, $\mathscr{Q} = [H^{1/2}_{00}(\Gamma_{\text{dyn}})]^*$ and constraint operator $\mathscr{B}(u, p) = u|_{\Gamma_{\text{dyn}}} - p$. Here, p denotes a
dummy variable modeling the dynamics on the boundary Γ_{dyn}. The constraint (5.4b)
couples the two variables u and p. This example fits into the framework of this paper
with $g \equiv 0$. Further, the nonlinearity satisfies the assumptions of the convergence
results in Theorems 3.2 and 4.2 due to well-known Sobolev embeddings, see [31,
p. 17f].

For the simulation we consider a spatial discretization by bilinear finite elements
on a uniform mesh with mesh size $h = 1/128$. The initial value of p is chosen
in a consistent manner, i.e., by $u_0|_{\Gamma_{\text{dyn}}}$. An illustration of the dynamics is given
in Fig. 1. The convergence results of the exponential Euler scheme of Sect. 3 and
the exponential integrator introduced in Sect. 4 are displayed in Fig. 2 and show
first and second-order convergence, respectively. Note that the smoothness of the
solution implies $\frac{d^i}{dt^i} f(t, u(t)) \in \mathscr{V}$, $i = 0, 1, 2$, which yields the full convergence
order.

Finally, we note that the computations remain stable for very coarse step sizes τ,
since we do not rely on a CFL condition here.

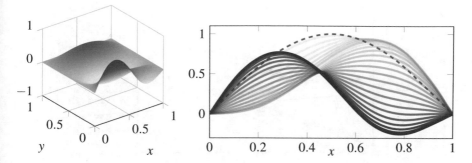

Fig. 1 Illustration of the solution (u, p). The left figure shows u at time $t = 0.7$, whereas the right figure includes several snapshots of p in the time interval $[0, 0.7]$. The dashed line shows the initial value of p. Both results are obtained for mesh size $h = 1/128$ and step size $\tau = 1/100$

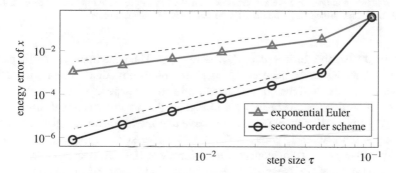

Fig. 2 Convergence history for the error in $x = [u; p]$, measured in the (discrete) $\mathscr{A}$-norm. The dashed lines show first and second-order rates

5.3 A Non-symmetric Example

In this final example we consider a case for which Assumption 2.2 is not satisfied. More precisely, we consider the coupled system

$$\dot{u} - \partial_{xx}u - \partial_{xx}v = -u^3 \quad \text{in } (0, 1),$$

$$\dot{v} + u - \partial_{xx}v = -v^3 \quad \text{in } (0, 1)$$

with initial value

$$u_0(x) = v_0(x) = \sum_{k=1}^{\infty} \frac{\sin(k\pi x)}{k^{1.55}}$$

and the constraint $u(t, 1) - v(t, 1) = g(t) = e^{2t} - 1$. At the other boundary point $x = 0$ we prescribe homogeneous Dirichlet boundary conditions. In this example, the

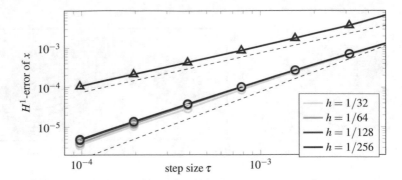

Fig. 3 Convergence history for the error in $x = [u; v]$, measured in the (discrete) $H^1(0, 1)$-norm, including Dirichlet boundary conditions in $x = 0$. The graphs show the results of the exponential Euler scheme (*triangle*) and the second order scheme (*circle*) for different values of h, displayed by its color. The dashed lines illustrate the orders 1 and 3/2, respectively

operator $\mathscr{A}$ has the form $-[\partial_{xx}, \partial_{xx}; -\text{id}, \partial_{xx}]$. Thus, the non-symmetric part $\mathscr{A}_2$ includes a second-order differential operator which contradicts Assumption 2.2. As a consequence, non of the convergence results in this paper apply.

The numerical results are shown in Fig. 3, using a finite element discretization with varying mesh sizes h. One can observe that the exponential Euler scheme still converges with order 1, whereas the second-order scheme introduced in Sect. 4 clearly converges with a reduced rate. Moreover, the rate decreases as the mesh size h gets smaller. By linear regression one can approximate the convergence rate as a value between 1.40 (coarsest mesh, $h = 1/32$) and 1.34 (finest mesh, $h = 1/256$). Thus, the convergence rate is strictly below 3/2. A deeper analysis with fractional powers of $\mathscr{A}$ may predict the exact convergence rate, cf. [17, Th. 4.2 & Th. 4.3]. However, this is a task for future work.

6   Conclusion

In this paper, we have introduced a novel class of time integration schemes for semi-linear parabolic equations restricted by a linear constraint. For this, we have combined exponential integrators for the dynamical part of the system with (stationary) saddle point problems for the 'algebraic part' of the solution. This results in exponential integrators for constrained systems of parabolic type for which we have proven convergence of first and second order, respectively. The theory is verified by numerical experiments.

Acknowledgments C. Zimmer acknowledges the support by the Deutsche Forschungsgemeinschaft (DFG, German Research Foundation) within the SFB 910, project number 163436311.

References

1. Altmann, R.: Regularization and simulation of constrained partial differential equations. Dissertation, Technische Universität Berlin (2015)
2. Altmann, R.: A PDAE formulation of parabolic problems with dynamic boundary conditions. Appl. Math. Lett. **90**, 202–208 (2019)
3. Altmann, R., Ostermann, A.: Splitting methods for constrained diffusion-reaction systems. Comput. Math. Appl. **74**(5), 962–976 (2017)
4. Altmann, R., Zimmer, C.: Runge-Kutta methods for linear semi-explicit operator differential-algebraic equations. Math. Comput. **87**(309), 149–174 (2018)
5. Altmann, R., Zimmer, C.: Time discretization schemes for hyperbolic systems on networks by ε-expansion. Preprint (2018). ArXiv:1810.04278
6. Birman, M.S., Solomjak, M.Z.: Spectral Theory of Self-Adjoint Operators in Hilbert Space. D. Reidel Publishing Company, Dordrecht (1987)
7. Brezzi, F., Fortin, M.: Mixed and Hybrid Finite Element Methods. Springer, New York (1991)
8. Certaine, J.: The solution of ordinary differential equations with large time constants. In: Mathematical Methods for Digital Computers, pp. 128–132. Wiley, New York (1960)
9. Caliari, M., Ostermann, A.: Implementation of exponential Rosenbrock-type integrators. Appl. Numer. Math. **59**(3–4), 568–581 (2009)
10. Calvo, M.P., Palencia, C.: A class of explicit multistep exponential integrators for semilinear problems. Numer. Math. **102**(3), 367–381 (2006)
11. Cox, S.M., Matthews, P.C.: Exponential time differencing for stiff systems. J. Comput. Phys. **176**(2), 430–455 (2002)
12. Eiermann, M., Ernst, O.G.: A restarted Krylov subspace method for the evaluation of matrix functions. SIAM J. Numer. Anal. **44**(6), 2481–2504 (2006)
13. Emmrich, E., Mehrmann, V.: Operator differential-algebraic equations arising in fluid dynamics. Comput. Methods Appl. Math. **13**(4), 443–470 (2013)
14. Egger, H., Kugler, T., Liljegren-Sailer, B., Marheineke, N., Mehrmann, V.: On structure-preserving model reduction for damped wave propagation in transport networks. SIAM J. Sci. Comput. **40**, A331–A365 (2018)
15. Goldberg, H., Kampowsky, W., Tröltzsch, F.: On Nemytskij operators in L_p-spaces of abstract functions. Math. Nachr. **155**, 127–140 (1992)
16. Hochbruck, M., Lubich, C.: On Krylov subspace approximations to the matrix exponential operator. SIAM J. Numer. Anal. **34**(5), 1911–1925 (1997)
17. Hochbruck, M., Ostermann, A.: Explicit exponential Runge–Kutta methods for semilinear parabolic problems. SIAM J. Numer. Anal. **43**(3), 1069–1090 (2005)
18. Hochbruck, M., Ostermann, A.: Exponential Runge–Kutta methods for parabolic problems. Appl. Numer. Math. **53**(2), 323–339 (2005)
19. Hochbruck, M., Ostermann, A.: Exponential integrators. Acta Numer. **19**, 209–286 (2010)
20. Hochbruck, M., Lubich, C., Selhofer, H.: Exponential integrators for large systems of differential equations. SIAM J. Sci. Comput. **19**(5), 1552–1574 (1998)
21. Hochbruck, M., Ostermann, A., Schweitzer, J.: Exponential Rosenbrock-type methods. SIAM J. Numer. Anal. **47**(1), 786–803 (2009)
22. Jansen, L., Tischendorf, C.: A unified (P)DAE modeling approach for flow networks. In: Progress in Differential-Algebraic Equations, pp. 127–151. Springer, Berlin, Heidelberg (2014)
23. Kooij, G.L., Botchev, M.A., Geurts, B.J.: An exponential time integrator for the incompressible Navier-Stokes equation. SIAM J. Sci. Comput. **40**(3), B684–B705 (2018)
24. Lawson, J.: Generalized Runge-Kutta processes for stable systems with large Lipschitz constants. SIAM J. Numer. Anal. **4**(3), 372–380 (1967)
25. Lamour, R., März, R., Tischendorf, C.: Differential-Algebraic Equations: A Projector Based Analysis. Springer, Berlin, Heidelberg (2013)
26. Luan, V.T., Ostermann, A.: Explicit exponential Runge–Kutta methods of high order for parabolic problems. J. Comput. Appl. Math. **256**, 168–179 (2014)

27. Moler, C., Van Loan, C.: Nineteen dubious ways to compute the exponential of a matrix. SIAM Rev. **20**(4), 801–836 (1978)
28. Newman, C.K.: Exponential integrators for the incompressible Navier-Stokes equations. Ph.D. thesis, Virginia Polytechnic Institute and State University, Blacksburg, Virginia (2003)
29. Niesen, J., Wright, W.M.: Algorithm 919: a Krylov subspace algorithm for evaluating the ϕ-functions appearing in exponential integrators. ACM Trans. Math. Software **38**(3), Art. 22 (2012)
30. Pazy, A.: emphSemigroups of Linear Operators and Applications to Partial Differential Equations. Springer, New York (1983)
31. Roubíček, T.: Nonlinear Partial Differential Equations with Applications. Birkhäuser Verlag, Basel (2005)
32. Saad, Y.: Analysis of some Krylov subspace approximations to the matrix exponential operator. SIAM J. Numer. Anal. **29**(1), 209–228 (1992)
33. Simeon, B.: DAEs and PDEs in elastic multibody systems. Numer. Algorithms **19**, 235–246 (1998)
34. Sprekels, J., Wu, H.: A note on parabolic equation with nonlinear dynamical boundary condition. Nonlinear Anal. **72**(6), 3028–3048 (2010)
35. Strehmel, K., Weiner, R., Podhaisky, H.: Numerik gewöhnlicher Differentialgleichungen: Nichtsteife, steife und differential-algebraische Gleichungen. Vieweg+Teubner Verlag, Wiesbaden (2012)
36. Tartar, L.: An Introduction to Navier-Stokes Equation and Oceanography. Springer, Berlin (2006)
37. Voulis, I., Reusken, A.: Discontinuous Galerkin time discretization methods for parabolic problems with linear constraints. J. Numer. Math. **27**(3), 155–182 (2019)
38. Zeidler, E.: Nonlinear Functional Analysis and its Applications IIa: Linear Monotone Operators. Springer, New York (1990)
39. Zeidler, E.: Nonlinear Functional Analysis and its Applications. I, Fixed-Point Theorems, 2nd edn. Springer, New York (1992)
40. Zimmer, C.: Theorie und Numerik von linearen Operator-differentiell-algebraischen Gleichungen mit zeitverzögertem Term. Master's thesis, Technische Universität Berlin, Germany (2015). In German

Improvement of Rosenbrock-Wanner Method RODASP

Enhanced Coefficient Set and MATLAB Implementation

Gerd Steinebach

Abstract Rosenbrock-Wanner methods for solving index-one DAEs usually suffer from order reduction to order $p = 1$ when the Jacobian matrix is not exactly computed in every time step. This may even happen when the Jacobian matrix is updated in every step, but numerically evaluated by finite differences. Recently, Jax (A rooted-tree based derivation of ROW-type methods with arbitrary jacobian entries for solving index-one DAEs, Dissertation, University Wuppertal (to appear)) could derive new order conditions for the avoidance of such order reduction phenomena. In this paper we present an improvement of the known Rosenbrock-Wanner method rodasp (Steinebach, Order-reduction of ROW-methods for DAEs and method of lines applications, Preprint-Nr. 1741. FB Mathematik, TH Darmstadt (1995)). It is possible to modify its coefficient set such that only an order reduction to $p = 2$ occurs. Several numerical tests indicate that the new method is more efficient than rodasp and the original method rodas from Hairer and Wanner (Solving Ordinary Differential Equations II, Stiff and differential algebraic Problems, 2nd edn. Springer, Berlin, Heidelberg (1996)). When additionally measures for the efficient evaluation of the Jacobian matrix are applied, the method can compete with the standard integrator ode15s of MATLAB.

Keywords Rosenbrock-Wanner methods · W methods · Index-one DAEs · Order reduction

Mathematics Subject Classification (2010) 65L80

G. Steinebach (✉)
Hochschule Bonn-Rhein-Sieg, Sankt Augustin, Germany
e-mail: Gerd.Steinebach@h-brs.de

T. Reis et al. (eds.), *Progress in Differential-Algebraic Equations II*,
Differential-Algebraic Equations Forum,
https://doi.org/10.1007/978-3-030-53905-4_6

1 Introduction

Rosenbrock-Wanner (ROW) methods for solving initial value problems of stiff ordinary differential equations (ODEs) are well known since the late seventies. Recently, Jens Lang [4] gave an excellent survey on the development of these methods.

Beside the simple implementation a major advantage of linear implicit ROW methods is the avoidance of solving nonlinear systems of equations. Instead, s linear systems must be solved per time step, where s describes the stage number of the method. A disadvantage in contrast to e.g. implicit Runge-Kutta methods is, that the Jacobian matrix must be recalculated at each time step. In order to save Jacobian evaluations W methods were introduced in [17]. In principle they can cope with any approximation of the Jacobian. In practice and due to stability issues the Jacobian is held constant for some time steps. In many technical applications the Jacobian matrix is not exactly known, but is approximated by finite differences. Here, W methods can also be advantageous, even if the Jacobian matrix is recalculated in each step.

However, the number of order conditions to be fulfilled is much higher for W methods than for ROW methods. For example, fourth-order ROW methods must satisfy 8 conditions, but W methods 21, see reference [1]. Therefore, W methods usually require a higher number of stages.

When applying ROW schemes to differential algebraic problems (DAEs), additional order conditions must be met. Roche [13] was able to derive these conditions using the Butcher tree theory for index-1 problems. Based on these results, a number of methods were constructed for DAEs. One of the best known methods is rodas by Hairer and Wanner [1]. It is a stiffly accurate method of order $p = 4$ with stage number $s = 6$.

Ostermann and Roche [8] were able to show that Rosenbrock schemes undergo order reduction for some problem classes. This occurs e.g. for semidiscretized parabolic problems and depends on the boundary conditions of the partial differential equations (PDEs). To avoid this order reduction, additional conditions have to be fulfilled. These agree with the conditions of Scholz, which he derived for the Prothero-Robinson model [15].

In [18] the coefficient set of rodas could be modified such that the conditions of Scholz were fulfilled. The corresponding procedure was called rodasp, where P stands for the suitability of the methods for semidiscretized PDEs.

Beside rodas and rodasp many other efficient ROW methods exist. Here only some well known methods are mentioned, that are related to rodas. The third order method ros3pl [5] with four stages is also stiffly accurate, but fulfills additional conditions of a W method for ODEs with $\mathcal{O}(h)$-approximations to the Jacobian. In order to avoid order reduction for the Prothero-Robinson model, Rang [10] enhanced this method to ros3prl2. Another modification leads to the method ros34prw [10] fulfilling more order conditions of W methods for ODEs. A more detailed review on ROW methods is given in reference [4].

In reference [3] W methods for DAEs were considered. At first only approximations to the Jacobian matrix for the differential part of the DAEs were allowed. Recently, Jax [2] could extend these results. DAE systems of the following form are considered:

$$y' = f(y, z); \quad y(t_0) = y_0, \tag{1.1}$$

$$0 = g(y, z); \quad z(t_0) = z_0. \tag{1.2}$$

Consistent initial values with $0 = g(y_0, z_0)$ are assumed and the index-1 condition guarantees a regular matrix g_z in the neighbourhood of the solution, where g_z denotes the Jacobian of function g of partial derivatives with respect to z. The ROW scheme with stage-number s considered by Roche for equations (1.1,1.2) is defined as follows:

$$y_1 = y_0 + \sum_{i=1}^{s} b_i k_i, \quad z_1 = z_0 + \sum_{i=1}^{s} b_i k_i^{alg}, \quad i = 1, \ldots, s \tag{1.3}$$

$$\begin{pmatrix} k_i \\ 0 \end{pmatrix} = h \begin{pmatrix} f(v_i, w_i) \\ g(v_i, w_i) \end{pmatrix} + h \begin{bmatrix} (f_y)_0 & (f_z)_0 \\ (g_y)_0 & (g_z)_0 \end{bmatrix} \sum_{j=1}^{i} \gamma_{ij} \begin{pmatrix} k_j \\ k_j^{alg} \end{pmatrix}, \tag{1.4}$$

$$v_i = y_0 + \sum_{j=1}^{i-1} \alpha_{ij} k_j, \quad w_i = z_0 + \sum_{j=1}^{i-1} \alpha_{ij} k_j^{alg}. \tag{1.5}$$

Here, y_1, z_1 denote approximations to the solution of equations (1.1,1.2) at time $t = t_0 + h$. The coefficients of the method are α_{ij}, γ_{ij} with $\gamma_{ii} = \gamma$ and weights are b_i. The Jacobian matrices are evaluated at time $t = t_0$, e.g. $(f_y)_0 = \frac{\partial f}{\partial y}(y_0, z_0)$.

Jax [2] now replaces f_y, f_z, g_y with arbitrary matrices A_y, A_z, B_y, only g_z remains exact. With the Butcher tree theory (see references [1, 13]) transferred to this case, he can derive new additional order conditions. The total number of conditions is e.g. $n = 26$ for such a new method of order $p = 3$ compared to $n = 13$ for a ROW method given by equations (1.3,1.4,1.5).

It is well known that rodas and rodasp suffer from an order reduction to $p = 1$ when implemented as a W method with inexact Jacobian. Even with an implementation as a ROW method with updated Jacobian matrix at every time step, problems may occur if the Jacobian matrix is calculated using finite differences and is therefore not exact. The aim of this paper is to improve the fourth order ROW method rodasp in such a way, that the order $p = 2$ is still maintained by using arbitrary approximations A_y, A_z, B_y to the Jacobian matrix. The free parameters are chosen such that the new conditions of Jax are fulfilled. In [20] the coefficient set of rodasp is given, but the preprint [18], where the construction of the method is described, has never been published in a journal. Therefore the construction of rodasp together with the optimized coefficients is outlined again in this paper.

Section 2 first summarises all the necessary conditions. In Sect. 3 the new method `rodasp2` is constructed. Since many integrators are implemented and tested in a MATLAB version, these aspects are treated in Sect. 4. For the efficient implementation the computation of the Jacobian plays a key role. Strategies for taking linear components into account and for vectorization are discussed. Subsequently, numerical tests are carried out and analysed in Sect. 5. The numerical tests cover problems from network simulation, because ROW methods are proven to be very well suited for such problems, especially in the context of fluid flow networks, see reference [19].

2 Order Conditions

We want to construct an L-stable ROW method of order $p = 4$ for index-1 DAE systems of type (1.1,1.2). According to [13] the order conditions 1 to 13 stated in Table 1 have to be fulfilled. The following abbreviations are used:

$$\beta_{ij} = \alpha_{ij} + \gamma_{ij} , \quad \text{with} \quad \beta_{ij} = 0 \text{ for } i < j \text{ and } \beta_{ii} = \gamma_{ii} = \gamma , \tag{2.1}$$

$$\beta_i = \sum_{j=1}^{i} \beta_{ij} , \quad \alpha_i = \sum_{j=1}^{i-1} \alpha_{ij} , \quad B = (\beta_{ij})_{i,j=1}^{s} , \quad W = B^{-1} = (w_{ij})_{i,j=1}^{s} . \tag{2.2}$$

The missing quantities in conditions 17 and 18 are explained below in the text.

In order to avoid order reduction to $p = 1$ when dealing with inexact Jacobians, we want to fulfill at least the conditions up to order $p = 2$ in this case. These are the conditions 14–16 in Table 1 which have been derived in [2]. Note, that nevertheless the Jacobian g_z must be exact.

Finally, we want to avoid severe order reduction when applying the method to the Prothero-Robinson model [9]

$$y' = \lambda(y - g) + g' \quad , \quad y(0) = g(0), \quad g(t) \text{ smooth and } Re\lambda < 0 \tag{2.3}$$

with the exact solution $y(t) = g(t)$. For $y(0) \neq g(0)$ and large stiffness $Re\lambda \ll 0$ the solution $y(t)$ attains $g(t)$ very quickly asymptotically. First, Scholz [15] studied ROW methods applied to this problem and derived additional order conditions for the stiff case. The Prothero-Robinson model is an important test problem in the context of parabolic partial differential equations, too. By semi-discretization of certain problems in space and diagonalization of the resulting matrix it can be shown, that a system of equations of type (2.3) will arise, see [18]. Ostermann and Roche [7, 8] investigated order reduction of Runge-Kutta and ROW methods when applied to semi-discretized parabolic problems. They could show that the conditions of Scholz also appear as additional requirements. Rang [10, 11] examined further the convergence of ROW methods for problem (2.3). He derived new order conditions which guarantee a global error of size $\mathcal{O}(\frac{h^k}{z^l})$ with $z = \lambda h$ and different exponents

Table 1 Order conditions to be fulfilled for the new method

No	Order	Tree	Condition
1	1	•	$\sum b_i = 1$
2	2		$\sum b_i \beta_i = 1/2$
3	3		$\sum b_i \alpha_i^2 = 1/3$
4			$\sum b_i \beta_{ij} \beta_j = 1/6$
5			$\sum b_i w_{ij} \alpha_j^2 = 1$
6	4		$\sum b_i \alpha_i^3 = 1/4$
7			$\sum b_i \alpha_i \alpha_{ij} \beta_j = 1/8$
8			$\sum b_i \beta_{ij} \alpha_j^2 = 1/12$
9			$\sum b_i \beta_{ij} \beta_{jk} \beta_k = 1/24$
10			$\sum b_i \alpha_i \alpha_{ij} w_{jk} \alpha_k^2 = 1/4$
11			$\sum b_i w_{ij} \alpha_j^3 = 1$
12			$\sum b_i w_{ij} \alpha_j \alpha_{jk} \beta_k = 1/2$
13			$\sum b_i w_{ij} \alpha_j \alpha_{jk} w_{kl} \alpha_l^2 = 1$
14	2		$\sum b_i \gamma_{ij} = 0$
15			$\sum b_i \alpha_{ij} w_{jk} \alpha_k = 1/2$
16			$\sum b_i w_{ij} \alpha_j = 1$
17	3		$C_2(H) = \sum_{i=0}^{s} A_i H^i = 0$
18	4		$C_3(H) = \sum_{i=0}^{s-1} B_i H^i = 0$

k and l. In the stiff case large values of $|z|$ are assumed even in the limit case $h \to 0$. Since the conditions of Scholz are included in those of Rang and `rodasp` has been derived with the Scholz conditions, we stick to these conditions.

In his paper [15] Scholz could show, that for strong A-stable methods fulfilling some conditions $C_1(H) \equiv \ldots \equiv C_{p-1}(H) \equiv 0$ with $H = \frac{z}{1-\gamma z}$ the global error is bounded by $C \cdot h^p$. This theorem is valid for small step sizes h and large stiffness parameter $|\lambda|$ such that $Re(z) \le \lambda_0 < 0$. Note, that condition $C_1(H) \equiv 0$ holds for every consistent ROW method. According to conditions 17 and 18 in Table 1 we want to have order $p = 4$. It should be mentioned that the estimation of the error constant C of the global error in the paper of Scholz is not sharp. It depends on H and behaves even like $C = \frac{1}{z} C_1$ for L-stable methods, see [10]. Therefore, for fixed h asymptotically exact results are obtained for $|\lambda| \to \infty$, but for fixed large stiffness $|\lambda|$ only order $p - 1$ must be visible in the numerical results.

The coefficients of polynomials $C_2(H) = \sum_{i=0}^{s} A_i H^i$, $C_3(H) = \sum_{i=0}^{s-1} B_i H^i$ are defined by, see [15]:

$$A_0 = -N^{(2)}(-1) + \gamma M(-1) + M(0) \tag{2.4}$$

$$A_i = -N^{(2)}(i-1) + 2\gamma M(i-1) + \gamma^2 M(i-2) + M(i) \tag{2.5}$$

$$\text{for } 0 < i < s$$

$$A_s = \gamma^2 M(s-2) \tag{2.6}$$

$$B_0 = -N^{(3)}(-1) + N^{(2)}(0) \tag{2.7}$$

$$B_i = -N^{(3)}(i-1) + \gamma N^{(2)}(i-1) + N^{(2)}(i) \tag{2.8}$$

$$\text{for } 0 < i < s-1$$

$$B_{s-1} = -N^{(3)}(s-2) + \gamma N^{(2)}(s-2) \tag{2.9}$$

$$M(\nu) = \sum_{i=1}^{s} b_i M_i(\nu), \quad N^{(\sigma)}(\nu) = \sum_{i=1}^{s} b_i N_i^{(\sigma)}(\nu) \quad \text{for } \sigma \geq 2 \tag{2.10}$$

with

$$M_i(\nu) = \begin{cases} 1 & \text{if } \nu < 0 \\ \beta_i' & \text{if } \nu = 0 \\ \sum \beta_{ij_1} \beta_{j_1 j_2} \cdots \beta_{j_{\nu-1} j_\nu} \beta_{j_\nu}' & \text{if } \nu = 1, \ldots, i-2 \\ 0 & \text{if } \nu \geq i-1 \end{cases} \tag{2.11}$$

$$\sigma! N_i^{(\sigma)}(\nu) = \begin{cases} 1 & \text{if } \nu < 0 \\ \alpha_i^\sigma & \text{if } \nu = 0 \\ \sum \beta_{ij_1} \beta_{j_1 j_2} \cdots \beta_{j_{\nu-1} j_\nu} \alpha_{j_\nu}^\sigma & \text{if } \nu = 1, \ldots, i-2 \\ 0 & \text{if } \nu \geq i-1 \end{cases} \tag{2.12}$$

and $\beta_i' = \sum_{j=1}^{i-1} \beta_{ij}$. The summation in (2.11), (2.12) is over $j_\nu < \cdots < j_1 < i$.

In order to fulfill conditions 17 and 18 in Table 1, all coefficients A_i and B_i must be zero.

An L-stable method is obtained, when $|R(z)| < 1$ for $Re(z) < 0$ and $R(\infty) = 0$ holds. The stability function is given by

$$R(z) = 1 + z b^T (I - zB)^{-1} e, \quad b^T = (b_1, \ldots, b_s), \quad e = (1, \ldots, 1)^T. \tag{2.13}$$

$R(z)$ can also be expressed in terms of $M(\nu)$ defined in (2.10):

$$R(z) = \sum_{i=0}^{s} M(i-2) H^i. \tag{2.14}$$

3   Construction of Coefficient Set

The aim is to construct an L-stable method that fulfills all conditions from Table 1. The construction is very close to that of `rodas` [1], and `rodasp` [18]. The first method meets conditions 1–13 and `rodasp` in addition conditions 17 and 18. Both are stiffly accurate methods with $s = 6$ stages. Stiffly accurate ROW-methods are characterized by

$$b_i = \beta_{si} \quad \text{for } i = 1, \ldots, s - 1, \quad b_s = \gamma, \quad \alpha_s = 1 . \tag{3.1}$$

The embedded method of order $\hat{p} = 3$ with stage number $\hat{s} = 5$ is stiffly accurate too:

$$\hat{b}_i = \beta_{s-1,i} \quad \text{for } i = 1, \ldots, s - 2, \quad \hat{b}_{s-1} = \gamma, \quad \alpha_{s-1} = 1 . \tag{3.2}$$

Moreover

$$\alpha_{si} = \beta_{s-1,i} \quad \text{for } i = 1, \ldots, s - 1 \tag{3.3}$$

is required with the consequence that the computation of the two last stages represent Newton-iteration steps, see [1].

If we choose in addition

$$\beta_2' = 0 \tag{3.4}$$

$$\sum \alpha_{s-1,i} \beta_i' = 1/2 - \gamma \tag{3.5}$$

$$\sum \alpha_{s-1,i} w_{ij} \alpha_j^2 = 1 \tag{3.6}$$

and since stiffly accurate ROW-methods lead to

$$\sum_{i=1}^{s} b_i w_{ij} = \begin{cases} 1 & \text{if } j = s \\ 0 & \text{otherwise,} \end{cases} \tag{3.7}$$

the order conditions are simplified significantly. From (3.1,3.2,3.5) and condition No.1 from Table 1 we get:

$$\alpha_5 = \alpha_6 = 1 , \quad \hat{b}_5 = b_6 = \gamma , \quad \beta_2' = 0 , \quad \beta_5' = \beta_6' = 1 - \gamma .$$

Conditions No.5, 11, 12, 13, 16 from Table 1 are fulfilled. Moreover, for the embedded method conditions No.1, 5 are met. The remaining conditions are shown in Table 2. For the Scholz conditions we require $A_3 = A_4 = A_5 = 0$ and $B_2 = B_3 = B_4 = B_5 = 0$, since $A_0 = A_1 = A_2 = A_6 = B_0 = B_1 = 0$ is already true. The conditions for the embedded method are marked with a hat ^.

Thus, we have to solve 23 equations for the 22 unknowns γ, α_{21}, α_{31}, α_{32}, α_{41}, α_{42}, α_{43}, α_{52}, α_{53}, α_{54}, β_{31}, β_{32}, β_{41}, β_{42}, β_{43}, β_{52}, β_{53}, β_{54}, β_{62}, β_{63}, β_{64}, β_{65}. This is only possible, when some conditions are redundant. We even can show, that the embedded method fulfills the Scholz conditions, too.

Table 2 Equations to be fulfilled for the new method

No	Previous No	Equation
1	1	$b_1 + b_2 + b_3 + b_4 + b_5 = 1 - \gamma$
2	2	$b_3\beta_3' + b_4\beta_4' + b_5(1 - \gamma) = \gamma^2 - 2\gamma + \frac{1}{2}$
3	3	$b_2\alpha_2^2 + b_3\alpha_3^2 + b_4\alpha_4^2 + b_5 = \frac{1}{3} - \gamma$
4	4	$b_4\beta_{43}\beta_3' + b_5(\frac{1}{2} - 2\gamma + \gamma^2) = -\gamma^3 + 3\gamma^2 - \frac{3}{2}\gamma + \frac{1}{6}$
5	6	$b_2\alpha_2^3 + b_3\alpha_3^3 + b_4\alpha_4^3 + b_5 = \frac{1}{4} - \gamma$
6	7	$b_4\alpha_4\alpha_{43}\beta_3 + b_5(\frac{1}{2} - \gamma) = \gamma^2 - \frac{5}{6}\gamma + \frac{1}{8}$
7	8	$b_3\beta_{32}\alpha_2^2 + b_4\beta_{42}\alpha_2^2 + b_4\beta_{43}\alpha_3^2 + b_5(\frac{1}{3} - \gamma) = \gamma^2 - \frac{2}{3}\gamma + \frac{1}{12}$
8	9	$b_5(\frac{1}{6} - \frac{3}{2}\gamma + 3\gamma^2 - \gamma^3) = \gamma^4 - 4\gamma^3 + 3\gamma^2 - \frac{2}{3}\gamma + \frac{1}{24}$
9	10	$b_3\alpha_3\alpha_{32}w_{22}\alpha_2^2 + b_4\alpha_4(\alpha_{42}w_{22}\alpha_2^2 + \alpha_{43}w_{32}\alpha_2^2 + \alpha_{43}w_{33}\alpha_3^2) + b_5 = \frac{1}{4} - \gamma$
10	$\hat{2}$	$\beta_{53}\beta_3 + \beta_{54}\beta_4 = \frac{1}{2} - 2\gamma + \gamma^2$
11	$\hat{3}$	$\beta_{52}\alpha_2^2 + \beta_{53}\alpha_3^2 + \beta_{54}\alpha_4^2 = \frac{1}{3} - \gamma$
12	$\hat{4}$	$\beta_{54}\beta_{43}\beta_3 = \frac{1}{6} - \frac{3}{2}\gamma + 3\gamma^2 - \gamma^3$
13	(3.5)	$\alpha_{53}\beta_3 + \alpha_{54}\beta_4 = \frac{1}{2} - \gamma$
14	(3.6)	$(\alpha_{52}w_{22} + \alpha_{53}w_{32} + \alpha_{54}w_{42})\alpha_2^2 + (\alpha_{53}w_{33} + \alpha_{54}w_{43})\alpha_3^2 + \alpha_{54}w_{44}\alpha_4^2 = 1$
15	17, A3	$N^{(2)}(2) = 2\gamma M(2) + \gamma^2 M(1) + M(3)$
16	17, A4	$N^{(2)}(3) = 2\gamma M(3) + \gamma^2 M(2)$
17	17, A5	$N^{(2)}(4) = \gamma^2 M(3)$
18	18, B2	$N^{(3)}(1) = \gamma N^{(2)}(1) + N^{(2)}(2)$
19	18, B3	$N^{(3)}(2) = \gamma N^{(2)}(2) + N^{(2)}(3)$
20	18, B4	$N^{(3)}(3) = \gamma N^{(2)}(3) + N^{(2)}(4)$
21	18, B5	$N^{(3)}(4) = \gamma N^{(2)}(4)$
22	14	$b_2\alpha_2 + b_3\alpha_3 + b_4\alpha_4 + b_5 = \frac{1}{2} - \gamma$
23	15	$\sum b_i\alpha_{ij}w_{jk}\alpha_k = 1/2$

The construction of the new method is according to `rodasp`, [18]. First, we choose $\gamma = 1/4$ according to `rodas` and α_3, α_4, β_4' as free parameters, see [1]. Then:

(a) b_5 is determined by No.8.
(b) No.21 is equivalent to

$$\frac{1}{6}b_6\beta_{65}\beta_{54}\beta_{43}\beta_{32}\alpha_2^3 = \frac{\gamma}{2}b_6\beta_{65}\beta_{54}\beta_{43}\beta_{32}\alpha_2^2$$

and yields $\alpha_2 = 3\gamma$.
(c) No.17 is equivalent to

$$\frac{1}{2}b_6\beta_{65}\beta_{54}\beta_{43}\beta_{32}\alpha_2^2 = \gamma^2 b_6\beta_{65}\beta_{54}\beta_{43}\beta_3',$$

it follows

$$\frac{1}{2}\beta_{32}\alpha_2^2 = \gamma^2\beta_3'. \tag{3.8}$$

(d) Putting No.16 and No.17 into No.20 yields $N^{(3)}(3) = 3\gamma^2 M(3) + \gamma^3 M(2)$. $M(2) = \sum b_i \beta_{ij} \beta_{jk} \beta_k' = -\gamma^3 + \frac{3}{2}\gamma^2 - \frac{\gamma}{2} + \frac{1}{24}$ is the order condition according to No.8. $M(3)$ is defined by

$$M(3) = \gamma^5 - 4\gamma^4 + 3\gamma^3 - \frac{2}{3}\gamma^2 + \frac{\gamma}{24}. \tag{3.9}$$

This can be seen by writing $M(3)$ as

$$\sum b_i \beta_{ij} \beta_{jk} \beta_{kl} \beta_l' = \sum_{i=1}^{5} \sum_{j,k,l} b_i \beta_{ij} \beta_{jk} \beta_{kl} \beta_l' + \gamma \sum_{i=1}^{5} \sum_{j,k} \beta_{ij} \beta_{jk} \beta_k'.$$

The term on the left side is zero. The right term can be successively computed by conditions No.8, No.4, No.2, No.1 resulting from Butcher trees without branches. No.20 is then equivalent to

$$\frac{1}{6}(b_5 \beta_{54} \beta_{43} \beta_{32} + \gamma b_5 \beta_{54} \beta_{42} + \gamma b_5 \beta_{53} \beta_{32} + \gamma b_4 \beta_{43} \beta_{32})\alpha_2^3$$

$$+\frac{1}{6}\gamma b_5 \beta_{54} \beta_{43} \alpha_3^3 = \gamma^3 (\frac{1}{6} - \frac{5}{2}\gamma + \frac{21}{2}\gamma^2 - 13\gamma^3 + 3\gamma^4) \tag{3.10}$$

and No.16 is equivalent to

$$\frac{1}{2}(b_5 \beta_{54} \beta_{43} \beta_{32} + \gamma b_5 \beta_{54} \beta_{42} + \gamma b_5 \beta_{53} \beta_{32} + \gamma b_4 \beta_{43} \beta_{32})\alpha_2^2 =$$

$$-\frac{1}{2}\gamma b_5 \beta_{54} \beta_{43} \alpha_3^2 + 2\gamma^2 b_5 \beta_{54} \beta_{43} \beta_3' + \gamma^2 (\frac{1}{24} - \frac{1}{2}\gamma + \frac{3}{2}\gamma^2 - \gamma^3) \tag{3.11}$$

Putting No.16 multiplied by α_2, b_5 from No.8 and $\beta_{54} \beta_{43} \beta_3'$ from No.12 into (3.10) yields

$$\beta_{54} \beta_{43} (-\frac{\gamma^2}{2}\alpha_3^2 + \frac{\gamma}{6}\alpha_3^3) = \gamma^3 (\frac{1}{6} - \frac{3}{2}\gamma + 3\gamma^2 - \gamma^3).$$

By comparing this equation with No.12 the following expression for β_3' is obtained

$$\beta_3' = \frac{\alpha_3^2}{\gamma^2}(-\frac{\gamma}{2} + \frac{\alpha_3}{6}). \tag{3.12}$$

β_{32} is then defined by (3.8).

(e) Expressions for $b_4\beta_{43}\beta_3'$ and $\beta_{54}\beta_{43}\beta_3'$ can be obtained from No.4 and No.12. Division of both terms yields an expression for β_{54} in linear dependence on b_4:

$$\beta_{54} = \frac{\beta_{54}\beta_{43}\beta_3'}{b_4\beta_{43}\beta_3'} b_4 . \tag{3.13}$$

With the help of (3.13), equations No.2, No.3, No.5, No.7 and No.10 define a linear system for the unknowns b_2, b_3, b_4, $b_4\beta_{42}$ and β_{53}.

Thereby β_{42} is also defined, and β_{43} is defined by No.4, β_{54} by No.12, b_1 by No.1, α_{43} by No.6 and β_{52} by No.13.

(f) Coefficients α_{52}, α_{53} and α_{54} have to be computed from No.13 and No.14. The additional degree of freedom is used to fulfill $\sum \alpha_{5i} w_{ij} = 1$. By this condition the error of the numerical solution is bounded by $O(h^2\delta)$ in case of inconsistent initial conditions (y_0, z_0) for DAEs of type (1.1,1.2) with $\|(g_z^{-1}g)(y_0, z_0)\| \leq \delta$, see [1].

(g) The remaining condition No.9 has to be considered for the computation of α_{32} and α_{42}. The additional degree of freedom is exploited to satisfy the extra order-condition $\sum b_i \alpha_i^2 \alpha_{ij} \alpha_j^2 = \frac{1}{18}$ for methods of order 6.

By this construction all conditions except No.22, 23 are fulfilled. Therefore, we use the free parameters α_3, α_4, β_4' to iterate this process by a nonlinear least-square method in order to satisfy these conditions. The extra degree of freedom is used to get a small truncation error. The computed coefficients of the new method named rodasp2 are given in Table 3.

Table 3 Coefficients for new method rodasp2

$\gamma = 0.25$	$\alpha_{21} = 7.500000000000000e-01$	$\beta_{21} = 0$
	$\alpha_{31} = 3.688749816109670e-01$	$\beta_{31} = -9.184372116108780e-02$
$b_1 = \beta_{61}$	$\alpha_{32} = -4.742684759792117e-02$	$\beta_{32} = -2.624106318888223e-02$
$b_2 = \beta_{62}$	$\alpha_{41} = 4.596170083041160e-01$	$\beta_{41} = -5.817702768270960e-02$
$b_3 = \beta_{63}$	$\alpha_{42} = 2.724432453018110e-01$	$\beta_{42} = -1.382129630513952e-01$
$b_4 = \beta_{64}$	$\alpha_{43} = -2.123145213282008e-01$	$\beta_{43} = 5.517478318046004e-01$
$b_5 = \beta_{65}$	$\alpha_{51} = 2.719770298548111e+00$	$\beta_{51} = -6.315720511779359e-01$
$b_6 = \gamma$	$\alpha_{52} = 1.358873794835473e+00$	$\beta_{52} = -3.326966988718489e-01$
	$\alpha_{53} = -2.838824065018641e+00$	$\beta_{53} = 1.154688683864917e+00$
$\hat{b}_1 = \beta_{51}$	$\alpha_{54} = -2.398200283649438e-01$	$\beta_{54} = 5.595800661848674e-01$
$\hat{b}_2 = \beta_{52}$	$\alpha_{61} = -6.315720511779362e-01$	$\beta_{61} = 1.464968119068509e-01$
$\hat{b}_3 = \beta_{53}$	$\alpha_{62} = -3.326966988718489e-01$	$\beta_{62} = 8.896159691002870e-02$
$\hat{b}_4 = \beta_{54}$	$\alpha_{63} = 1.154688683864918e+00$	$\beta_{63} = 1.648843942975147e-01$
$\hat{b}_5 = \gamma$	$\alpha_{64} = 5.595800661848674e-01$	$\beta_{64} = 4.568000540284631e-01$
	$\alpha_{65} = 2.500000000000000e-01$	$\beta_{65} = -1.071428571428573e-01$

For the stability function we get

$$R(z) = \frac{-4(7z^4 + 8z^3 - 96z^2 - 192z + 768)}{3(z-4)^5} .$$

It can be shown that thereby the method is A-stable an hence L-stable.

4   Implementation Issues

In contrast to implicit Runge-Kutta or BDF schemes, ROW methods must recompute the Jacobian matrix J in every time step. Therefore, it is essential to do this computation as efficient as possible. Usually, the user is not willing or even not able to supply an analytical Jacobian and the integrator must compute a numerical approximation by finite differences. When nothing is known about the structure of J, this computation requires n evaluations of the right-hand side of the DAE-system (1.1,1.2). This system is usually summarized into its linear implicit form

$$My' = f(t, y), \quad y(t_0) = y_0 \tag{4.1}$$

with singular $n \times n$ matrix M and right-hand side f consisting of both functions f and g from equations (1.1,1.2). The application of a ROW scheme to non-autonomous systems like (4.1) is given in [1]. A common way to compute J is given below in MATLAB notation.

```
f0 = f(t,y);
for i=1:n
    y1 = y;  y1(i) = y1(i) + del;
    J(:,i) = ( f(t,y1) - f0 )/del;
end
```

The first function evaluation of f is required for the integrator as well, independent of J. By each of the subsequent evaluations a whole column of J is computed and *del* is an appropriate increment for the finite difference approximation. For problems with large dimension n of equations it is therefore important to provide the integrator with the pattern of non zeros of J, which is often sparse. By this knowledge several components of y can be altered at once and large reduction of function evaluations may be possible. E.g. for a tridiagonal matrix J only three function evaluations are necessary regardless of dimension n.

Further reductions are possible when at least parts of J are constant, i.e. function $f(t, y)$ has linear components with respect to some components of y. Then the corresponding entries of J must be calculated only once in the first time step. In order to supply the integrator with such information, beside the JPATTERN option, the user can specify columns and/or rows of J which are constant.

Finally, in MATLAB vectorization can be applied. In this case, the function which evaluates $f(t, y)$ must be modified, such that a whole matrix y instead of a vector can be given as input parameter and the resulting output value of function $f(t, y)$ is a matrix as well. By this option only one function call to f is required in order to compute the whole Jacobian J.

5 Numerical Tests

This section discusses numerical results obtained with the new method `rodasp2`. The results of `rodas` and `rodasp` are used for comparison. Also `ode15s`, the standard BDF integrator for DAEs in MATLAB [16] is used.

In the first example, the order of the methods is determined numerically. It is the DAE problem

$$y' = z \tag{5.1}$$

$$0 = y^2 + z^2 - 1 \tag{5.2}$$

with initial conditions $y(0) = 0$, $z(0) = 1$ and analytical solution $y(t) = \sin(t)$, $z(t) = \cos(t)$. In Table 4 the numerical results are summarized. The problem was solved in the time interval $t \in [0, 1]$ with constant time steps $h = \frac{0.1}{2^n}, n = 0, \ldots, 6$. For the three methods `rodas`, `rodasp`, `rodasp2` and its embedded methods the absolute error $err = max(|y(1) - \sin(1)|, |y(2) - \cos(1)|)$ and the numerical obtained order p of convergence is shown. The problem was solved with the exact Jacobian $J = \begin{pmatrix} 0 & 1 \\ 2y & 2z \end{pmatrix}$ and with the inexact Jacobian $J_{inexact} = \begin{pmatrix} 0 & 0 \\ 0 & 2z \end{pmatrix}$. All results are in agreement with the theoretical expectations. For the exact Jacobian all methods reach the order $p = 4$ and $\hat{p} = 3$ for their embedded formulas. When using the inexact Jacobian, `rodasp2` suffers an order reduction to $p = 2$, while the other methods show a reduction to $p = 1$.

The second example treats the Prothero-Robinson equation (2.3) with function $g(t) = 10 - (10 + t)e^{-t}$ and stiffness $\lambda = -1$ and $\lambda = -10^5$ in time interval $t \in [0, 2]$, see [15]. The calculations were carried out with the exact Jacobian matrix. The results are presented in the same manner as in Table 4. Again, the theoretical expectations are confirmed. `rodas` shows an order reduction to $p = 1$ in case of high stiffness, whereas the order reduction of `rodasp` and `rodasp2` is only to $p = 3$. Additionally to Table 5 it can be shown, that due to their stiffly accuracy all methods show asymptotic convergence with $\frac{1}{|\lambda|}$ regarding high stiffness for fixed step size h.

Table 4 Numerical results for DAE problem (5.1, 5.2)

J exact	rodasp2		rodasp		rodas	
h	err	p	err	p	err	p
1.00e−01	7.94e−06		2.91e−05		8.93e−06	
5.00e−02	3.93e−07	4.34	1.51e−06	4.27	4.46e−07	4.32
2.50e−02	1.70e−08	4.53	6.69e−08	4.49	1.88e−08	4.57
1.25e−02	7.24e−10	4.56	2.88e−09	4.54	7.47e−10	4.65
6.25e−03	3.25e−11	4.48	1.30e−10	4.47	3.05e−11	4.61
3.13e−03	1.60e−12	4.35	6.44e−12	4.34	1.35e−12	4.49
1.56e−03	8.84e−14	4.18	3.46e−13	4.22	6.58e−14	4.36
	Embedded methods					
1.00e−01	6.43e−05		2.41e−04		8.00e−05	
5.00e−02	6.56e−06	3.29	2.53e−05	3.25	8.64e−06	3.21
2.50e−02	6.39e−07	3.36	2.50e−06	3.34	8.84e−07	3.29
1.25e−02	6.46e−08	3.31	2.54e−07	3.30	9.30e−08	3.25
6.25e−03	6.98e−09	3.21	2.75e−08	3.21	1.03e−08	3.17
3.13e−03	7.98e−10	3.13	3.15e−09	3.13	1.20e−09	3.10
1.56e−03	9.50e−11	3.07	3.75e−10	3.07	1.45e−10	3.06
J inexact	rodasp2		rodasp		rodas	
h	err	p	err	p	err	p
1.00e−01	6.35e−03		7.53e−03		7.32e−03	
5.00e−02	1.77e−03	1.84	3.86e−03	0.96	2.12e−03	1.79
2.50e−02	4.69e−04	1.92	2.14e−03	0.85	1.53e−03	0.46
1.25e−02	1.20e−04	1.96	1.07e−03	0.99	1.03e−03	0.57
6.25e−03	3.05e−05	1.98	5.31e−04	1.02	6.02e−04	0.77
3.13e−03	7.69e−06	1.99	2.63e−04	1.01	3.26e−04	0.88
1.56e−03	1.93e−06	1.99	1.31e−04	1.01	1.69e−04	0.94
	Embedded methods					
1.00e−01	1.69e−02		3.90e−02		2.36e−01	
5.00e−02	1.10e−02	0.62	9.60e−03	2.02e	2.94e−02	3.01
2.50e−02	6.14e−03	0.83	1.86e−03	2.37e	1.15e−02	1.35
1.25e−02	3.24e−03	0.92	3.27e−04	2.51e	5.85e−03	0.98
6.25e−03	1.66e−03	0.96	2.02e−04	0.69e	3.15e−03	0.89
3.13e−03	8.41e−04	0.98	1.33e−04	0.60e	1.63e−03	0.95
1.56e−03	4.23e−04	0.99	8.05e−05	0.72e	8.27e−04	0.97

The third and fourth examples are PDE problems. First, the parabolic problem

$$\frac{\partial u}{\partial t} = \frac{\partial^2 u}{\partial x^2} - \frac{(x+\frac{1}{2})(\frac{3}{2}-x)}{(1+t)^2} + \frac{2}{1+t} , \quad t \in [0, 0.1] , \quad x \in [0, 1] \qquad (5.3)$$

Table 5 Numerical results for Prothero-Robinson problem (2.3) for different stiffness parameters λ

$\lambda = -1$	rodasp2		rodasp		rodas	
h	err	p	err	p	err	p
1.00e−01	1.25e−07		1.19e−07		6.77e−07	
5.00e−02	7.90e−09	3.98	7.65e−09	3.96	3.20e−08	4.40
2.50e−02	4.97e−10	3.99	4.85e−10	3.98	1.48e−09	4.43
1.25e−02	3.11e−11	4.00	3.05e−11	3.99	7.16e−11	4.37
6.25e−03	1.94e−12	4.00	1.91e−12	4.00	3.73e−12	4.26
	Embedded methods					
1.00e−01	1.12e−07		9.85e−08		5.08e−05	
5.00e−02	3.43e−08	1.71	3.33e−08	1.57	7.12e−06	2.83
2.50e−02	5.69e−09	2.59	5.63e−09	2.56	9.54e−07	2.90
1.25e−02	8.06e−10	2.82	8.01e−10	2.81	1.24e−07	2.95
6.25e−03	1.07e−10	2.91	1.07e−10	2.91	1.58e−08	2.97
$\lambda = -10^5$	rodasp2		rodasp		rodas	
h	err	p	err	p	err	p
1.00e−01	7.46e−11		2.61e−11		1.76e−08	
5.00e−02	9.22e−12	3.02	3.33e−12	2.97	1.14e−08	0.626
2.50e−02	1.15e−12	3.01	4.19e−13	2.99	6.30e−09	0.856
1.25e−02	1.43e−13	3.00	5.11e−14	3.04	3.28e−09	0.942
6.25e−03	1.91e−14	2.91	4.88e−15	3.39	1.66e−09	0.985
	Embedded methods					
1.00e−01	2.26e−10		2.10e−10		1.63e−07	
5.00e−02	2.76e−11	3.03	2.57e−11	3.03	7.55e−08	1.11
2.50e−02	3.41e−12	3.01	3.18e−12	3.02	3.62e−08	1.06
1.25e−02	4.24e−13	3.01	3.96e−13	3.01	1.77e−08	1.03
6.25e−03	5.46e−14	2.96	5.11e−14	2.95	8.72e−09	1.02

is considered, see [18]. Initial values and inhomogeneous boundary conditions are taken from the analytical solution

$$u(x,t) = \frac{(x + \frac{1}{2})(\frac{3}{2} - x)}{1 + t}.$$

Since $u(x,t)$ is quadratic with respect to x, the space discretization of (5.3) by standard finite differences does not introduce a numerical error. In Table 6 the numerical results are given. The problem was solved by using $n_x = 1000$ discretization points x_i in space with time steps $h = \frac{0.1}{2^n}$, $n = 0, \ldots, 7$. The error at time $t_{end} = 0.1$ is measured by

$$err = \sqrt{\frac{1}{n_x} \sum_{i=1}^{n_x} (u(x_i, t_{end}) - y_i)^2}$$

Table 6 Numerical results for parabolic problem (5.3)

h	rodasp2		rodasp		rodas	
	err	p	err	p	err	p
1.00e−01	1.01e−06		9.71e−07		3.49e−06	
5.00e−02	6.48e−08	3.97	6.24e−08	3.96	2.53e−07	3.79
2.50e−02	4.11e−09	3.98	3.99e−09	3.97	3.19e−08	2.99
1.25e−02	2.61e−10	3.98	2.54e−10	3.97	9.87e−09	1.69
6.25e−03	1.65e−11	3.98	1.61e−11	3.98	2.50e−09	1.98
3.13e−03	1.04e−12	3.98	1.02e−12	3.98	5.70e−10	2.13
1.56e−03	6.66e−14	3.97	6.55e−14	3.97	1.25e−10	2.19
7.81e−04	5.78e−15	3.53	5.72e−15	3.52	2.67e−11	2.22
	Embedded methods					
1.00e−01	1.73e−06		1.66e−06		5.69e−05	
5.00e−02	3.36e−07	2.37	3.31e−07	2.33	9.21e−06	2.63
2.50e−02	5.08e−08	2.73	5.05e−08	2.71	1.57e−06	2.55
1.25e−02	6.96e−09	2.87	6.94e−09	2.86	2.86e−07	2.46
6.25e−03	9.10e−10	2.94	9.09e−10	2.93	5.47e−08	2.39
3.13e−03	1.16e−10	2.97	1.16e−10	2.97	1.09e−08	2.33
1.56e−03	1.47e−11	2.98	1.47e−11	2.98	2.21e−09	2.30
7.81e−04	1.85e−12	2.99	1.85e−12	2.99	4.57e−10	2.28

where y_i denotes the numerical solution in space point x_i at time t_{end}. Methods rodasp and rodasp2 do not suffer from order reduction, whereas the order of rodas is reduced to approximately $p = 2.25$. This is in agreement to the theory of Ostermann and Roche [8] for parabolic problems.

Similar observations can even be made for the hyperbolic problem, [14]:

$$\frac{\partial u}{\partial t} = -\frac{\partial u}{\partial x} + \frac{t - x}{(1 + t)^2}, \quad t \in [0, 1], \quad x \in [0, 1] \tag{5.4}$$

Again, initial values and left boundary condition are taken from the analytical solution

$$u(x, t) = \frac{1 + x}{1 + t}.$$

Discretization in space is made by using the first order upwind finite difference and $n_x = 1000$ space points and the error is measured in the same way as in the parabolic problem. Here, time steps $h = \frac{1}{10 \cdot 2^n}$, $n = 0, \ldots, 7$ are applied. Method rodas shows an order reduction to approximately $p = 3.25$, all other methods including the embedded formulas show no order reduction.

In practical applications, all methods are used with variable time step control. Therefore, the numerical results obtained with step size control for the parabolic and hyperbolic problem are shown in Fig. 1. For different tolerances $atol = rtol =$

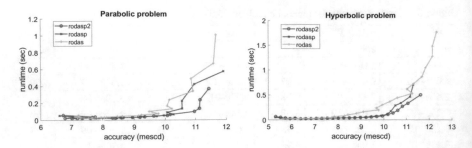

Fig. 1 Numerical results (runtime versus accuracy) for parabolic problem (5.3) and hyperbolic problem (5.4)

$10^{-(4+\frac{m}{4})}$, $m = 0, \ldots, m_{max}$ the achieved accuracy and the CPU runtime is plotted. For the parabolic problem (5.3) $m_{max} = 24$ and for the hyperbolic problem (5.4) $m_{max} = 26$ is used. The accuracy is measured as the mixed error significant digits value (mescd):

$$mescd = -\log_{10}\left(rtol \cdot \max_i \frac{|y_i^{true} - y_i^{num}|}{atol + rtol \cdot |y_i^{true}|} \right)_{t=t_{end}},$$

where y_i^{true} denotes the analytical solution $u(x_i, t_{end})$ and y_i^{num} the corresponding numerical solution. The mescd value should reflect the value q of the tolerance $rtol = 10^{-q}$, see [6].

Due to the order reduction of rodas the differences between the standard and embedded method are small, which leads to difficulties in the step size control. This results in the increased computing times of rodas compared to the other methods shown in Fig. 1. Although the results of rodasp and rodasp2 are very similar in Tables 6 and 7, the new method rodasp2 is slightly more efficient.

With ROW methods good results could be achieved in the network simulation, see [19]. Therefore the next numerical tests refer to such problems. First, the water tube system described in [6] is considered. It is a an index-2 system of 49 non-linear DAEs of type (4.1). In Fig. 2, the effects of the optimized implementation described in Sect. 4 are first considered. The results of the MATLAB standard integrator ode15s are compared with those of rodasp with and without optimizations. The optimizations refers to the vectorization of the right side $f(t, y)$ of the DAE System. In addition, 13 columns of the Jacobian matrix of function $f(t, y)$ are linear and need to be evaluated only once. It can be shown that these optimization measures lead to a significant increase in performance. The comparison between the different rodas methods in Fig. 2 shows no significant differences here.

Next, an electrical circuit is considered. The two transistor amplifier was originally treated in [12] and is also part of the test set [6]. It is a system of eight stiff DAEs of index-one of type (4.1). Consistent initial values must be carefully determined. In addition, large nonlinearities occur due to the modelling

Table 7 Numerical results for hyperbolic problem (5.4)

	rodasp2		rodasp		rodas	
h	err	p	err	p	err	p
1.00e−01	8.16e−07		8.25e−07		1.45e−06	
5.00e−02	5.03e−08	4.02	5.05e−08	4.03	7.04e−08	4.37
2.50e−02	3.12e−09	4.01	3.12e−09	4.02	5.59e−09	3.65
1.25e−02	1.94e−10	4.01	1.94e−10	4.01	6.43e−10	3.12
6.25e−03	1.21e−11	4.00	1.21e−11	4.00	7.45e−11	3.11
3.13e−03	7.57e−13	4.00	7.57e−13	4.00	8.11e−12	3.20
1.56e−03	4.74e−14	4.00	4.72e−14	4.00	8.65e−13	3.23
7.81e−04	2.82e−15	4.07	2.81e−15	4.07	8.76e−14	3.30
	Embedded methods					
1.00e−01	8.27e−06		8.27e−06		2.22e−05	
5.00e−02	1.11e−06	2.90	1.11e−06	2.90	2.83e−06	2.97
2.50e−02	1.44e−07	2.94	1.44e−07	2.94	3.63e−07	2.97
1.25e−02	1.84e−08	2.97	1.84e−08	2.97	4.68e−08	2.95
6.25e−03	2.33e−09	2.98	2.33e−09	2.98	6.10e−09	2.94
3.13e−03	2.92e−10	2.99	2.92e−10	2.99	8.02e−10	2.93
1.56e−03	3.67e−11	3.00	3.67e−11	3.00	1.06e−10	2.92
7.81e−04	4.59e−12	3.00	4.59e−12	3.00	1.40e−11	2.92

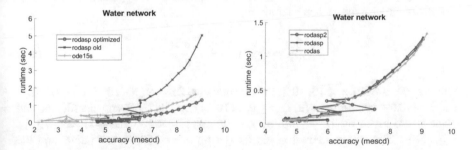

Fig. 2 Numerical results (runtime versus accuracy) for water tube system described in [6]. Left: Comparison of `ode15s` and `rodasp` with or without optimized implementation. Right: Comparison of the different `rodas` methods

of the transistors. The results in Fig. 3 show that `rodasp2` performs better than `rodas` and `rodasp`. Using the analytical Jacobian matrix, the accuracy of `rodas` is slightly higher compared to `rodasp2`. However, if the numerically calculated Jacobian matrix is used, this accuracy advantage disappears.

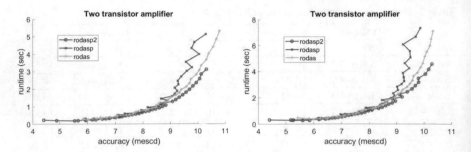

Fig. 3 Numerical results (runtime versus accuracy) for two transistor amplifier described in [6, 12]. Left: Analytical Jacobian matrix. Right: Numerically computed Jacobian matrix

Table 8 Numerical efforts for convection diffusion problem (5.5). NSUCC = number of successful time steps, NFAIL = number of failed steps, NFCN = number of funciton evaluations, CPU = CPU time

	NSUCC	NFAIL	NFCN	CPU
ode15s	1672	18	3279	1.7
rodasp2	413	10	2941	1.7

Finally, the new method `rodasp2` is compared again with `ode15s`. A convection diffusion problem is considered:

$$\frac{\partial c}{\partial t} = -u(t)\frac{\partial c}{\partial x} + D\frac{\partial^2 c}{\partial x^2} \tag{5.5}$$

with $t \in (0, 10800]$, $x \in [0, 10^5]$, initial value $c(x, 0) = \exp(-10^{-6}(x - 10000)^2)$ and boundary conditions $c(0, t) = 0$, $c(10^5, t) = 0$. Diffusion coefficient and velocity are given by $D = 100$, $u(t) = 15$ for $t < 5400$ and $u(t) = -15$ for $t \geq 5400$. The space interval is discretized by $n = 1000$ space points and the second order derivative is approximated by standard second order central finite differences. The convection term was discretized with a WENO scheme of fifth order. The ODE system resulting from the semidiscretization in space was solved with the standard tolerances $rtol = atol = 10^{-4}$. In Table 8 the numerical efforts for both methods `ode15s` and `rodasp2` are given. It turns out that both methods are similarly efficient in terms of computing times. However, if one compares the quality of the solution in Fig. 4, `rodasp2` gives better results.

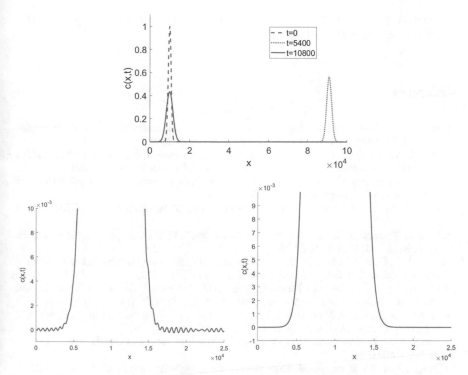

Fig. 4 Numerical results for convection diffusion problem. Above: Solution at different times. Below: Zoom into solutions obtained at $t = 10800$, left with ode15s, right with rodasp2

6 Conclusion

A disadvantage of Rosenbrock-Wanner methods for solving index-one DAEs is that the Jacobian matrix has to be re-evaluated in every time step. Jax [2] was able to derive new order conditions for Rosenbrock-Wanner methods with the help of the Butcher tree theory, so that this disadvantage is mostly eliminated. Only the part $(g_z)_0$ in equation (1.4) has to be updated.

In this paper, the rodasp method could be modified on the basis of Jax's new conditions. If the exact Jacobian matrix is used in each time step, the new method rodasp2 is still a 4th-order scheme with all the properties of rodasp. In the case of the inexact Jacobian matrix, the order reduction could be limited to $p = 2$ instead of $p = 1$. Various numerical tests have shown that the new method is efficient. By specific additional measures for the efficient evaluation of the Jacobian matrix, further computing time reductions can be achieved. These measures are e.g. vectorization or the avoidance of repeated evaluation of linear components. Thus, the new rodasp2 method can be an alternative to the MATLAB standard integrators.

Example programs and the new method rodasp2 are given as supplementary material or can be requested from the author.

References

1. Hairer, E., Wanner, G.: Solving Ordinary Differential Equations II, Stiff and differential algebraic Problems, 2nd edn. Springer, Berlin, Heidelberg (1996)
2. Jax, T.: A rooted-tree based derivation of ROW-type methods with arbitrary jacobian entries for solving index-one DAEs. Dissertation, Bergische Universität Wuppertal (2019)
3. Jax, T., Steinebach, G.: Generalized ROW-type methods for solving semi-explicit DAEs of index-1. J. Comput. Appl. Math. **316**, 213–228 (2017)
4. Lang, J.: Rosenbrock-Wanner Methods: Construction and Mission, Preprint, TU Darmstadt (2019)
5. Lang, J., Teleaga, D.: Towards a fully space-time adaptive FEM for magnetoquasistatics. IEEE Trans. Magn. **44**, 1238–1241 (2008)
6. Mazzia, F., Cash, J.R., Soetaert, K.: A test set for stiff initial value problem solvers in the open source software R. J. Comput. Appl. Math. **236**, 4119–4131 (2012)
7. Ostermann, A., Roche, M.: Runge-Kutta methods for partial differential equations and fractional orders of convergence. Math. Comput. **59**, 403–420 (1992)
8. Ostermann, A., Roche, M.: Rosenbrock methods for partial differential equations and fractional orders of convergence. SIAM J. Numer. Anal. **30**, 1084–1098 (1993)
9. Prothero, A., Robinson, A.: The stability and accuracy of one-step methods. Math. Comput. **28**, 145–162 (1974)
10. Rang, J.: Improved traditional Rosenbrock-Wanner methods for stiff ODEs and DAEs. J. Comput. Appl. Math. **286**, 128–144 (2015)
11. Rang, J.: The Prothero and Robinson example: Convergence studies for Runge–Kutta and Rosenbrock–Wanner methods. Appl. Numer. Math. **108**, 37–56 (2016)
12. Rentrop, P., Roche, M. Steinebach, G.: The application of Rosenbrock-Wanner type methods with stepsize control in differential-algebraic equations. Numer. Math. **55**, 545–563 (1989)
13. Roche, M.: Rosenbrock methods for differential algebraic equations. Numer. Math. **52**, 45–63 (1988)
14. Sanz-Serna, J.M., Verwer, J.G., Hundsdorfer, W.H.: Convergence and order reduction of Runge-Kutta schemes applied to evolutionary problems in partial differential eqautions. Numer. Math. **50**, 405–418 (1986)
15. Scholz, S.: Order barriers for the B-convergence of ROW methods. Computing **41**, 219–235 (1989)
16. Shampine, L.E., Gladwell, I., Thompson, S.: ODEs with MATLAB. Cambridge University Press, Cambridge (2003)
17. Steihaug, T., Wolfbrandt, A.: An attempt to avoid exact Jacobian and nonlinear equations in the numerical solution of stiff differential equations. Math. Comput. **33**, 521–534 (1979)
18. Steinebach, G.: Order-reduction of ROW-methods for DAEs and method of lines applications. Preprint-Nr. 1741. FB Mathematik, TH Darmstadt (1995)
19. Steinebach, G.: From river Rhine alarm model to water supply network simulation by the method of lines. In: Russo, G., Capasso, V., Nicosia, G., Romano, V. (eds.) Progress in Industrial Mathematics at ECMI 2014, pp. 783–792. Springer International Publishing (2016)
20. Steinebach, G., Rentrop, P.: An adaptive method of lines approach for modelling flow and transport in rivers. In: Vande Wouver, A., Sauces, Ph., Schiesser, W.E. (eds.) Adaptive Method of Lines, pp. 181–205. Chapman & Hall/CRC (2001)

Data-Driven Model Reduction for a Class of Semi-Explicit DAEs Using the Loewner Framework

Athanasios C. Antoulas, Ion Victor Gosea, and Matthias Heinkenschloss

Abstract This paper introduces a modified version of the recently proposed data-driven Loewner framework to compute reduced order models (ROMs) for a class of semi-explicit differential algebraic equation (DAE) systems, which include the semi-discretized linearized Navier–Stokes/Oseen equations. The modified version estimates the polynomial part of the original transfer function from data and incorporates this estimate into the Loewner ROM construction. Without this proposed modification the transfer function of the Loewner ROM is strictly proper, i.e., goes to zero as the magnitude of the frequency goes to infinity, and therefore may have a different behavior for large frequencies than the transfer function of the original system. The modification leads to a Loewner ROM with a transfer function that has a strictly proper and a polynomial part, just as the original model. This leads to better approximations for transfer function components in which the coefficients in the polynomial part are not too small. The construction of the improved Loewner ROM is described and the improvement is demonstrated on a large-scale system governed by the semi-discretized Oseen equations.

Keywords Model reduction · Loewner framework · Rational interpolation · Transfer function · Semi-explicit DAE · Oseen equations

A. C. Antoulas
Department of Electrical an Computer Engineering, Rice University, Houston, TX, USA

Max Planck Institute for Dynamics of Complex Technical Systems, Magdeburg, Germany
e-mail: aca@rice.edu

I. V. Gosea (✉)
Max Planck Institute for Dynamics of Complex Technical Systems, Magdeburg, Germany
e-mail: gosea@mpi-magdeburg.mpg.de

M. Heinkenschloss
Department of Computational and Applied Mathematics, Rice University, Houston, TX, USA
e-mail: heinken@rice.edu

© The Editor(s) (if applicable) and The Author(s), under exclusive licence   185
to Springer Nature Switzerland AG 2020
T. Reis et al. (eds.), *Progress in Differential-Algebraic Equations II*,
Differential-Algebraic Equations Forum,
https://doi.org/10.1007/978-3-030-53905-4_7

Mathematics Subject Classification (2010) 37M99, 37N10, 41A20, 93A15, 93B15, 93C15

1 Introduction

This paper introduces a modified version of the data-driven Loewner framework to compute reduced order models (ROMs) for a class of semi-explicit differential algebraic equation (DAE) systems, which includes systems arising from semi-discretized linearized Navier–Stokes/Oseen equations . The improvement is in the estimation of the polynomial part of the transfer function from measurements and in the incorporation of this estimate into the Loewner ROM construction, which in many cases leads to ROMs with better approximation properties.

Most ROM approaches first compute subspaces that contain the important dynamics of the system and then generate a ROM by applying a Galerkin or Petrov–Galerkin projection of the original full order model (FOM) onto these subspaces. These projection based ROM approaches include balanced truncation, interpolation based methods, proper orthogonal decomposition, reduced basis methods, and others. See, e.g., the books [1, 3, 5, 9, 12]. All of these ROM approaches require explicit access to the system matrices to apply the projection and generate the ROM. In contrast, the Loewner framework computes a ROM directly from measurements of the transfer function and does not require explicit knowledge of the system matrices. Thus, the Loewner framework can be applied even if the mathematical model of the system is not available, e.g., because proprietary software is used or measurements are generated directly from the physical system. The Loewner framework is described, e.g., in the book [3, Chapter 4] and in the recent survey [2].

The Loewner framework computes a ROM directly from transfer function measurements in such a way that the ROM transfer function approximately interpolates the transfer function of the original FOM at the measurements. However, the Loewner ROM generated with the original approach has a strictly proper transfer function. In particular, the ROM transfer function goes to zero as the magnitude of the frequency goes to infinity. In contrast, the transfer function of the original model may have a polynomial part which is bounded away from zero, or is even unbounded as the magnitude of the frequency goes to infinity. In this case, this substantially different behavior of transfer functions generates substantial differences away from the measurements, which means that the ROM may not capture important features of the original problem. As mentioned before, this paper shows how to estimate the polynomial part from transfer function measurements and how to incorporate these estimates into the Loewner ROM construction to generate better ROMs. In principle, there is no difference between the computation of a Loewner ROM for an ordinary differential equation (ODE) system and for a DAE system. However, for ODE systems the structure of the ODE system allows one to directly identify the polynomial part, especially assessing whether it is non-zero. Unfortunately, this is more involved for DAE systems. For theoretical purposes we derive the analytical forms of the strictly proper and polynomial parts of the transfer function for our class of semi-explicit DAE systems. If available, the analytical form of the polynomial part of the transfer function could be used. However this requires access to the system matrices. As an alternative, we propose to estimate the polynomial part of the transfer function from measurements. We then show how to incorporate this estimate into the Loewner ROM construction to generate better ROMs. This paper specifically focuses on the structure of semi-explicit DAE systems arising, e.g., from semi-discretized Oseen equations and complements [7].

The class of semi-explicit DAE systems is given by

$$\mathbf{E}_{11}\frac{d}{dt}\mathbf{v}(t) = \mathbf{A}_{11}\mathbf{v}(t) + \mathbf{A}_{12}\mathbf{p}(t) + \mathbf{B}_{1,0}\mathbf{g}(t) + \mathbf{B}_{1,1}\frac{d}{dt}\mathbf{g}(t), \qquad t \in (0, T),$$
(1.1a)

$$\mathbf{0} = \mathbf{A}_{12}^T\mathbf{v}(t) + \mathbf{B}_{2,0}\mathbf{g}(t), \qquad t \in (0, T),$$
(1.1b)

$$\mathbf{v}(0) = \mathbf{0},$$
(1.1c)

$$\mathbf{y}(t) = \mathbf{C}_1\mathbf{v}(t) + \mathbf{C}_2\mathbf{p}(t) + \mathbf{D}_0\mathbf{g}(t) + \mathbf{D}_1\frac{d}{dt}\mathbf{g}(t) \qquad t \in (0, T).$$
(1.1d)

Here $\mathbf{v}$, $\mathbf{p}$ are the states (velocities and pressures in the Oseen system), $\mathbf{g}$ are the inputs, and $\mathbf{y}$ are the outputs. The matrix $\mathbf{E}_{11} \in \mathbb{R}^{n_v \times n_v}$ is symmetric positive definite, $\mathbf{A}_{11} \in \mathbb{R}^{n_v \times n_v}$, $\mathbf{A}_{12}^T \in \mathbb{R}^{n_p \times n_v}$, $n_p < n_v$, is a matrix with rank n_p, $\mathbf{B}_{1,0}, \mathbf{B}_{1,1} \in \mathbb{R}^{n_v \times n_g}$, $\mathbf{B}_{2,0} \in \mathbb{R}^{n_p \times n_g}$, $\mathbf{C}_1 \in \mathbb{R}^{n_y \times n_v}$, $\mathbf{C}_2 \in \mathbb{R}^{n_y \times n_p}$, and $\mathbf{D}_0, \mathbf{D}_1 \in \mathbb{R}^{n_y \times n_g}$. See, e.g., the books [6, 10]. Derivatives $\frac{d}{dt}\mathbf{g}$ of the inputs appear in the semi-discretized equations, e.g., when inputs on the partial differential equation (PDE) level are given as Dirichlet conditions on the velocities (e.g., the input corresponds to suction/blowing actuation on the boundary).

Often it will be convenient to define $n = n_v + n_p$,

$$\mathbf{x}(t) = \begin{pmatrix} \mathbf{v}(t) \\ \mathbf{p}(t) \end{pmatrix}, \qquad \mathbf{E} = \begin{pmatrix} \mathbf{E}_{11} & \mathbf{0} \\ \mathbf{0} & \mathbf{0} \end{pmatrix}, \qquad \mathbf{A} = \begin{pmatrix} \mathbf{A}_{11} & \mathbf{A}_{12} \\ \mathbf{A}_{12}^T & \mathbf{0} \end{pmatrix},$$
(1.2a)

$$\mathbf{B}_0 = \begin{pmatrix} \mathbf{B}_{1,0} \\ \mathbf{B}_{2,0} \end{pmatrix}, \qquad \mathbf{B}_1 = \begin{pmatrix} \mathbf{B}_{1,1} \\ \mathbf{0} \end{pmatrix}, \qquad \mathbf{C} = \begin{pmatrix} \mathbf{C}_1 & \mathbf{C}_2 \end{pmatrix},$$
(1.2b)

and write (1.1) in the compact notation

$$\mathbf{E}\frac{d}{dt}\mathbf{x}(t) = \mathbf{A}\mathbf{x}(t) + \mathbf{B}_0\mathbf{g}(t) + \mathbf{B}_1\frac{d}{dt}\mathbf{g}(t), \qquad t \in (0, T),$$
(1.3a)

$$\mathbf{E}\mathbf{x}(0) = \mathbf{0},$$
(1.3b)

$$\mathbf{y}(t) = \mathbf{C}\mathbf{x}(t) + \mathbf{D}_0\mathbf{g}(t) + \mathbf{D}_1\frac{d}{dt}\mathbf{g}(t), \qquad t \in (0, T).$$
(1.3c)

This paper is organized as follows. In the next Sect. 2 we derive the analytical representations of the strictly proper and polynomial parts of the transfer function. Section 3 reviews the Loewner approach. Our approach for estimating the polynomial part of the transfer function from data is introduced in Sect. 4. Section 5 applies the Loewner approach with identification of the polynomial part of the transfer function to the Oseen equation.

2   Transfer Function

As mentioned before, the Loewner framework constructs a ROM such that its transfer function approximates the transfer function of the FOM. The transfer function $\mathbf{H}(s)$ of the FOM additively splits into a so-called strictly proper part $\mathbf{H}_{\mathrm{spr}}(s)$, which is a rational function in s with $\|\mathbf{H}_{\mathrm{spr}}(s)\| \rightarrow 0$ as $|s| \rightarrow \infty$, and a polynomial part $\mathbf{H}_{\mathrm{poly}}(s)$. Depending on the transfer function measurements available it can be difficult to obtain a good approximation of the combined transfer function

$$\mathbf{H}(s) = \mathbf{C}\big(s\,\mathbf{E} - \mathbf{A}\big)^{-1}(\mathbf{B}_0 + s\,\mathbf{B}_1) + \mathbf{D}_0 + s\,\mathbf{D}_1 \tag{2.1}$$

associated with (1.3), and in these cases a separate approximation of the strictly proper and of the polynomial part can yield much better results. This section computes $\mathbf{H}_{\mathrm{spr}}(s)$ and $\mathbf{H}_{\mathrm{poly}}(s)$.

2.1   Transfer Function of an ODE System

First consider (1.3) with an invertible matrix $\mathbf{E}$, i.e., consider an ODE system. Since

$$\begin{aligned}
\big(s\,\mathbf{E} - \mathbf{A}\big)^{-1}(\mathbf{B}_0 + s\,\mathbf{B}_1) &= \big(s\,\mathbf{E} - \mathbf{A}\big)^{-1}(\mathbf{B}_0 + \mathbf{A}\mathbf{E}^{-1}\mathbf{B}_1 + (s\,\mathbf{E} - \mathbf{A})\,\mathbf{E}^{-1}\mathbf{B}_1) \\
&= \big(s\,\mathbf{E} - \mathbf{A}\big)^{-1}(\mathbf{B}_0 + \mathbf{A}\mathbf{E}^{-1}\mathbf{B}_1) + \mathbf{E}^{-1}\mathbf{B}_1,
\end{aligned}$$

the transfer function (2.1) can be written as

$$\mathbf{H}(s) = \underbrace{\mathbf{C}\big(s\,\mathbf{E} - \mathbf{A}\big)^{-1}(\mathbf{B}_0 + \mathbf{A}\mathbf{E}^{-1}\mathbf{B}_1)}_{=\mathbf{H}_{\mathrm{spr}}(s)} + \underbrace{\mathbf{C}\mathbf{E}^{-1}\mathbf{B}_1 + \mathbf{D}_0 + s\,\mathbf{D}_1}_{\mathbf{H}_{\mathrm{poly}}(s)}\,.$$

If $\mathbf{E}$ is invertible, the strictly proper part and the polynomial part of the transfer function can be determined directly from the matrices in (1.3). Specifically, the polynomial part is at most linear,

$$\mathbf{H}_{\text{poly}}(s) = \mathbf{P}_0 + s\,\mathbf{P}_1 \quad \text{with} \quad \mathbf{P}_0 = \mathbf{C}\mathbf{E}^{-1}\mathbf{B}_1 + \mathbf{D}_0, \quad \mathbf{P}_1 = \mathbf{D}_1,$$

and the polynomial part is zero if $\mathbf{B}_1, \mathbf{D}_0, \mathbf{D}_1$ are zero.

2.2 Transfer Function of the Semi-Explicit DAE System

Now consider (1.1). Because the corresponding $\mathbf{E}$ in (1.2) is singular, the representation (2.1) does not directly expose the strictly proper part and the polynomial part of the transfer function. We proceed as in [8] and transform (1.1) into an ODE system.

We write

$$\mathbf{v}(t) = \mathbf{v}_0(t) + \mathbf{v}_g(t), \tag{2.2}$$

where

$$\mathbf{v}_g(t) = -\mathbf{E}_{11}^{-1}\mathbf{A}_{12}(\mathbf{A}_{12}^T\mathbf{E}_{11}^{-1}\mathbf{A}_{12})^{-1}\mathbf{B}_{2,0}\mathbf{g}(t) \tag{2.3}$$

is a particular solution of (1.1b) and $\mathbf{v}_0(t)$ satisfies $\mathbf{0} = \mathbf{A}_{12}^T\mathbf{v}_0(t)$. Furthermore, we define the projection

$$\boldsymbol{\Pi} = \mathbf{I} - \mathbf{A}_{12}(\mathbf{A}_{12}^T\mathbf{E}_{11}^{-1}\mathbf{A}_{12})^{-1}\mathbf{A}_{12}^T\mathbf{E}_{11}^{-1}.$$

It can be verified that $\boldsymbol{\Pi}^2 = \boldsymbol{\Pi}$, $\boldsymbol{\Pi}\mathbf{E}_{11} = \mathbf{E}_{11}\boldsymbol{\Pi}^T$, $\text{null}(\boldsymbol{\Pi}) = \text{range}(\mathbf{A}_{12})$ and $\text{range}(\boldsymbol{\Pi}) = \text{null}(\mathbf{A}_{12}^T\mathbf{E}_{11}^{-1})$, i.e., $\boldsymbol{\Pi}$ is an $\mathbf{E}_{11}$-orthogonal projection. For (1.1) derived from a finite element discretization, $\boldsymbol{\Pi}$ is a discrete version of the Leray projector [4]. The properties of $\boldsymbol{\Pi}$ imply that

$$\mathbf{A}_{12}^T\mathbf{v}_0(t) = \mathbf{0} \quad \text{if and only if} \quad \boldsymbol{\Pi}^T\mathbf{v}_0(t) = \mathbf{v}_0(t). \tag{2.4}$$

Inserting (2.2), (2.3) into (1.1) gives

$$\mathbf{E}_{11}\frac{d}{dt}\mathbf{v}_0(t) = \mathbf{A}_{11}\mathbf{v}_0(t) + \mathbf{A}_{12}\mathbf{p}(t) + \mathbf{B}_3\mathbf{g}(t)$$

$$+ \left(\mathbf{B}_{1,1} + \mathbf{A}_{12}(\mathbf{A}_{12}^T\mathbf{E}_{11}^{-1}\mathbf{A}_{12})^{-1}\mathbf{B}_{2,0}\right)\frac{d}{dt}\mathbf{g}(t) \tag{2.5a}$$

$$\mathbf{0} = \mathbf{A}_{12}^T\mathbf{v}_0(t), \tag{2.5b}$$

$$\mathbf{v}_0(0) = -\mathbf{v}_g(0), \tag{2.5c}$$

$$\mathbf{y}(t) = \mathbf{C}_1\mathbf{v}_0(t) + \mathbf{C}_2\mathbf{p}(t) + \left(\mathbf{D}_0 - \mathbf{C}_1\mathbf{E}_{11}^{-1}\mathbf{A}_{12}(\mathbf{A}_{12}^T\mathbf{E}_{11}^{-1}\mathbf{A}_{12})^{-1}\mathbf{B}_{2,0}\right)\mathbf{g}(t)$$

$$+ \mathbf{D}_1\frac{d}{dt}\mathbf{g}(t), \tag{2.5d}$$

where

$$\mathbf{B}_3 := \mathbf{B}_{1,0} - \mathbf{A}_{11}\mathbf{E}_{11}^{-1}\mathbf{A}_{12}(\mathbf{A}_{12}^T\mathbf{E}_{11}^{-1}\mathbf{A}_{12})^{-1}\mathbf{B}_{2,0}. \tag{2.6}$$

Next we express $\mathbf{p}$ in terms of $\mathbf{v}_0$ and project onto the constraint (2.5b). Specifically, we multiply (2.5a) by $\mathbf{A}_{12}^T\mathbf{E}_{11}^{-1}$, then use (2.5b) and finally solve the resulting equation for $\mathbf{p}$ to get

$$\mathbf{p}(t) = -(\mathbf{A}_{12}^T\mathbf{E}_{11}^{-1}\mathbf{A}_{12})^{-1}\mathbf{A}_{12}^T\mathbf{E}_{11}^{-1}\mathbf{A}_{11}\mathbf{v}_0(t)$$

$$- (\mathbf{A}_{12}^T\mathbf{E}_{11}^{-1}\mathbf{A}_{12})^{-1}\mathbf{A}_{12}^T\mathbf{E}_{11}^{-1}\mathbf{B}_3\,\mathbf{g}(t)$$

$$- (\mathbf{A}_{12}^T\mathbf{E}_{11}^{-1}\mathbf{A}_{12})^{-1}\left(\mathbf{A}_{12}^T\mathbf{E}_{11}^{-1}\mathbf{B}_{1,1} + \mathbf{B}_{2,0}\right)\frac{d}{dt}\mathbf{g}(t). \tag{2.7}$$

Now we insert (2.7) into (2.5d), apply (2.4), and use $\boldsymbol{\Pi}\mathbf{A}_{12}(\mathbf{A}_{12}^T\mathbf{E}_{11}^{-1}\mathbf{A}_{12})^{-1} = \mathbf{0}$ to write (2.5) as

$$\boldsymbol{\Pi}\mathbf{E}_{11}\boldsymbol{\Pi}^T\frac{d}{dt}\mathbf{v}_0(t) = \boldsymbol{\Pi}\mathbf{A}_{11}\boldsymbol{\Pi}^T\mathbf{v}_0(t) + \boldsymbol{\Pi}\mathbf{B}_3\mathbf{g}(t) + \boldsymbol{\Pi}\mathbf{B}_{1,1}\frac{d}{dt}\mathbf{g}(t), \quad t \in (0, T), \tag{2.8a}$$

$$\boldsymbol{\Pi}^T\mathbf{v}_0(0) = -\boldsymbol{\Pi}^T\mathbf{v}_g(0), \tag{2.8b}$$

$$\mathbf{y}(t) = \mathbf{C}_3\boldsymbol{\Pi}^T\mathbf{v}_0(t) + \widetilde{\mathbf{P}}_0\,\mathbf{g}(t) + \mathbf{P}_1\frac{d}{dt}\mathbf{g}(t), \qquad t \in (0, T), \tag{2.8c}$$

where $\mathbf{B}_3$ is given by (2.6) and

$$\mathbf{C}_3 := \mathbf{C}_1 - \mathbf{C}_2(\mathbf{A}_{12}^T\mathbf{E}_{11}^{-1}\mathbf{A}_{12})^{-1}\mathbf{A}_{12}^T\mathbf{E}_{11}^{-1}\mathbf{A}_{11}, \tag{2.9a}$$

$$\widetilde{\mathbf{P}}_0 := \mathbf{D}_0 - \mathbf{C}_1\mathbf{E}_{11}^{-1}\mathbf{A}_{12}(\mathbf{A}_{12}^T\mathbf{E}_{11}^{-1}\mathbf{A}_{12})^{-1}\mathbf{B}_{2,0}$$

$$- \mathbf{C}_2(\mathbf{A}_{12}^T\mathbf{E}_{11}^{-1}\mathbf{A}_{12})^{-1}\mathbf{A}_{12}^T\mathbf{E}_{11}^{-1}\mathbf{B}_3, \tag{2.9b}$$

$$\mathbf{P}_1 := \mathbf{D}_1 - \mathbf{C}_2(\mathbf{A}_{12}^T\mathbf{E}_{11}^{-1}\mathbf{A}_{12})^{-1}\left(\mathbf{A}_{12}^T\mathbf{E}_{11}^{-1}\mathbf{B}_{1,1} + \mathbf{B}_{2,0}\right). \tag{2.9c}$$

The system (2.8) is a dynamical system in the $n_v - n_p$ dimensional subspace null($\boldsymbol{\Pi}$) and (2.8a,b) has to be solved for $\boldsymbol{\Pi}^T \mathbf{v} = \mathbf{v}$. This can be made more explicit by decomposing

$$\boldsymbol{\Pi} = \boldsymbol{\Theta}_l \boldsymbol{\Theta}_r^T \tag{2.10a}$$

with $\boldsymbol{\Theta}_l, \boldsymbol{\Theta}_r \in \mathbb{R}^{n_v \times (n_v - n_p)}$ satisfying

$$\boldsymbol{\Theta}_l^T \boldsymbol{\Theta}_r = \mathbf{I}. \tag{2.10b}$$

Substituting this decomposition into (2.8) shows that $\tilde{\mathbf{v}}_0 = \boldsymbol{\Theta}_l^T \mathbf{v}_0 \in \mathbb{R}^{n_v - n_p}$ must satisfy

$$\boldsymbol{\Theta}_r^T \mathbf{E}_{11} \boldsymbol{\Theta}_r \frac{d}{dt} \tilde{\mathbf{v}}_0(t) = \boldsymbol{\Theta}_r^T \mathbf{A}_{11} \boldsymbol{\Theta}_r \tilde{\mathbf{v}}_0(t)$$

$$+ \boldsymbol{\Theta}_r^T \mathbf{B}_3 \mathbf{g}(t) + \boldsymbol{\Theta}_r^T \mathbf{B}_{1,1} \frac{d}{dt} \mathbf{g}(t), \quad t \in (0, T), \tag{2.11a}$$

$$\tilde{\mathbf{v}}_0(0) = -\boldsymbol{\Theta}_l^T \mathbf{v}_g(0), \tag{2.11b}$$

$$\mathbf{y}(t) = \mathbf{C}_3 \boldsymbol{\Theta}_r \tilde{\mathbf{v}}_0(t) + \tilde{\mathbf{P}}_0 \mathbf{g}(t) + \mathbf{P}_1 \frac{d}{dt} \mathbf{g}(t), \quad t \in (0, T). \tag{2.11c}$$

The systems (1.1) and (2.11) are equivalent. Again we refer to [8] for details. Specifically, the transfer function of (1.1) is identical to the transfer function of (2.11). Since the $(n_v - n_p) \times (n_v - n_p)$ matrix $\boldsymbol{\Theta}_r^T \mathbf{E}_{11} \boldsymbol{\Theta}_r$ has full rank, we can proceed as in Sect. 2.1 to read off the strictly proper part and the polynomial part of the transfer function from the system representation (2.11),

$$\mathbf{H}(s) = \mathbf{H}_{\mathrm{spr}}(s) + \mathbf{H}_{\mathrm{poly}}(s), \tag{2.12a}$$

where

$$\mathbf{H}_{\mathrm{spr}}(s) = \mathbf{C}_3 \boldsymbol{\Theta}_r \left(s \, \boldsymbol{\Theta}_r^T \mathbf{E}_{11} \boldsymbol{\Theta}_r - \boldsymbol{\Theta}_r^T \mathbf{A}_{11} \boldsymbol{\Theta}_r \right)^{-1}$$

$$\times \left(\boldsymbol{\Theta}_r^T \mathbf{B}_3 + \boldsymbol{\Theta}_r^T \mathbf{A}_{11} \boldsymbol{\Theta}_r (\boldsymbol{\Theta}_r^T \mathbf{E}_{11} \boldsymbol{\Theta}_r)^{-1} \boldsymbol{\Theta}_r^T \mathbf{B}_{1,1} \right), \tag{2.12b}$$

$$\mathbf{H}_{\mathrm{poly}}(s) = \underbrace{\mathbf{C}_3 \boldsymbol{\Theta}_r (\boldsymbol{\Theta}_r^T \mathbf{E}_{11} \boldsymbol{\Theta}_r)^{-1} \boldsymbol{\Theta}_r^T \mathbf{B}_{1,1} + \tilde{\mathbf{P}}_0}_{=\mathbf{P}_0} + s \, \mathbf{P}_1. \tag{2.12c}$$

Thus the polynomial part of the transfer function of (1.1) is again at most linear, but the matrices $\mathbf{P}_0$ and $\mathbf{P}_1$ are more involved.

If the system matrices $\mathbf{E}_{11}, \ldots$ in (1.1) are available then the matrices in (2.6) and (2.9) and the matrices $\mathbf{P}_0$ and $\mathbf{P}_1$ in (2.12c) can be computed using results already applied in [8]. We summarize these results next. However, if one does not have access to the system matrices $\mathbf{E}_{11}, \ldots$ one needs to estimate the polynomial parts $\mathbf{P}_0$ and $\mathbf{P}_1$ from transfer function measurements, as we will describe in Sect. 4.

2.3 Computational Details

If

$$\begin{pmatrix} \mathbf{E}_{11} & \mathbf{A}_{12} \\ \mathbf{A}_{12}^T & \mathbf{0} \end{pmatrix} \begin{pmatrix} \mathbf{X}_1 \\ \mathbf{Z}_1 \end{pmatrix} = \begin{pmatrix} \mathbf{0} \\ \mathbf{B}_{2,0} \end{pmatrix}, \quad \begin{pmatrix} \mathbf{E}_{11} & \mathbf{A}_{12} \\ \mathbf{A}_{12}^T & \mathbf{0} \end{pmatrix} \begin{pmatrix} \mathbf{X}_2 \\ \mathbf{Z}_2 \end{pmatrix} = \begin{pmatrix} \mathbf{0} \\ \mathbf{C}_2 \end{pmatrix}, \quad (2.13)$$

then $\mathbf{X}_1^T = \mathbf{B}_{2,0}^T (\mathbf{A}_{12}^T \mathbf{E}_{11}^{-1} \mathbf{A}_{12})^{-1} \mathbf{A}_{12}^T \mathbf{E}_{11}^{-1}$, $\mathbf{Z}_1^T = -\mathbf{B}_{2,0}(\mathbf{A}_{12}^T \mathbf{E}_{11}^{-1} \mathbf{A}_{12})^{-1}$, and $\mathbf{X}_2^T = \mathbf{C}_2^T (\mathbf{A}_{12}^T \mathbf{E}_{11}^{-1} \mathbf{A}_{12})^{-1} \mathbf{A}_{12}^T \mathbf{E}_{11}^{-1}$, $\mathbf{Z}_2^T = -\mathbf{C}_2(\mathbf{A}_{12}^T \mathbf{E}_{11}^{-1} \mathbf{A}_{12})^{-1}$. Hence, the matrices in (2.6) and (2.9) can be written as

$$\mathbf{B}_3 = \mathbf{B}_{1,0} - \mathbf{A}_{11}\mathbf{X}_1, \qquad \mathbf{C}_3 = \mathbf{C}_1 - \mathbf{X}_2^T \mathbf{A}_{11},$$

and

$$\widetilde{\mathbf{P}}_0 = \mathbf{D}_0 - \mathbf{C}_1 \mathbf{X}_1 - \mathbf{X}_2^T \mathbf{B}_3, \qquad \mathbf{P}_1 = \mathbf{D}_1 - \mathbf{X}_2^T \mathbf{B}_{1,1} + \mathbf{Z}_2^T \mathbf{B}_{2,0}.$$

If

$$\begin{pmatrix} \mathbf{E}_{11} & \mathbf{A}_{12} \\ \mathbf{A}_{12}^T & \mathbf{0} \end{pmatrix} \begin{pmatrix} \mathbf{X}_3 \\ \mathbf{Z}_3 \end{pmatrix} = \begin{pmatrix} \mathbf{B}_{1,1} \\ \mathbf{0} \end{pmatrix}, \quad (2.14)$$

then $\mathbf{P}_0$ in (2.12c) can be written as

$$\mathbf{P}_0 = \widetilde{\mathbf{P}}_0 + \mathbf{C}_3 \mathbf{X}_3.$$

In fact, $\mathbf{A}_{12}^T \mathbf{X}_3 = \mathbf{0}$ implies $\mathbf{X}_3 = \boldsymbol{\Pi}^T \mathbf{X}_3 = \boldsymbol{\Theta}_r \boldsymbol{\Theta}_l^T \mathbf{X}_3$ by (2.4) and (2.10a). Hence, with $\widetilde{\mathbf{X}}_3 = \boldsymbol{\Theta}_l^T \mathbf{X}_3$ the first block in (2.14) reads $\mathbf{E}_{11} \boldsymbol{\Theta}_r \widetilde{\mathbf{X}}_3 + \mathbf{A}_{12} \mathbf{Z}_3 = \mathbf{B}_{1,1}$. Since $\text{null}(\boldsymbol{\Theta}_r^T) = \text{null}(\boldsymbol{\Pi}) = \text{range}(\mathbf{A}_{12})$, $\boldsymbol{\Theta}_r^T \mathbf{E}_{11} \boldsymbol{\Theta}_r \widetilde{\mathbf{X}}_3 = \boldsymbol{\Theta}_r^T \mathbf{B}_{1,1}$. This gives $\mathbf{C}_3 \boldsymbol{\Theta}_r (\boldsymbol{\Theta}_r^T \mathbf{E}_{11} \boldsymbol{\Theta}_r)^{-1} \boldsymbol{\Theta}_r^T \mathbf{B}_{1,1} = \mathbf{C}_3 \boldsymbol{\Theta}_r \widetilde{\mathbf{X}}_3 = \mathbf{C}_3 \mathbf{X}_3$.

3 Loewner Framework Applied to the Oseen Equations

We review the Loewner framework applied to (1.3). The presentation is standard and follows the recent tutorial paper [2] and book [3, Chapter 4]. In the next Sect. 4 we modify it to better account for the presence of a polynomial part in the transfer function (2.12).

The Loewner framework (LF) is a data-driven model identification and reduction technique that was originally introduced in [11] and was continuously developed, improved and extended to different problems and system classes during the last decade. It is an interpolation-based method that produces ROMs that (approximately) interpolate the transfer function corresponding to the underlying FOM at the given interpolation frequencies. Unlike other interpolation-based methods the LF computes the ROM from measurements of the transfer function rather than by projection of the original system.

Let $m = n_g$ be the number of inputs and $p = n_y$ be the number of outputs, so that $\mathbf{H}(s) \in \mathbb{C}^{p \times m}$. We assume that given frequencies

$$\mu_j, \lambda_j \in \mathbb{C}, \qquad j = 1, \ldots, N, \tag{3.1a}$$

left tangential directions

$$\boldsymbol{\ell}_j \in \mathbb{C}^p, \qquad j = 1, \ldots, N, \tag{3.1b}$$

and right tangential directions

$$\mathbf{r}_j \in \mathbb{C}^m \qquad j = 1, \ldots, N, \tag{3.1c}$$

we have transfer function measurements

$$\mathbf{v}_j^* := \boldsymbol{\ell}_j^* \mathbf{H}(\mu_j) \in \mathbb{C}^{1 \times m}, \qquad \mathbf{w}_j := \mathbf{H}(\lambda_j) \mathbf{r}_j \in \mathbb{C}^{p \times 1}, \qquad j = 1, \ldots, N. \tag{3.1d}$$

We seek a ROM of the form[1]

$$\widehat{\mathbf{E}} \frac{d}{dt} \widehat{\mathbf{x}}(t) = \widehat{\mathbf{A}} \widehat{\mathbf{x}}(t) + \widehat{\mathbf{B}}_0 \mathbf{g}(t) + \widehat{\mathbf{B}}_1 \frac{d}{dt} \mathbf{g}(t), \qquad t \in (0, T), \tag{3.2a}$$

$$\widehat{\mathbf{E}} \widehat{\mathbf{x}}(0) = \mathbf{0}, \tag{3.2b}$$

$$\widehat{\mathbf{y}}(t) = \widehat{\mathbf{C}} \widehat{\mathbf{x}}(t) + \widehat{\mathbf{P}}_0 \mathbf{g}(t) + \widehat{\mathbf{P}}_1 \frac{d}{dt} \mathbf{g}(t), \qquad t \in (0, T), \tag{3.2c}$$

where $\widehat{\mathbf{E}}$ and $\widehat{\mathbf{A}}$ are of size $r \times r$ with small r, $\widehat{\mathbf{B}}_0, \widehat{\mathbf{B}}_0$ have r rows, and $\widehat{\mathbf{C}}$ has r columns, such that the corresponding transfer function $\widehat{H}$ is an approximate tangential interpolant to the original transfer function $\mathbf{H}$, i.e., such that

$$\boldsymbol{\ell}_j^* \widehat{H}(\mu_j) \approx \boldsymbol{\ell}_j^* \mathbf{H}(\mu_j) = \mathbf{v}_j^* \qquad \text{for} \quad j = 1, \ldots, N,$$
$$\widehat{H}(\lambda_j) \mathbf{r}_j \approx \mathbf{H}(\lambda_j) \mathbf{r}_j = \mathbf{w}_j \qquad \text{for} \quad j = 1, \ldots, N. \tag{3.3}$$

---

[1] The matrices $\widehat{\mathbf{E}}$ and $\widehat{\mathbf{A}}$ do not have the block 2×2 structure of $\mathbf{E}$ and $\mathbf{A}$ in (1.2).

Because of the left and right tangential interpolation conditions $\{\mu_j\}_{j=1}^N \subset \mathbb{C}$ are called the left interpolation points, $\{\mathbf{v}_j\}_{j=1}^N \subset \mathbb{C}^m$ are called the left sample values, $\{\boldsymbol{\ell}_j\}_{j=1}^N \subset \mathbb{C}^p$ are called the left tangential directions and $\{\lambda_j\}_{j=1}^N \subset \mathbb{C}$ are called the right interpolation points, $\{\mathbf{w}_j\}_{j=1}^N \subset \mathbb{C}^p$ are called the right sample values, $\{\mathbf{r}_j\}_{j=1}^N \subset \mathbb{C}^m$ are called the right tangential directions.

We assume that the left interpolation points and the right interpolation points are distinct, i.e. that

$$\{\mu_j\}_{j=1}^N \cap \{\lambda_j\}_{j=1}^N = \emptyset.$$

The measured data are arranged into matrix format as follows[2]

$$\mathbf{M} = \mathrm{diag}(\mu_1, \mu_2, \ldots, \mu_N) \in \mathbb{C}^{N \times N}, \quad \boldsymbol{\Lambda} = \mathrm{diag}(\lambda_1, \lambda_2, \ldots, \lambda_N) \in \mathbb{C}^{N \times N},$$

$$\mathbf{L}^* = \begin{bmatrix} \boldsymbol{\ell}_1 & \boldsymbol{\ell}_2 & \cdots & \boldsymbol{\ell}_N \end{bmatrix} \in \mathbb{C}^{p \times N}, \qquad \mathbf{R} = \begin{bmatrix} \mathbf{r}_1 & \mathbf{r}_2 & \cdots & \mathbf{r}_N \end{bmatrix} \in \mathbb{C}^{m \times N}, \qquad (3.4)$$

$$\mathbf{V}^* = \begin{bmatrix} \mathbf{v}_1 & \mathbf{v}_2 & \cdots & \mathbf{v}_N \end{bmatrix} \in \mathbb{C}^{m \times N}, \qquad \mathbf{W} = \begin{bmatrix} \mathbf{w}_1 & \mathbf{w}_2 & \cdots & \mathbf{w}_N \end{bmatrix} \in \mathbb{C}^{p \times N}.$$

The Loewner matrix is given by

$$\mathbb{L} = \begin{bmatrix} \frac{\mathbf{v}_1^* \mathbf{r}_1 - \boldsymbol{\ell}_1^* \mathbf{w}_1}{\mu_1 - \lambda_1} & \cdots & \frac{\mathbf{v}_1^* \mathbf{r}_N - \boldsymbol{\ell}_1^* \mathbf{w}_N}{\mu_1 - \lambda_N} \\ \vdots & \ddots & \vdots \\ \frac{\mathbf{v}_N^* \mathbf{r}_1 - \boldsymbol{\ell}_N^* \mathbf{w}_1}{\mu_N - \lambda_1} & \cdots & \frac{\mathbf{v}_N^* \mathbf{r}_N - \boldsymbol{\ell}_N^* \mathbf{w}_N}{\mu_N - \lambda_N} \end{bmatrix} \in \mathbb{C}^{N \times N}. \qquad (3.5)$$

Using (3.4) it can be verified that the Loewner matrix (3.5) solves the Sylvester equation

$$\mathbf{M}\mathbb{L} - \mathbb{L}\boldsymbol{\Lambda} = \mathbf{V}\mathbf{R} - \mathbf{L}\mathbf{W}.$$

The shifted Loewner matrix is given by

$$\mathbb{L}_s = \begin{bmatrix} \frac{\mathbf{v}_1^T \mathbf{r}_1 \mu_1 - \boldsymbol{\ell}_1^* \mathbf{w}_1 \lambda_1}{\mu_1 - \lambda_1} & \cdots & \frac{\mathbf{v}_1^* \mathbf{r}_N \mu_1 - \boldsymbol{\ell}_1^* \mathbf{w}_N \lambda_N}{\mu_1 - \lambda_N} \\ \vdots & \ddots & \vdots \\ \frac{\mathbf{v}_N^* \mathbf{r}_1 \mu_N - \boldsymbol{\ell}_N^* \mathbf{w}_1 \lambda_1}{\mu_N - \lambda_1} & \cdots & \frac{\mathbf{v}_N^* \mathbf{r}_N \mu_N - \boldsymbol{\ell}_N^* \mathbf{w}_N \lambda_N}{\mu_N - \lambda_N} \end{bmatrix} \in \mathbb{C}^{N \times N}. \qquad (3.6)$$

---

[2]Note that the matrices $\mathbf{V}^* \in \mathbb{C}^{m \times N}$ and $\mathbf{W} \in \mathbb{C}^{p \times N}$ contain transfer function measurements (3.1) and are not projection matrices.

Using (3.4) it can be verified that the shifted Loewner matrix (3.6) solves the Sylvester equation

$$\mathbf{M}\mathbb{L}_s - \mathbb{L}_s \mathbf{\Lambda} = \mathbf{MVR} - \mathbf{LW\Lambda}.$$

If the 'right' amount of data is given,[3] then the ROM computed with the (classical) Loewner method is (3.2) with

$$\widehat{\mathbf{E}} = -\mathbb{L}, \quad \widehat{\mathbf{A}} = -\mathbb{L}_s, \quad \widehat{\mathbf{B}}_0 = \mathbf{V}, \quad \widehat{\mathbf{B}}_1 = \mathbf{0}, \quad \widehat{\mathbf{C}} = \mathbf{W}, \quad \widehat{\mathbf{P}}_0 = \widehat{\mathbf{P}}_1 = \mathbf{0}. \qquad (3.7)$$

The ROM (3.2) with (3.7) is in general complex. However, if the data (3.1) contain also the conjugate complex data ($\{\mu_j\}_{j=1}^N = \{\overline{\mu}_j\}_{j=1}^N$, $\{\lambda_j\}_{j=1}^N = \{\overline{\lambda}_j\}_{j=1}^N$, etc.), then the complex ROM (3.7) can be transformed into a real ROM with the same transfer function, as shown in [2, p. 360]. The transfer function $\widehat{H}$ corresponding to (3.7) satisfies the interpolation conditions (3.3) with equality. However, while it satisfies the interpolation conditions (3.3), the transfer function $\widehat{H}$ by design is strictly proper, $\widehat{H}_{\text{poly}} = \mathbf{0}$, and therefore the error $\mathbf{H} - \widehat{H}$ is large, especially for large frequency. We will address this deficiency in Sect. 4.

Often more data than necessary are provided and the pencil ($\mathbb{L}_s, \mathbb{L}$) is singular. In this case we use the singular value decomposition (SVD) to extract the important information. Specifically, we compute the (short) SVDs of the augmented Loewner matrices as

$$\begin{bmatrix} \mathbb{L} & \mathbb{L}_s \end{bmatrix} = \mathbf{Y}_1 \mathbf{S}_1 \mathbf{X}_1^*, \quad \begin{bmatrix} \mathbb{L} \\ \mathbb{L}_s \end{bmatrix} = \mathbf{Y}_2 \mathbf{S}_2 \mathbf{X}_2^*, \qquad (3.8)$$

where $\mathbf{S}_1 = \text{diag}(\sigma_1^{(1)}, \ldots, \sigma_N^{(1)}) \in \mathbb{R}^{N \times N}$ and $\mathbf{S}_2 = \text{diag}(\sigma_1^{(2)}, \ldots, \sigma_N^{(2)}) \in \mathbb{R}^{N \times N}$ are the matrices with singular values on the diagonal, and $\mathbf{Y}_1, \mathbf{X}_2 \in \mathbb{C}^{N \times N}$, $\mathbf{Y}_2, \mathbf{X}_1 \in \mathbb{C}^{2N \times N}$ are the matrices of singular vectors.

The size r of the ROM can be chosen as follows. Given a tolerance $\tau > 0$ the truncation order r is the smallest integer such that the normalized singular values satisfy $\sigma_j^{(1)}/\sigma_1^{(1)} < \tau, \sigma_j^{(2)}/\sigma_1^{(2)} < \tau, j = r + 1, \ldots, N$.

The matrices $\mathbf{Y}, \mathbf{X} \in \mathbb{C}^{N \times r}$ are obtained by selecting the first r columns of the matrices $\mathbf{Y}_1$ and $\mathbf{X}_2$. The reduced Loewner system is constructed by multiplying the matrices $\mathbb{L}, \mathbb{L}_s, \mathbf{V}, \mathbf{W}$ with $\mathbf{Y}^*$ and $\mathbf{X}$ to the left and respectively, to the right, as

$$\widehat{\mathbb{L}} = \mathbf{Y}^* \mathbb{L} \mathbf{X}, \quad \widehat{\mathbb{L}}_s = \mathbf{Y}^* \mathbb{L}_s \mathbf{X}, \quad \widehat{\mathbf{V}} = \mathbf{Y}^* \mathbf{V}, \quad \widehat{\mathbf{W}} = \mathbf{W} \mathbf{X}. \qquad (3.9)$$

---

[3]What the 'right' amount of data is depends on the transfer function. Since we typically have more data, the case we describe below, we omit specification of the 'right' amount of data.

The ROM computed with the (classical) Loewner method is (3.2) with

$$\widehat{\mathbf{E}} = -\widehat{\mathbb{L}}, \quad \widehat{\mathbf{A}} = -\widehat{\mathbb{L}}_s, \quad \widehat{\mathbf{B}}_0 = \widehat{\mathbf{V}}, \quad \widehat{\mathbf{B}}_1 = \mathbf{0}, \quad \widehat{\mathbf{C}} = \widehat{\mathbf{W}}, \quad \widehat{\mathbf{P}}_0 = \widehat{\mathbf{P}}_1 = \mathbf{0}. \qquad (3.10)$$

As before, if the data $\{\mu_j\}_{j=1}^N, \{\lambda_i\}_{i=1}^N, \{\mathbf{v}_j\}_{j=1}^N, \{\mathbf{w}_j\}_{i=1}^N$ contain also the conjugate complex data, then the complex ROM (3.2) with (3.10) can be transformed into a real ROM with the same transfer function, as shown in [2, p. 360].

The transfer function $\widehat{H}$ corresponding to (3.10) satisfies the approximate interpolation conditions (3.3). However, by design, the transfer function $\widehat{H}$ is strictly proper, $\widehat{H}_{\text{poly}} = \mathbf{0}$, and therefore the error $\mathbf{H} - \widehat{H}$ is large, especially for large frequency. We will address this deficiency next.

4 Accounting for the Polynomial Part of the Transfer Function

As we have seen in Sect. 2.2, the transfer function is composed of a strictly proper part and a polynomial part. The exact structure of these parts is shown in (2.12). We write $\mathbf{H}(s) = \mathbf{H}_{\text{spr}}(s) + \mathbf{H}_{\text{poly}}(s)$ with

$$\mathbf{H}_{\text{poly}}(s) = \mathbf{P}_0 + s\mathbf{P}_1.$$

Especially for the DAE system (1.1), the exact form (2.12) of $\mathbf{P}_0, \mathbf{P}_1 \in \mathbb{R}^{p \times m}$ is complicated. Here, as in Sect. 3, $m = n_g$ is the number of inputs and $p = n_y$ is the number of outputs, so that $\mathbf{H}(s) \in \mathbb{C}^{p \times m}$. Even if all system matrices in (1.1) were available, the computation of $\mathbf{P}_0, \mathbf{P}_1 \in \mathbb{R}^{p \times m}$ from (2.12) is tedious. More importantly, if only transfer function $\mathbf{H}(s)$ measurements are available, it is impossible to compute $\mathbf{P}_0, \mathbf{P}_1 \in \mathbb{R}^{p \times m}$ from (2.12). In this section we explain how we can estimate $\mathbf{H}_{\text{poly}}$ and account for it in the Loewner framework. The key assumption is that information about the transfer function is known at high frequency bands. More precisely, we assume that $\mathbf{H}(\iota\,\omega)$ is known for large positive real numbers ω. Here, denote the imaginary unit with $\iota = \sqrt{-1}$. Since $\lim_{|\omega| \to \infty} |\mathbf{H}_{\text{spr}}(\iota\omega)| = 0$, the contribution of the strictly proper part $\mathbf{H}_{\text{spr}}(s)$ to the transfer function $\mathbf{H}(s)$ becomes negligible for high frequency ranges.

4.1 Estimation from One and Two Data Points

Assume that the transfer function $\mathbf{H}(s)$ is known at one point $\iota\,\eta$ located on the imaginary axis where $\eta \in \mathbb{R}$ and $\eta \gg 1$. Since $\lim_{\eta \to \infty} |\mathbf{H}_{\text{spr}}(\iota\,\eta)| = 0$,

$$\mathbf{H}(\iota\,\eta) = \mathbf{H}_{\text{spr}}(\iota\,\eta) + \mathbf{P}_0 + \iota\,\eta\mathbf{P}_1 \approx \mathbf{P}_0 + \iota\,\eta\mathbf{P}_1.$$

This gives the estimates

$$\widehat{\mathbf{P}}_0 = \mathrm{Re}\left(\mathbf{H}(\imath\,\eta)\right), \qquad \widehat{\mathbf{P}}_1 = \eta^{-1}\,\mathrm{Im}\left(\mathbf{H}(\imath\,\eta)\right). \tag{4.1}$$

Next, assume that the transfer function $\mathbf{H}(s)$ is known for two points $\imath\,\eta$ and $\imath\,\theta$ on the imaginary axis with $\eta, \theta \in \mathbb{R}$ and $\theta > \eta \gg 1$. We have

$$\mathbf{H}(\imath\,\theta) - \mathbf{H}(\imath\,\eta) = \left(\mathbf{H}_{\mathrm{spr}}(\imath\,\theta) + \mathbf{P}_0 + \imath\,\theta\mathbf{P}_1\right) - \left(\mathbf{H}_{\mathrm{spr}}(\imath\,\eta) + \mathbf{P}_0 + \imath\,\eta\mathbf{P}_1\right)$$

$$= \mathbf{H}_{\mathrm{spr}}(\imath\,\theta) - \mathbf{H}_{\mathrm{spr}}(\imath\,\eta) + (\imath\,\theta - \imath\,\eta)\mathbf{P}_1 \approx (\imath\,\theta - \imath\,\eta)\mathbf{P}_1. \tag{4.2}$$

Hence, we can estimate $\mathbf{P}_1$ in terms of a divided difference value that appears in the Loewner matrix with $\lambda = \imath\,\eta$ and $\mu = \imath\,\theta$ (that is approximating the derivative when $\theta \to \eta$), as follows

$$\widehat{\mathbf{P}}_1 = \mathrm{Re}\left(\frac{\mathbf{H}(\imath\,\theta) - \mathbf{H}(\imath\,\eta)}{\imath\,\theta - \imath\,\eta}\right). \tag{4.3a}$$

We also have

$$\imath\,\theta\mathbf{H}(\imath\,\theta) - \imath\,\eta\mathbf{H}(\imath\,\eta)$$

$$= \left(\imath\,\theta\mathbf{H}_{\mathrm{spr}}(i\theta) + \imath\,\theta\mathbf{P}_0 - \theta^2\mathbf{P}_1\right) - \left(\imath\,\eta\mathbf{H}_{\mathrm{spr}}(\imath\,\eta) + \imath\,\eta\mathbf{P}_0 - \eta^2\mathbf{P}_1\right)$$

$$= \imath\,\theta\mathbf{H}_{\mathrm{spr}}(\imath\,\theta) - \imath\,\eta\mathbf{H}_{\mathrm{spr}}(\imath\,\eta) + (\imath\,\theta - \imath\,\eta)\mathbf{P}_0 + (\eta^2 - \theta^2)\mathbf{P}_1,$$

which implies

$$\frac{\imath\,\theta\mathbf{H}(\imath\,\theta) - \imath\,\eta\mathbf{H}(\imath\,\eta)}{\imath\,\theta - \imath\,\eta} = \frac{\imath\,\theta\mathbf{H}_{\mathrm{spr}}(\imath\,\theta) - \imath\,\eta\mathbf{H}_{\mathrm{spr}}(\imath\,\eta)}{\imath\,\theta - \imath\,\eta} + \mathbf{P}_0 + \imath(\eta + \theta)\mathbf{P}_1$$

$$\approx \mathbf{P}_0 + \imath(\eta + \theta)\mathbf{P}_1.$$

The previous approximation gives the following estimate for $\mathbf{P}_0$,

$$\widehat{\mathbf{P}}_0 = \mathrm{Re}\left(\frac{\imath\,\theta\mathbf{H}(\imath\,\theta) - \imath\,\eta\mathbf{H}(\imath\,\eta)}{\imath\,\theta - \imath\,\eta}\right). \tag{4.3b}$$

Example 4.1 Consider the order $n = 3$ DAE system

$$\underbrace{\begin{bmatrix} 1 & 0 \\ 0 & 1 \end{bmatrix}}_{\mathbf{E}_{11}} \frac{d}{dt}\mathbf{v}(t) = \underbrace{\begin{bmatrix} 1 & 0 \\ 0 & 2 \end{bmatrix}}_{\mathbf{A}_{11}} \mathbf{v}(t) + \underbrace{\begin{bmatrix} 1 \\ 0 \end{bmatrix}}_{\mathbf{A}_{12}} \mathbf{p}(t) + \underbrace{\begin{bmatrix} 1 & -1 \\ 1 & 1 \end{bmatrix}}_{\mathbf{B}_{1,0}} \mathbf{g}(t),$$

$$\mathbf{0} = \underbrace{\begin{bmatrix} 1 & 0 \end{bmatrix}}_{\mathbf{A}_{12}^T} \mathbf{v}(t) + \underbrace{\begin{bmatrix} 1 & 2 \end{bmatrix}}_{\mathbf{B}_{2,0}} \mathbf{g}(t),$$

$$\mathbf{y}(t) = \underbrace{\begin{bmatrix} 2 & 1 \\ 0 & 1 \end{bmatrix}}_{\mathbf{C}_1} \mathbf{v}(t) + \underbrace{\begin{bmatrix} 3 \\ 1 \end{bmatrix}}_{\mathbf{C}_2} \mathbf{p}(t),$$

which is of the form (1.1), with $\mathbf{B}_{1,1} = \mathbf{D}_0 = \mathbf{D}_1 = \mathbf{0}_{2\times2}$. For this small example we can compute the transfer function explicitly, e.g., using the symbolic toolbox in Matlab applied to (2.1), to get

$$\mathbf{H}_{\text{spr}}(s) = \frac{1}{s-2}\begin{bmatrix} 1 & 1 \\ 1 & 1 \end{bmatrix}, \qquad \mathbf{H}_{\text{poly}}(s) = \underbrace{\begin{bmatrix} -2 & 5 \\ 0 & 3 \end{bmatrix}}_{=\mathbf{P}_0} + s\underbrace{\begin{bmatrix} -3 & -6 \\ -1 & -2 \end{bmatrix}}_{=\mathbf{P}_1}. \qquad (4.5)$$

First, we estimate $\mathbf{P}_0$ and $\mathbf{P}_1$ in (4.5) from one measurement pair $(\imath\eta, \mathbf{H}(\imath\eta))$ using (4.1). In this simple example, these errors can be computed analytically from (4.5) and happen to be nearly identical,

$$\mathbf{P}_0 - \widehat{\mathbf{P}}_0 = \mathbf{P}_0 - \text{Re}\left(\mathbf{H}(\imath\eta)\right) = \frac{2}{\eta^2+4}\begin{bmatrix} 1 & 1 \\ 1 & 1 \end{bmatrix} = O(\eta^{-2}),$$

$$\mathbf{P}_1 - \widehat{\mathbf{P}}_1 = \mathbf{P}_1 - \eta^{-1}\text{Im}\left(\mathbf{H}(\imath\eta)\right) = \frac{1}{\eta^2+4}\begin{bmatrix} 1 & 1 \\ 1 & 1 \end{bmatrix} = O(\eta^{-2}).$$

The errors for different η are depicted by the black curves with crosses in Fig. 1.

Next, we estimate the values of $\mathbf{P}_0$ and $\mathbf{P}_1$ in (4.5) from two measurement pairs $(\imath\eta, \mathbf{H}(\imath\eta))$ and $(\imath\theta, \mathbf{H}(\imath\theta))$ using the estimates (4.3). Specifically, we use the second frequency $\theta = 10\eta$ or $\theta = 100\eta$. The errors $\|\mathbf{P}_0 - \widehat{\mathbf{P}}_0\|_2$ are shown in the left plot in Fig. 1, while the errors $\|\mathbf{P}_1 - \widehat{\mathbf{P}}_1\|_2$ are shown in the right plot. The red curves with circles correspond to the estimates (4.3) with $\theta = 10\eta$ and green curves with diamonds correspond to the estimates (4.3) with $\theta = 100\eta$. Again, the errors behave like $O(\eta^{-2})$. Adding a second frequency $\theta = 10^k\eta, k = 1, 2$, reduces the error approximately by 10^{-k}.

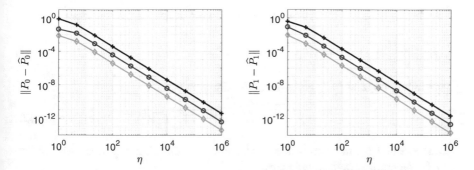

Fig. 1 Errors $\|\mathbf{P}_0 - \widehat{\mathbf{P}}_0\|_2$ (left plot) and $\|\mathbf{P}_1 - \widehat{\mathbf{P}}_1\|_2$ (right plot) for $\widehat{\mathbf{P}}_0, \widehat{\mathbf{P}}_1$ estimated from (4.1) and (4.3). The black curves with crosses show the error for the estimates obtained from (4.1) for $\eta \in [10^0, 10^6]$. The red curves with circles and green curves with diamonds show the error for the estimates obtained from (4.3) for $\eta \in [10^0, 10^6]$ and $\theta = 10\,\eta$ (red curves with circles) or $\theta = 100\,\eta$ (green curves with diamonds). The errors behave like $O(\eta^{-2})$ and adding a second frequency $\theta = 10^k \eta, k = 1, 2$, reduces the error by approximately by 10^{-k}

4.2 Estimation from 2L Data Points—The General Case

Now assume that we have $2L$ measurements available with sampling points located in high frequency bands, i.e., on the imaginary axis with high absolute value. We will extend the formulas in (4.3) to the general case $L \geq 1$ using the definitions of the Loewner matrices in (3.5) and (3.6).

The set-up is as in Sect. 3. The left interpolation points $\{\iota\,\theta_i\}_{i=1}^L$ and right interpolation points $\{\iota\,\eta_j\}_{j=1}^L$ are chosen on the imaginary axis $\iota\,\mathbb{R}$ with $\min\{\theta_i\}, \min\{\eta_j\} \gg 1$. The goal is to estimate the coefficient matrices $\mathbf{P}_0, \mathbf{P}_1$ taking into account all $2L$ measurements, and not only two of them as in (4.3).

We begin by extending (4.3a) for the estimation of $\mathbf{P}_1$. We write the (i, j) entry of the Loewner matrix $\mathbb{L}$ (3.5) with $\lambda = \iota\,\eta$ and $\mu = \iota\,\theta$. Instead of the generic notation $\mathbb{L}$ for the Loewner matrix, we now use the notation $\mathbb{L}^{\text{hi}}$ to indicate that this Loewner matrix is computed with data located in high frequency bands, and to later differentiate it from the Loewner matrix $\mathbb{L}^{\text{lo}}$ obtained from the remaining data in low frequency band.

Using the equalities (3.3) and (4.2), it follows that the (i, j) entry of the Loewner matrix $\mathbb{L}^{\text{hi}}$ (3.5) with $\lambda = \iota\,\eta$ and $\mu = \iota\,\theta$ has the expression

$$
\begin{aligned}
\mathbb{L}^{\text{hi}}_{(i,j)} &= \frac{\mathbf{v}_i^* \mathbf{r}_j - \boldsymbol{\ell}_i^* \mathbf{w}_j}{\iota\,\theta_i - \iota\,\eta_j} = \frac{\boldsymbol{\ell}_i^* \mathbf{H}(\iota\,\theta_i)\mathbf{r}_j - \boldsymbol{\ell}_i^* \mathbf{H}(\iota\,\eta_j)\mathbf{r}_j}{\iota\,\theta_i - \iota\,\eta_j} = \boldsymbol{\ell}_i^* \left(\frac{\mathbf{H}(\iota\,\theta_i) - \mathbf{H}(\iota\,\eta_j)}{\iota\,\theta_i - \iota\,\eta_j} \right) \mathbf{r}_j \\
&= \boldsymbol{\ell}_i^* \left(\frac{\mathbf{H}_{\text{spr}}(\iota\,\theta_i) - \mathbf{H}_{\text{spr}}(\iota\,\eta_j) + (\iota\,\theta_i - \iota\,\eta_j)\mathbf{P}_1}{\iota\,\theta_i - \iota\,\eta_j} \right) \mathbf{r}_j \\
&= \underbrace{\boldsymbol{\ell}_i^* \left(\frac{\mathbf{H}_{\text{spr}}(\iota\,\theta_i) - \mathbf{H}_{\text{spr}}(\iota\,\eta_j)}{\iota\,\theta_i - \iota\,\eta_j} \right) \mathbf{r}_j}_{:=\mathbb{L}^{\text{hi,spr}}_{(i,j)}} + \boldsymbol{\ell}_i^* \mathbf{P}_1 \mathbf{r}_j = \mathbb{L}^{\text{hi,spr}}_{(i,j)} + \boldsymbol{\ell}_i^* \mathbf{P}_1 \mathbf{r}_j. \quad (4.6)
\end{aligned}
$$

As in (3.4), the directional vectors $\boldsymbol{\ell}_i$ and $\mathbf{r}_j$ are collected into matrices

$$\left(\mathbf{L}^{\text{hi}}\right)^* = \begin{bmatrix} \boldsymbol{\ell}_1 & \boldsymbol{\ell}_2 & \cdots & \boldsymbol{\ell}_L \end{bmatrix} \in \mathbb{C}^{p \times L}, \qquad \mathbf{R}^{\text{hi}} = \begin{bmatrix} \mathbf{r}_1 & \mathbf{r}_2 & \cdots & \mathbf{r}_L \end{bmatrix} \in \mathbb{C}^{m \times L}. \qquad (4.7)$$

Combining (4.6) and (4.7) gives the approximation formula

$$\mathbb{L}^{\text{hi}} = \mathbb{L}^{\text{hi,spr}} + \mathbf{L}^{\text{hi}} \mathbf{P}_1 \mathbf{R}^{\text{hi}} \approx \mathbf{L}^{\text{hi}} \mathbf{P}_1 \mathbf{R}^{\text{hi}}, \qquad (4.8)$$

again obtained by neglecting the contribution of the strictly proper part of the transfer function at high frequencies.

Provided that $L \geq \max\{p, m\}$ (recall that here m is the number of inputs and p is the number of outputs), one can write the estimated linear polynomial coefficient matrix as follows

$$\widehat{\mathbf{P}}_1 = \text{Re}\left(\left(\mathbf{L}^{\text{hi}}\right)^{\dagger} \mathbb{L}^{\text{hi}} \left(\mathbf{R}^{\text{hi}}\right)^{\dagger}\right), \qquad (4.9a)$$

where $\mathbf{X}^{\dagger} \in \mathbb{C}^{v \times u}$ is the Moore-Penrose pseudo-inverse of $\mathbf{X} \in \mathbb{C}^{u \times v}$.

Similarly to the procedure used for estimating $\mathbf{P}_1$, one can extend the formula in (4.3b) for estimating $\mathbf{P}_0$ from the shifted Loewner matrix $\mathbb{L}_s^{\text{hi}}$ computed from L sampling points located in high frequency bands as follows

$$\widehat{\mathbf{P}}_0 = \text{Re}\left(\left(\mathbf{L}^{\text{hi}}\right)^{\dagger} \mathbb{L}_s^{\text{hi}} \left(\mathbf{R}^{\text{hi}}\right)^{\dagger}\right). \qquad (4.9b)$$

4.3   The Proposed Procedure

Assume that we have samples of the transfer function evaluated at high frequencies (to capture the polynomial part) as well as at low frequencies (to capture the strictly proper part). Algorithm 1 below adapts the Loewner framework for DAE systems by preserving the polynomial structure of the underlying transfer function. The ROM constructed with Algorithm 1 has the form

$$\widehat{\mathbf{E}} \frac{d}{dt} \widehat{\mathbf{x}}(t) = \widehat{\mathbf{A}} \widehat{\mathbf{x}}(t) + \widehat{\mathbf{B}}_0 \, \mathbf{g}(t), \qquad\qquad t \in (0, T), \qquad (4.10a)$$

$$\widehat{\mathbf{E}} \widehat{\mathbf{x}}(0) = \mathbf{0}, \qquad (4.10b)$$

$$\widehat{\mathbf{y}}(t) = \widehat{\mathbf{C}} \widehat{\mathbf{x}}(t) + \widehat{\mathbf{P}}_0 \, \mathbf{g}(t) + \widehat{\mathbf{P}}_1 \frac{d}{dt} \mathbf{g}(t), \qquad t \in (0, T). \qquad (4.10c)$$

The derivative $\frac{d}{dt} \mathbf{g}(t)$ is not an explicit input into the dynamics (4.10a), i.e., $\widehat{\mathbf{B}}_1 = \mathbf{0}$, but its influence on the output is modeled by the feed-through term $\widehat{\mathbf{P}}_1 \frac{d}{dt} \mathbf{g}(t)$ in the output equation (4.10c). While some structural details of the ROM (4.10) are different from the original FOM (1.3), the transfer function

$$\widehat{\mathbf{H}}(s) = \widehat{\mathbf{H}}_{\mathrm{spr}}(s) + \widehat{\mathbf{H}}_{\mathrm{poly}}(s) \tag{4.11a}$$

of the ROM (4.10), now has a strictly proper part and a polynomial part,

$$\widehat{\mathbf{H}}_{\mathrm{spr}}(s) = \widehat{\mathbf{C}}\big(s\,\widehat{\mathbf{E}} - \widehat{\mathbf{A}}\big)^{-1}\widehat{\mathbf{B}}_0, \quad \widehat{\mathbf{H}}_{\mathrm{poly}}(s) = \widehat{\mathbf{P}}_0 + s\,\widehat{\mathbf{P}}_1. \tag{4.11b}$$

Numerical experiments indicate that each of these match the ones of the FOM (2.12) well, provided enough transfer measurements are available.

Instead of the generic $\lambda_j, \mu_j \in \mathbb{C}$ used in Sect. 3 we now specify $\lambda_j = \iota\,\eta_j$ and $\mu_j = \iota\,\theta_j$ in Algorithm 1 with $\eta_j, \theta_j \in \mathbb{R}$.

Algorithm 1 Modified Loewner method with identification of polynomial terms in transfer function

---

Input: A data set composed of $2(N + L)$ sample points, $2(N + L)$ tangential directions, and
$\quad 2(N + L)$ measured values of the transfer function as introduced in (3.1).
Output: Loewner ROM specified by $\widehat{\mathbf{E}}, \widehat{\mathbf{A}}, \widehat{\mathbf{B}}, \widehat{\mathbf{C}}, \widehat{\mathbf{P}}_0, \widehat{\mathbf{P}}_1$.
1: Split the data into $2N$ data corresponding to the low frequency range and into $2L$ data
$\quad$ corresponding to the high frequency range.
2: Use the $2L$ data corresponding to the high frequency range to estimate $\widehat{\mathbf{P}}_0, \widehat{\mathbf{P}}_1$ using (4.9).
3: Adjust the $2N$ transfer function measurements corresponding to the low frequency range, by
$\quad$ subtracting the estimated polynomial part $\widehat{\mathbf{H}}_{\mathrm{poly}}(\omega) = \widehat{\mathbf{P}}_0 + \iota\,\omega\widehat{\mathbf{P}}_1$ for $\omega \in \{\theta_i \mid 1 \leq i \leq$
$\quad N\} \cup \{\eta_j \mid 1 \leq j \leq N\}$ from the original measurement values, i.e., compute

$$\begin{aligned} \text{left}: \quad & \big(\iota\,\theta_j, \boldsymbol{\ell}_j, \mathbf{v}_i - \widehat{\mathbf{H}}_{poly}(\iota\,\theta_j)^*\boldsymbol{\ell}_j\big), \quad j = 1, \dots, N, \\ \text{right}: \quad & \big(\iota\,\eta_j, \mathbf{r}_j, \mathbf{w}_j - \widehat{\mathbf{H}}_{poly}(\iota\,\eta_j)\mathbf{r}_j\big), \quad j = 1, \dots, N. \end{aligned} \tag{4.39}$$

4: Use the $2N$ data (4.39) to construct data matrices $\mathbf{V}^{\mathrm{lo}} \in \mathbb{C}^{N \times m}$, $\mathbf{W}^{\mathrm{lo}} \in \mathbb{C}^{p \times N}$ as in (3.4), and
$\quad$ Loewner matrices $\mathbb{L}^{\mathrm{lo}}, \mathbb{L}_s{}^{\mathrm{lo}} \in \mathbb{C}^{N \times N}$ as in (3.5) and (3.6).
5: Compute the SVD of the augmented Loewner matrices obtained with $\mathbb{L}^{\mathrm{lo}}, \mathbb{L}_s{}^{\mathrm{lo}}$ and project as
$\quad$ in (3.9) to construct $\widehat{\mathbf{E}} = -\widehat{\mathbb{L}}^{\mathrm{lo}} = -\mathbf{Y}^*\mathbb{L}^{\mathrm{lo}}\mathbf{X}$, $\widehat{\mathbf{A}} = -\widehat{\mathbb{L}}_s^{\mathrm{lo}} = -\mathbf{Y}^*\mathbb{L}_s^{\mathrm{lo}}\mathbf{X}$, $\widehat{\mathbf{B}}_0 = \widehat{\mathbf{V}}^{\mathrm{lo}} =$
$\quad \mathbf{Y}^*\mathbf{V}^{\mathrm{lo}}$, $\widehat{\mathbf{C}} = \widehat{\mathbf{W}}^{\mathrm{lo}} = \mathbf{W}^{\mathrm{lo}}\mathbf{X}$.

---

5 Numerical Example—Oseen Equations

In this section we apply the Loewner framework to the Oseen equations. The example specifications are from [8].

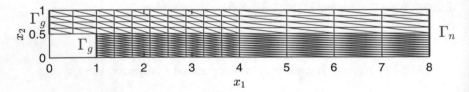

Fig. 2 The channel geometry and coarse grid

5.1 Problem Specification

For completeness we first review the main problem specifications. Let $\Omega \subset \mathbb{R}^2$ be the backward facing step geometry shown in Fig. 2. The boundary is decomposed into segments $\Gamma_n, \Gamma_d, \Gamma_g$, where $\Gamma_n = \{8\} \times (0, 1)$ is the outflow boundary, inputs are applied on $\Gamma_g = \{0\} \times (1/2, 1) \cup \{1\} \times (0, 1/2)$, and the velocities are set to zero on $\Gamma_d = \partial \Omega \setminus (\Gamma_g \cup \Gamma_n)$.

We consider the Oseen equations

$$\frac{\partial}{\partial t} v(x, t) + (a(x) \cdot \nabla) v(x, t) - \nu \Delta v(x, t) + \nabla p(x, t) = 0 \qquad \text{in } \Omega \times (0, T),$$

$$\nabla \cdot v(x, t) = 0 \qquad \text{in } \Omega \times (0, T),$$

$$(-p(x, t)I + \nu \nabla v(x, t))n(x) = 0 \qquad \text{on } \Gamma_n \times (0, T),$$

$$v(x, t) = 0 \qquad \text{on } \Gamma_d \times (0, T),$$

$$v(x, t) = g_\Gamma(x, t) \qquad \text{on } \Gamma_g \times (0, T),$$

$$v(x, 0) = 0 \qquad \text{in } \Omega,$$

where $\nu = 1/50$ is the dynamic viscosity and where $n(x)$ is the unit outward normal to Ω at x. Here v, p are the velocity and pressure of the fluid respectively, and g_Γ denotes the boundary input. The advection field a is computed as in [8, Sec. 7.2] by solving the steady-state Stokes equation with velocity $8(x_2 - 1/2)(1 - x_2)$ on the inflow boundary segment $\Gamma_{in} = \{0\} \times (1/2, 1)$ and and zero velocity boundary conditions on $\partial \Omega \setminus (\Gamma_n \cup \Gamma_{in})$.

Our boundary inputs are given as in [8] by

$$g_\Gamma(x, t) = \sum_{k=1}^{n_g} \mathbf{g}_k(t) \boldsymbol{\gamma}_k(x) \qquad (5.1)$$

with $n_g = 6$ boundary control functions $\boldsymbol{\gamma}_j : \mathbb{R}^2 \to \mathbb{R}^2$ given as follows. The first three functions are defined on the inflow boundary segment $\{0\} \times (1/2, 1)$ and are given by

$$\boldsymbol{\gamma}_k(x) = \begin{pmatrix} \sin(2j\pi(x_2 - 1/2)) \\ 0 \end{pmatrix}, \quad k = 1, 2, 3;$$

the remaining three are defined on the backstep boundary segment $\{1\} \times (0, 1/2)$ and are of the form

$$\boldsymbol{\gamma}_{3+k}(x) = \begin{pmatrix} \sin(2j\pi x_2) \\ 0 \end{pmatrix}, \quad k = 1, 2, 3.$$

We use a $P1 - P2$ Taylor-Hood discretization to arrive at the semi-discrete equations (1.1a–c). (Note that the $\mathbf{B}_{1,1}$ term has accidentally been dropped in [8, Sec. 7.2].) We use a mesh that is obtained from a uniform refinement of the coarse mesh shown in Fig. 2.

We consider the second output from [8, Sec. 7.2], which is the integral of the stress force on the boundary segment $\Gamma_{\text{obs}} = (1, 8) \times \{0\}$,

$$y(t) = \int_{\Gamma_{\text{obs}}} \left(- p(x, t)I + v\nabla v(x, t)\right)n(x)ds, \tag{5.2}$$

approximated using the weak form (see [8] for details). This leads to (1.1d) with $\mathbf{C}_1 \in \mathbb{R}^{2\times n_v}$, $\mathbf{C}_2 \in \mathbb{R}^{2\times n_p}$, $\mathbf{D}_0 \in \mathbb{R}^{2\times n_g}$, and $\mathbf{D}_1 = \mathbf{0}$. Note that the output matrices represent derivatives of the finite element approximations of velocity v and pressure p and therefore scale with the mesh size h; the finite element approximation of the output $y(t)$ itself does not.

In summary, the semi-discretized DAE model is of dimension $n = n_v + n_p$ with $m = n_g = 6$ inputs and $p = n_y = 2$ outputs.

5.2 Numerical Experiments

We report numerical experiments for a discretization with $n_v = 12,504$ velocity degrees of freedom and $n_p = 1,669$ pressure degrees of freedom. Other discretization sizes gave similar results.

The polynomial coefficient matrices are explicitly computed using the approach in Sect. 2.3 and given by (four digits are shown)

$$\mathbf{P}_0 = \begin{bmatrix} -7.088 & -1.124\cdot10^{-4} & -2.363 & -7.731 & -4.172\cdot10^{-1} & -2.724 \\ 4.845\cdot10^1 & -2.940\cdot10^{-4} & 1.615\cdot10^1 & 4.927\cdot10^1 & 8.727\cdot10^{-3} & 1.656\cdot10^1 \end{bmatrix},$$

$$\mathbf{P}_1 = \begin{bmatrix} -5.484\cdot10^{-17} & -2.242\cdot10^{-22} & -1.828\cdot10^{-17} \\ 7.814 & 1.997\cdot10^{-5} & 2.605 \end{bmatrix}$$

$$\begin{bmatrix} -5.576\cdot10^{-17} & -4.468\cdot10^{-19} & -1.866\cdot10^{-17} \\ 7.889 & 3.275\cdot10^{-2} & 2.632 \end{bmatrix}.$$

Next we compute the Loewner ROM using the classical Loewner approach (3.7) and the modified Loewner approach. For the modified Loewner approach we first modify the transfer function measurements using the true polynomial part $\widehat{\mathbf{P}}_0 = \mathbf{P}_0$, $\widehat{\mathbf{P}}_1 = \mathbf{P}_1$ computed using the approach in Sect. 2.3. Thus the modified Loewner uses Algorithm 1, with Steps 1 and 2 replaced by the computation of $\widehat{\mathbf{P}}_0 = \mathbf{P}_0$, $\widehat{\mathbf{P}}_1 = \mathbf{P}_1$ using the approach in Sect. 2.3. We assume that we have $2N = 200$ measurements logarithmically spaced in the low frequency range $[10^{-2}, 10^1]$ $\imath$. The left $\boldsymbol{\ell}_j$ and right $\mathbf{r}_j$ tangential vectors are chosen randomly.

The singular value decay of the Loewner matrices (3.8) computed using measurements in the low frequency range is shown in Fig. 3. The ROM size r is chosen as the largest integer such that $\sigma_r/\sigma_1 > \tau = 10^{-10}$ and is $r = 24$ for the classical Loewner ROM. In the modified Loewner approach we compute the Loewner matrices from the shifted transfer function measurements (Steps 3+4 in Algorithm 1). The singular value decay of these Loewner matrices is similar the one shown in Fig. 3 and are not plotted. The ROM size r is again chosen as the largest integer such that $\sigma_r/\sigma_1 > \tau = 10^{-10}$ and is $r = 23$ for the modified Loewner

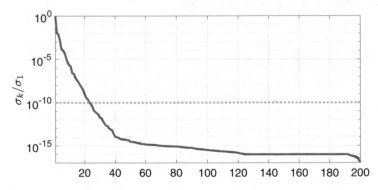

Fig. 3 Singular value decay of the Loewner matrices (3.8) computed using measurements in the low frequency range and tolerance $\tau = 10^{-10}$ used to determine the ROM size. The normalized singular values for the two Loewner matrices (3.8) are visually identical

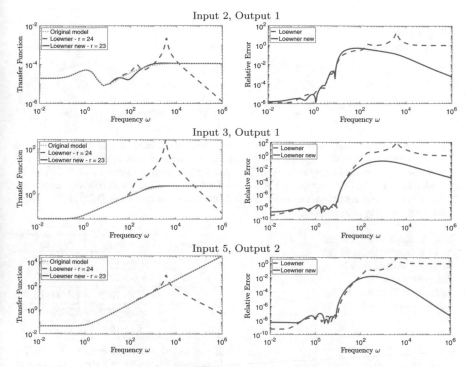

Fig. 4 Left plots: absolute values of frequency responses of the original system (yellow dotted lines) of the reduced system computed with the classical Loewner approach (blue dashed line), and of the reduced system computed with the modified Loewner approach with true $\widehat{\mathbf{P}}_0 = \mathbf{P}_0$, $\widehat{\mathbf{P}}_1 = \mathbf{P}_1$ (red solid line) for various components of the 2×6 transfer function. Loewner ROMs computed using $2N = 200$ measurements logarithmically spaced in the low frequency range $[10^{-2}, 10^1]\, \iota$. Right plots: corresponding relative errors

ROM. The left plots in Fig. 4 show the absolute values of frequency responses of the original system (yellow dotted lines) of the reduced system computed with the classical Loewner approach (blue dashed line), and of the reduced system computed with the modified Loewner approach (red solid line) for various components of the 2×6 transfer function at 300 logarithmically spaced frequencies $\omega\, \iota$ in $[10^{-2}, 10^6]\,\iota$. The right plots in Fig. 4 show the corresponding relative errors. We have picked three transfer function components which well represent the overall behavior of the Loewner approach.

The modified Loewner approach generally leads to ROMs with transfer functions that better approximate the true transfer function. The approximation of the transfer function for large frequencies ω is always substantially better when the modified Loewner approach is used. For the transfer function component $\mathbf{H}(i\omega)_{12}$ corresponding to input 2 and output 1 the modified Loewner approach leads to a slightly larger error for frequencies roughly between 10^1 and 10^2. This is due to the fact that we only use measurements in $[10^{-2}, 10^1]\,\iota$. If instead we use $2N = 200$

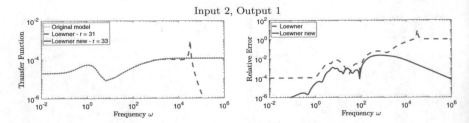

Fig. 5 Left plots: absolute values of frequency responses of the original system (yellow dotted lines) of the reduced system computed with the classical Loewner approach (blue dashed line), and of reduced system computed with the modified Loewner approach with true $\widehat{\mathbf{P}}_0 = \mathbf{P}_0$, $\widehat{\mathbf{P}}_1 = \mathbf{P}_1$ (red solid line) for the $(1, 2)$ component of the transfer function. Loewner ROMs computed using $2N = 200$ measurements logarithmically spaced in the low frequency range $[10^{-2}, 10^2]\,\imath$. Right plots: corresponding relative errors

Table 1 Estimation error for $\widehat{\mathbf{P}}_0, \widehat{\mathbf{P}}_1$ computed using (4.9a) and (4.9b) with $2L = 20$ measurements logarithmically spaced in the high frequency range $[10^f, 10^{f+2}]\,\imath$ for $f = 3, \dots, 7$

Freq. range	$\|\mathbf{P}_0 - \widehat{\mathbf{P}}_0\|_2$	$\|\mathbf{P}_1 - \widehat{\mathbf{P}}_1\|_2$
$[10^3, 10^5]$	$6.0161 \cdot 10^{-2}$	$2.7859 \cdot 10^{-4}$
$[10^4, 10^6]$	$2.5535 \cdot 10^{-4}$	$1.1163 \cdot 10^{-6}$
$[10^5, 10^7]$	$3.0575 \cdot 10^{-6}$	$1.3111 \cdot 10^{-8}$
$[10^6, 10^8]$	$2.8019 \cdot 10^{-8}$	$1.2303 \cdot 10^{-10}$
$[10^7, 10^9]$	$5.8920 \cdot 10^{-10}$	$2.6729 \cdot 10^{-12}$

The observed estimation error for $\widehat{\mathbf{P}}_0$ and for $\widehat{\mathbf{P}}_1$ behaves like $O(10^{-2f})$ and in this example the $\mathbf{P}_1$ estimation error is two orders of magnitude smaller than the $\mathbf{P}_0$ estimation error

measurements logarithmically spaced in the low frequency range $[10^{-2}, 10^2]\,\imath$, we get the frequency response in Fig. 5. Approximations for the other transfer function components are also improved when the modified Loewner approach is used, but not plotted because of space limitations. However, note that the classical and modified Loewer ROMs computed using these data are of larger sizes $r = 31$ and $r = 33$. (The ROM size r is again chosen as the largest integer such that $\sigma_r/\sigma_1 > \tau = 10^{-10}$.)

Next we estimate the polynomial part using (4.9a) and (4.9b). Assume that we have $2L = 20$ measurements logarithmically spaced in the high frequency range $[10^f, 10^{f+2}]\,\imath$. The left $\boldsymbol{\ell}_j$ and right $\mathbf{r}_j$ tangential vectors are chosen randomly. Table 1 shows the estimation error for varying frequency ranges. The observed estimation error for both $\widehat{\mathbf{P}}_0, \widehat{\mathbf{P}}_1$ behaves like $O(10^{-2f})$.

In our last experiments we compute the Loewner ROM using the classical Loewner approach (3.7) and the modified Loewner approach, Algorithm 1. Thus, in contrast to the experiments shown in Figs. 4 and 5 we now estimate the polynomial part. Again we assume that we have $2N = 200$ measurements logarithmically spaced in the low frequency range $[10^{-2}, 10^1]\,\imath$. In addition we assume that we have $2L = 20$ measurements logarithmically spaced in the high frequency range

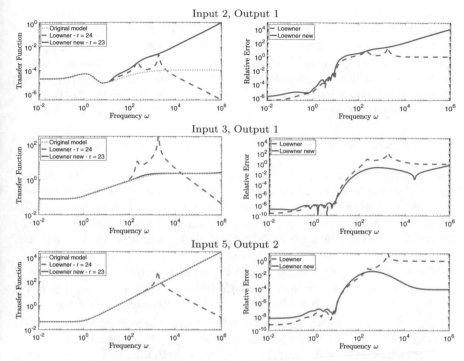

Fig. 6 Left plots: absolute values of frequency responses of the original system (yellow dotted lines) of the reduced system computed with the classical Loewner approach (blue dashed line), and of the reduced system computed with the modified Loewner approach with estimated $\widehat{\mathbf{P}}_0$, $\widehat{\mathbf{P}}_1$ (red solid line) for various components of the 2×6 transfer function. Right plots: corresponding relative errors

$[10^4, 10^6]\,\iota$ to compute estimates $\widehat{\mathbf{P}}_0$ and $\widehat{\mathbf{P}}_1$. In all cases the left $\boldsymbol{\ell}_j$ and right $\mathbf{r}_j$ tangential vectors are chosen randomly.

The left plots in Fig. 6 show the absolute values of frequency responses of the original system (yellow dotted lines) of the reduced system computed with the classical Loewner approach (blue dashed line), and of reduced system computed with the modified Loewner approach (red solid line) for various components of the 2×6 transfer function at 300 logarithmically spaced frequencies $\omega\iota$ in $[10^{-2}, 10^6]\iota$. The right plots in Fig. 6 show the corresponding relative errors.

In most cases the modified Loewner approach improves the approximation properties of the ROM transfer function. For large frequencies $\omega \gg 1$, the estimation error $\omega\,|(\widehat{\mathbf{P}}_1)_{jk} - (\mathbf{P}_1)_{jk}|$ starts to dominate the overall error in transfer function approximation. The beginning of this can be seen in Fig. 6 for Input 3 and Output 1, where the error between FOM transfer function and modified Loewner ROM transfer function begins to grow linearly in ω for $\omega > 10^5$. As indicted by Table 1 the errors $\|\mathbf{P}_0 - \widehat{\mathbf{P}}_0\|_2$, $\|\mathbf{P}_1 - \widehat{\mathbf{P}}_1\|_2$ when $2L$ measurements at higher frequencies are available to compute $\widehat{\mathbf{P}}_0$, $\widehat{\mathbf{P}}_1$. Thus while a linear growth in error between FOM and ROM transfer function is unavoidable when $\mathbf{P}_1$ is present, the impact can be delayed by using measurements at higher frequencies.

The behavior of modified Loewner ROM for the transfer function component corresponding to Input 2 and Output 1 is worse than that of the classical Loewner ROM. Note that this component of the transfer function is substantially smaller than all other components. Moreover, this component of the transfer function has a constant polynomial part, i.e.,

$$\mathbf{H}(\iota\,\omega)_{1,2} = \mathbf{H}_{\mathrm{spr}}(\iota\,\omega)_{1,2} + (\mathbf{P}_0)_{1,2}, \quad (\mathbf{P}_0)_{1,2} \approx 10^{-4}, \ (\mathbf{P}_1)_{1,2} = 0,$$

but is estimated by $\widehat{\mathbf{H}}(\iota\,\omega)_{1,2} = \widehat{\mathbf{H}}_{\mathrm{spr}}(\iota\,\omega)_{1,2} + (\widehat{\mathbf{P}}_0)_{1,2} + \iota\,\omega\,(\widehat{\mathbf{P}}_1)_{1,2}$. The errors in the transfer functions for the modified Loewner and the classical Loewner are nearly identical in the range $[10^{-2}, 10^1]\iota$ where measurements were taken, but both ROM transfer functions have the wrong asymptotics for large frequencies. The difficulty for the modified Loewner approach is that both $(\mathbf{P}_0)_{1,2}$ and $(\mathbf{P}_1)_{1,2}$ are small (in fact $(\mathbf{P}_1)_{1,2} = 0$).

The modified Loewner ROM can be improved somewhat by thresholding. If there is an error estimate τ_0 and τ_1 available such that $|(\mathbf{P}_0)_{j,k} - (\widehat{\mathbf{P}}_0)_{j,k}| \le \tau_0$ and $|(\mathbf{P}_1)_{j,k} - (\widehat{\mathbf{P}}_1)_{j,k}| \le \tau_1$, then for small polynomials components with $|(\widehat{\mathbf{P}}_0)_{j,k}| \le \tau_0$ or $|(\widehat{\mathbf{P}}_1)_{j,k}| \le \tau_1$, respectively, the estimation error may be as large as the estimated quantity itself. Hence for components with $|(\widehat{\mathbf{P}}_0)_{j,k}| \le \tau_0$ we set $(\widehat{\mathbf{P}}_0)_{j,k} = 0$, and for components with $|(\widehat{\mathbf{P}}_1)_{j,k}| \le \tau_1$ we set $(\widehat{\mathbf{P}}_1)_{j,k} = 0$. Unfortunately, currently there is no rigorous error estimate τ_0 and τ_1 available. Motivated by Table 1 we set $\tau_0 = \tau_1 = 10^{-f}$ when the polynomial part is estimated from measurements in the high frequency range $[10^f, 10^{f+2}]\iota$. Specifically, since we have $2L = 20$ measurements logarithmically spaced in the high frequency range $[10^4, 10^6]\iota$ to compute estimates $\widehat{\mathbf{P}}_0$ and $\widehat{\mathbf{P}}_1$ we set $\tau_0 = \tau_1 = 10^{-4}$. With this thresholding $(\widehat{\mathbf{P}}_1)_{1,k} = 0$, $k = 1, \ldots 6$, and $(\widehat{\mathbf{P}}_1)_{2,2} = 0$. The absolute values of frequency responses for the (1,2) component of the transfer function and corresponding relative errors are shown in Fig. 7. The error in transfer function

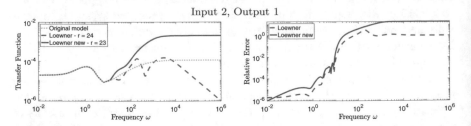

Fig. 7 Left plots: Absolute values of frequency responses of the original system (yellow dotted lines) of the reduced system computed with the classical Loewner approach (blue dashed line), and of the reduced system computed with the modified Loewner approach with estimated $\widehat{\mathbf{P}}_0$, $\widehat{\mathbf{P}}_1$ and thresholding (red solid line) for the (1,2) component of the transfer function. Right plots: corresponding relative errors

for the modified Loewner and the classical Loewner are again nearly identical in the range $[10^{-2}, 10^1] \imath$ where measurements were taken. For large frequencies the observed relative error in the transfer function for the modified Loewner approach is approximately $|(\mathbf{P}_0)_{1,2} - (\widehat{\mathbf{P}}_0)_{1,2}|/|(\mathbf{P}_0)_{1,2}|$, whereas the relative error in the transfer function for the classical Loewner is always asymptotically equal to one. The fundamental issue is that small polynomial components $|(\mathbf{P}_0)_{j,k}| \ll 1$ and especially $|(\mathbf{P}_1)_{j,k}| \ll 1$ need to be estimated with even smaller absolute errors. This is difficult and requires more measurements at higher frequencies.

6 Conclusions

This paper has provided a detailed description of the analytical form of the transfer function for a class of semi-explicit DAE systems, which includes the semi-discretized Oseen equations, and it has introduced a modified version of the data-driven Loewner framework to compute reduced order models (ROMs) for these DAE systems The algorithmic improvement is in the estimation of the polynomial part of the transfer function from measurements and in the incorporation of this estimate into the Loewner ROM construction, which in many cases lead to ROMs with better approximation properties. The modified Loewner approach uses measurements of the transfer function at high frequencies to estimate the polynomial part, and then applies the standard Loewner approach to measurement contributions from the strictly proper part of the transfer function. In particular, the split of the transfer function into a strictly proper and a polynomial part is explicit in the construction of the Loewner ROM to ensure that the resulting ROM transfer function has the same structure. Numerical experiments on the semi-discretized Oseen equations indicate that the modified Loewner approach generates ROMs that better approximate the transfer function if a linear polynomial part is present. In cases, where the polynomial part is linear with a small linear term, the modified Loewner approach can introduce a spurious polynomial part, which then leads to large errors for large frequencies. This can be somewhat avoided by thresholding, but the estimation of small components in the polynomial parts, especially in the linear part remains a difficulty. For the modified Loewner approach precise theoretical error estimates and improvement bounds are not yet available, and are part of future work.

Acknowledgments We thank the referee for carefully reading the paper and providing constructive comments that have lead to a better presentation.

This research was supported in part by NSF grants CCF-1816219 and DMS-1819144.

References

1. Antoulas, A.C.: Approximation of large-scale dynamical systems. In: Advances in Design and Control, vol. 6. Society for Industrial and Applied Mathematics (SIAM), Philadelphia (2005). https://doi.org/10.1137/1.9780898718713
2. Antoulas, A.C., Lefteriu, S., Ionita, A.C.: Chapter 8: a tutorial introduction to the Loewner framework for model reduction. In: Benner, P., Cohen, A., Ohlberger, M., Willcox, K. (eds.) Model Reduction and Approximation: Theory and Algorithms, pp. 335–376. SIAM, Philadelphia (2017). https://doi.org/10.1137/1.9781611974829.ch8
3. Antoulas, A.C., Beattie, C.A., Gugercin, S.: Interpolatory model reduction. In: Computational Science & Engineering, vol. 21. Society for Industrial and Applied Mathematics (SIAM), Philadelphia (2020). https://doi.org/10.1137/1.9781611976083
4. Bänsch, E., Benner, P., Saak, J., Weichelt, H.K.: Riccati-based boundary feedback stabilization of incompressible Navier-Stokes flows. SIAM J. Sci. Comput. **37**(2), A832–A858 (2015). http://dx.doi.org/10.1137/140980016
5. Benner, P., Cohen, A., Ohlberger, M., Willcox, K. (eds.): Model Reduction and Approximation: Theory and Algorithms. Computational Science and Engineering. SIAM, Philadelphia (2017). https://doi.org/10.1137/1.9781611974829
6. Elman, H.C., Silvester, D.J., Wathen, A.J.: Finite Elements and Fast Iterative Solvers with Applications in Incompressible Fluid Dynamics. Numerical Mathematics and Scientific Computation, 2nd edn. Oxford University Press, Oxford (2014). http://dx.doi.org/10.1093/acprof:oso/9780199678792.001.0001
7. Gosea, I.V., Zhang, Q., Antoulas, A.C.: Preserving the DAE structure in the Loewner model reduction and identification framework. Adv. Comput. Math. **46**(1), 3 (2020). https://doi.org/10.1007/s10444-020-09752-8
8. Heinkenschloss, M., Sorensen, D.C., Sun, K.: Balanced truncation model reduction for a class of descriptor systems with application to the Oseen equations. SIAM J. Sci. Comput. **30**(2), 1038–1063 (2008). http://doi.org/10.1137/070681910
9. Hesthaven, J.S., Rozza, G., Stamm, B.: Certified Reduced Basis Methods for Parametrized Partial Differential Equations. Springer Briefs in Mathematics. Springer, New York (2015). http://doi.org/10.1007/978-3-319-22470-1
10. Layton, W.: Introduction to the Numerical Analysis of Incompressible Viscous Flows. Computational Science & Engineering, vol. 6. Society for Industrial and Applied Mathematics (SIAM), Philadelphia (2008). https://doi.org/10.1137/1.9780898718904
11. Mayo, A.J., Antoulas, A.C.: A framework for the solution of the generalized realization problem. Linear Algebra Appl. **425**(2–3), 634–662 (2007). https://doi.org/10.1016/j.laa.2007.03.008
12. Quarteroni, A., Manzoni, A., Negri, F.: Reduced basis methods for partial differential equations. An introduction. In: Unitext, vol. 92. Springer, Cham (2016). https://doi.org/10.1007/978-3-319-15431-2

Part III
Closed-Loop and Optimal Control

Vector Relative Degree and Funnel Control for Differential-Algebraic Systems

Thomas Berger, Huy Hoàng Lê, and Timo Reis

Abstract We consider tracking control for multi-input multi-output differential-algebraic systems. First, the concept of vector relative degree is generalized for linear systems and we arrive at the novel concept of "truncated vector relative degree", and we derive a new normal form. Thereafter, we consider a class of nonlinear functional differential-algebraic systems which comprises linear systems with truncated vector relative degree. For this class we introduce a feedback controller which achieves that, for a given sufficiently smooth reference signal, the tracking error evolves within a pre-specified performance funnel.

Keywords Adaptive control · Differential-algebraic equations · Funnel control · Relative degree

Mathematics Subject Classification (2010) 34A09, 93C05, 93C10, 93C23, 93C40

T. Berger (✉)
Institut für Mathematik, Universität Paderborn, Paderborn, Germany
e-mail: thomas.berger@math.upb.de

H. H. Lê
Department of Mathematics, National University of Civil Engineering, Hai Ba Trung, Hanoi, Vietnam
e-mail: hoanglh@nuce.edu.vn

T. Reis
Fachbereich Mathematik, Universität Hamburg, Hamburg, Germany
e-mail: timo.reis@uni-hamburg.de

1 Introduction

Funnel control has been introduced in [23] almost two decades ago. Meanwhile, plenty of articles have been published in which funnel control from both a theoretical and an applied perspective are considered, see e.g. [3–5, 9, 10, 16, 17, 20, 26, 29] to mention only a few.

A typical assumption in funnel control is that the system has a *strict relative degree*, which means that the input-output behavior can be described by a differential equation which has the same order for all outputs. However, multi-input, multi-output systems that appear in real-world applications do not always have a strict relative degree. Instead, the input-output behavior is described by a collection of differential equations of different order for each output, which is referred to as *vector relative degree*.

The subject of this article is twofold: First we consider linear (not necessarily regular) systems described by differential-algebraic equations (DAEs). We generalize the notion of vector relative degree as given in [1, Def. 5.3.4] for regular DAEs, see [24, 27] for systems of ordinary differential equations (ODEs). Furthermore, we develop a normal form for linear DAE systems which allows to read off this new *truncated* vector relative degree as well as the zero dynamics. Thereafter, we consider a class of nonlinear functional DAE systems which encompasses linear systems in this normal form, and we introduce a new funnel controller for this system class.

Our results generalize, on the one hand, the results of [9], where systems with strict relative degree are considered. On the other hand, concerning funnel control, the results in this article generalize those of [3, 8] for linear and nonlinear DAEs, where the truncated vector relative degree (although this notion does not appear in these articles) is restricted to be component-wise less or equal to one. Note that [3] already encompasses the results found in [7] for linear DAE systems with properly invertible transfer function. DAEs with higher relative degree have been considered in [6], and even this article is comprised by the present results. Therefore, the present article can be seen as a unification of the funnel control results presented in the previous works [3, 6–9] to a fairly general class of nonlinear DAE systems. Parts of our results have been published in the doctoral thesis [25] by one of the authors.

1.1 Nomenclature

Throughout this article, $\mathbb{R}_{\geq 0} = [0, \infty)$ and $\|x\|$ is the Euclidean norm of $x \in \mathbb{R}^n$. The symbols $\mathbb{N}$ denotes the set of natural numbers and $\mathbb{N}_0 = \mathbb{N} \cup \{0\}$. The ring of real polynomials is denoted by $\mathbb{R}[s]$, and $\mathbb{R}(s)$ is its quotient field. In other words, $\mathbb{R}(s)$ is the field of real rational functions. Further, $\mathbf{Gl}_n(\mathbb{R})$ stands for the group of invertible matrices in $\mathbb{R}^{n \times n}$.

The restriction of a function $f : V \to \mathbb{R}^n$ to $W \subseteq V$ is denoted by $f|_W$, $V \subseteq W$. For $p \in [1, \infty]$, $L^p(I \to \mathbb{R}^n)$ $(L^p_{\mathrm{loc}}(I \to \mathbb{R}^n))$ stands for the space of measurable and (locally) p-th power integrable functions $f : I \to \mathbb{R}^n$, $I \subseteq \mathbb{R}$ an interval. Likewise $L^\infty(I \to \mathbb{R}^n)$ $(L^\infty_{\mathrm{loc}}(I \to \mathbb{R}^n))$ is the space of measurable and (locally) essentially bounded functions $f : I \to \mathbb{R}^n$, and $\|f\|_\infty$ stands for the essential supremum of f. Note that functions which agree almost everywhere are identified. Further, for $p \in [1, \infty]$ and $k \in \mathbb{N}_0$, $W^{k,p}(I \to \mathbb{R}^n)$ is the Sobolev space of elements of $L^p(I \to \mathbb{R}^n)$ $(L^p_{\mathrm{loc}}(I \to \mathbb{R}^n))$ with the property that the first k weak derivatives exist and are elements of $L^p(I \to \mathbb{R}^n)$ $(L^p_{\mathrm{loc}}(I \to \mathbb{R}^n))$. Moreover, $C^k(V \to \mathbb{R}^n)$ is the set of k-times continuously differentiable functions $f : V \to \mathbb{R}^n$, $V \subseteq \mathbb{R}^m$, and we set $C(V \to \mathbb{R}^n) := C^0(V \to \mathbb{R}^n)$.

2 Linear Systems and the Truncated Vector Relative Degree

In this section, we consider linear constant coefficient DAE systems

$$\frac{\mathrm{d}}{\mathrm{d}t} Ex(t) = Ax(t) + Bu(t),$$
$$y(t) = Cx(t), \tag{2.1}$$

where $E, A \in \mathbb{R}^{l \times n}$, $B \in \mathbb{R}^{l \times m}$, $C \in \mathbb{R}^{p \times n}$. We denote the class of these systems by $\Sigma_{l,n,m,p}$ and write $[E, A, B, C] \in \Sigma_{l,n,m,p}$. We stress that these systems are not required to be *regular*, which would mean that $l = n$ and $\det(sE - A) \in \mathbb{R}[s] \setminus \{0\}$. The functions $u : \mathbb{R} \to \mathbb{R}^m$, $x : \mathbb{R} \to \mathbb{R}^n$, and $y : \mathbb{R} \to \mathbb{R}^p$ are called *input, (generalized) state variable,* and *output* of the system, respectively. We introduce the *behavior* of system (2.1) as

$$\mathfrak{B}_{[E,A,B,C]} := \left\{ (x, u, y) \in L^1_{\mathrm{loc}}(\mathbb{R} \to \mathbb{R}^n \times \mathbb{R}^m \times \mathbb{R}^p) \; \middle| \right.$$

$$\left. Ex \in W^{1,1}_{\mathrm{loc}}(\mathbb{R} \to \mathbb{R}^l) \wedge \tfrac{\mathrm{d}}{\mathrm{d}t} Ex = Ax + Bu \wedge y = Cx + Du \right\}.$$

Note that the equalities in the above definition are to be understood as equalities of functions in L^1_{loc}. For a regular system $[E, A, B, C] \in \Sigma_{n,n,m,p}$, the *transfer function* is defined by

$$G(s) = C(sE - A)^{-1}B \in \mathbb{R}(s)^{p \times m}.$$

2.1 Zero Dynamics and Right-Invertibility

To specify the class that we consider, we introduce the *zero dynamics* which are the set of solutions resulting in a trivial output. For more details on the concept of zero dynamics and a literature survey we refer to [1].

Definition 2.1 The zero dynamics of $[E, A, B, C] \in \Sigma_{l,n,m,p}$ are the set

$$\mathscr{Z}\mathscr{D}_{[E,A,B,C]} := \left\{ (x, u, y) \in \mathfrak{B}_{[E,A,B,C]} \mid y = 0 \right\}.$$

We call $\mathscr{Z}\mathscr{D}_{[E,A,B,C]}$ *autonomous*, if

$$\forall \omega \in \mathscr{Z}\mathscr{D}_{[E,A,B,C]} \ \forall I \subseteq \mathbb{R} \text{ open interval: } \omega|_I = 0 \ \Rightarrow \ \omega = 0,$$

and *asymptotically stable*, if

$$\forall (x, u, y) \in \mathscr{Z}\mathscr{D}_{[E,A,B,C]} : \lim_{t \to \infty} \left\| (x, u)\big|_{[t,\infty)} \right\|_\infty = 0.$$

Remark 2.1 Let $[E, A, B, C] \in \Sigma_{l,n,m,p}$.

(a) It has been shown in [3, Prop. 3.5] that

$$\mathscr{Z}\mathscr{D}_{[E,A,B,C]} \text{ are autonomous } \iff \ker_{\mathbb{R}(s)} \begin{bmatrix} -sE+A & B \\ C & 0 \end{bmatrix} = \{0\}.$$

In particular, $\begin{bmatrix} -sE+A & B \\ C & 0 \end{bmatrix}$ is left invertible over $\mathbb{R}(s)$ if, and only if, $\mathscr{Z}\mathscr{D}_{[E,A,B,C]}$ are autonomous. If $[E, A, B, C]$ is regular, then its transfer function $G(s)$ satisfies

$$\begin{bmatrix} -sE + A & B \\ C & 0 \end{bmatrix} \begin{bmatrix} I_n & (sE - A)^{-1}B \\ 0 & I_m \end{bmatrix} = \begin{bmatrix} -sE + A & 0 \\ C & G(s) \end{bmatrix}, \qquad (2.2)$$

hence autonomy of the zero dynamics is equivalent to $G(s)$ having full column rank over $\mathbb{R}(s)$, cf. [3, Prop. 4.8].

(b) It has been shown in [3, Lem. 3.11] that

$$\mathscr{Z}\mathscr{D}_{[E,A,B,C]} \text{ are asymptotically stable}$$

$$\iff \ker_{\mathbb{C}} \begin{bmatrix} -\lambda E+A & B \\ C & 0 \end{bmatrix} = \{0\} \text{ for all } \lambda \in \mathbb{C}_+ \text{with } \operatorname{Re}(\lambda) \geq 0.$$

We will consider systems with autonomous zero dynamics throughout this article. We will furthermore assume that the system is *right-invertible*, which is defined in the following.

Definition 2.2 The system $[E, A, B, C] \in \Sigma_{l,n,m,p}$ is called *right-invertible*, if

$$\forall y \in C^\infty(\mathbb{R} \to \mathbb{R}^p) \; \exists (x, u) \in L^1_{\text{loc}}(\mathbb{R} \to \mathbb{R}^n \times \mathbb{R}^m) : (x, u, y) \in \mathfrak{B}_{[E,A,B,C]}.$$

The notion of right-invertibility has been used in [30, Sec. 8.2] for systems governed by ordinary differential equations and in [2, 3] for the differential-algebraic case. The concept is indeed motivated by tracking control: Namely, right-invertibility means that any smooth signal can be tracked by the output on a right-invertible system.

Remark 2.2 Consider a regular system $[E, A, B, C] \in \Sigma_{n,n,m,p}$ with transfer function $G(s)$. It has been shown in [3, Prop. 4.8] that

$$[E, A, B, C] \text{ is right-invertible } \iff \text{im}_{\mathbb{R}(s)} G(s) = \mathbb{R}(s)^p,$$

whence, by (2.2),

$$[E, A, B, C] \text{ is right-invertible } \iff \text{im}_{\mathbb{R}(s)} \begin{bmatrix} -sE+A & B \\ C & 0 \end{bmatrix} = \mathbb{R}(s)^{n+p},$$

Combining this with Remark 2.1 (a), we can infer from the dimension formula that for regular square systems $[E, A, B, C] \in \Sigma_{n,n,m,m}$ (i.e., the dimensions of input and output coincide) with transfer function $G(s) \in \mathbb{R}(s)^{m \times m}$, the following statements are equivalent:

(i) $\mathscr{L}\mathscr{D}_{[E,A,B,C]}$ are autonomous,
(ii) $[E, A, B, C]$ is right-invertible,
(iii) $G(s) \in \mathbb{R}(s)^{m \times m}$ is invertible over $\mathbb{R}(s)$,
(iv) $\begin{bmatrix} -sE+A & B \\ C & 0 \end{bmatrix}$ is invertible over $\mathbb{R}(s)$.

For general right-invertible systems with autonomous zero dynamics, we can derive a certain normal form under state space transformation. The following result is a straightforward combination of [3, Lem. 4.2 & Thm. 4.3 & Prop. 4.6].

Theorem 2.1 *Let a right-invertible system $[E, A, B, C] \in \Sigma_{l,n,m,p}$ with autonomous zero dynamics be given. Then there exist $W \in \mathbf{Gl}_l(\mathbb{R})$, $T \in \mathbf{Gl}_n(\mathbb{R})$ such that*

$$W(sE - A)T = \begin{bmatrix} sI_{n_1} - Q & -A_{12} & 0 \\ -A_{21} & sE_{22} - A_{22} & sE_{23} \\ 0 & sE_{32} & sN - I_{n_3} \\ 0 & 0 & -sE_{43} \end{bmatrix}, \quad WB = \begin{bmatrix} 0 \\ I_m \\ 0 \\ 0 \end{bmatrix},$$

$$CT = \begin{bmatrix} 0 & I_p & 0 \end{bmatrix}, \tag{2.3}$$

where $n_1, n_3, n_4 \in \mathbb{N}_0$, $N \in \mathbb{R}^{n_3 \times n_3}$ is nilpotent and

$$Q \in \mathbb{R}^{n_1 \times n_1}, \quad A_{12} \in \mathbb{R}^{n_1 \times p}, \quad A_{21} \in \mathbb{R}^{m \times n_1},$$
$$E_{22}, A_{22} \in \mathbb{R}^{m \times p}, \quad E_{23} \in \mathbb{R}^{m \times n_3},$$
$$E_{32} \in \mathbb{R}^{n_3 \times p}, \quad E_{43} \in \mathbb{R}^{n_4 \times n_3}$$

are such that $E_{43} N^j E_{32} = 0$ for all $j \in \mathbb{N}_0$.

Remark 2.3 Let $[E, A, B, C] \in \Sigma_{l,n,m,p}$ be right-invertible and have autonomous zero dynamics. Using the form (2.3), we see that $(x, u, y) \in \mathfrak{B}_{[E,A,B,C]}$ if, and only if,

$$Tx = (\eta^\top, y^\top, x_3^\top)^\top \in L^1_{\text{loc}}(\mathbb{R} \to \mathbb{R}^{n_1 + p + n_3})$$

satisfies

$$\begin{bmatrix} E_{22} \\ E_{32} \end{bmatrix} y \in W^{1,1}_{\text{loc}}(\mathbb{R} \to \mathbb{R}^{m+n_3}), \quad \begin{bmatrix} E_{23} \\ N \\ E_{43} \end{bmatrix} x_3 \in W^{1,1}_{\text{loc}}(\mathbb{R} \to \mathbb{R}^{m+n_3+n_4})$$

$$(2.4)$$

and the equations

$$\dot{\eta} = Q\eta + A_{12}y, \tag{2.5a}$$

$$0 = - \sum_{i=0}^{n_3-1} E_{23} N^i E_{32} y^{(i+2)} - E_{22}\dot{y} + A_{22}y + A_{21}\eta + u, \tag{2.5b}$$

$$x_3 = \sum_{i=0}^{n_3-1} N^i E_{32} y^{(i+1)} \tag{2.5c}$$

hold in the distributional sense. In particular, the zero dynamics of $[E, A, B, C]$ are asymptotically stable if, and only if, any eigenvalue of Q has negative real part. Further note that $\eta \in L^1_{\text{loc}}(\mathbb{R} \to \mathbb{R}^{n_1})$, $y \in L^1_{\text{loc}}(\mathbb{R} \to \mathbb{R}^p)$ together with (2.5a) imply that $\eta \in W^{1,1}_{\text{loc}}(\mathbb{R} \to \mathbb{R}^{n_1})$.

2.2 Truncated Vector Relative Degree

Our aim in this section is to present a suitable generalization of the concept of vector relative degree to differential-algebraic systems which are not necessarily regular. For regular systems a definition of this concept is given in [3, Def. B.1].

Definition 2.3 Let a regular system $[E, A, B, C] \in \Sigma_{n,n,m,p}$ with transfer function $G(s) \in \mathbb{R}(s)^{p \times m}$ be given. We say that $[E, A, B, C]$ has *vector relative degree* $(r_1, \ldots, r_p) \in \mathbb{Z}^{1 \times p}$, if there exists a matrix $\Gamma \in \mathbb{R}^{p \times m}$ with rk $\Gamma = p$ and

$$\lim_{\lambda \to \infty} \mathrm{diag}(\lambda^{r_1}, \ldots, \lambda^{r_p}) G(\lambda) = \Gamma.$$

If the above holds with $r_1 = \ldots = r_p =: r$, then we say that $[E, A, B, C]$ has *strict relative degree r*.

Since this definition involves the transfer function, it is only applicable to regular systems. To avoid this limitation, we introduce a novel concept. Let us start by introducing the notion of column degree of a rational matrix. This generalizes the concept of column degree for polynomial matrices in [15, Sec. 2.4].

Definition 2.4 For a rational function $r(s) = \frac{p(s)}{q(s)} \in \mathbb{R}(s)$ we define

$$\deg r(s) := \deg p(s) - \deg q(s).$$

Further, for $r(s) = (r_1(s), r_2(s), \ldots, r_p(s))^\top \in \mathbb{R}(s)^p$ we define

$$\deg r(s) = \max_{1 \le i \le p} \deg r_i(s).$$

Note that the degree of a rational function $r(s) = \frac{p(s)}{q(s)}$ is independent of the choice of $p(s)$ and $q(s)$, i.e., they do not need to be coprime.

If $[E, A, B, C] \in \Sigma_{l,n,m,p}$ has autonomous zero dynamics, then we can conclude from Remark 2.1 that $\begin{bmatrix} -sE+A & B \\ C & 0 \end{bmatrix} \in \mathbb{R}(s)^{(l+p) \times (n+m)}$ possesses a left inverse $L(s) \in \mathbb{R}(s)^{(n+m) \times (l+p)}$. Then we set

$$H(s) := -\begin{bmatrix} 0 & I_m \end{bmatrix} L(s) \begin{bmatrix} 0 \\ I_p \end{bmatrix} \in \mathbb{R}(s)^{m \times p}. \tag{2.6}$$

Remark 2.4

(a) Assume that $[E, A, B, C] \in \Sigma_{l,n,m,p}$ has autonomous zero dynamics and is right-invertible. Then it has been shown in [3, Lem. A.1] that the rational matrix $H(s) \in \mathbb{R}(s)^{m \times p}$ is uniquely determined by $[E, A, B, C]$. Moreover, with the notation from Theorem 2.1, we have

$$H(s) = sE_{22} - A_{22} - A_{21}(sI_{n_1} - Q)^{-1} A_{12} - s^2 E_{23}(sN - I_{n_3})^{-1} E_{32}. \tag{2.7}$$

We stress that the above representation is independent of the transformation matrices W and T in (2.3).

(b) If $[E, A, B, C] \in \Sigma_{n,n,m,m}$ has autonomous zero dynamics and is regular with transfer function $G(s) \in \mathbb{R}(s)^{m \times m}$, then, invoking (2.2) and Remark 2.2, it can be shown that $H(s) = G(s)^{-1}$, see also [3, Rem. A.4].

In view of Remark 2.4, we see that for any regular system $[E, A, B, C] \in \Sigma_{n,n,m,m}$ with transfer function $G(s)$ and vector relative degree $(r_1, \ldots, r_m)$, we have

$$\lim_{\lambda \to \infty} \mathrm{diag}(\lambda^{r_1}, \ldots, \lambda^{r_m}) G(\lambda) = \Gamma \in \mathbf{Gl}_m(\mathbb{R})$$

$$\iff \quad \lim_{\lambda \to \infty} H(\lambda) \, \mathrm{diag}(\lambda^{-r_1}, \ldots, \lambda^{-r_m}) = \Gamma^{-1} \in \mathbf{Gl}_m(\mathbb{R}), \tag{2.8}$$

with $H(s)$ as in (2.6). This motivates to use $H(s)$ instead of the transfer function $G(s)$ to define a generalization of the vector relative degree to DAE systems which are not necessarily regular.

Definition 2.5 Assume that $[E, A, B, C] \in \Sigma_{l,n,m,p}$ is right-invertible and has autonomous zero dynamics. Let $H(s) \in \mathbb{R}(s)^{m \times p}$ be defined as in (2.6) (which is well-defined by Remark 2.4 (a), $h_i(s) = H(s)e_i \in \mathbb{R}(s)^m$ for $i = 1, \ldots, p$ and set $r_i = \max\{\deg h_i(s), 0\}$. Let q be the number of nonzero entries of $(r_1, \ldots, r_p)$,

$$\hat{\Gamma} := \lim_{\lambda \to \infty} H(\lambda) \, \mathrm{diag}(\lambda^{-r_1}, \ldots, \lambda^{-r_p}) \in \mathbb{R}^{m \times p}, \tag{2.9}$$

and $\hat{\Gamma}_q \in \mathbb{R}^{m \times q}$ be the matrix which is obtained from $\hat{\Gamma}$ by deleting all the columns corresponding to $r_i = 0$. Then we call $r = (r_1, \ldots, r_p) \in \mathbb{N}_0^{1 \times p}$ the *truncated vector relative degree* of the system $[E, A, B, C]$, if $\mathrm{rk}\,\hat{\Gamma}_q = q$.
A truncated vector relative degree $(r_1, \ldots, r_p)$ is called *ordered*, if $r_1 \geq \ldots \geq r_p$.

Remark 2.5 Let the system $[E, A, B, C] \in \Sigma_{l,n,m,p}$ be right invertible and have autonomous zero dynamics.

(a) Assume that $[E, A, B, C]$ has ordered truncated vector relative degree $(r_1, \ldots, r_q, 0, \ldots, 0)$ with $r_q > 0$. Then the matrices $\hat{\Gamma}$ and $\hat{\Gamma}_q$ in Definition 2.5 are related by

$$\hat{\Gamma}_q = \hat{\Gamma} \begin{bmatrix} I_q \\ 0 \end{bmatrix}.$$

(b) Assume that $[E, A, B, C]$ has truncated vector relative degree $(r_1, \ldots, r_p) \in \mathbb{N}_0^{1 \times p}$. Consider a permutation matrix $P_\sigma \in \mathbb{R}^{p \times p}$ induced by the permutation $\sigma : \{1, \ldots, p\} \to \{1, \ldots, p\}$. A straightforward calculation shows that $H_\sigma(s)$ as in (2.6) corresponding to $[E, A, B, P_\sigma C]$ satisfies $H_\sigma(s) = H(s)P_\sigma$, thus the system $[E, A, B, P_\sigma C]$ has truncated vector relative degree $(r_{\sigma(1)}, \ldots, r_{\sigma(p)})$. In particular, there exists a permutation σ such that the output-permuted system $[E, A, B, P_\sigma C]$ has ordered truncated vector relative degree.

(c) Assume that $[E, A, B, C]$ has ordered truncated vector relative degree $(r_1, \ldots, r_p) \in \mathbb{N}_0^{1 \times p}$. Using the notation from Theorem 2.1 and (2.7), we obtain that

$$
\begin{aligned}
\hat{\Gamma} &= \lim_{\lambda \to \infty} H(\lambda) \operatorname{diag}(\lambda^{-r_1}, \ldots, \lambda^{-r_p}) \\
&= \lim_{\lambda \to \infty} \left[(\lambda E_{22} - A_{22}) - A_{21}(\lambda I - Q)^{-1} A_{12} - \lambda^2 E_{23}(\lambda N - I_{n_3})^{-1} E_{32} \right] \cdot \\
&\qquad\qquad \cdot \operatorname{diag}(\lambda^{-r_1}, \ldots, \lambda^{-r_p}) \\
&= \lim_{\lambda \to \infty} \left[\lambda E_{22} - A_{22} + \sum_{k=0}^{n_3-1} \lambda^{k+2} E_{23} N^k E_{32} \right] \operatorname{diag}(\lambda^{-r_1}, \ldots, \lambda^{-r_p}).
\end{aligned}
$$

(d) Consider a regular system $[E, A, B, C] \in \Sigma_{n,n,m,p}$. If $m > p$, then, in view of Remark 2.1 (a), the zero dynamics of $[E, A, B, C]$ are not autonomous, because $\begin{bmatrix} -sE+A & B \\ C & 0 \end{bmatrix}$ has a non-trivial kernel over $\mathbb{R}(s)$. Therefore, such a system does not have a truncated vector relative degree, but a vector relative degree may exist. As an example consider the system $[E, A, B, C] \in \Sigma_{1,1,2,1}$ with $E = C = [1]$, $A = [0]$ and $B = [1, 1]$, for which a truncated relative degree does not exist. However, the transfer function is given by $G(s) = s^{-1}[1, 1]$ and hence the system even has strict relative degree $r = 1$.

If $m \le p$ and $[E, A, B, C]$ has a vector relative degree, then also a truncated vector relative degree exists. This can be seen as follows: First observe that, as a consequence of Definition 2.3, $p \le m$ and hence we have $p = m$. Therefore, the matrix $\Gamma \in \mathbb{R}^{m \times m}$ in Definition 2.3 is invertible. Let $F(s) := \operatorname{diag}(\lambda^{r_1}, \ldots, \lambda^{r_m}) G(s)$, then $F(s) = \Gamma + G_{\mathrm{sp}}(s)$ for some *strictly proper* $G_{\mathrm{sp}}(s) \in \mathbb{R}(s)^{m \times m}$, i.e., $\lim_{\lambda \to \infty} G_{\mathrm{sp}}(\lambda) = 0$. Then $\tilde{G}(s) := -\Gamma^{-1} G_{\mathrm{sp}}(s)$ is strictly proper as well. Let $p(s) \in \mathbb{R}(s)^m$ be such that $F(s)p(s) = 0$, then $p(s) = \tilde{G}(s)p(s)$. A component-wise comparison of the degrees yields that

$$
\forall i = 1, \ldots, m : \ \deg p_i(s) = \deg \sum_{j=1}^m \tilde{G}_{ij}(s) p_j(s) \le \max_{j=1,\ldots,m} \left(\deg p_j(s) - 1 \right),
$$

because $\deg \tilde{G}_{ij}(s) \le -1$ for all $i, j = 1, \ldots, m$. Therefore,

$$
\max_{i=1,\ldots,m} \deg p_i(s) \le \max_{j=1,\ldots,m} \left(\deg p_j(s) - 1 \right) = \left(\max_{j=1,\ldots,m} \deg p_j(s) \right) - 1,
$$

a contradiction. This shows that $F(s)$ is invertible over $\mathbb{R}(s)$ and hence $G(s)$ is invertible over $\mathbb{R}(s)$. Then Remark 2.2 yields that $[E, A, B, C]$ is right-invertible and has autonomous zero dynamics. Moreover, Remark 2.4 (b) gives that $H(s) = -G(s)^{-1}$ and hence it follows that a truncated vector relative degree exists with $\hat{\Gamma} = -\Gamma^{-1}$ as in (2.9).

(e) If $[E, A, B, C] \in \Sigma_{n,n,m,m}$ is regular and has autonomous zero dynamics, then $[E, A, B, C]$ has truncated vector relative degree $(0, \ldots, 0)$ if, any only if, the transfer function $G(s) \in \mathbb{R}(s)^{m \times m}$ of $[E, A, B, C]$ is *proper*, i.e., $\lim_{\lambda \to \infty} G(\lambda) \in \mathbb{R}^{m \times m}$ exists. This is an immediate consequence of the fact that, by Remark 2.4 (b), the matrix $H(s)$ in (2.6) satisfies $G(s)^{-1}$.

(f) A motivation for the definition of the truncated vector relative degree, even when only regular systems are considered, is given by output feedback control: Whilst the regular system $[E, A, B, C] \in \Sigma_{2,2,1,1}$ with

$$
E = \begin{bmatrix} 0 & 1 \\ 0 & 0 \end{bmatrix}, \quad A = \begin{bmatrix} 1 & 0 \\ 0 & 1 \end{bmatrix}, \quad B = \begin{bmatrix} 0 \\ 1 \end{bmatrix}, \quad C = \begin{bmatrix} 1 & 0 \end{bmatrix}
$$

has transfer function $G(s) = -s$ and thus vector relative degree $(r_1) = (-1)$, application of the static output feedback $u(t) = Ky(t) + v(t)$ with new input v leads to the system $[E, A+BKC, B, C]$ with transfer function $G_K(s) = \frac{-s}{1+Ks}$. We may infer that the vector relative degree of $[E, A + BKC, B, C]$ is zero unless $K = 0$, thus the vector relative degree is not invariant under output feedback in general.

In the following we show that the truncated vector relative degree is however invariant under static output feedback.

Proposition 2.2 *Let $[E, A, B, C] \in \Sigma_{l,n,m,p}$ and $K \in \mathbb{R}^{m \times p}$ be given. Then the following statements hold:*

(a) $\mathscr{L}\mathscr{D}_{[E,A,B,C]}$ *are autonomous if, and only if, $\mathscr{L}\mathscr{D}_{[E,A+BKC,B,C]}$ are autonomous.*

(b) $[E, A, B, C]$ *is right-invertible if, and only if, $[E, A + BKC, B, C]$ is right-invertible.*

(c) $[E, A, B, C]$ *has a truncated vector relative degree if, and only if, $[E, A + BKC, B, C]$ has a truncated vector relative degree. In this case, the truncated vector relative degrees of $[E, A, B, C]$ and $[E, A + BKC, B, C]$ coincide.*

Proof

(a) This follows from Remark 2.1 (a) together with

$$
\begin{bmatrix} -sE + A + BKC & B \\ C & 0 \end{bmatrix} = \begin{bmatrix} I_l & BK \\ 0 & I_p \end{bmatrix} \begin{bmatrix} -sE + A & B \\ C & 0 \end{bmatrix}. \tag{2.10}
$$

(b) Since $[E, A + BKC, B, C]$ is obtained from $[E, A, B, C]$ by output feedback $u(t) = Ky(t) + v(t)$ with new input $v \in L^1_{\text{loc}}(\mathbb{R} \to \mathbb{R}^m)$, we obtain that $(x, u, y) \in \mathfrak{B}_{[E,A,B,C]}$ if, and only if, $(x, u - Ky, y) \in \mathfrak{B}_{[E,A+BKC,B,C]}$. In particular, the set of generated outputs of $[E, A, B, C]$ and $[E, A+BKC, B, C]$

are the same, whence $[E, A, B, C]$ is right-invertible if, and only if, $[E, A + BKC, B, C]$ is right-invertible.

(c) Since $[E, A, B, C]$ is obtained from $[E, A + BKC, B, C]$ by applying the feedback $-K$, it suffices to prove one implication. In view of Remark 2.5 (b), it is no loss of generality to assume that $[E, A, B, C]$ has ordered truncated vector relative degree $(r_1, \ldots, r_q, 0 \ldots, 0) \in \mathbb{N}_0^{1 \times p}$ with $r_q > 0$. Let $L(s), L_K(s) \in \mathbb{R}(s)^{(n+m) \times (l+p)}$ be left inverses of

$$\begin{bmatrix} -sE + A & B \\ C & 0 \end{bmatrix} \quad \text{and} \quad \begin{bmatrix} -sE + A + BKC & B \\ C & 0 \end{bmatrix}, \quad \text{resp.,}$$

and partition

$$L(s) = \begin{bmatrix} L_{11}(s) & L_{12}(s) \\ L_{21}(s) & H(s) \end{bmatrix}.$$

From (2.10) it follows that $L(s) \begin{bmatrix} I_l & -BK \\ 0 & I_p \end{bmatrix}$ is a left inverse of $\begin{bmatrix} -sE+A+BKC & B \\ C & 0 \end{bmatrix}$. Since $H_K(s) = [0, I_m] L_K(s) \begin{bmatrix} 0 \\ I_p \end{bmatrix}$ is independent of the choice of the left inverse $L_K(s)$ by Remark 2.4 (a), we may infer that

$$\begin{aligned} H_K(s) &= \begin{bmatrix} 0 & I_m \end{bmatrix} L(s) \begin{bmatrix} I_n & -BK \\ 0 & I_m \end{bmatrix} \begin{bmatrix} 0 \\ I_p \end{bmatrix} \\ &= \begin{bmatrix} 0 & I_m \end{bmatrix} \begin{bmatrix} L_{11}(s) & L_{12}(s) \\ L_{21}(s) & H(s) \end{bmatrix} \begin{bmatrix} -BK \\ I_p \end{bmatrix} \\ &= H(s) - L_{21}(s)BK. \end{aligned}$$

The relation $L(s) \begin{bmatrix} -sE+A & B \\ C & 0 \end{bmatrix} = I_{n+m}$ leads to $L_{21}(s)B = I_m$. Therefore, $H_K(s) = H(s) - K$ and we find

$$\begin{aligned} \hat{\Gamma}_K &= \lim_{\lambda \to \infty} H_K(\lambda) \operatorname{diag}(\lambda^{-r_1}, \ldots, \lambda^{-r_q}, 1, \ldots, 1) \\ &= \hat{\Gamma} - \lim_{\lambda \to \infty} K \operatorname{diag}(\lambda^{-r_1}, \ldots, \lambda^{-r_q}, 1, \ldots, 1). \end{aligned}$$

This implies that $\hat{\Gamma}_K \begin{bmatrix} I_q \\ 0 \end{bmatrix} = \hat{\Gamma}_q$, and thus

$$\operatorname{rk} \hat{\Gamma}_K \begin{bmatrix} I_q \\ 0 \end{bmatrix} = \operatorname{rk} \hat{\Gamma}_q = q.$$

Therefore, the truncated vector relative degree of the feedback system $[E, A + BKC, B, C]$ is $(r_1, \ldots, r_q, 0 \ldots, 0)$, i.e., that of $[E, A, B, C]$. $\square$

Remark 2.6

(a) The truncated vector relative degree of a right-invertible system with autonomous zero dynamics does not necessarily exist: For instance, consider $[E, A, B, C] \in \Sigma_{4,4,2,2}$ with

$$E = \begin{bmatrix} 1 & 0 & 0 & 0 \\ 0 & 1 & 1 & 0 \\ 0 & 1 & 1 & 0 \\ 0 & 0 & 0 & 0 \end{bmatrix}, \quad A = \begin{bmatrix} -1 & 0 & 0 & 0 \\ 0 & 1 & -1 & 0 \\ 0 & 1 & 2 & 0 \\ 0 & 0 & 0 & 1 \end{bmatrix}, \quad B = \begin{bmatrix} 0 & 0 \\ 1 & 0 \\ 0 & 1 \\ 0 & 0 \end{bmatrix}, \quad C = \begin{bmatrix} 0 & 1 & 0 & 0 \\ 0 & 0 & 1 & 0 \end{bmatrix}.$$

For this system, we have

$$H(s) = \begin{bmatrix} s - 1 & s + 1 \\ s - 1 & s - 2 \end{bmatrix}.$$

Moreover,

$$\hat{\Gamma} = \lim_{\lambda \to \infty} H(\lambda) \, \mathrm{diag}(\lambda^{-1}, \lambda^{-1}) = \begin{bmatrix} 1 & 1 \\ 1 & 1 \end{bmatrix} = \hat{\Gamma}_q.$$

Since $\mathrm{rk}\,\hat{\Gamma}_q = 1 < 2$, which is the number of columns of $H(s)$ with positive degree. Hence, this system does not have a truncated vector relative degree.

(b) There exist right-invertible regular systems with autonomous zero dynamics with the property that the truncated vector relative degree exists, but the vector relative degree according to Definition 2.3 does not exist. For instance, consider $[E, A, B, C] \in \Sigma_{5,5,2,2}$ with

$$E = \begin{bmatrix} 1 & 0 & 0 & 0 & 0 \\ 0 & 1 & 0 & 1 & 0 \\ 0 & -1 & 0 & 0 & 0 \\ 0 & 0 & 0 & 0 & 1 \\ 0 & 1 & 0 & 0 & 0 \end{bmatrix}, \quad A = \begin{bmatrix} -1 & 1 & -2 & 0 & 0 \\ 3 & 5 & 0 & 0 & 0 \\ 0 & 0 & 0 & 0 & 0 \\ 0 & 0 & 0 & 1 & 0 \\ 0 & 0 & 0 & 0 & 1 \end{bmatrix}, \quad B = \begin{bmatrix} 0 & 0 \\ 1 & 0 \\ 0 & 1 \\ 0 & 0 \\ 0 & 0 \end{bmatrix}, \quad C = \begin{bmatrix} 0 & 1 & 0 & 0 & 0 \\ 0 & 0 & 1 & 0 & 0 \end{bmatrix}.$$

(2.11)

Then

$$G(s) = C(sE - A)^{-1}B = \begin{bmatrix} 0 & -\frac{1}{s} \\ \frac{s+1}{6} & \frac{s^4 + s^3 + s^2 - 4s - 8}{6s} \end{bmatrix}.$$

We have

$$\Gamma := \lim_{\lambda \to \infty} \operatorname{diag}(\lambda, \lambda^{-3}) G(\lambda) = \begin{bmatrix} 0 & -1 \\ 0 & \frac{1}{6} \end{bmatrix}, \text{ and } \operatorname{rk} \Gamma = 1 < 2.$$

This implies that the system does not have vector relative degree in the sense of Definition 2.3. Invoking Remark 2.4 (b), we obtain

$$H(s) = G(s)^{-1} = \begin{bmatrix} \frac{s^4 + s^3 + s^2 - 4s - 8}{s+1} & \frac{6}{s+1} \\ -s & 0 \end{bmatrix},$$

and

$$\hat{\Gamma} := \lim_{\lambda \to \infty} H(\lambda) \operatorname{diag}(\lambda^{-3}, 1) = \begin{bmatrix} 1 & 0 \\ 0 & 0 \end{bmatrix} \text{ and } \hat{\Gamma}_q = \begin{bmatrix} 1 \\ 0 \end{bmatrix}.$$

Then $\operatorname{rk} \hat{\Gamma}_q = 1 = q$, and consequently this system has truncated vector relative degree $(3, 0)$.

2.3 A Representation for Systems with Truncated Vector Relative Degree

For ODE systems, Byrnes and Isidori have introduced a normal form under state space transformation which allows to read off the relative degree and internal dynamics [11, 24]. This normal form plays an important role in designing local and global stabilizing feedback controllers for nonlinear systems [12–14], adaptive observers [28], and adaptive controllers [19, 22]. A normal form for linear ODE systems with vector relative degree has been developed in [27]. Further, a normal form for regular linear DAE systems with strict relative degree has been derived in [6], whereas a normal form for regular linear differential-algebraic systems with proper inverse transfer function in [7]. The latter has been extended to (not necessarily regular) DAE systems with truncated vector relative degree pointwise less or equal to one in [3], although this notion was not used there. Note that the concept of truncated vector relative degree encompasses systems governed by ODEs with strict or vector relative degree as well as regular DAE systems with strict relative degree (up to some extent, cf. Remark 2.5 (d)) or proper inverse transfer function, and we introduce a novel representation which comprises all the aforementioned results.

Assume that $[E, A, B, C] \in \Sigma_{l,n,m,p}$ is right-invertible, has autonomous zero dynamics and possesses a truncated vector relative degree $(r_1, \ldots, r_p) \in \mathbb{N}_0^{1 \times p}$. By Remark 2.5 (b), it is further no loss of generality to assume that the latter is ordered, i.e., $r_1 \geq \ldots \geq r_q > 0 = r_{q+1} = \ldots = r_p$. Introduce the polynomial matrix

$$F(s) := sE_{22} - A_{22} + \sum_{k=0}^{n_3-1} s^{k+2} E_{23} N^k E_{32} \in \mathbb{R}(s)^{m \times p}.$$

By Remark 2.5 (c) we have

$$\hat{\Gamma} = \lim_{\lambda \to \infty} H(\lambda) \operatorname{diag}(\lambda^{-r_1}, \ldots, \lambda^{-r_q}, 1, \ldots, 1) = \lim_{\lambda \to \infty} F(\lambda) \operatorname{diag}(\lambda^{-r_1}, \ldots, \lambda^{-r_q}, 1, \ldots, 1)$$

$$= \begin{bmatrix} \hat{\Gamma}_{11} & \hat{\Gamma}_{12} \\ \hat{\Gamma}_{21} & \hat{\Gamma}_{22} \end{bmatrix} \in \mathbb{R}^{m \times p},$$

where the latter partition is with $\hat{\Gamma}_{11} \in \mathbb{R}^{q \times q}$, $\hat{\Gamma}_{12} \in \mathbb{R}^{q \times (p-q)}$, $\hat{\Gamma}_{21} \in \mathbb{R}^{(m-q) \times q}$ and $\hat{\Gamma}_{22} \in \mathbb{R}^{(m-q) \times (p-q)}$. Then Definition 2.5 yields

$$\operatorname{rk} \begin{bmatrix} \hat{\Gamma}_{11} \\ \hat{\Gamma}_{21} \end{bmatrix} = \operatorname{rk} \hat{\Gamma} \begin{bmatrix} I_q \\ 0 \end{bmatrix} = q.$$

Let $h \in \mathbb{N}$ be such that $r_h > 1$ and $r_{h+1} = 1$. Denote the jth column of a matrix M by $M^{(j)}$. Then

$$\hat{\Gamma} = \lim_{\lambda \to \infty} F(\lambda) \operatorname{diag}(\lambda^{-r_1}, \ldots, \lambda^{-r_q}, 1, \ldots, 1)$$

$$= \left[E_{23} N^{r_1-2} E_{32}^{(1)} \ \ldots \ E_{23} N^{r_h-2} E_{32}^{(h)} \ E_{22}^{(h+1)} \ \ldots \ E_{22}^{(q)} \ -A_{22}^{(q+1)} \ \ldots \ -A_{22}^{(p)} \right],$$

and thus

$$\hat{\Gamma}_q = \begin{bmatrix} \hat{\Gamma}_{11} \\ \hat{\Gamma}_{21} \end{bmatrix} = \left[E_{23} N^{r_1-2} E_{32}^{(1)} \ \ldots \ E_{23} N^{r_h-2} E_{32}^{(h)} \ E_{22}^{(h+1)} \ \ldots \ E_{22}^{(q)} \right] \in \mathbb{R}^{m \times q}.$$

$$(2.12)$$

Since $\operatorname{rk} \hat{\Gamma}_q = q$, by reordering the inputs and—accordingly—reording the rows of A_{21}, E_{22}, A_{22} and E_{23}, it is no loss of generality to assume that the first q rows of $\hat{\Gamma}_q$ are linearly independent, thus $\hat{\Gamma}_{11} \in \mathbf{Gl}_q(\mathbb{R})$. Consider the matrix

$$\Gamma := \begin{bmatrix} \Gamma_{11} & 0 \\ \Gamma_{21} & I_{m-q} \end{bmatrix} \in \mathbf{Gl}_m(\mathbb{R}), \qquad (2.13)$$

where $\Gamma_{11} = \hat{\Gamma}_{11}^{-1} \in \mathbf{Gl}_q(\mathbb{R})$, $\Gamma_{21} = -\hat{\Gamma}_{21} \hat{\Gamma}_{11}^{-1} \in \mathbb{R}^{(m-q) \times q}$, then

$$\Gamma \hat{\Gamma}_q = \begin{bmatrix} I_q \\ 0 \end{bmatrix}. \qquad (2.14)$$

On the other hand, using the notation from Theorem 2.1 and invoking Remark 2.3, we have that $(x, u, y) \in \mathfrak{B}_{[E,A,B,C]}$ if, and only if, $Tx = (\eta^\top, y^\top, x_3^\top)^\top \in L^1_{\text{loc}}(\mathbb{R} \to \mathbb{R}^{n_1+p+n_3})$ solves (2.5) in the distributional sense, and the components satisfy (2.4). Since (2.5b) can be written as $F(\frac{d}{dt})y = A_{21}\eta + u$, by construction of $\hat{\Gamma}_q$ and (2.12) we may rewrite this as

$$
\hat{\Gamma}_q \begin{pmatrix} y_1^{(r_1)} \\ \vdots \\ y_q^{(r_q)} \end{pmatrix} = M_1 \begin{pmatrix} y_1 \\ \vdots \\ y_1^{(r_1-1)} \end{pmatrix} + \ldots + M_q \begin{pmatrix} y_q \\ \vdots \\ y_q^{(r_q-1)} \end{pmatrix} + M \begin{pmatrix} y_{q+1} \\ \vdots \\ y_m \end{pmatrix} + A_{21}\eta + u
$$

$$(2.15)$$

for some $M_1 \in \mathbb{R}^{m \times r_1}, \ldots, M_q \in \mathbb{R}^{m \times r_q}$, $M \in \mathbb{R}^{m \times (p-q)}$ which can be constructed from the columns of $E_{23}N^i E_{32}$, E_{22} and A_{22}, $i = 0, \ldots, r_1$. Define $R_{j,1} \in \mathbb{R}^{q \times r_j}$, $R_{j,2} \in \mathbb{R}^{(m-q) \times r_j}$ for $j = 1, \ldots, q$ and $S_1 \in \mathbb{R}^{q \times (p-q)}$, $S_2 \in \mathbb{R}^{(m-q) \times (p-q)}$, $P_1 \in \mathbb{R}^{q \times n_1}$, $P_2 \in \mathbb{R}^{(m-q) \times n_1}$ by

$$
\begin{bmatrix} R_{j,1} \\ R_{j,2} \end{bmatrix} := \Gamma M_j, \quad j = 1, \ldots, q, \qquad \begin{bmatrix} S_1 \\ S_2 \end{bmatrix} := \Gamma M, \qquad \begin{bmatrix} P_1 \\ P_2 \end{bmatrix} := \Gamma A_{21}.
$$

$$(2.16)$$

By a multiplication of (2.15) from the left with $\Gamma \in \mathbf{Gl}_m(\mathbb{R})$, we obtain that, also invoking (2.5a) and (2.5c),

$$
\dot{\eta} = Q\eta + A_{12}y,
$$

$$
\begin{pmatrix} y_1^{(r_1)} \\ \vdots \\ y_q^{(r_q)} \end{pmatrix} = R_{1,1} \begin{pmatrix} y_1 \\ \vdots \\ y_1^{(r_1-1)} \end{pmatrix} + \ldots + R_{q,1} \begin{pmatrix} y_q \\ \vdots \\ y_q^{(r_q-1)} \end{pmatrix} + S_1 \begin{pmatrix} y_{q+1} \\ \vdots \\ y_m \end{pmatrix} + P_1\eta + \Gamma_{11} \begin{pmatrix} u_1 \\ \vdots \\ u_q \end{pmatrix},
$$

$$
0 = R_{1,2} \begin{pmatrix} y_1 \\ \vdots \\ y_1^{(r_1-1)} \end{pmatrix} + \ldots + R_{q,2} \begin{pmatrix} y_q \\ \vdots \\ y_q^{(r_q-1)} \end{pmatrix} + S_2 \begin{pmatrix} y_{q+1} \\ \vdots \\ y_m \end{pmatrix}
$$

$$
+ P_2\eta + \Gamma_{21} \begin{pmatrix} u_1 \\ \vdots \\ u_q \end{pmatrix} + \begin{pmatrix} u_{q+1} \\ \vdots \\ u_m \end{pmatrix},
$$

$$
x_3 = \sum_{i=0}^{n_3-1} N^i E_{32} y^{(i+1)}.
$$

$$(2.17)$$

We have thus derived a representation for systems with truncated vector relative degree and summarize the findings in the following result.

Theorem 2.3 *Let a right-invertible system* $[E, A, B, C] \in \Sigma_{l,n,m,p}$ *with autonomous zero dynamics be given. Assume that* $[E, A, B, C]$ *has ordered truncated vector relative degree* $(r_1, \ldots, r_q, 0, \ldots, 0)$ *with* $r_q > 0$. *Use the notation from Theorem 2.1, (2.13) and (2.16). Then* $(x, u, y) \in \mathcal{B}_{[E,A,B,C]}$, *if, and only if, after a reordering of the inputs so that* $\hat{\Gamma}_{11}$ *in (2.12) is invertible,*

$$Tx = (\eta^\top, y^\top, x_3^\top)^\top \in L^1_{loc}(\mathbb{R} \to \mathbb{R}^{n_1+p+n_3})$$

satisfies the smoothness conditions in (2.4) and solves (2.17) in the distributional sense.

Remark 2.7 Consider a regular and right-invertible system $[E, A, B, C] \in \Sigma_{n,n,m,m}$ with autonomous zero dynamics and ordered truncated vector relative degree $(r_1, \ldots, r_q, 0, \ldots, 0) \in \mathbb{N}_0^{1 \times p}$ such that $r_q > 0$.

(a) If $[E, A, B, C]$ has strict relative degree $r > 0$, then $q = m$ and $r_1 = \ldots = r_m = r$. In this case, the representation (2.17) simplifies to

$$\dot{\eta} = Q\eta + A_{12}y,$$

$$y^{(r)} = R_{1,1} \begin{pmatrix} y_1 \\ \vdots \\ y_1^{(r-1)} \end{pmatrix} + \ldots + R_{m,1} \begin{pmatrix} y_m \\ \vdots \\ y_m^{(r-1)} \end{pmatrix} + P_1\eta + \Gamma_{11}u,$$

$$x_3 = \sum_{i=0}^{n_3-1} N^i E_{32} y^{(i+1)}.$$

Since the second equation can be rewritten as

$$y^{(r)} = Q_{r-1}y^{(r-1)} + \ldots + Q_0 y + P_1\eta + \Gamma_{11}u$$

for matrices $Q_0, \ldots, Q_{r-1}$, this is exactly the form which has been developed in [6].

(b) If the transfer function $G(s) \in \mathbb{R}(s)^{m \times m}$ of $[E, A, B, C]$ has a proper inverse, then we have that $H(s) = G(s)^{-1}$ (see Remark 2.4 (b)) is proper, hence $q = 0$ and the truncated vector relative degree is $(0, \ldots, 0)$. In this case, the representation (2.17) simplifies to

$$\dot{\eta} = Q\eta + A_{12}y,$$

$$0 = S_2 y + P_2 \eta + u,$$

$$x_3 = \sum_{i=0}^{n_3-1} N^i E_{32} y^{(i+1)},$$

which is exactly the form developed in [7].

(c) If the system is an ODE, that is $E = I_n$, then its transfer function $G(s)$ is strictly proper, i.e., $\lim_{\lambda \to \infty} G(\lambda) = 0$. We can further infer from Remark 2.2 that the transfer function $G(s) \in \mathbb{R}(s)^{m \times m}$ is invertible. Then (2.8) implies $q = m$, i.e., the truncated vector relative degree (which coincides with the vector relative degree by Remark 2.5 (d)) is $(r_1, \dots, r_m) \in \mathbb{N}^{1 \times m}$. In this case, (2.17) simplifies to

$$\dot{\eta} = Q\eta + A_{12}y,$$

$$\begin{pmatrix} y_1^{(r_1)} \\ \vdots \\ y_m^{(r_m)} \end{pmatrix} = R_{1,1} \begin{pmatrix} y_1 \\ \vdots \\ y_1^{(r_1-1)} \end{pmatrix} + \dots + R_{m,1} \begin{pmatrix} y_m \\ \vdots \\ y_m^{(r_m-1)} \end{pmatrix} + P_1 \eta + \Gamma_{11} u,$$

$$x_3 = \sum_{i=0}^{n_3-1} N^i E_{32} y^{(i+1)}.$$

This form comprises the one presented in [27], where, additionally,

$$x_3 = (\dot{y}_1, \dots, y_1^{(r_1-1)}, \dots, \dot{y}_m, \dots, y_m^{(r_m-1)})^\top \in \mathbb{R}^{n_3},$$

$$N = \operatorname{diag}(N_1, \dots, N_m) \in \mathbb{R}^{n_3 \times n_3} \text{ with } N_i = \begin{bmatrix} 0 \\ 1 & \ddots \\ & \ddots & \ddots \\ & & 1 & 0 \end{bmatrix} \in \mathbb{R}^{(r_i-1) \times (r_i-1)},$$

$$E_{32} = \operatorname{diag}(e_1^{[r_1-1]}, \dots, e_1^{[r_1-1]}) \in \mathbb{R}^{n_3 \times m},$$

where $e_1^{[k]} \in \mathbb{R}^k$ is the first canonical unit vector. We note that the above nilpotent matrix N has index $\nu = \max_{1 \le i \le m} (r_i - 1)$.

3 Nonlinear Systems with Truncated Vector Relative Degree

In this section, we consider a class of nonlinear DAE systems which comprises the class of linear DAE systems which have a truncated vector relative degree and the same number of inputs and outputs. More precisely, we consider nonlinear functional differential-algebraic systems of the form

$$
\begin{pmatrix} y_1^{(r_1)}(t) \\ y_2^{(r_2)}(t) \\ \vdots \\ y_q^{(r_q)}(t) \end{pmatrix} = f_1\Big(d_1(t), T_1\big(y_1, \ldots, y_1^{(r_1-1)}, \ldots, y_q^{(r_q-1)}, y_{q+1}, \ldots, y_m\big)(t)\Big)
$$
$$
\qquad\qquad + \Gamma_I\Big(d_2(t), T_1\big(y_1, \ldots, y_1^{(r_1-1)}, \ldots, y_q^{(r_q-1)}, y_{q+1}, \ldots, y_m\big)(t)\Big) u_I(t),
$$

$$
0 = f_2\Big(y_1(t), \ldots, y_1^{(r_1-1)}(t), \ldots, y_q^{(r_q-1)}(t), y_{q+1}(t), \ldots, y_m(t)\Big)
$$
$$
\qquad + f_3\Big(d_3(t), (T_2 y)(t)\Big) + \Gamma_{II}\Big(d_4(t), (T_2 y)(t)\Big) u_I(t) + f_4\Big(d_5(t), (T_2 y)(t)\Big) u_{II}(t),
$$

$$
y|_{[-h,0]} = y^0
$$

$$\tag{3.1}$$

with initial data

$$
y^0 = (y_1^0, y_2^0, \ldots, y_m^0)^\top, \quad y_i^0 \in C^{r_i-1}([-h, 0] \to \mathbb{R}), \quad i = 1, \ldots, q,
$$
$$
y_i^0 \in C([-h, 0] \to \mathbb{R}), \quad i = q+1, \ldots, m,
$$

$$\tag{3.2}$$

where $f_1, \ldots, f_4, \Gamma_I, \Gamma_{II}, d_1, \ldots, d_5$ are functions and T_1, T_2 are operators with properties being specified in the sequel. The output is $y = (y_1, \ldots, y_m)^\top$ and the input of the system is $u = (u_1, \ldots, u_m)^\top$, for which we set

$$
u_I = (u_1, \ldots, u_q)^\top, \quad u_{II} = (u_{q+1}, \ldots, u_m)^\top,
$$

i.e., $u = (u_I^\top, u_{II}^\top)^\top$. The functions $d_1, \ldots, d_5 : \mathbb{R}_{\geq 0} \to \mathbb{R}^s$ play the roles of disturbances. We denote $\bar{r} = r_1 + \ldots + r_p$ and call—in virtue of Sect. 2.3—the tuple $(r_1, \ldots, r_p, 0, \ldots, 0) \in \mathbb{N}_0^{1 \times m}$ with $r_i > 0$ for $i = 1, \ldots, q$ the *truncated vector relative degree* of (3.1). We will later show that linear DAE systems which have a truncated vector relative degree belong to this class. Similar to [8], we introduce the following classes of operators.

Definition 3.1 For $m, k \in \mathbb{N}$ and $h \geq 0$ the set $\mathbb{T}_{m,k,h}$ denotes the class of operators $T : C([-h, \infty) \to \mathbb{R}^m) \to L^\infty_{\text{loc}}(\mathbb{R}_{\geq 0} \to \mathbb{R}^k)$ with the following properties:

(i) T is causal, i.e, for all $t \geq 0$ and all $\zeta, \xi \in C([-h, \infty) \to \mathbb{R}^m)$,

$$
\zeta|_{[-h,t)} = \xi|_{[-h,t)} \implies T(\zeta)|_{[0,t)} = T(\xi)|_{[0,t)}.
$$

(ii) T is locally Lipschitz continuous in the following sense: for all $t \geq 0$ and all $\xi \in C([-h, t] \to \mathbb{R}^m)$ there exist $\tau, \delta, c > 0$ such that, for all $\zeta_1, \zeta_2 \in$

$\mathscr{C}([-h, \infty) \rightarrow \mathbb{R}^m)$ with $\zeta_i|_{[-h,t]} = \xi$ and $\|\zeta_i(s) - \xi(t)\| < \delta$ for all $s \in [t, t + \tau]$ and $i = 1, 2$, we have

$$\left\| (T(\zeta_1) - T(\zeta_2))|_{[t,t+\tau]} \right\|_\infty \leq c \left\| (\zeta_1 - \zeta_2)|_{[t,t+\tau]} \right\|_\infty.$$

(iii) T maps bounded trajectories to bounded trajectories, i.e, for all $c_1 > 0$, there exists $c_2 > 0$ such that for all $\zeta \in \mathscr{C}([-h, \infty) \rightarrow \mathbb{R}^m)$

$$\|\zeta|_{[-h,\infty)}\|_\infty \leq c_1 \quad \Longrightarrow \quad \|T(\zeta)|_{[0,\infty)}\|_\infty \leq c_2.$$

Furthermore, the set $\mathbb{T}^{\mathrm{DAE}}_{m,k,h}$ denotes the subclass of operators

$$T : \ C([-h, \infty) \rightarrow \mathbb{R}^m) \rightarrow C^1(\mathbb{R}_{\geq 0} \rightarrow \mathbb{R}^k)$$

such that $T \in \mathbb{T}_{m,k,h}$ and, additionally,

(iv) there exist $z \in C(\mathbb{R}^m \times \mathbb{R}^k \rightarrow \mathbb{R}^k)$ and $\tilde{T} \in \mathbb{T}_{m,k,h}$ such that

$$\forall \zeta \in C([-h, \infty) \rightarrow \mathbb{R}^m) \ \forall t \geq 0 : \ \tfrac{\mathrm{d}}{\mathrm{d}t}(T\zeta)(t) = z\big(\zeta(t), (\tilde{T}\zeta)(t)\big).$$

Assumption 3.1 *We assume that the functional differential-algebraic system* (3.1) *has the following properties:*

(i) *the gain* $\Gamma_I \in C(\mathbb{R}^s \times \mathbb{R}^k \rightarrow \mathbb{R}^{q \times q})$ *satisfies* $\Gamma_I(d, \eta) + \Gamma_I(d, \eta)^\top > 0$ *for all* $(d, \eta) \in \mathbb{R}^s \times \mathbb{R}^k$, *and* $\Gamma_{II} \in C^1(\mathbb{R}^s \times \mathbb{R}^k \rightarrow \mathbb{R}^{(m-q) \times q})$.
(ii) *the disturbances satisfy* $d_1, d_2 \in L^\infty(\mathbb{R}_{\geq 0} \rightarrow \mathbb{R}^s)$ *and* $d_3, d_4, d_5 \in W^{1,\infty}(\mathbb{R}_{\geq 0} \rightarrow \mathbb{R}^s)$.
(iii) $f_1 \in C(\mathbb{R}^s \times \mathbb{R}^k \rightarrow \mathbb{R}^q)$, $f_2 \in C^1(\mathbb{R}^{\bar{r}+m-q} \rightarrow \mathbb{R}^{m-q})$, $f_3 \in C^1(\mathbb{R}^s \times \mathbb{R}^k \rightarrow \mathbb{R}^{m-q})$, *and* $f_2' \begin{bmatrix} 0 \\ I_{m-q} \end{bmatrix}$ *is bounded.*
(iv) $f_4 \in C^1(\mathbb{R}^s \times \mathbb{R}^k \rightarrow \mathbb{R})$ *and there exists* $\alpha > 0$ *such that* $f_4(d, v) \geq \alpha$ *for all* $(d, v) \in \mathbb{R}^s \times \mathbb{R}^k$.
(v) $T_1 \in \mathbb{T}_{\bar{r}+m-q,k,h}$ *and* $T_2 \in \mathbb{T}^{\mathrm{DAE}}_{m,k,h}$.

In the remainder of this section we show that any right-invertible system $[E, A, B, C] \in \Sigma_{l,n,m,m}$ with truncated vector relative degree $(r_1, \ldots, r_q, 0, \ldots, 0)$, where $r_1, \ldots, r_q \in \mathbb{N}$, belongs to the class of systems (3.1) which satisfy Assumption 3.1 as long as $[E, A, B, C]$ has asymptotically stable zero dynamics and the matrix Γ_{11} in (2.17) satisfies $\Gamma_{11} + \Gamma_{11}^\top > 0$. We have seen in Remark 2.3 that asymptotic stability of the zero dynamics is equivalent to the matrix Q in (2.17) having only eigenvalues with negative real part.

Consider the three first equations in (2.17) and the operator

$$T_2: \quad C([0, \infty) \to \mathbb{R}^m) \to C^1(\mathbb{R}_{\geq 0} \to \mathbb{R}^{n_3})$$

$$y \mapsto \left(t \mapsto (T_2 y)(t) := \eta(t) = e^{Qt} \eta^0 + \int_0^t e^{Q(t-\tau)} A_{12} y(\tau) d\tau \right),$$

which is parameterized by the initial value $\eta^0 \in \mathbb{R}^{n_1}$. This operator is clearly causal, locally Lipschitz, and, since all eigenvalues of Q have negative real part, T satisfies property (iii) in Definition 3.1. The derivative is given by

$$\frac{d}{dt}(T_2 y)(t) = Q e^{Qt} \eta^0 + A_{12} y(t) + Q \int_0^t e^{Q(t-\tau)} A_{12} y(\tau) d\tau =: (\tilde{T} y)(t), \quad t \geq 0,$$

and it is straightforward to check that $\tilde{T} \in \mathbb{T}_{m,n_3,0}$. Therefore, we obtain that $T_2 \in \mathbb{T}_{m,n_3,0}^{DAE}$. Further consider the operator $T_1 : C([0, \infty) \to \mathbb{R}^{\bar{r}+m-q}) \to L^\infty_{loc}(\mathbb{R}_{\geq 0} \to \mathbb{R}^q)$ defined by

$$T_1(\zeta_{1,1}, \ldots, \zeta_{1,r_1}, \ldots, \zeta_{q,r_q}, \zeta_{q+1,1}, \ldots, \zeta_{m,1})$$

$$= R_{1,1} \begin{pmatrix} \zeta_{1,1} \\ \vdots \\ \zeta_{1,r_1} \end{pmatrix} + \ldots + R_{q,1} \begin{pmatrix} \zeta_{q,1} \\ \vdots \\ \zeta_{q,r_q} \end{pmatrix} + S_1 \begin{pmatrix} \zeta_{q+1,1} \\ \vdots \\ \zeta_{m,1} \end{pmatrix} + P_1 T_2(\zeta_{1,1}, \zeta_{2,1}, \ldots, \zeta_{m,1}),$$

then, likewise, we obtain that $T_1 \in \mathbb{T}_{\bar{r}+m-q,q,0}$. The remaining functions are given by

$$f_1(d, \eta) = \eta, \quad \Gamma_I(d, \eta) = \Gamma_{11}, \quad f_3(d, \eta) = P_2 \eta, \quad \Gamma_{II}(d, \eta) = \Gamma_{21}, \quad f_4(d, \eta) = 1$$

and

$$f_2(\zeta_{1,1}, \ldots, \zeta_{1,r_1}, \ldots, \zeta_{q,r_q}, \zeta_{q+1,1}, \ldots, \zeta_{m,1})$$

$$= R_{1,2} \begin{pmatrix} \zeta_{1,1} \\ \vdots \\ \zeta_{1,r_1} \end{pmatrix} + \ldots + R_{q,2} \begin{pmatrix} \zeta_{q,1} \\ \vdots \\ \zeta_{q,r_q} \end{pmatrix} + S_2 \begin{pmatrix} \zeta_{q+1,1} \\ \vdots \\ \zeta_{m,1} \end{pmatrix}.$$

The function f_2 satisfies condition (iii) in Assumption 3.1 since

$$f_2' \begin{bmatrix} 0 \\ I_{m-q} \end{bmatrix} = S_2 \in \mathbb{R}^{(m-q) \times (m-q)}.$$

Note that system (2.17) does not entirely belong to the class (3.1) since the fourth equation in (2.17) is not included. However, the control objective formulated in the following section can also be achieved for (2.17), see also Remark 4.1 (e).

4 Funnel Control

4.1 Control Objective

Let a reference signal $y_{ref} = (y_{ref,1}, \ldots, y_{ref,m})^\top$ with $y_{ref,i} \in W^{r_i,\infty}(\mathbb{R}_{\geq 0} \to \mathbb{R})$ for $i = 1, \ldots, q$ and $y_{ref,i} \in W^{1,\infty}(\mathbb{R}_{\geq 0} \to \mathbb{R})$ for $i = q + 1, \ldots, m$ be given, and let $e = y - y_{ref}$ be the *tracking error*. The objective is to design an output error feedback of the form

$$u(t) = F\left(t, e_1(t), \ldots, e_1^{(r_1-1)}(t), \ldots, e_q^{(r_q-1)}(t), e_{q+1}(t), \ldots, e_m(t)\right),$$

such that in the closed-loop system the tracking error evolves within a prescribed performance funnel

$$\mathscr{F}_\varphi^m := \left\{ (t, e) \in \mathbb{R}_{\geq 0} \times \mathbb{R}^m \;\middle|\; \varphi(t)\|e\| < 1 \right\}, \tag{4.1}$$

which is determined by a function φ belonging to

$$\Phi_k := \left\{ \varphi \in C^k(\mathbb{R}_{\geq 0} \to \mathbb{R}) \;\middle|\; \begin{array}{l} \varphi, \dot{\varphi}, \ldots, \varphi^{(k)} \text{ are bounded,} \\ \varphi(\tau) > 0 \text{ for all } \tau > 0, \text{ and } \liminf_{\tau \to \infty} \varphi(\tau) > 0 \end{array} \right\}. \tag{4.2}$$

A further objective is that all signals $u, e_1, \ldots, e_1^{(r_1-1)}, \ldots, e_q^{(r_q-1)}, e_{q+1}, \ldots, e_m :$ $\mathbb{R}_{\geq 0} \to \mathbb{R}^m$ should remain bounded.

The funnel boundary is given by the reciprocal of φ, see Fig. 1. It is explicitly allowed that $\varphi(0) = 0$, meaning that no restriction on the initial value is imposed since $\varphi(0)\|e(0)\| < 1$; the funnel boundary $1/\varphi$ has a pole at $t = 0$ in this case. Since every $\varphi \in \Phi_k$ is bounded, the boundary of the associated performance funnel $\mathscr{F}_\varphi^m$ is bounded away from zero, which means that there exists $\lambda > 0$ with $1/\varphi(t) \geq \lambda$ for all $t > 0$. Further note that the funnel boundary is not necessarily monotonically decreasing, but it might be beneficial to choose a wider funnel over some later time interval, for instance in the presence of periodic disturbance or when the reference signal varies strongly. Various different funnel boundaries are possible, see e.g. [18, Sec. 3.2].

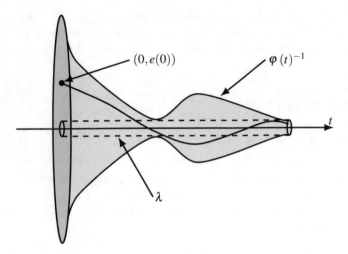

Fig. 1 Error evolution in a funnel $\mathscr{F}_\varphi^1$ with boundary $\varphi(t)^{-1}$ for $t > 0$

4.2 Controller Design

The funnel controller for systems of the form (3.1) satisfying Assumption 3.1 is of the following form:

$$
\begin{aligned}
&\text{For } i = 1, \ldots, q : \\
&\quad e_{i0}(t) = e_i(t) = y_i(t) - y_{\text{ref},i}(t), \\
&\quad e_{i1}(t) = \dot{e}_{i0}(t) + k_{i0}(t)e_{i0}(t), \\
&\quad e_{i2}(t) = \dot{e}_{i1}(t) + k_{i1}(t)e_{i1}(t), \\
&\qquad\qquad \vdots \\
&\quad e_{i,r_i-1}(t) = \dot{e}_{i,r_i-2}(t) + k_{i,r_i-2}(t)e_{i,r_i-2}(t), \\
&\qquad k_{ij}(t) = \frac{1}{1 - \varphi_{ij}^2(t)|e_{ij}(t)|^2}, \quad j = 0, \ldots, r_i - 2. \\
&\text{For } i = q+1, \ldots, m : \quad e_i(t) = y_i(t) - y_{\text{ref},i}(t), \\
&e_I(t) = (e_{1,r_1-1}(t), \ldots, e_{q,r_q-1}(t))^\top, \quad e_{II}(t) = (e_{q+1}(t), \ldots, e_m(t))^\top, \\
&k_I(t) = \frac{1}{1 - \varphi_I(t)^2 \|e_I(t)\|^2}, \qquad\qquad k_{II}(t) = \frac{\hat{k}}{1 - \varphi_{II}(t)^2 \|e_{II}(t)\|^2}, \\
&u(t) = \begin{pmatrix} u_I(t) \\ u_{II}(t) \end{pmatrix} = \begin{pmatrix} -k_I(t)e_I(t) \\ -k_{II}(t)e_{II}(t) \end{pmatrix},
\end{aligned}
$$

$$\tag{4.3}$$

where we impose the following conditions on the reference signal and funnel functions:

$$y_{\text{ref}} = (y_{\text{ref},1}, \ldots, y_{\text{ref},m})^\top, \quad y_{\text{ref},i} \in W^{r_i,\infty}(\mathbb{R}_{\geq 0} \to \mathbb{R}), \ i = 1, \ldots, q$$

$$y_{\text{ref},i} \in W^{1,\infty}(\mathbb{R}_{\geq 0} \to \mathbb{R}), \ i = q+1, \ldots, m$$

$$\varphi_I, \varphi_{II} \in \Phi_1, \ \varphi_{ij} \in \Phi_{r_i - j}, \ i = 1, \ldots, q, \ j = 0, \ldots, r_i - 2.$$

$$(4.4)$$

We further assume that $\hat{k}$ satisfies

$$\hat{k} > \alpha^{-1} \sup_{Y \in \mathbb{R}^{\bar{r}+m-q}} \left\| f_2'(Y) \begin{bmatrix} 0 \\ I_{m-q} \end{bmatrix} \right\|. \tag{4.5}$$

Remark 4.1

(a) By a solution of the closed-loop system (3.1) and (4.3) on $[-h, \omega)$, $\omega \in (0, \infty]$, with initial data y^0 as in (3.2) we mean a function $y = (y_1, \ldots, y_m)^\top$ such that $y|_{[-h,0]} = y^0$, $y_i \in C^{r_i-1}([-h, \omega) \to \mathbb{R})$ and $y_i^{(r_i-1)}|_{[0,\omega)}$ is weakly differentiable for $i = 1, \ldots, q$, $y_i \in C([-h, \omega] \to \mathbb{R})$ and $y_i|_{[0,\omega)}$ is weakly differentiable for $i = q + 1, \ldots, m$, and y satisfies the differential-algebraic equation in (3.1) with u defined in (4.3) in the weak sense. The solution y is called *maximal*, if it has no right extension that is also a solution, and *global*, if $\omega = \infty$.

(b) Assumption 3.1 (iii) together with condition (4.5) are essential for the solvability of the closed-loop system (3.1) and (4.3), since they guarantee the invertibility of $\alpha \hat{k} I_{m-q} - f_3'(Y) \begin{bmatrix} 0 \\ I_{m-q} \end{bmatrix}$. This property is crucial for the explicit solution of the algebraic constraint in the closed-loop system (3.1) and (4.3).

(c) If the system (3.1) has strict relative degree, i.e., $q = m$ and $r_1 = \ldots = r_m =: r > 0$, then it satisfies the assumptions of [9, Thm. 3.1]. In this case, the funnel controller (4.3) simplifies to

$$
\begin{aligned}
&\text{For } i = 1, \ldots, m, \\
&\quad e_{i0}(t) = e_i(t) = y_i(t) - y_{\text{ref},i}(t), \\
&\quad e_{i1}(t) = \dot{e}_{i0}(t) + k_{i0}(t)e_{i0}(t), \\
&\quad e_{i2}(t) = \dot{e}_{i1}(t) + k_{i1}(t)e_{i1}(t), \\
&\qquad\qquad \vdots \\
&\quad e_{i,r-1}(t) = \dot{e}_{i,r-2}(t) + k_{i,r-2}(t)e_{i,r-2}(t), \\
&\quad k_{ij}(t) = \frac{1}{1 - \varphi_{ij}^2(t)|e_{ij}(t)|^2}, \ j = 0, \ldots, r - 2, \\
&\quad e_{r-1}(t) = (e_{1,r-1}(t), \ldots, e_{m,r-1}(t))^\top \\
&\quad k_{r-1}(t) = \frac{1}{1 - \varphi_{r-1}(t)^2 \|e_{r-1}(t)\|^2}, \\
&\quad u(t) = -k_{r-1}(t)e_{r-1}(t).
\end{aligned}
$$

This controller slightly differs from the one presented in [9] for systems with strict relative degree (even when we choose $\varphi_{ij} = \varphi_{1j}$ for all $i = 1, \ldots, m$), which reads

$$
\begin{aligned}
e_0(t) &= e(t) = y(t) - y_{\text{ref}}(t), \\
e_1(t) &= \dot{e}_0(t) + k_0(t)e_0(t), \\
e_2(t) &= \dot{e}_1(t) + k_1(t)e_1(t), \\
&\;\;\vdots \\
e_{r-1}(t) &= \dot{e}_{r-2}(t)k_{r-2}(t)e_{r-2}(t), \\
k_i(t) &= \frac{1}{1-\varphi_i(t)^2\|e_i(t)\|^2}, \quad i = 0, \ldots, r-1, \\
u(t) &= -k_{r-1}(t)e_{r-1}(t).
\end{aligned}
\tag{4.6}
$$

(d) If the system (3.1) satisfies $q = 0$, then the funnel controller (4.3) simplifies to

$$
e(t) = y(t) - y_{\text{ref}}(t), \qquad k(t) = \frac{\hat{k}}{1-\varphi(t)^2\|e(t)\|^2},
$$

$$
u(t) = -k(t)e(t),
$$

and feasibility follows from the results in [8] where funnel control for this type has been considered.

(e) Let us stress again that a linear system of the form (2.17) does not completely belong to the class (3.1) as the fourth equation in (2.17) is not included. However, we like to emphasize that in

$$
x_3(t) = \sum_{i=0}^{n_3-1} N^i E_{32} y^{(i+1)}(t),
$$

the output y is required smooth enough for x_3 to be well defined. Nevertheless, the funnel controller (4.3) can also be applied to systems of the form (2.17). To see this, assume that there exists a solution to (4.3) applied to (2.17) except for the fourth equation. If the funnel functions φ_I, φ_{II} and $\varphi_{ij}, i = 1, \ldots, q, j = 0, \ldots, r_i - 2$ are additionally in $C^{n_3+1}(\mathbb{R}_{\geq 0} \to \mathbb{R})$ and y_{ref} is additionally in $W^{n_3+2,\infty}(\mathbb{R}_{\geq 0} \to \mathbb{R}^m)$, then the solution $y|_{[0,\infty)}$ will be at least in $C^{n_3+1}(\mathbb{R}_{\geq 0} \to \mathbb{R}^m)$, so that x_3 is well defined and continuously differentiable. The proof of this statement is similar to Step 2 of the proof of [3, Thm. 5.3]. Furthermore, using $y_{\text{ref}} \in W^{n_3+2,\infty}(\mathbb{R}_{\geq 0} \to \mathbb{R}^m)$ also yields boundedness of x_3, cf. Step 4 of the proof of [3, Thm. 5.3].

Remark 4.2 Consider a system (3.1) which satisfies Assumption 3.1 and let the reference signal and funnel functions be as in (4.4). Since the second equation in (3.1) is an algebraic equation we need to guarantee that it is initially satisfied for a solution to exist. Since $T_2 \in \mathbb{T}_{m,k,h}^{\text{DAE}}$ is causal it "localizes", in a natural way,

to an operator $\hat{T}_2 : C([-h, \omega] \to \mathbb{R}^n) \to C^1([0, \omega] \to \mathbb{R}^k)$, cf. [21, Rem. 2.2].
With some abuse of notation, we will henceforth not distinguish between T_2 and its
"localization" $\hat{T}_2$. Note that for $\omega = 0$ we have that $\hat{T}_2 : C([-h, 0] \to \mathbb{R}^n) \to \mathbb{R}^k$.
Hence, an initial value y^0 as in (3.2) is called *consistent* for the closed loop
system (3.1), (4.3), if

$$f_2\left(y_1^0(0), \ldots, \left(\tfrac{d}{dt}\right)^{r_1-1}(y_1^0)(0), \ldots, \left(\tfrac{d}{dt}\right)^{r_q-1}(y_q^0)(0), y_{q+1}^0(0), \ldots, y_m^0(0)\right)$$

$$+ f_3\big(d_3(0), T_2(y^0)\big) + \Gamma_{II}\big(d_4(0), T_2(y^0)\big)u_I(0) + f_4\big(d_5(0), T_2(y^0)\big)u_{II}(0) = 0,$$
$$(4.7)$$

where $u_I(0)$, $u_{II}(0)$ are defined by (4.3).

4.3 Feasibility of Funnel Control

We show feasibility of the funnel controller (4.3) for systems of the form (3.1)
satisfying Assumption 3.1. The following theorem unifies and extends the funnel
control results from [3, 6–9], which are all special cases of it.

Theorem 4.1 *Consider a system (3.1) satisfying Assumption 3.1. Let y_{ref} and
$\varphi_I, \varphi_{II}, \varphi_{ij}, i = 1, \ldots, q, j = 0, \ldots, r_i - 2$ be as in (4.4) and $\hat{k} > 0$ such that (4.5)
holds. Then for any consistent initial value y^0 as in (3.2) (i.e., y^0 satisfies (4.7))
such that $e_I, e_{II}, e_{ij}, i = 1, \ldots, q, j = 0, \ldots, r_i - 2$ defined in (4.3) satisfy*

$$\varphi_I(0)\|e_I(0)\| < 1, \quad \varphi_{II}(0)\|e_{II}(0)\| < 1,$$
$$\varphi_{ij}(0)|e_{ij}(0)| < 1, \quad i = 1, \ldots, q, \; j = 0, \ldots, r_i - 2,$$
$$(4.8)$$

*the application of the funnel controller (4.3) to (3.1) yields a closed-loop initial
value problem that has a solution and every solution can be extended to a global
solution. Furthermore, for every global solution $y(\cdot)$,*

(i) *the input $u : \mathbb{R}_{\geq 0} \to \mathbb{R}^m$ and the gain functions $k_I, k_{II}, k_{ij} : \mathbb{R}_{\geq 0} \to \mathbb{R}$,
$i = 1, \ldots, q, j = 0, \ldots, r_i - 2$ are bounded;*
(ii) *the functions $e_I : \mathbb{R}_{\geq 0} \to \mathbb{R}^q$, $e_{II} : \mathbb{R}_{\geq 0} \to \mathbb{R}^{m-q}$ and $e_{ij} : \mathbb{R}_{\geq 0} \to \mathbb{R}$,
$i = 1, \ldots, q, j = 0, \ldots, r_i - 2$ evolve in their respective performance funnels,
i.e., for all $i = 1, \ldots, q, j = 0, \ldots, r_i - 2$ and $t \geq 0$ we have*

$$(t, e_I(t)) \in \mathscr{F}_{\varphi_I}^q, \; (t, e_{II}(t)) \in \mathscr{F}_{\varphi_{II}}^{m-q}, \; (t, e_{ij}(t)) \in \mathscr{F}_{\varphi_{ij}}^1.$$

Furthermore, the signals $e_I(\cdot), e_{II}(\cdot), e_{ij}(\cdot)$ are uniformly bounded away from the funnel boundaries in the following sense:

$$\exists\, \varepsilon_I > 0\ \forall t > 0: \qquad\qquad\qquad \|e_I(t)\| \le \varphi_I(t)^{-1} - \varepsilon_I,$$

$$\exists\, \varepsilon_{II} > 0\ \forall t > 0: \qquad\qquad\qquad \|e_{II}(t)\| \le \varphi_{II}(t)^{-1} - \varepsilon_{II},$$

$$\forall\, i = 1, \ldots, q,\ j = 0, \ldots, r_i - 2\ \exists\, \varepsilon_{ij} > 0\ \forall t > 0: |e_{ij}(t)| \le \varphi_{ij}(t)^{-1} - \varepsilon_{ij}.$$

$$(4.9)$$

In particular, each error component $e_i(t) = y_i - y_{\text{ref},i}(t)$ evolves in the funnel $\mathscr{F}^1_{\varphi_{i0}}$, for $i = 1, \ldots, q$, or $\mathscr{F}^1_{\varphi_{II}}$, for $i = q+1, \ldots, m$, resp., and stays uniformly away from its boundary.

The proof of this theorem is similar to the one of [9, Thm. 3.1], where the feasibility of the funnel controller (4.6) for ODE systems with strict relative degree has been treated. However, one of the additional difficulties in proving this theorem is that the closed-loop system (3.1) and (4.3) is now a DAE because of the second equation in (3.1).

Proof We proceed in several steps.

Step 1 We show that a maximal solution $y : [-h, \omega) \to \mathbb{R}^m$, $\omega \in (0, \infty]$, of the closed-loop system (3.1) and (4.3) exists. To this end, we seek to reformulate (3.1) and (4.3) as an initial value problem of the form

$$\dot{X}_I(t) = F_I\left(t, \begin{pmatrix} X_I(t) \\ X_{II}(t) \end{pmatrix}, T_1\begin{pmatrix} X_I \\ X_{II} \end{pmatrix}(t)\right),$$

$$0 = F_{II}\left(t, \begin{pmatrix} X_I(t) \\ X_{II}(t) \end{pmatrix}, \hat{T}_2\begin{pmatrix} X_I \\ X_{II} \end{pmatrix}(t)\right)$$

$$(4.10)$$

with

$$X_I|_{[-h,0]} = \left(y_1^0, \ldots, \left(\tfrac{d}{dt}\right)^{r_1-1} y_1^0, \ldots, \left(\tfrac{d}{dt}\right)^{r_q-1} y_q^0\right)^\top,$$

$$X_{II}|_{[-h,0]} = \left(y_{q+1}^0, \ldots, y_m^0\right)^\top.$$

$$(4.11)$$

Step 1a Define, for $i = 1, \ldots, q$, and $j = 0, \ldots, r_i - 2$, the sets

$$\mathscr{D}_{ij} := \left\{ (t, e_{i0}, \ldots, e_{ij}) \in \mathbb{R}_{\ge 0} \times \mathbb{R} \times \cdots \times \mathbb{R} \,\middle|\, (t, e_{i\ell}) \in \mathscr{F}^1_{\varphi_{i\ell}}, \ell = 0, \ldots, j \right\},$$

where $\mathscr{F}_{\varphi_{i\ell}}^1$ is as in (4.1), and the functions $K_{ij} : \mathscr{D}_{ij} \to \mathbb{R}$ recursively by

$$K_{i0}(t, e_{i0}) := \frac{e_{i0}}{1-\varphi_{i0}^2(t)|e_{i0}|^2},$$

$$K_{ij}(t, e_{i0}, \dots, e_{ij}) := \frac{e_{ij}}{1-\varphi_{ij}^2(t)|e_{ij}|^2} + \frac{\partial K_{i,j-1}}{\partial t}(t, e_{i0}, \dots, e_{i,j-1})$$

$$+ \sum_{\ell=0}^{j-1} \frac{\partial K_{i,j-1}}{\partial e_{\ell j}}(t, e_{i0}, \dots, e_{i,j-1}) \left(e_{i,\ell+1} - \frac{e_{i\ell}}{1-\varphi_{i\ell}^2(t)|e_{i\ell}|^2} \right).$$

Now recall that $\bar{r} = r_1 + \dots + r_q$ and set

$$\mathscr{D}_I := \left\{ (t, e_{10}, \dots, e_{1,r_1-1}, \dots, e_{q,r_q-1}) \in \mathbb{R}_{\geq 0} \times \mathbb{R}^{\bar{r}} \; \right|$$

$$\forall i = 1, \dots, q : \left(t, e_{i0}, \dots, e_{i,r_i-2} \right) \in \mathscr{D}_{i,r_i-2} \wedge (t, e_{1,r_1-1}, \dots, e_{q,r_q-1}) \in \mathscr{F}_{\varphi_I}^q \left. \right\},$$

$$\mathscr{D}_{II} := \mathscr{F}_{\varphi_{II}}^{m-q},$$

$$\mathscr{D} := \left\{ (t, e_I, e_{II}) \in \mathbb{R}_{\geq 0} \times \mathbb{R}^{\bar{r}} \times \mathbb{R}^{m-q} \; \middle| \; (t, e_I) \in \mathscr{D}_I \wedge (t, e_{II}) \in \mathscr{D}_{II} \; \right\}.$$

Choose some interval $I \subseteq \mathbb{R}_{\geq 0}$ with $0 \in I$ and let

$$(e_{10}, \dots, e_{1,r_1-1}, \dots, e_{q,r_q-1}) : I \to \mathbb{R}^{\bar{r}}$$

be sufficiently smooth such that for all $t \in I$ we have

$$\left(t, e_{10}(t), \dots, e_{1,r_1-1}(t), \dots, e_{q,r_q-1}(t) \right) \in \mathscr{D}_I,$$

$$(t, e_{q+1}(t), \dots, e_m(t)) \in \mathscr{D}_{II}$$

and $(e_{i0}, \dots, e_{i,r_i-1})$, $i = 1, \dots, q$, satisfies the relations in (4.3). Then $e_i = e_{i0}$ satisfies, on the interval I,

$$e_i^{(j)} = e_{ij} - \sum_{\ell=0}^{j-1} \left(\tfrac{d}{dt} \right)^{j-1-\ell} k_{i\ell} e_{i\ell}, \quad i = 1, \dots, q, \; j = 1, \dots, r_i - 1. \tag{4.12}$$

Step 1b We show by induction that for all $i = 1, \dots, q$, and $j = 0, \dots, r_i - 2$ we have

$$\forall t \in I : \sum_{\ell=0}^{j} \left(\tfrac{d}{dt} \right)^{j-\ell} \left(k_{i\ell}(t) e_{i\ell}(t) \right) = K_{ij} \left(t, e_{i0}(t), \dots, e_{ij}(t) \right). \tag{4.13}$$

Fix $t \in I$. Equation (4.13) is obviously true for $j = 0$. Assume that $j \in \{1, \ldots, r_i - 2\}$ and the statement holds for $j - 1$. Then

$$\sum_{\ell=0}^{j} \left(\tfrac{d}{dt}\right)^{j-\ell} \left(k_{i\ell}(t)e_{i\ell}(t)\right) = k_{ij}(t)e_{ij}(t) + \tfrac{d}{dt}\left(\sum_{\ell=0}^{j-1} \left(\tfrac{d}{dt}\right)^{j-\ell-1}\left(k_{i\ell}(t)e_{i\ell}(t)\right)\right)$$

$$= k_{ij}(t)e_{ij}(t) + \tfrac{d}{dt}K_{i,j-1}\Big(t, e_{i0}(t), \ldots, e_{i,j-1}(t)\Big)$$

$$= K_{ij}\Big(t, e_{i0}(t), \ldots, e_{ij}(t)\Big).$$

Therefore, (4.13) is shown and, invoking (4.12), we have for all $i = 1, \ldots, q$ and $t \in I$ that

$$e_i^{(j)}(t) = e_{ij}(t) - K_{i,j-1}\big(t, e_{i0}(t), \ldots, e_{i,j-1}(t)\big), \quad j = 1, \ldots, r_i - 1. \quad (4.14)$$

Step 1c Define, for $i = 1, \ldots, q$,

$$\tilde{K}_{i0} : \mathbb{R}_{\geq 0} \times \mathbb{R} \to \mathbb{R}, \quad (t, y_{i0}) \mapsto y_{i0} - y_{\text{ref},i}(t)$$

and the set

$$\tilde{\mathscr{D}}_{i0} := \left\{ (t, y_i) \in \mathbb{R}_{\geq 0} \times \mathbb{R} \ \middle| \ \big(t, \tilde{K}_{i0}(t, y_i)\big) \in \mathscr{D}_{i0} \right\}.$$

Furthermore, recursively define the maps

$$\tilde{K}_{ij} : \tilde{\mathscr{D}}_{i,j-1} \times \mathbb{R} \to \mathbb{R},$$

$$(t, y_{i0}, \ldots, y_{ij}) \mapsto y_{ij} - y_{\text{ref},i}^{(j)}(t) + K_{i,j-1}\big(t, \tilde{K}_{i0}(t, y_{i0}), \ldots, \tilde{K}_{i,j-1}(t, y_{i0}, \ldots, y_{i,j-1})\big),$$

for $j = 1, \ldots, r_i - 1$ and the sets

$$\tilde{\mathscr{D}}_{ij} := \left\{ (t, y_{i0}, \ldots, y_{ij}) \in \tilde{\mathscr{D}}_{i,j-1} \times \mathbb{R} \ \middle| \ \big(t, \tilde{K}_{i0}(t, y_{i0}), \ldots, \tilde{K}_{ij}(t, y_{i0}, \ldots, y_{ij})\big) \in \mathscr{D}_{ij} \right\}$$

for $j = 1, \ldots, r_i - 2$. Then it follows from (4.14) and a simple induction that for all $t \in I$, $i = 1, \ldots, q$, and $j = 0, \ldots, r_i - 1$ we have

$$e_{ij}(t) = \tilde{K}_{ij}\big(t, y_i(t), \ldots, y_i^{(j)}(t)\big).$$

Now, define

$$\tilde{\mathscr{D}}_I := \Big\{ (t, y_{10}, \ldots, y_{1,r_1-1}, \ldots, y_{q,r_q-1}) \in \mathbb{R}_{\geq 0} \times \mathbb{R}^{\bar r} \ \Big|$$

$$\forall i = 1, \ldots, q : \ (t, y_{i0}, \ldots, y_{i,r_i-1}) \in \tilde{\mathscr{D}}_{i,r_i-2} \times \mathbb{R}$$

$$\wedge \left(t, \tilde K_{1,r_1-1}(t, y_{10}, \ldots, y_{1,r_1-1}), \ldots, \tilde K_{q,r_q-1}(t, y_{q0}, \ldots, y_{q,r_q-1}) \right) \in \mathscr{F}_{\varphi_I}^q \Big\},$$

$$\tilde{\mathscr{D}}_{II} := \Big\{ (t, y_{q+1}, \ldots, y_m) \in \mathbb{R}_{\geq 0} \times \mathbb{R}^{m-q} \ \Big| \ (t, y_{q+1} - y_{\mathrm{ref},q+1}(t), \ldots, y_m - y_{\mathrm{ref},m}(t)) \in \mathscr{D}_{II} \Big\},$$

and the map

$$\tilde K_I : \tilde{\mathscr{D}}_I \to \mathbb{R}^q, \ (t, y_{10}, \ldots, y_{1,r_1-1}, \ldots, y_{q,r_q-1})$$

$$\mapsto \left(\tilde K_{1,r_1-1}(t, y_{10}, \ldots, y_{1,r_1-1}), \ldots, \tilde K_{q,r_q-1}(t, y_{q0}, \ldots, y_{q,r_q-1}) \right)^\top,$$

then we find that, for all $t \in I$,

$$e_I(t) := \left(e_{1,r_1-1}(t), \ldots, e_{q,r_q-1}(t) \right)^\top = \tilde K_I \left(t, y_1(t), \ldots, y_1^{(r_1-1)}(t), \ldots, y_q^{(r_q-1)}(t) \right).$$

Further denote, for $t \in I$,

$$X_I(t) = \left(y_1(t), \ldots, y_1^{(r_1-1)}(t) \ldots, y_q^{(r_q-1)}(t) \right)^\top, \quad X_{II}(t) = (y_{q+1}(t), \ldots, y_m(t))^\top,$$

$$X_{\mathrm{ref},II}(t) = (y_{\mathrm{ref},q+1}(t), \ldots, y_{\mathrm{ref},m}(t))^\top,$$

then

$$e_I(t) = \tilde K_I(t, X_I(t)),$$

$$e_{II}(t) := (y_{q+1}(t) - y_{\mathrm{ref},q+1}(t), \ldots, y_m(t) - y_{\mathrm{ref},m}(t))^\top = X_{II}(t) - X_{\mathrm{ref},II}(t)$$

and the feedback u in (4.3) reads

$$u(t) = \begin{pmatrix} \dfrac{-\tilde K_I(t, X_I(t))}{1 - \varphi_I(t)^2 \| \tilde K_I(t, X_I(t)) \|^2} \\[2mm] \dfrac{-\hat k(X_{II}(t) - X_{\mathrm{ref},II}(t))}{1 - \varphi_{II}(t)^2 \| X_{II}(t) - X_{\mathrm{ref},II}(t) \|^2} \end{pmatrix}.$$

Step 1d Now, we set

$$
H = \mathrm{diag}\left((e_1^{[r_1]})^\top, \ldots, (e_1^{[r_q]})^\top \right) \in \mathbb{R}^{q \times \bar{r}},
$$

$$
S = \begin{bmatrix} H & 0 \\ 0 & I_{m-q} \end{bmatrix} \in \mathbb{R}^{m \times (\bar{r}+m-q)},
\tag{4.15}
$$

where $e_1^{[k]} \in \mathbb{R}^k$ is the first canonical unit vector. This construction yields

$$
\forall t \in I: \quad S \begin{pmatrix} X_I(t) \\ X_{II}(t) \end{pmatrix} = y(t).
$$

We define an operator $\hat{T}_2 : C([-h,\infty) \to \mathbb{R}^{\bar{r}+m-q}) \to C^1(\mathbb{R}_{\geq 0} \to \mathbb{R}^k)$ such that for $\zeta_1 \in C([-h,\infty) \to \mathbb{R}^{\bar{r}})$, $\zeta_2 \in C([-h,\infty) \to \mathbb{R}^{m-q})$ we have

$$
\hat{T}_2 \begin{pmatrix} \zeta_1 \\ \zeta_2 \end{pmatrix}(t) := T_2 \left(S \begin{pmatrix} \zeta_1 \\ \zeta_2 \end{pmatrix} \right)(t), \quad t \geq 0.
$$

Since $T_2 \in \mathbb{T}_{m,k,h}^{\mathrm{DAE}}$ we obtain that $\hat{T}_2 \in \mathbb{T}_{\bar{r}+m-q,k,h}^{\mathrm{DAE}}$. Set

$$
\tilde{\mathscr{D}} := \left\{ (t, X_I, X_{II}) \in \mathbb{R}_{\geq 0} \times \mathbb{R}^{\bar{r}} \times \mathbb{R}^{m-q} \ \middle|\ (t, X_I) \in \tilde{\mathscr{D}}_I \text{ and } (t, X_{II}) \in \tilde{\mathscr{D}}_{II} \right\}.
$$

We rewrite f_1, and Γ_I from system (3.1) in vector form

$$
f_1 = \begin{pmatrix} f_1^1 \\ \vdots \\ f_1^q \end{pmatrix}, \quad \Gamma_I = \begin{pmatrix} \Gamma_I^1 \\ \vdots \\ \Gamma_I^q \end{pmatrix}
$$

with components $f_1^i \in C(\mathbb{R}^s \times \mathbb{R}^k \to \mathbb{R})$ and $\Gamma_I^i \in C(\mathbb{R}^s \times \mathbb{R}^k \to \mathbb{R}^{1 \times q})$ for $i = 1, \ldots, q$. We now define functions $F_I : \tilde{\mathscr{D}} \times \mathbb{R}^k \to \mathbb{R}^{\bar{r}}$, $F_{II} : \tilde{\mathscr{D}} \times \mathbb{R}^k \to \mathbb{R}^{m-q}$ with

$$
F_I : \quad (t, \underbrace{y_{10}, \ldots, y_{1,r_1-1}, \ldots, y_{q,r_q-1}}_{=X_I}, \underbrace{y_{q+1}, \ldots, y_m}_{=X_{II}}, \eta) \mapsto
$$

$$
\left(y_{11}, \ldots, y_{1,r_1-1}, f_1^1(d_1(t), \eta) - \frac{\Gamma_I^1(d_2(t), \eta) \tilde{K}_I(t, X_I)}{1 - \varphi_I(t)^2 \|\tilde{K}_I(t, X_I)\|^2}, \right.
$$

$$\ldots, y_{q,r_q-1}, f_1^q(d_1(t), \eta) - \frac{\Gamma_I^q(d_2(t), \eta)\tilde{K}_I(t, X_I)}{1 - \varphi_I(t)^2\|\tilde{K}_I(t, X_I)\|^2}\Bigg),$$

$$F_{II}: \quad (t, \underbrace{y_{10}, \ldots, y_{1,r_1-1}, \ldots, y_{q,r_q-1}}_{=X_I}, \underbrace{y_{q+1}, \ldots, y_m}_{=X_{II}}, \eta) \mapsto$$

$$\left(f_2(X_I, X_{II}) + f_3(d_3(t), \eta) - \frac{\Gamma_{II}(d_4(t), \eta)\tilde{K}_I(t, X_I)}{1 - \varphi_I(t)^2\|\tilde{K}_I(t, X_I)\|^2} \right.$$

$$\left. - f_4(d_5(t), \eta)\frac{\hat{k}(X_{II} - X_{\mathrm{ref},II}(t))}{1 - \varphi_{II}(t)^2\|X_{II} - X_{\mathrm{ref},II}(t)\|^2} \right).$$

Then the closed-loop system (3.1) and (4.3) is equivalent to (4.10).

Step 1e In order to show that (4.10) has a solution we take the derivative of the second equation and rewrite it appropriately. First observe that since $T_2 \in \mathbb{T}_{m,k,h}^{\mathrm{DAE}}$ there exist $z \in C(\mathbb{R}^m \times \mathbb{R}^k \to \mathbb{R}^k)$ and $\tilde{T}_2 \in \mathbb{T}_{m,k,h}$ such that

$$\forall \zeta \in C([-h, \infty) \to \mathbb{R}^m)\ \forall\, t \geq 0: \tfrac{\mathrm{d}}{\mathrm{d}t}(T_2\zeta)(t) = z\big(\zeta(t), (\tilde{T}_2\zeta)(t)\big).$$

Now define the operator $\hat{T}_3 : C([-h, \infty) \to \mathbb{R}^{\bar{r}+m-q}) \to C(\mathbb{R}_{\geq 0} \to \mathbb{R}^k)$ by the property that for $\zeta_1 \in C([-h, \infty) \to \mathbb{R}^{\bar{r}})$, $\zeta_2 \in C([-h, \infty) \to \mathbb{R}^{m-q})$ we have

$$\hat{T}_3\begin{pmatrix}\zeta_1\\\zeta_2\end{pmatrix}(t) := \tilde{T}_2\left(S\begin{pmatrix}\zeta_1\\\zeta_2\end{pmatrix}\right)(t), \quad t \geq 0,$$

then $\hat{T}_3 \in \mathbb{T}_{\bar{r}+m-q,k,h}$. A differentiation of the second equation in (4.10) yields

$$0 = \frac{\partial F_{II}}{\partial t}\left(t, \begin{pmatrix}X_I(t)\\X_{II}(t)\end{pmatrix}, \hat{T}_2\begin{pmatrix}X_I\\X_{II}\end{pmatrix}(t)\right) + \frac{\partial F_{II}}{\partial X_I}\left(t, \begin{pmatrix}X_I(t)\\X_{II}(t)\end{pmatrix}, \hat{T}_2\begin{pmatrix}X_I\\X_{II}\end{pmatrix}(t)\right)\dot{X}_I(t)$$

$$+ \frac{\partial F_{II}}{\partial X_{II}}\left(t, \begin{pmatrix}X_I(t)\\X_{II}(t)\end{pmatrix}, \hat{T}_2\begin{pmatrix}X_I\\X_{II}\end{pmatrix}(t)\right)\dot{X}_{II}(t)$$

$$+ \frac{\partial F_{II}}{\partial \eta}\left(t, \begin{pmatrix}X_I(t)\\X_{II}(t)\end{pmatrix}, \hat{T}_2\begin{pmatrix}X_I\\X_{II}\end{pmatrix}(t)\right)\frac{\mathrm{d}}{\mathrm{d}t}\left(\hat{T}_2\begin{pmatrix}X_I\\X_{II}\end{pmatrix}\right)(t),$$

by which, using the first equation in (4.10) and

$$\frac{d}{dt}\left(\hat{T}_2\begin{pmatrix}X_I\\X_{II}\end{pmatrix}\right)(t) = z\left(S\begin{pmatrix}X_I(t)\\X_{II}(t)\end{pmatrix}, \hat{T}_3\begin{pmatrix}X_I\\X_{II}\end{pmatrix}(t)\right),$$

we obtain

$$\frac{\partial F_{II}}{\partial X_{II}}\left(t, \begin{pmatrix}X_I(t)\\X_{II}(t)\end{pmatrix}, \hat{T}_2\begin{pmatrix}X_I\\X_{II}\end{pmatrix}(t)\right) \dot{X}_{II}(t)$$

$$= \hat{F}_{II}\left(t, \begin{pmatrix}X_I(t)\\X_{II}(t)\end{pmatrix}, T_1\begin{pmatrix}X_I\\X_{II}\end{pmatrix}(t), \hat{T}_2\begin{pmatrix}X_I\\X_{II}\end{pmatrix}(t), \hat{T}_3\begin{pmatrix}X_I\\X_{II}\end{pmatrix}(t)\right)$$

for some $\hat{F}_{II} : \tilde{\mathscr{D}} \times \mathbb{R}^{3k} \to \mathbb{R}^{m-q}$. We show that the matrix

$$\frac{\partial F_{II}}{\partial X_{II}}(t, X_I, X_{II}, \eta) = \frac{\partial f_2(X_I, X_{II})}{\partial X_{II}} - \frac{\hat{k} f_4(d_5(t), \eta)}{1 - \varphi_{II}(t)^2 \|X_{II} - X_{\mathrm{ref}, II}(t)\|^2} \cdot$$

$$\cdot \left(I_{m-q} + \frac{2\varphi_{II}(t)^2(X_{II} - X_{\mathrm{ref}, II}(t))(X_{II} - X_{\mathrm{ref}, II}(t))^\top}{1 - \varphi_{II}(t)^2 \|X_{II} - X_{\mathrm{ref}, II}(t)\|^2}\right) \qquad (4.16)$$

is invertible for all $(t, X_I, X_{II}, \eta) \in \tilde{\mathscr{D}} \times \mathbb{R}^k$: The symmetry and positive semi-definiteness of

$$\mathscr{G}(t, X_{II}) := \frac{2\varphi_{II}(t)^2(X_{II} - X_{\mathrm{ref}, II}(t))(X_{II} - X_{\mathrm{ref}, II}(t))^\top}{1 - \varphi_{II}(t)^2 \|X_{II} - X_{\mathrm{ref}, II}(t)\|^2}$$

implies positive definiteness (and hence invertibility) of $I_{m-q} + \mathscr{G}(t, X_{II})$ for all $(t, X_{II}) \in \tilde{\mathscr{D}}_{II}$, and by Berger et al. [8, Lem. 3.3] we further have

$$\left\|\left(I_{m-q} + \mathscr{G}(t, X_{II})\right)^{-1}\right\| \le 1.$$

Therefore, according to (4.5) and Assumption 3.1 (iv), we have for all $(t, X_I, X_{II}, \eta) \in \tilde{\mathscr{D}} \times \mathbb{R}^k$ that

$$\left\|(1 - \varphi_{II}(t)^2 \|X_{II} - X_{\mathrm{ref}, II}(t)\|^2)\hat{k}^{-1}\left[f_4(d_5(t), \eta)\right]^{-1}(I_{m-q} + \mathscr{G}(t, X_{II}))^{-1}\frac{\partial f_2(X_I, X_{II})}{\partial X_{II}}\right\|$$

$$\le \hat{k}^{-1}\alpha^{-1}\left\|\frac{\partial f_2(X_I, X_{II})}{\partial X_{II}}\right\| \overset{(4.5)}{<} 1.$$

This implies invertibility of $\frac{\partial F_{II}}{\partial X_{II}}(t, X_I, X_{II}, \eta)$ for all $(t, X_I, X_{II}, \eta) \in \tilde{\mathscr{D}} \times \mathbb{R}^k$.
With $\tilde{F}_{II} : \tilde{\mathscr{D}} \times \mathbb{R}^{3k} \to \mathbb{R}^{m-q}$ defined by

$$\tilde{F}_{II}(t, X_I, X_{II}, \eta_1, \eta_2, \eta_3) := \left(\frac{\partial F_{II}}{\partial X_{II}}(t, X_I, X_{II}, \eta_2) \right)^{-1} \hat{F}_{II}(t, X_I, X_{II}, \eta_1, \eta_2, \eta_3)$$

and the first equation in (4.10) we obtain the ODE

$$\dot{X}_I(t) = F_I \left(t, \begin{pmatrix} X_I(t) \\ X_{II}(t) \end{pmatrix}, T_1 \begin{pmatrix} X_I \\ X_{II} \end{pmatrix}(t) \right),$$

$$\dot{X}_{II}(t) = \tilde{F}_{II} \left(t, \begin{pmatrix} X_I(t) \\ X_{II}(t) \end{pmatrix}, T_1 \begin{pmatrix} X_I \\ X_{II} \end{pmatrix}(t), \hat{T}_2 \begin{pmatrix} X_I \\ X_{II} \end{pmatrix}(t), \hat{T}_3 \begin{pmatrix} X_I \\ X_{II} \end{pmatrix}(t) \right),$$

$$(4.17)$$

with initial conditions (4.11).

Step 1f Consider the initial value problem (4.17), (4.11), then we have
$(0, X_I(0), X_{II}(0)) \in \tilde{\mathscr{D}}$, F_I is measurable in t, continuous in (X_I, X_{II}, η),
and locally essentially bounded, and $\tilde{F}_{II}$ is measurable in t, continuous in
$(X_I, X_{II}, \eta_1, \eta_2, \eta_3)$, and locally essentially bounded. Therefore, by [21, Theorem
B.1][1] we obtain existence of solutions to (4.17), and every solution can be extended
to a maximal solution. Furthermore, for a maximal solution $(X_I, X_{II}) : [-h, \omega) \to$
$\mathbb{R}^{\bar{r}+m-q}$, $\omega \in (0, \infty]$, of (4.17), (4.11) the closure of the graph of this solution is
not a compact subset of $\tilde{\mathscr{D}}$.

We show that (X_I, X_{II}) is also a maximal solution of (4.10). Since (X_I, X_{II}) is
particular satisfies, by construction,

$$\forall t \in [0, \omega) : \quad 0 = \frac{d}{dt} F_{II} \left(t, \begin{pmatrix} X_I(t) \\ X_{II}(t) \end{pmatrix}, \hat{T}_2 \begin{pmatrix} X_I \\ X_{II} \end{pmatrix}(t) \right),$$

there exists $c \in \mathbb{R}^{m-q}$ such that

$$\forall t \in [0, \omega) : \quad c = F_{II} \left(t, \begin{pmatrix} X_I(t) \\ X_{II}(t) \end{pmatrix}, \hat{T}_2 \begin{pmatrix} X_I \\ X_{II} \end{pmatrix}(t) \right).$$

Invoking (4.11), the definition of F_{II} and $\hat{T}_2$, and the consistency condition (4.7)
we may infer that $c = 0$. Therefore, (X_I, X_{II}) is a solution of (4.10). Furthermore,

---

[1] In [21] a domain $\mathscr{D} \subseteq \mathbb{R}_{\geq 0} \times \mathbb{R}$ is considered, but the generalization to the higher dimensional
case is straightforward.

(X_I, X_{II}) is also a maximal solution of (4.10), since any right extension would be a solution of (4.17) following the procedure in Step 1e, a contradiction.

Recall $X_I(t) = \left(y_1(t), \ldots, y_1^{(r_1-1)}(t) \ldots, y_q^{(r_q-1)}(t) \right)^\top$, $X_{II}(t) = (y_{q+1}(t),$
$\ldots, y_m(t))^\top$ and define

$$(e_{10}, \ldots, e_{1,r_1-1}, \ldots, e_{q,r_q-1}, e_{q+1}, \ldots, e_m) : [0, \omega) \to \mathbb{R}^{\bar{r}+m-q} \qquad (4.18)$$

by

$$e_{ij}(t) = \tilde{K}_{ij}(t, y_i(t), \ldots, y_i^{(j)}(t)), \quad \text{for } i = 1, \ldots, q \text{ and } j = 0, \ldots, r_i - 1,$$

$$e_i(t) = y_i(t) - y_{\text{ref},i}(t), \qquad\qquad \text{for } i = q + 1, \ldots, m,$$

then the closure of the graph of the function in (4.18) is not a compact subset of $\mathscr{D}$.

Step 2 We show boundedness of the gain functions $k_I(\cdot)$, $k_{II}(\cdot)$ and $k_{ij}(\cdot)$ as in (4.3) on $[0, \omega)$. This also proves (4.9).

Step 2a The proof of boundedness of $k_{ij}(\cdot)$ for $i = 1, \ldots, q$, $j = 0, \ldots, r_i - 2$ on $[0, \omega)$ is analogous to Step 2a of the proof of [9, Thm. 3.1] and hence omitted.

Step 2b We prove by induction that there exist constants M_{ij}^ℓ, N_{ij}^ℓ, $K_{ij}^\ell > 0$ such that, for all $t \in [0, \omega)$,

$$\left| \left(\tfrac{d}{dt} \right)^\ell \left[k_{ij}(t) e_{ij}(t) \right] \right| \leq M_{ij}^\ell, \quad \left| \left(\tfrac{d}{dt} \right)^\ell e_{ij}(t) \right| \leq N_{ij}^\ell, \quad \left| \left(\tfrac{d}{dt} \right)^\ell k_{ij}(t) \right| \leq K_{ij}^\ell,$$

$$(4.19)$$

for $i = 1, \ldots, q$, $j = 0, \ldots, r_i - 2$, and $\ell = 0, \ldots, r_i - 1 - j$.

First, we may infer from *Step 2a* that $k_{ij}(\cdot)$, for $i = 1, \ldots, q$, $j = 0, \ldots, r_i - 2$, are bounded. Furthermore, e_{ij} are bounded since they evolve in the respective performance funnels. Therefore, for each $i = 1, \ldots, q$ and $j = 0, \ldots, r_i - 2$, (4.19) is true whenever $\ell = 0$. Fix $i \in \{1, \ldots, q\}$. We prove (4.19) for $j = r_i - 2$ and $\ell = 1$:

$$\dot{e}_{i,r_i-2}(t) = e_{i,r_i-1}(t) - k_{i,r_i-2}(t) e_{i,r_i-2}(t),$$

$$\dot{k}_{i,r_i-2}(t) = 2k_{i,r_i-2}^2(t) \big(\varphi_{i,r_i-2}^2(t) e_{i,r_i-2}(t) \dot{e}_{i,r_i-2}(t)$$

$$+ \varphi_{i,r_i-2}(t) \dot{\varphi}_{i,r_i-2}(t) |e_{i,r_i-2}(t)|^2 \big),$$

$$\tfrac{d}{dt} \left[k_{i,r_i-2}(t) e_{i,r_i-2}(t) \right] = \dot{k}_{i,r_i-2}(t) e_{i,r_i-2}(t) + k_{i,r_i-2}(t) \dot{e}_{i,r_i-2}(t).$$

Boundedness of k_{i,r_i-2}, φ_{i,r_i-2}, $\dot{\varphi}_{i,r_i-2}$, e_{i,r_i-2} together with the above equations implies that $\dot{e}_{i,r_i-2}(t)$, $\dot{k}_{i,r_i-2}(t)$ and $\tfrac{d}{dt} \left[k_{i,r_i-2}(t) e_{i,r_i-2}(t) \right]$ are bounded. Now consider indices $s \in \{0, \ldots, r_i-3\}$ and $l \in \{0, \ldots, r_i-1-s\}$ and assume that (4.19) is true for all $j = s + 1, \ldots, r_i - 2$ and all $\ell = 0, \ldots, r_i - 1 - j$ as well as for

$j = s$ and all $\ell = 0, \ldots, l-1$. We show that it is true for $j = s$ and $\ell = l$:

$$\left(\tfrac{d}{dt}\right)^l e_{is}(t) = \left(\tfrac{d}{dt}\right)^{l-1}\left[e_{i,s+1}(t) - k_{is}(t)e_{is}(t)\right]$$

$$= \left(\tfrac{d}{dt}\right)^{l-1} e_{i,s+1}(t) - \left(\tfrac{d}{dt}\right)^{l-1}\left[k_{is}(t)e_{is}(t)\right],$$

$$\left(\tfrac{d}{dt}\right)^l k_{is}(t) = \left(\tfrac{d}{dt}\right)^{l-1}\left(2k_{is}^2(t)\left(\varphi_{is}^2(t)e_{is}(t)\dot{e}_{is}(t) + \varphi_{is}(t)\dot{\varphi}_{is}(t)|e_{is}(t)|^2\right)\right),$$

$$\left(\tfrac{d}{dt}\right)^l \left[k_{is}(t)e_{is}(t)\right] = \left(\tfrac{d}{dt}\right)^{l-1}\left(\dot{k}_{is}(t)e_{is}(t) + k_{is}(t)\dot{e}_{is}(t)\right).$$

Then, successive application of the product rule and using the induction hypothesis as wells as the fact that $\varphi_{is}, \dot{\varphi}_{is}, \ldots, \varphi_{is}^{(r_i-s)}$ are bounded, yields that the above terms are bounded. Therefore, the proof of (4.19) is complete.

It follows from (4.19) and (4.12) that, for all $i = 1, \ldots, q$ and $j = 0, \ldots, r_i - 1$, $e_i^{(j)}$ is bounded on $[0, \omega)$.

Step 2c We show that $k_I(\cdot)$ as in (4.3) is bounded. It follows from (4.12) that, for $i = 1, \ldots, q$,

$$e_i^{(r_i)}(t) = \dot{e}_{i,r_i-1}(t) - \sum_{j=0}^{r_i-2} \left(\tfrac{d}{dt}\right)^{r_i-1-j}\left[k_{ij}(t)e_{ij}(t)\right].$$

Then we find that by (4.10)

$$\dot{e}_I(t) = f_1\Big(d_1(t), T_1\big(y_1, \ldots, y_1^{(r_1-1)}, \ldots, y_q^{(r_q-1)}, y_{q+1}, \ldots, y_m\big)(t)\Big)$$

$$- \Gamma_1\Big(d_1(t), T_1\big(y_1, \ldots, y_1^{(r_1-1)}, \ldots, y_q^{(r_q-1)}, y_{q+1}, \ldots, y_m\big)(t)\Big) k_I(t)e_I(t)$$

$$+ \begin{pmatrix} \sum_{j=0}^{r_1-2} \left(\tfrac{d}{dt}\right)^{r_1-1-j} k_{1j}(t)e_{1j}(t) \\ \sum_{j=0}^{r_2-2} \left(\tfrac{d}{dt}\right)^{r_2-1-j} k_{2j}(t)e_{2j}(t) \\ \vdots \\ \sum_{j=0}^{r_q-2} \left(\tfrac{d}{dt}\right)^{r_q-1-j} k_{qj}(t)e_{qj}(t) \end{pmatrix} - \begin{pmatrix} y_{\text{ref},1}^{(r_1)}(t) \\ \vdots \\ y_{\text{ref},q}^{(r_q)}(t) \end{pmatrix}.$$

$$(4.20)$$

Again we use $X_I(t) = \left(y_1(t), \ldots, y_1^{(r_1-1)}(t) \ldots, y_q^{(r_q-1)}(t)\right)^\top$, $X_{II}(t) = (y_{q+1}(t), \ldots, y_m(t))^\top$ and we set, for $t \in [0, \omega)$,

$$\hat{F}_I(t) := f_1\left(d_1(t), T_1(X_I, X_{II})(t)\right)$$

$$+ \begin{pmatrix} \sum_{j=0}^{r_1-2} \left(\frac{d}{dt}\right)^{r_1-1-j} k_{1j}(t)e_{1j}(t) \\ \sum_{j=0}^{r_2-2} \left(\frac{d}{dt}\right)^{r_2-1-j} k_{2j}(t)e_{2j}(t) \\ \vdots \\ \sum_{j=0}^{r_q-2} \left(\frac{d}{dt}\right)^{r_q-1-j} k_{qj}(t)e_{qj}(t) \end{pmatrix} - \begin{pmatrix} y_{\text{ref},1}^{(r_1)}(t) \\ \vdots \\ y_{\text{ref},q}^{(r_q)}(t) \end{pmatrix}. \tag{4.21}$$

We obtain from (4.19) and (4.12) that $e_i^{(j)}$ is bounded on the interval $[0, \omega)$ for $i = 1, \ldots, q$ and $j = 0, \ldots, r_i - 2$. Furthermore, e_I evolves in the performance funnel $\mathcal{F}_{\varphi_I}^q$, thus $|e_{i,r_i-1}(t)|^2 \leq \|e_I(t)\|^2 < \varphi_I(t)^{-1}$ for all $t \in [0, \omega)$, so e_{i,r_i-1} is bounded on $[0, \omega)$ for $i = 1, \ldots, q$. Invoking boundedness of $y_{\text{ref},i}^{(j)}$ yields boundedness of $y_i^{(j)}$ for $i = 1, \ldots, q$, $j = 0, \ldots, r_i - 1$. Then the bounded-input, bounded-output property of T_1 in Definition 3.1 (iii) implies that $T_1(X_I, X_{II})$ is bounded by

$$M_{T_1} := \|T_1(X_I, X_{II})|_{[0,\omega)}\|_\infty.$$

This property together with (4.19), continuity of f_1 and boundedness of d_1 yields that $\hat{F}_I(\cdot)$ is bounded on $[0, \omega)$. In other words, there exists some $M_{\hat{F}_I} > 0$ such that $\|\hat{F}_I|_{[0,\omega)}\|_\infty \leq M_{\hat{F}_I}$. Now define the compact set

$$\Omega = \left\{ (\delta, \eta, e_I) \in \mathbb{R}^s \times \mathbb{R}^k \times \mathbb{R}^q \,\middle|\, \|\delta\| \leq \|d_2|_{[0,\omega)}\|_\infty, \ \|\eta\| \leq M_{T_1}, \ \|e_I\| = 1 \right\},$$

then, since $\Gamma_I + \Gamma_I^\top$ is pointwise positive definite by Assumption 3.1 (i) and the map

$$\Omega \ni (\delta, \eta, e_I) \mapsto e_I^\top \left(\Gamma_I(\delta, \eta) + \Gamma_I(\delta, \eta)^\top\right) e_I \in \mathbb{R}_{>0}$$

is continuous, it follows that there exists $\gamma > 0$ such that

$$\forall (\delta, \eta, e_I) \in \Omega : \quad e_I^\top \left(\Gamma_I(\delta, \eta) + \Gamma_I(\delta, \eta)^\top\right) e_I \geq \gamma.$$

Therefore, we have for all $t \in [0, \omega)$ that

$$e_I(t)^\top \left(\Gamma_I\big(d_1(t), T_1(X_I, X_{II})(t)\big) + \Gamma_I\big(d_1(t), T_1(X_I, X_{II})(t)\big)^\top \right) e_I(t) \geq \gamma \|e_I(t)\|^2.$$

Now, set $\psi_I(t) := \varphi_I(t)^{-1}$ for $t \in (0, \omega)$, let $T_I \in (0, \omega)$ be arbitrary but fixed and set $\lambda_I := \inf_{t \in (0, \omega)} \psi_I(t)$. Since $\dot{\varphi}_I$ is bounded and $\liminf_{t \to \infty} \varphi_I(t) > 0$ we find that $\frac{d}{dt}\psi_I|_{[0, \omega)}$ is bounded and hence $\psi_I|_{[0, \omega)}$ is Lipschitz continuous with Lipschitz bound $L_I > 0$. Choose $\varepsilon_I > 0$ small enough such that

$$\varepsilon_I \leq \min \left\{ \frac{\lambda_I}{2}, \inf_{t \in (0, T_I]} (\psi_I(t) - \|e_I(t)\|) \right\}$$

$$\text{and} \quad L_I \leq \frac{\lambda_I^2}{8\varepsilon_I}\gamma - M_{\hat{F}_I}, \tag{4.22}$$

We show that

$$\forall t \in (0, \omega) : \ \psi_I(t) - \|e_I(t)\| \geq \varepsilon_I. \tag{4.23}$$

By definition of ε_I this holds on $(0, T_I]$. Seeking a contradiction suppose that

$$\exists t_{I,1} \in [T_I, \omega) : \ \psi_I(t_{I,1}) - \|e_I(t_{I,1})\| < \varepsilon_I.$$

Set $t_{I,0} = \max\{t \in [T_I, t_{I,1}) \mid \psi_I(t) - \|e_I(t)\| = \varepsilon_I\}$. Then, for all $t \in [t_{I,0}, t_{I,1}]$, we have

$$\psi_I(t) - \|e_I(t)\| \leq \varepsilon_I,$$

$$\|e_I(t)\| \geq \psi_I(t) - \varepsilon_I \geq \frac{\lambda_I}{2},$$

$$k_I(t) = \frac{1}{1 - \varphi_I^2(t)\|e_I(t)\|^2} \geq \frac{\lambda_I}{2\varepsilon_I}.$$

Then it follows from (4.20) and (4.21) that for all $t \in [t_{I,0}, t_{I,1}]$,

$$\frac{1}{2}\frac{d}{dt}\|e_I(t)\|^2 = \frac{1}{2}\left(e_I^\top(t)\dot{e}_I(t) + \dot{e}_I^\top(t)e_I(t) \right)$$

$$= e_I^\top(t)\left(\hat{F}_I(t) - \frac{1}{2}\big(\Gamma_I\big(d_1(t), T_1(X_I, X_{II})(t)\big) + \Gamma_I\big(d_1(t), T_1(X_I, X_{II})(t)\big)^\top\big) k_I(t)e_I(t) \right)$$

$$\leq \left(M_{\hat{F}_I} - \frac{\lambda_I^2}{8\varepsilon_I}\gamma \right) \|e_I(t)\| \overset{(4.22)}{\leq} -L_I\|e_I(t)\|.$$

Then, using $\|e_I(t)\| \geq \frac{\lambda_I}{2} > 0$ for all $t \in [t_{I,0}, t_{I,1}]$,

$$\|e_I(t_{I,1})\| - \|e_I(t_{I,0})\| = \int_{t_{I,0}}^{t_{I,1}} \frac{1}{2} \|e_I(t)\|^{-1} \frac{\mathrm{d}}{\mathrm{d}t} \|e_I(t)\|^2 \, \mathrm{d}t$$

$$\leq -L_I(t_{I,1} - t_{I,0}) \leq -|\psi_I(t_{I,1}) - \psi_I(t_{I,0})| \leq \psi_I(t_{I,1}) - \psi_I(t_{I,0}),$$

and thus we obtain $\varepsilon_I = \psi_I(t_{I,0}) - \|e_I(t_{I,0})\| \leq \psi_I(t_{I,1}) - \|e_I(t_{I,1})\| < \varepsilon_I$, a contradiction.

Step 2d We show that $k_{II}(\cdot)$ as in (4.3) is bounded. Seeking a contradiction, we assume that $k_{II}(t) \to \infty$ for $t \to \omega$. Set, for $t \in [0, \omega)$,

$$\check{F}_{II}(t) := f_2\big(X_I(t), X_{II}(t)\big) + f_3\big(d_3(t), (T_2y)(t)\big) - \Gamma_{II}\big(d_4(t), (T_2y)(t)\big) k_I(t)e_I(t). \tag{4.24}$$

Since k_I is bounded on $[0\omega)$ by Step 2c, it follows from Step 2b, boundedness of $T_2(y)$, d_3 and d_4 and continuity of f_2, f_3 and Γ_{II} that $\check{F}_{II}(\cdot)$ is bounded on $[0, \omega)$. By (4.10) we have

$$0 = \check{F}_{II}(t) - f_4\big(d_5(t), (T_2y)(t)\big) k_{II}(t)e_{II}(t). \tag{4.25}$$

We show that $e_{II}(t) \to 0$ for $t \to \omega$. Seeking a contradiction, assume that there exist $\kappa > 0$ and a sequence $(t_n) \subset \mathbb{R}_{\geq 0}$ with $t_n \nearrow \omega$ such that $\|e_{II}(t_n)\| \geq \kappa$ for all $n \in \mathbb{N}$. Then, from (4.25) we obtain, for all $t \in [0, \omega)$,

$$\|\check{F}_{II}(t)\| = \|f_4\big(d_5(t), (T_2y)(t)\big) k_{II}(t)e_{II}(t)\| = |f_4\big(d_5(t), (T_2y)(t)\big)| \cdot |k_{II}(t)| \cdot \|e_{II}(t)\|.$$

Since $k_{II}(t) \to \infty$ for $t \to \omega$, $\|e_{II}(t_n)\| \geq \kappa$ and $f_4\big(d_5(t_n), (T_2y)(t_n)\big) \geq \alpha$, we find that

$$\|\check{F}_{II}(t_n)\| \geq \alpha \kappa k_{II}(t_n) \to \infty \quad \text{for } n \to \infty,$$

which contradicts boundedness of $\check{F}_{II}(\cdot)$.

Hence, we have $e_{II}(t) \to 0$ for $t \to \omega$, by which $\lim_{t\to\infty} \varphi_{II}(t)^2 \|e_{II}(t)\|^2 = 0$ because $\varphi_{II}(\cdot)$ is bounded. This leads to the contradiction $\lim_{t\to\infty} k_{II}(t) = \hat{k}$, thus $k_{II}(\cdot)$ is bounded.

Step 3 We show that $\omega = \infty$. Seeking a contradiction, we assume that $\omega < \infty$. Then, since e_I, e_{II}, k_I, k_{II} and e_{ij}, k_{ij} are bounded for $i = 1, \ldots, q$, $j = 0, \ldots, r_i - 2$ by Step 2, it follows that the closure of the graph of the function in (4.18) is a compact subset of $\mathscr{D}$, which is a contradiction. This finishes the proof of the theorem. $\qquad\square$

5 Simulations

In this section we illustrate the application of the funnel controller (4.3) by considering the following academic example:

$$
\begin{aligned}
\ddot{y}_1(t) = & - \sin y_1(t) + y_1(t)\dot{y}_1(t) + y_2(t)^2 \\
& + \dot{y}_1(t)^2 T(y_1, y_2)(t) + (y_1(t)^2 + y_2(t)^4 + 1)u_I(t), \\
0 = & \; y_1(t)^3 + y_1(t)\dot{y}_1(t)^3 + y_2(t) + T(y_1, y_2)(t) + \\
& + T(y_1, y_2)(t)u_I(t) + u_{II}(t),
\end{aligned}
\tag{5.1}
$$

where $T : C(\mathbb{R}_{\geq 0} \to \mathbb{R}^m) \to C^1(\mathbb{R}_{\geq 0} \to \mathbb{R})$ is given by

$$
T(y_1, y_2)(t) := e^{-2t}\eta^0 + \int_0^t e^{-2(t-s)}\Big(2y_1(s) - y_2(s)\Big)ds, \quad t \geq 0,
$$

for any fix $\eta^0 \in \mathbb{R}$. Similar as we have calculated for the operator T_2 on page 232, we may calculate that $T \in \mathbb{T}^{\mathrm{DAE}}_{2,1,0}$. Define

$$
T_1(y_1, y_1^d, y_2)(t) := \begin{pmatrix} y_1(t) \\ y_1^d(t) \\ y_2(t) \\ T(y_1, y_2)(t) \end{pmatrix}, \quad t \geq 0,
$$

then $T_1 \in \mathbb{T}_{2,3,0}$, and set $T_2 := T$. Furthermore, define the functions

$$
\begin{aligned}
& f_1 : \mathbb{R}^4 \to \mathbb{R}, \; (\eta_1, \eta_2, \eta_3, \eta_4) \mapsto - \sin \eta_1 + \eta_3\eta_2 + \eta_3^2 + \eta_2^2\eta_4, \\
& \Gamma_I : \mathbb{R}^4 \to \mathbb{R}, \; (\eta_1, \eta_2, \eta_3, \eta_4) \mapsto \eta_1^2 + \eta_3^4 + 1, \\
& f_2 : \mathbb{R}^4 \to \mathbb{R}, \; (y_1, y_1^d, y_2) \mapsto y_1^3 + y_1(y_1^d)^3 + y_2, \\
& f_3 : \mathbb{R} \to \mathbb{R}, \; \eta \mapsto \eta, \\
& \Gamma_{II} : \mathbb{R} \to \mathbb{R}, \; \eta \mapsto \eta, \\
& f_4 : \mathbb{R} \to \mathbb{R}, \; \eta \mapsto 1.
\end{aligned}
$$

Then system (5.1) is of the form (3.1) with $m = 2$, $q = 1$ and $r_1 = 2$. It is straightforward to check that Assumption 3.1 is satisfied. In particular, condition (iii) is satisfied, because

$$f_2'(y_1, y_1^d, y_2) \begin{bmatrix} 0 \\ I_{m-q} \end{bmatrix} = \frac{\partial f_2}{\partial y_2}(y_1, y_1^d, y_2) = 1$$

is bounded. Furthermore, $f_4(\eta) \geq 1 =: \alpha$ for all $\eta \in \mathbb{R}$, and hence we may choose $\hat{k} = 2$, with which condition (4.5) is satisfied.

For the simulation we choose the reference signal $y_{ref}(t) = (\cos 2t, \sin t)^\top$, and initial values

$$y_1(0) = \dot{y}_1(0) = y_2(0) = 0 \quad \text{and} \quad \eta^0 = 0.$$

For the controller (4.3) we choose the funnel functions $\varphi_{10} = \varphi_I = \varphi_{II} = \varphi$ with

$$\varphi : \mathbb{R}_{\geq 0} \to \mathbb{R}_{\geq 0}, \ t \mapsto \tfrac{1}{2}te^{-t} + 2\arctan t.$$

It is straightforward to check that $\dot{\varphi}$ and $\ddot{\varphi}$ are bounded, thus $\varphi \in \Phi_2$. Moreover, since $\varphi(0) = 0$, no restriction is put on the initial error and we find that (4.8) is satisfied and $k_{10}(0) = k_I(0) = 1$ and $k_{II}(0) = 2$. Furthermore,

$$e_I(0) = e_{1,1}(0) = \dot{e}_{10}(0) + k_{10}(0)e_{10}(0)$$

$$= \dot{y}_1(0) - \dot{y}_{ref,1}(0) + k_{10}(0)\big(y_1(0) - y_{ref,1}(0)\big) = -1,$$

$$e_{II}(0) = y_2(0) - y_{ref,2}(0) = 0,$$

and hence we obtain

$$u_I(0) = -k_I(0)e_I(0) = 1, \quad u_{II}(0) = -k_{II}(0)e_{II}(0) = 0.$$

Since $h = 0$, we find that in view of Remark 4.2 the localization of T_2 satisfies $T_2(0, 0) = 0$. With this finally find that the initial value is indeed consistent, i.e., condition (4.7) is satisfied. We have now verified all assumptions of Theorem 4.1, by which funnel control via (4.3) is feasible for the system (5.1).

The simulation of the controller (4.3) applied to (5.1) has been performed in MATLAB (solver: **ode15s**, rel. tol.: 10^{-14}, abs. tol.: 10^{-10}) over the time interval [0,10] and is depicted in Fig. 2.

Figure 2a shows the tracking error components, which stay uniformly within the funnel boundaries. The components of the generated input functions are shown in Fig. 2b, which exhibit an acceptable performance.

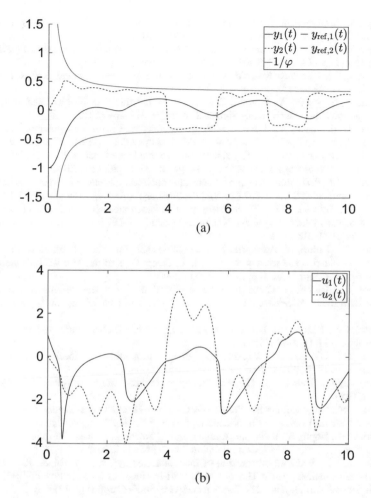

Fig. 2 Simulation of the controller (4.3) for the system (5.1). (**a**) Funnel and tracking errors. (**b**) Input functions

Acknowledgement This work was supported by the German Research Foundation (Deutsche Forschungsgemeinschaft) via the grant BE 6263/1-1.

References

1. Berger, T.: On differential-algebraic control systems. Ph.D. Thesis, Institut für Mathematik, Technische Universität Ilmenau, Universitätsverlag Ilmenau (2014). http://www.db-thueringen.de/servlets/DocumentServlet?id=22652
2. Berger, T.: Zero dynamics and stabilization for linear DAEs. In: Schöps, S., Bartel, A., Günther, M., ter Maten, E.J.W., Müller, P.C. (eds.) Progress in Differential-Algebraic Equations. Differential-Algebraic Equations Forum, pp. 21–45. Springer, Berlin (2014)

3. Berger, T.: Zero dynamics and funnel control of general linear differential-algebraic systems. ESAIM Control Optim. Calc. Var. **22**(2), 371–403 (2016)
4. Berger, T., Rauert, A.L.: A universal model-free and safe adaptive cruise control mechanism. In: Proceedings of the MTNS 2018, pp. 925–932. Hong Kong (2018)
5. Berger, T., Reis, T.: Zero dynamics and funnel control for linear electrical circuits. J. Franklin Inst. **351**(11), 5099–5132 (2014)
6. Berger, T., Ilchmann, A., Reis, T.: Zero dynamics and funnel control of linear differential-algebraic systems. Math. Control Signals Syst. **24**(3), 219–263 (2012)
7. Berger, T., Ilchmann, A., Trenn, S.: The quasi-Weierstraß form for regular matrix pencils. Linear Algebra Appl. **436**(10), 4052–4069 (2012). https://doi.org/10.1016/j.laa.2009.12.036
8. Berger, T., Ilchmann, A., Reis, T.: Funnel control for nonlinear functional differential-algebraic systems. In: Proceedings of the MTNS 2014, pp. 46–53. Groningen (2014)
9. Berger, T., Lê, H.H., Reis, T.: Funnel control for nonlinear systems with known strict relative degree. Automatica **87**, 345–357 (2018). https://doi.org/10.1016/j.automatica.2017.10.017
10. Berger, T., Otto, S., Reis, T., Seifried, R.: Combined open-loop and funnel control for underactuated multibody systems. Nonlinear Dynam. **95**, 1977–1998 (2019). https://doi.org/10.1007/s11071-018-4672-5
11. Byrnes, C.I., Isidori, A.: A frequency domain philosophy for nonlinear systems, with application to stabilization and to adaptive control. In: Proceedings of the 23rd IEEE Conference on Decision and Control, vol. 1, pp. 1569–1573 (1984)
12. Byrnes, C.I., Isidori, A.: Global feedback stabilization of nonlinear systems. In: Proceedings of the 24th IEEE Conference on Decision and Control, Ft. Lauderdale, vol. 1, pp. 1031–1037 (1985)
13. Byrnes, C.I., Isidori, A.: Local stabilization of minimum-phase nonlinear systems. Syst. Control Lett. **11**(1), 9–17 (1988)
14. Byrnes, C.I., Isidori, A.: New results and examples in nonlinear feedback stabilization. Syst. Control Lett. **12**(5), 437–442 (1989)
15. Fuhrmann, P.A., Helmke, U.: The Mathematics of Networks of Linear Systems. Springer, New York (2015)
16. Hackl, C.M.: Non-identifier Based Adaptive Control in Mechatronics–Theory and Application. Lecture Notes in Control and Information Sciences, vol. 466. Springer, Cham (2017)
17. Hackl, C.M., Hopfe, N., Ilchmann, A., Mueller, M., Trenn, S.: Funnel control for systems with relative degree two. SIAM J. Control Optim. **51**(2), 965–995 (2013)
18. Ilchmann, A.: Decentralized tracking of interconnected systems. In: Hüper, K., Trumpf, J. (eds.) Mathematical System Theory - Festschrift in Honor of Uwe Helmke on the Occasion of his Sixtieth Birthday, pp. 229–245. CreateSpace, South Carolina (2013)
19. Ilchmann, A., Ryan, E.P.: Universal λ-tracking for nonlinearly-perturbed systems in the presence of noise. Automatica **30**(2), 337–346 (1994)
20. Ilchmann, A., Ryan, E.P.: High-gain control without identification: a survey. GAMM Mitt. **31**(1), 115–125 (2008)
21. Ilchmann, A., Ryan, E.P.: Performance funnels and tracking control. Int. J. Control **82**(10), 1828–1840 (2009)
22. Ilchmann, A., Townley, S.B.: Simple adaptive stabilization of high-gain stabilizable systems. Syst. Control Lett. **20**(3), 189–198 (1993)
23. Ilchmann, A., Ryan, E.P., Sangwin, C.J.: Tracking with prescribed transient behaviour. ESAIM Control Optim Calc Var **7**, 471–493 (2002)
24. Isidori, A.: Nonlinear Control Systems. Communications and Control Engineering Series, 3rd edn. Springer, Berlin (1995)
25. Lê, H.H.: Funnel control for systems with known vector relative degree. Ph.D. Thesis, Universität Hamburg, Hamburg (2019). https://ediss.sub.uni-hamburg.de/volltexte/2019/9711/pdf/Dissertation.pdf
26. Liberzon, D., Trenn, S.: The bang-bang funnel controller for uncertain nonlinear systems with arbitrary relative degree. IEEE Trans. Autom. Control **58**(12), 3126–3141 (2013)

27. Mueller, M.: Normal form for linear systems with respect to its vector relative degree. Linear Algebra Appl. **430**(4), 1292–1312 (2009)
28. Nicosia, S., Tornambè, A.: High-gain observers in the state and parameter estimation of robots having elastic joints. Syst. Control Lett. **13**(4), 331–337 (1989)
29. Senfelds, A., Paugurs, A.: Electrical drive DC link power flow control with adaptive approach. In: Proceedings of the 55th International Scientific Conference on Power and Electrical Engineering of Riga Technical University, Latvia, pp. 30–33 (2014)
30. Trentelman, H.L., Stoorvogel, A.A., Hautus, M.L.J.: Control Theory for Linear Systems. Communications and Control Engineering. Springer, London (2001). https://doi.org/10.1007/978-1-4471-0339-4

Observers for Differential-Algebraic Systems with Lipschitz or Monotone Nonlinearities

Thomas Berger and Lukas Lanza

Abstract We study state estimation for nonlinear differential-algebraic systems, where the nonlinearity satisfies a Lipschitz condition or a generalized monotonicity condition or a combination of these. The presented observer design unifies earlier approaches and extends the standard Luenberger type observer design. The design parameters of the observer can be obtained from the solution of a linear matrix inequality restricted to a subspace determined by the Wong sequences. Some illustrative examples and a comparative discussion are given.

Keywords Differential-algebraic system · Nonlinear system · Observer · Wong sequence · Linear matrix inequality

Mathematics Subject Classification (2010) 34A09, 93B07, 93C05

1 Introduction

The description of dynamical systems using *differential-algebraic equations* (DAEs), which are a combination of differential equations with algebraic constraints, arises in various relevant applications, where the dynamics are algebraically constrained, for instance by tracks, Kirchhoff laws, or conservation laws. To name but a few, DAEs appear naturally in mechanical multibody dynamics [16], electrical networks [36] and chemical engineering [23], but also in non-natural scientific contexts such as economics [33] or demography [13]. The aforementioned problems often cannot be modeled by ordinary differential equations (ODEs) and hence it is of practical interest to investigate the properties of DAEs. Due to their power in applications, nowadays DAEs are an established

T. Berger (✉) · L. Lanza
Institut für Mathematik, Universität Paderborn, Paderborn, Germany
e-mail: thomas.berger@math.upb.de; lukas.lanza@math.upb.de

© The Editor(s) (if applicable) and The Author(s), under exclusive licence
to Springer Nature Switzerland AG 2020
T. Reis et al. (eds.), *Progress in Differential-Algebraic Equations II*,
Differential-Algebraic Equations Forum,
https://doi.org/10.1007/978-3-030-53905-4_9

257

field in applied mathematics and subject of various monographs and textbooks, see e.g. [12, 24, 25].

In the present paper we study state estimation for a class of nonlinear differential-algebraic systems. Nonlinear DAE systems seem to have been first considered by Luenberger [32]; cf. also the textbooks [24, 25] and the recent works [3, 4]. Since it is often not possible to directly measure the state of a system, but only the external signals (input and output) and an internal model are available, it is of interest to construct an "observing system" which approximates the original system's state. Applications for observers are for instance error detection and fault diagnosis, disturbance (or unknown input) estimation and feedback control, see e.g. [14, 42].

Several results on observer design for nonlinear DAEs are available in the literature. Lu and Ho [29] developed a Luenberger type observer for square systems with Lipschitz continuous nonlinearities, utilising solutions of a certain linear matrix inequality (LMI) to construct the observer. This is more general than the results obtained in [19], where the regularity of the linear part was assumed. Extensions of the work from [29] are discussed in [15], where non-square systems are treated, and in [43, 45], inter alia considering nonlinearities in the output equation. We stress that the approach in [11] and [22], where ODE systems with unknown inputs are considered, is similar to the aforementioned since these systems may be treated as DAEs as well. Further but different approaches are taken in [1], where completely nonlinear DAEs which are semi-explicit and index-1 are investigated, in [41], where a nonlinear generalized PI observer design is used, and in [44], where the Lipschitz condition is avoided by regularizing the system via an injection of the output derivatives.

Recently, Gupta et al. [20] presented a reduced-order observer design which is applicable to non-square DAEs with generalized monotone nonlinearities. Systems with nonlinearities which satisfy a more general monotonicity condition are considered in [40], but the results found there are applicable to square systems only.

A novel observer design using so called innovations has been developed in [34, 37] and considered for linear DAEs in [6] and for DAEs with Lipschitz continuous nonlinearities in [5]. Roughly speaking, the innovations are "[...] a measure for the correctness of the overall internal model at time t" [6]. This approach extends the classical Luenberger type observer design and allows for non-square systems.

It is our aim to present an observer design framework which unifies the above mentioned approaches. To this end, we use the approach from [6] for linear DAEs (which can be non-square) and extend it to incorporate both nonlinearities which are Lipschitz continuous as in [5, 29] and nonlinearities which are generalized monotone as in [20, 40], or combinations thereof. We show that if a certain LMI restricted to a subspace determined by the Wong sequences is solvable, then there exists a state estimator (or observer) for the original system, where the gain matrices corresponding to the innovations in the observer are constructed out of the solution of the LMI. We will distinguish between an *(asymptotic) observer* and a *state estimator*, cf. Sect. 2. To this end, we speak of an *observer candidate* before such a system is found to be an observer or a state estimator. We stress that such an observer candidate is a DAE system in general; for the investigation of the existence of ODE observers see e.g. [5, 7, 15, 20].

This paper is organised as follows: We briefly state the basic definitions and some preliminaries on matrix pencils in Sect. 2. The unified framework for the observer design is presented in Sect. 3. In Sects. 4 and 5 we state and prove the main results of this paper. Subsequent to the proofs we give some instructive examples for the theorems in Sect. 6. A discussion as well as a comparison to the relevant literature is provided in Sect. 7 and computational aspects are discussed in Sect. 8.

1.1 Nomenclature

$A \in \mathbb{R}^{n \times m}$	The matrix A is in the set of real $n \times m$ matrices;
rk A, im A, ker A	The rank, image and kernel of $A \in \mathbb{R}^{n \times m}$, resp.;
$\mathscr{C}^k(X \to Y)$	The set of $k-$times continuously differentiable functions $f : X \to Y, k \in \mathbb{N}_0$;
dom(f)	The domain of the function f;
$A >_V 0$	$: \iff \forall x \in V \setminus \{0\} : x^\top A x > 0, V \subseteq \mathbb{R}^n$ a subspace;
$\mathbb{R}[s]$	The ring of polynomials with coefficients in $\mathbb{R}$.

2 Preliminaries

We consider nonlinear DAE systems of the form

$$\frac{\mathrm{d}}{\mathrm{d}t} E x(t) = f(x(t), u(t), y(t))$$
$$y(t) = h(x(t), u(t)), \tag{2.1}$$

with $E \in \mathbb{R}^{l \times n}$, $f \in \mathscr{C}(\mathscr{X} \times \mathscr{U} \times \mathscr{Y} \to \mathbb{R}^l)$ and $h \in \mathscr{C}(\mathscr{X} \times \mathscr{U} \to \mathbb{R}^p)$, where $\mathscr{X} \subseteq \mathbb{R}^n$, $\mathscr{U} \subseteq \mathbb{R}^m$ and $\mathscr{Y} \subseteq \mathbb{R}^p$ are open. The functions $x : I \to \mathbb{R}^n$, $u : I \to \mathbb{R}^m$ and $y : I \to \mathbb{R}^p$ are called the *state*, *input* and *output* of (2.1), resp. Since solutions not necessarily exist globally we consider local solutions of (2.1), which leads to the following solution concept, cf. [5].

Definition 2.1 Let $I \subseteq \mathbb{R}$ be an open interval. A trajectory $(x, u, y) \in \mathscr{C}(I \to \mathscr{X} \times \mathscr{U} \times \mathscr{Y})$ is called *solution* of (2.1), if $x \in \mathscr{C}^1(I \to \mathscr{X})$ and (2.1) holds for all $t \in I$. The set

$$\mathfrak{B}_{(2.1)} := \left\{ (x, u, y) \in \mathscr{C}(I \to \mathscr{X} \times \mathscr{U} \times \mathscr{Y}) \mid I \subseteq \mathbb{R} \text{ open intvl., } (x, u, y) \text{ is a solution of (2.1)} \right\}$$

of all possible solution trajectories is called the *behavior* of system (2.1).

We stress that the interval of definition I of a solution of (2.1) does not need to be maximal and, moreover, it depends on the choice of the input u. Next we introduce

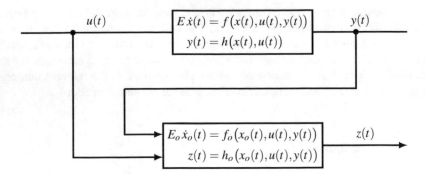

Fig. 1 Interconnection with an acceptor

the concepts of an acceptor, an (asymptotic) observer and a state estimator. These definitions follow in essence the definitions given in [5].

Definition 2.2 Consider a system (2.1). The system

$$\frac{\mathrm{d}}{\mathrm{d}t} E_o x_o(t) = f_o(x_o(t), u(t), y(t)),$$

$$z(t) = h_o(x_o(t), u(t), y(t)),$$
(2.2)

where $E_o \in \mathbb{R}^{l_o \times n_o}$, $f_o \in \mathscr{C}(\mathscr{X}_o \times \mathscr{U} \times \mathscr{Y} \to \mathbb{R}^{l_o})$, $h_o \in \mathscr{C}(\mathscr{X}_o \times \mathscr{U} \times \mathscr{Y} \to \mathbb{R}^{p_o})$, $\mathscr{X}_o \subseteq \mathbb{R}^{n_o}$ open, is called *acceptor for* (2.1) , if for all $(x, u, y) \in \mathfrak{B}_{(2.1)}$ with $I = \mathrm{dom}(x)$, there exist $x_o \in \mathscr{C}^1(I \to \mathscr{X}_o)$, $z \in \mathscr{C}(I \to \mathbb{R}^{p_o})$ such that

$$\left(x_o, \left(\begin{smallmatrix} u \\ y \end{smallmatrix}\right), z\right) \in \mathfrak{B}_{(2.2)}.$$

The definition of an acceptor shows that the original system influences, or may influence, the acceptor but not vice-versa, i.e., there is a directed signal flow from (2.1) to (2.2), see Fig. 1.

Definition 2.3 Consider a system (2.1). Then a system (2.2) with $p_o = n$ is called

(a) an *observer for* (2.1), if it is an acceptor for (2.1), and

$$\forall I \subseteq \mathbb{R} \text{ open interval } \forall t_0 \in I \ \forall (x, u, y, x_o, z) \in \mathscr{C}(I \to \mathscr{X} \times \mathscr{U} \times \mathscr{Y} \times \mathscr{X}_o \times \mathbb{R}^n):$$

$$\left((x, u, y) \in \mathfrak{B}_{(2.1)} \ \wedge \ (x_o, \left(\begin{smallmatrix} u \\ y \end{smallmatrix}\right), z) \in \mathfrak{B}_{(2.2)} \ \wedge \ Ez(t_0) = Ex(t_0)\right) \implies z = x;$$
(2.3)

(b) a *state estimator for* (2.1), if it is an acceptor for (2.1), and

$$\forall t_0 \in \mathbb{R} \, \forall \, (x, u, y, x_o, z) \in \mathscr{C}([t_0, \infty) \to \mathscr{X} \times \mathscr{U} \times \mathscr{Y} \times \mathscr{X}_o \times \mathbb{R}^n) :$$

$$\left((x, u, y) \in \mathfrak{B}_{(2.1)} \wedge (x_o, \left(\begin{smallmatrix} u \\ y \end{smallmatrix}\right), z) \in \mathfrak{B}_{(2.2)} \right) \implies \lim_{t \to \infty} z(t) - x(t) = 0;$$

(2.4)

(c) an *asymptotic observer for* (2.1), if it is an observer and a state estimator for (2.1).

The property of being a state estimator is much weaker than being an asymptotic observer. Since there is no requirement such as (2.3) it might even happen that the state estimator's state matches the original system's state for some time, but eventually evolves in a different direction.

Concluding this section we recall some important concepts for matrix pencils. First, a matrix pencil $sE - A \in \mathbb{R}[s]^{l \times n}$ is called *regular*, if $l = n$ and $\det(sE - A) \neq 0 \in \mathbb{R}[s]$. An important geometric tool are the *Wong sequences*, named after Wong [39], who was the first to use both sequences for the analysis of matrix pencils. The Wong sequences are investigated and utilized for the decomposition of matrix pencils in [8–10].

Definition 2.4 Consider a matrix pencil $sE - A \in \mathbb{R}[s]^{l \times n}$. The *Wong sequences* are sequences of subspaces, defined by

$$\mathscr{V}^0_{[E,A]} := \mathbb{R}^n, \quad \mathscr{V}^{i+1}_{[E,A]} := A^{-1}(E\mathscr{V}^i_{[E,A]}) \subseteq \mathbb{R}^n, \quad \mathscr{V}^*_{[E,A]} := \bigcap_{i \in \mathbb{N}_0} \mathscr{V}^i_{[E,A]},$$

$$\mathscr{W}^0_{[E,A]} := \{0\}, \quad \mathscr{W}^{i+1}_{[E,A]} := E^{-1}(A\mathscr{W}^i_{[E,A]}) \subseteq \mathbb{R}^n, \quad \mathscr{W}^*_{[E,A]} := \bigcup_{i \in \mathbb{N}_0} \mathscr{W}^i_{[E,A]},$$

where $A^{-1}(S) = \{x \in \mathbb{R}^n \mid Ax \in S\}$ is the preimage of $S \subseteq \mathbb{R}^l$ under A. The subspaces $\mathscr{V}^*_{[E,A]}$ and $\mathscr{W}^*_{[E,A]}$ are called the *Wong limits*.

As shown in [8] the Wong sequences terminate, are nested and satisfy

$$\exists k^* \in \mathbb{N} \, \forall j \in \mathbb{N} : \mathscr{V}^0_{[E,A]} \supsetneq \mathscr{V}^1_{[E,A]} \supsetneq \cdots \supsetneq \mathscr{V}^{k^*}_{[E,A]} = \mathscr{V}^{k^*+j}_{[E,A]} = \mathscr{V}^*_{[E,A]} \supseteq \ker(A),$$

$$\exists l^* \in \mathbb{N} \, \forall j \in \mathbb{N} : \mathscr{W}^0_{[E,A]} \subseteq \ker(E) = \mathscr{W}^1_{[E,A]} \subsetneq \cdots \subsetneq \mathscr{W}^{l^*}_{[E,A]} = \mathscr{W}^{l^*+j}_{[E,A]} = \mathscr{W}^*_{[E,A]}.$$

Remark 2.1 Let $sE - A \in \mathbb{R}[s]^{l \times n}$ and consider the associated DAE $\frac{d}{dt}Ex(t) = Ax(t)$. In view of Definition 2.1 we may associate with the previous equation the behavior

$$\mathfrak{B}_{[E,A]} = \left\{ x \in \mathscr{C}^1(I \to \mathbb{R}^n) \mid E\dot{x} = Ax, \, I \subseteq \mathbb{R} \text{ open interval} \right\}.$$

We have that all trajectories in $\mathfrak{B}_{[E,A]}$ evolve in $\mathscr{V}^*_{[E,A]}$, that is

$$\forall x \in \mathfrak{B}_{[E,A]} \ \forall t \in \operatorname{dom}(x): \ x(t) \in \mathscr{V}^*_{[E,A]}. \tag{2.5}$$

This can be seen as follows: For $x \in \mathfrak{B}_{[E,A]}$ we have that $x(t) \in \mathbb{R}^n = \mathscr{V}^0_{[E,A]}$ for all $t \in \operatorname{dom}(x)$. Since the linear spaces $\mathscr{V}^i_{[E,A]}$ are closed they are invariant under differentiation and hence $\dot{x}(t) \in \mathscr{V}^0_{[E,A]}$. Due to the fact that $x \in \mathfrak{B}_{[E,A]}$ it follows for all $t \in \operatorname{dom}(x)$ that $x(t) \in A^{-1}(E\mathscr{V}^0_{[E,A]}) = \mathscr{V}^1_{[E,A]}$. Now assume that $x(t) \in \mathscr{V}^i_{[E,A]}$ for some $i \in \mathbb{N}_0$ and all $t \in \operatorname{dom}(x)$. By the previous arguments we find that $x(t) \in A^{-1}(E\mathscr{V}^i_{[E,A]}) = \mathscr{V}^{i+1}_{[E,A]}$.

An important concept in the context of DAEs is the index of a matrix pencil, which is based on the (quasi-)Weierstraß form (QWF), cf. [10, 18, 24, 25].

Definition 2.5 Consider a regular matrix pencil $sE - A \in \mathbb{R}[s]^{n \times n}$ and let $S, T \in \mathbb{R}^{n \times n}$ be invertible such that

$$S(sE - A)T = s\begin{bmatrix} I_r & 0 \\ 0 & N \end{bmatrix} - \begin{bmatrix} J & 0 \\ 0 & I_{n-r} \end{bmatrix}$$

for some $J \in \mathbb{R}^{r \times r}$ and nilpotent $N \in \mathbb{R}^{(n-r) \times (n-r)}$. Then

$$\nu := \begin{cases} 0, & \text{if } r = n, \\ \min\left\{ k \in \mathbb{N} \mid N^k = 0 \right\}, & \text{if } r < n \end{cases}$$

is called the *index* of $sE - A$.

The index is independent of the choice of S, T and can be computed via the Wong sequences as shown in [10].

3 System, Observer Candidate and Error Dynamics

In this section we present the observer design used in this paper, which invokes so called innovations and was developed in [34, 37] for linear behavioral systems. It is an extension of the classical approach to observer design which goes back to Luenberger, see [30, 31].

We consider nonlinear DAE systems of the form

$$\frac{\mathrm{d}}{\mathrm{d}t} Ex(t) = Ax(t) + B_L f_L(x(t), u(t), y(t)) + B_M f_M(Jx(t), u(t), y(t)),$$

$$y(t) = Cx(t) + h(u(t)),$$

$$\tag{3.1}$$

where $E, A \in \mathbb{R}^{l \times n}$ with $0 \leq r = \mathrm{rk}(E) \leq n$, $B_L \in \mathbb{R}^{l \times q_L}$, $B_M \in \mathbb{R}^{l \times q_M}$, $J \in \mathbb{R}^{q_M \times n}$ with $\mathrm{rk}\, J = q_M$, $C \in \mathbb{R}^{p \times n}$ and $h \in \mathscr{C}(\mathcal{U} \to \mathbb{R}^p)$ with $\mathcal{U} \subseteq \mathbb{R}^m$ open. Furthermore, for some open sets $\mathscr{X} \subseteq \mathbb{R}^n$, $\mathscr{Y} \subseteq \mathbb{R}^p$ and $\hat{\mathscr{X}} := J\mathscr{X} \subseteq \mathbb{R}^{q_M}$, the nonlinear function $f_L : \mathscr{X} \times \mathcal{U} \times \mathscr{Y} \to \mathbb{R}^{q_L}$ satisfies a Lipschitz condition in the first variable

$$\forall\, x, z \in \mathscr{X} \;\, \forall\, u \in \mathcal{U} \;\, \forall\, y \in \mathscr{Y} : \;\; \| f_L(z, u, y) - f_L(x, u, y) \| \leq \| F(z - x) \| \tag{3.2}$$

with $F \in \mathbb{R}^{j \times n}$, $j \in \mathbb{N}$; and $f_M : \hat{\mathscr{X}} \times \mathcal{U} \times \mathscr{Y} \to \mathbb{R}^{q_M}$ satisfies a generalized monotonicity condition in the first variable

$$\forall\, x, z \in \hat{\mathscr{X}} \;\, \forall\, u \in \mathcal{U} \;\, \forall\, y \in \mathscr{Y} : \;\; (z - x)^\top \Theta \big(f_M(z, u, y) - f_M(x, u, y) \big) \geq \frac{1}{2} \mu \| z - x \|^2 \tag{3.3}$$

for some $\Theta \in \mathbb{R}^{q_M \times q_M}$ and $\mu \in \mathbb{R}$. We stress that $\mu < 0$ is explicitly allowed and Θ can be singular, i.e., in particular Θ does not necessarily satisfy any definiteness conditions as in [40]. We set $B := [B_L, B_M] \in \mathbb{R}^{l \times (q_L + q_M)}$ and

$$f : \mathscr{X} \times \mathcal{U} \times \mathscr{Y} \to \mathbb{R}^{q_L} \times \mathbb{R}^{q_M}, \quad (x, u, y) \mapsto \begin{pmatrix} f_L(x, u, y) \\ f_M(Jx, u, y) \end{pmatrix}.$$

Let us consider a system (3.1) and assume that $n = l$. Then another system driven by the external variables u and y of (3.1) of the form

$$\begin{aligned} \frac{\mathrm{d}}{\mathrm{d}t} E z(t) &= A z(t) + B f(z(t), u(t), y(t)) + L(y(t) - \hat{y}(t)) \\ &= A z(t) + B f(z(t), u(t), y(t)) + L(C x(t) - C z(t)) \\ &= (A - LC) z(t) + B f(z(t), u(t), y(t)) + L \underbrace{C x(t)}_{= y(t) - h(u(t))} \end{aligned} \tag{3.4}$$

with $\hat{y}(t) = C z(t) + h(u(t))$

is a Luenberger type observer, where $L \in \mathbb{R}^{n \times p}$ is the observer gain. The dynamics for the error state $e(t) = z(t) - x(t)$ read

$$\frac{\mathrm{d}}{\mathrm{d}t} E e(t) = (A - LC) e(t) + B \big(f(x(t), u(t), y(t)) - f(z(t), u(t), y(t)) \big).$$

The observer (3.4) incorporates a copy of the original system, and in addition the outputs' difference $\hat{y}(t) - y(t)$, the influence of which is weighted with the observer gain L.

In this paper we consider a generalization of the design (3.4) which incorporates an extra variable d that takes the role of the innovations. The innovations are used to describe "the difference between what we actually observe and what we had expected to observe" [34], and hence they generalize the effect of the observer gain L in (3.4). We consider the following observer candidate, which is an additive composition of an internal model of the system (3.1) and a further term which involves the innovations:

$$\tfrac{\mathrm{d}}{\mathrm{d}t} E z(t) = A z(t) + B f(z(t), u(t), y(t)) + L_1 d(t)$$
$$0 = C z(t) - y(t) + h(u(t)) + L_2 d(t),$$
(3.5)

where $x_o(t) = \begin{pmatrix} z(t) \\ d(t) \end{pmatrix}$ is the observer state and $L_1 \in \mathbb{R}^{l \times k}$, $L_2 \in \mathbb{R}^{p \times k}$, $\mathscr{X}_o = \mathscr{X} \times \mathbb{R}^k$. From the second line in (3.5) we see that the innovations term balances the difference between the system's and the observer's output. In a sense, the smaller the variable d, the better the approximate state z in (3.5) matches the state x of the original system (3.1).

We stress that $n \neq l$ is possible in general, and if L_2 is invertible, then the observer candidate reduces to

$$\tfrac{\mathrm{d}}{\mathrm{d}t} E z(t) = A z(t) + B f(z(t), u(t), y(t)) + L_1 L_2^{-1} (y(t) - C z(t) - h(u(t)))$$

$$= (A - L_1 L_2^{-1} C) z(t) + B f(z(t), u(t), y(t)) + L_1 L_2^{-1} \underbrace{(y(t) - h(u(t)))}_{=C x(t)},$$
(3.6)

which is a Luenberger type observer of the form (3.4) with gain $L = L_1 L_2^{-1}$. Hence the Luenberger type observer is a special case of the observer design (3.5). Being square is a necessary condition for invertibility of L_2, i.e., $k = p$.

For later use we consider the dynamics of the error state $e(t) := z(t) - x(t)$ between systems (3.1) and (3.5),

$$\tfrac{\mathrm{d}}{\mathrm{d}t} E e(t) = A e(t) + B \phi(t) + L_1 d(t)$$
$$0 = C e(t) + L_2 d(t),$$
(3.7)

where

$$\phi(t) := f(z(t), u(t), y(t)) - f(x(t), u(t), y(t)) = \begin{pmatrix} f_L(z(t),u(t),y(t)) - f_L(x(t),u(t),y(t)) \\ f_M(Jz(t),u(t),y(t)) - f_M(Jx(t),u(t),y(t)) \end{pmatrix},$$

and rewrite (3.7) as

$$\tfrac{\mathrm{d}}{\mathrm{d}t} \mathscr{E} \begin{pmatrix} e(t) \\ d(t) \end{pmatrix} = \mathscr{A} \begin{pmatrix} e(t) \\ d(t) \end{pmatrix} + \mathscr{B} \phi(t),$$
(3.8)

where

$$\mathscr{E} = \begin{bmatrix} E & 0 \\ 0 & 0 \end{bmatrix} \in \mathbb{R}^{(l+p)\times(n+k)}, \quad \mathscr{A} = \begin{bmatrix} A & L_1 \\ C & L_2 \end{bmatrix} \in \mathbb{R}^{(l+p)\times(n+k)}$$

$$and \quad \mathscr{B} = \begin{bmatrix} B \\ 0 \end{bmatrix} \in \mathbb{R}^{(l+p)\times(q_L+q_M)}.$$

The following lemma is a consequence of (2.5).

Lemma 3.1 *Consider a system* (3.1) *and the observer candidate* (3.5). *Then* (3.5) *is an acceptor for* (3.1). *Furthermore, for all open intervals* $I \subseteq \mathbb{R}$, *all* $(x, u, y) \in \mathfrak{B}_{(3.1)}$ *and all* $\left(\binom{z}{d}, \binom{u}{y}, z \right) \in \mathfrak{B}_{(3.5)}$ *with* $\mathrm{dom}(x) = \mathrm{dom}\binom{z}{d} = I$ *we have:*

$$\forall t \in I : \begin{pmatrix} e(t) \\ d(t) \\ \phi(t) \end{pmatrix} \in \mathscr{V}^*_{[[\mathscr{E},0],[\mathscr{A},\mathscr{B}]]}. \tag{3.9}$$

Proof Let $I \subseteq \mathbb{R}$ be an open interval and $(x, u, y) \in \mathfrak{B}_{(3.1)}$. For any $(x, u, y) \in \mathfrak{B}_{(3.1)}$ it holds $\left(\binom{x}{0}, \binom{u}{y}, x \right) \in \mathfrak{B}_{(3.5)}$, hence (3.5) is an acceptor for (3.1).
Now let $(x, u, y) \in \mathfrak{B}_{(3.1)}$ and $\left(\binom{z}{d}, \binom{u}{y}, z \right) \in \mathfrak{B}_{(3.5)}$, with $I = \mathrm{dom}(x) = \mathrm{dom}\binom{z}{d}$ and rewrite (3.8) as

$$\tfrac{\mathrm{d}}{\mathrm{d}t}[\mathscr{E}, 0] \begin{pmatrix} e(t) \\ d(t) \\ \phi(t) \end{pmatrix} = [\mathscr{A}, \mathscr{B}] \begin{pmatrix} e(t) \\ d(t) \\ \phi(t) \end{pmatrix}.$$

Then (3.9) is immediate from Remark 2.1. $\qquad\square$

In the following lemma we show that for a state estimator to exist, it is necessary that the system (3.1) does not contain free state variables, i.e., solutions (if they exist) are unique.

Lemma 3.2 *Consider a system* (3.1) *and the observer candidate* (3.5). *If* (3.5) *is a state estimator for* (3.1), *then either*

$$\left(\forall (x, u, y) \in \mathfrak{B}_{(3.1)} \, \exists t_0 \in \mathbb{R} : \ \mathrm{dom}(x) \cap [t_0, \infty) = \emptyset \right)$$

$$\vee \ \left(\forall \left(\binom{z}{d}, \binom{u}{y}, z \right) \in \mathfrak{B}_{(3.5)} \, \exists t_0 \in \mathbb{R} : \ \mathrm{dom}(z, d) \cap [t_0, \infty) = \emptyset \right), \tag{3.10}$$

or we have $\mathrm{rk}_{\mathbb{R}(s)} \begin{bmatrix} sE-A \\ C \end{bmatrix} = n.$

Proof Let (3.5) be a state estimator for (3.1) and assume that (3.10) is not true. Set
$E' := \begin{bmatrix} E \\ 0 \end{bmatrix}$, $A' := \begin{bmatrix} A \\ C \end{bmatrix}$ and let $(x, u, y) \in \mathfrak{B}_{(3.1)}$ with $[t_0, \infty) \subseteq \text{dom}(x)$ for some
$t_0 \in \mathbb{R}$. Then we have that, for all $t \geq t_0$,

$$\tfrac{d}{dt} E'x(t) = \begin{bmatrix} A \\ C \end{bmatrix} x(t) + \begin{bmatrix} B & L_1 \\ 0 & L_2 \end{bmatrix} \begin{pmatrix} f(x(t), u(t), y(t)) \\ d(t) \end{pmatrix} =: A'x(t) + g(x(t), u(t), y(t), d(t))$$

(3.11)

with $d(t) \equiv 0$. Using [8, Thm. 2.6] we find matrices $S \in Gl_{l+p}(\mathbb{R})$, $T \in Gl_n(\mathbb{R})$
such that

$$S\left(sE' - A'\right)T = s \begin{bmatrix} E_P & 0 & 0 \\ 0 & E_R & 0 \\ 0 & 0 & E_Q \end{bmatrix} - \begin{bmatrix} A_P & 0 & 0 \\ 0 & A_R & 0 \\ 0 & 0 & A_Q \end{bmatrix}, \quad (3.12)$$

where

(i) $E_P, A_P \in \mathbb{R}^{m_P \times n_P}$, $m_P < n_P$, are such that $\text{rk}_{\mathbb{C}}(\lambda E_P - A_P) = m_P$ for all
$\lambda \in \mathbb{C} \cup \{\infty\}$,
(ii) $E_R, A_R \in \mathbb{R}^{m_R \times n_R}$, $m_R = n_R$, with $sE_R - A_R$ regular,
(iii) $E_Q, A_Q \in \mathbb{R}^{m_Q \times n_Q}$, $m_Q > n_Q$, are such that $\text{rk}_{\mathbb{C}}(\lambda E_Q - A_Q) = n_Q$ for all
$\lambda \in \mathbb{C} \cup \{\infty\}$.

We consider the underdetermined pencil $sE_P - A_P$ in (3.12) and the corresponding
DAE. If $n_P = 0$, then [8, Lem. 3.1] implies that $\text{rk}_{\mathbb{R}(s)} sE_Q - A_Q = n_Q$ and
invoking $\text{rk}_{\mathbb{R}(s)} sE_R - A_R = n_R$ gives that $\text{rk}_{\mathbb{R}(s)} sE' - A' = n$. So assume that
$n_p > 0$ in the following and set

$$\begin{pmatrix} x_P \\ x_R \\ x_Q \end{pmatrix} := T^{-1}x, \qquad \begin{pmatrix} g_P \\ g_R \\ g_Q \end{pmatrix} := Sg.$$

If $m_p = 0$, then x_P can be chosen arbitrarily. Otherwise, we have

$$\tfrac{d}{dt} E_P x_P(t) = A_P x_P(t) + g_P \left(T \begin{pmatrix} x_P(t) \\ x_R(t) \\ x_Q(t) \end{pmatrix}, u(t), y(t), d(t) \right). \quad (3.13)$$

As a consequence of [8, Lem. 4.12] we may w.l.o.g. assume that $sE_P - A_P = s[I_{m_p}, 0] - [N, R]$ with $R \in \mathbb{R}^{m_P \times (n_P - m_P)}$ and nilpotent $N \in \mathbb{R}^{m_P \times m_P}$. Partition
$x_P = \begin{pmatrix} x_P^1 \\ x_P^2 \end{pmatrix}$, then (3.13) is equivalent to

$$\dot{x}_P^1(t) = N x_P^1(t) + R x_P^2(t) + g_P\left(T(x_P^1(t)^\top, x_P^2(t)^\top, x_R(t)^\top, x_Q(t)^\top)^\top, u(t), y(t), d(t)\right)$$

(3.14)

for all $t \geq t_0$, and hence $x_P^2 \in \mathcal{C}([t_0, \infty) \to \mathbb{R}^{n_P - m_P})$ can be chosen arbitrarily and every choice preserves $[t_0, \infty) \subseteq \mathrm{dom}(x)$. Similarly, if $\left(\left(\begin{smallmatrix} z \\ d \end{smallmatrix} \right), \left(\begin{smallmatrix} u \\ y \end{smallmatrix} \right), z \right) \in \mathfrak{B}_{(3.5)}$ with $[t_0, \infty) \subseteq \mathrm{dom}(z)$—w.l.o.g. the same t_0 can be chosen—then (3.11) is satisfied for $x = z$ and, proceeding in an analogous way, z_P^2 can be chosen arbitrarily, in particular such that $\lim_{t \to \infty} z_P^2(t) \neq \lim_{t \to \infty} x_P^2(t)$. Therefore, $\lim_{t \to \infty} z(t) - x(t) = \lim_{t \to \infty} e(t) \neq 0$, which contradicts that (3.5) is a state estimator for (3.1). Thus $n_P = 0$ and $\mathrm{rk}_{\mathbb{R}(s)} sE' - A' = n$ follows. $\square$

As a consequence of Lemma 3.2, a necessary condition for (3.5) to be a state estimator for (3.1) is that $n \leq l + p$. This will serve as a standing assumption in the subsequent sections.

4 Sufficient Conditions for State Estimators

In this section we show that if certain matrix inequalities are satisfied, then there exists a state estimator for system (3.1) which is of the form (3.5). The design parameters of the latter can be obtained from a solution of the matrix inequalities. The proofs of the subsequent theorems are inspired by the work of Lu and Ho [29] and by [5], where LMIs are considered on the Wong limits only.

Theorem 4.1 *Consider a system* (3.1) *with* $n \leq l + p$ *which satisfies conditions* (3.2) *and* (3.3). *Let* $k \in \mathbb{N}_0$ *and denote with* $\mathcal{V}^*_{[[\mathcal{E}, 0], [\mathcal{A}, \mathcal{B}]]}$ *the Wong limit of the pencil* $s[\mathcal{E}, 0] - [\mathcal{A}, \mathcal{B}] \in \mathbb{R}[s]^{(l+p) \times (n+k+q_L+q_M)}$, *and* $\overline{\mathcal{V}}^*_{[[\mathcal{E}, 0], [\mathcal{A}, \mathcal{B}]]} := [I_{n+k}, 0] \mathcal{V}^*_{[[\mathcal{E}, 0], [\mathcal{A}, \mathcal{B}]]}$. *Further let* $\hat{A} = \left[\begin{smallmatrix} A & 0 \\ C & 0 \end{smallmatrix} \right]$,

$$H = \begin{bmatrix} 0_{n \times n} & 0 \\ 0 & I_k \end{bmatrix} = H^{\mathsf{T}}, \quad \mathcal{F} = [F, 0] \in \mathbb{R}^{j \times (n+k)}, \quad j \in \mathbb{N},$$

$$\hat{\Theta} = \begin{bmatrix} 0 & J^{\mathsf{T}} \Theta \\ 0 & 0 \end{bmatrix} \in \mathbb{R}^{(n+k) \times (q_L + q_M)}, \quad \mathcal{J} = \begin{bmatrix} J^{\mathsf{T}} J & 0 \\ 0 & 0 \end{bmatrix} \in \mathbb{R}^{(n+k) \times (n+k)}$$

and $\Lambda_{q_L} := \begin{bmatrix} I_{q_L} & 0 \\ 0 & 0 \end{bmatrix} \in \mathbb{R}^{(q_L + q_M) \times (q_L + q_M)}$.

If there exist $\delta > 0$, $\mathcal{P} \in \mathbb{R}^{(l+p) \times (n+k)}$ *and* $\mathcal{K} \in \mathbb{R}^{(n+k) \times (n+k)}$ *such that*

$$\mathcal{Q} := \begin{bmatrix} \hat{A}^{\mathsf{T}} \mathcal{P} + \mathcal{P}^{\mathsf{T}} \hat{A} + H^{\mathsf{T}} \mathcal{K}^{\mathsf{T}} + \mathcal{K} H + \delta \mathcal{F}^{\mathsf{T}} \mathcal{F} - \mu \mathcal{J} & \mathcal{P}^{\mathsf{T}} \mathcal{B} + \hat{\Theta} \\ \mathcal{B}^{\mathsf{T}} \mathcal{P} + \hat{\Theta}^{\mathsf{T}} & -\delta \Lambda_{q_L} \end{bmatrix} <_{\mathcal{V}^*_{[[\mathcal{E}, 0], [\mathcal{A}, \mathcal{B}]]}} 0 \tag{4.1}$$

and

$$\mathcal{P}^{\mathsf{T}} \mathcal{E} = \mathcal{E}^{\mathsf{T}} \mathcal{P} >_{\overline{\mathcal{V}}^*_{[[\mathcal{E}, 0], [\mathcal{A}, \mathcal{B}]]}} 0, \tag{4.2}$$

then for all $L_1 \in \mathbb{R}^{l \times k}$, $L_2 \in \mathbb{R}^{p \times k}$ *such that* $\mathscr{P}^\top \begin{bmatrix} 0 & L_1 \\ 0 & L_2 \end{bmatrix} = \mathscr{K} H$ *the system* (3.5) *is a state estimator for* (3.1).

Furthermore, there exists at least one such pair L_1, L_2 *if, and only if,* im $\mathscr{K} H \subseteq$ im $\mathscr{P}^\top$.

Proof Using Lemma 3.1, we have that (3.5) is an acceptor for (3.1). To show that (3.5) satisfies condition (2.4) let $t_0 \in \mathbb{R}$ and $(x, u, y, x_o, z) \in \mathscr{C}([t_0, \infty) \to \mathscr{X} \times \mathscr{U} \times \mathscr{Y} \times \mathscr{X}_o \times \mathbb{R}^n)$ such that $(x, u, y) \in \mathfrak{B}_{(3.1)}$ and $(x_o, \binom{u}{y}, z) \in \mathfrak{B}_{(3.5)}$, with $x_o(t) = \binom{z(t)}{d(t)}$ and $\mathscr{X}_o = \mathscr{X} \times \mathbb{R}^k$.

The last statement of the theorem is clear. Let $\hat{L} = [0_{(l+p) \times n}, *]$ be a solution of $\mathscr{P}^\top \hat{L} = \mathscr{K} H$ and $\mathscr{A} = \hat{A} + \hat{L}$, further set $\eta(t) := \binom{e(t)}{d(t)}$, where $e(t) = z(t) - x(t)$. Recall that

$$\phi(t) = f(z(t), u(t), y(t)) - f(x(t), u(t), y(t))$$

$$= \begin{pmatrix} f_L(z(t), u(t), y(t)) - f_L(x(t), u(t), y(t)) \\ f_M(Jz(t), u(t), y(t)) - f_M(Jx(t), u(t), y(t)) \end{pmatrix} =: \begin{pmatrix} \phi_L(t) \\ \phi_M(t) \end{pmatrix}.$$

In view of condition (3.2) we have for all $t \geq t_0$ that

$$\delta(\eta^\top(t)\mathscr{F}^\top \mathscr{F} \eta(t) - \phi_L^\top(t)\phi_L(t)) \geq 0 \tag{4.3}$$

and by (3.3)

$$([J, 0]\eta(t))^\top \Theta \phi_M(t) + \phi_M^\top(t)\Theta^\top[J, 0]\eta(t) - \mu([J, 0]\eta(t))^\top ([J, 0]\eta(t)) \geq 0. \tag{4.4}$$

Now assume that (4.1) and (4.2) hold. Consider a Lyapunov function candidate

$$\tilde{V} : \mathbb{R}^{n+k} \to \mathbb{R}, \quad \eta \mapsto \eta^\top \mathscr{E}^\top \mathscr{P} \eta$$

and calculate the derivative along solutions for $t \geq t_0$:

$$\frac{d}{dt}\tilde{V}(\eta(t)) = \dot{\eta}^\top(t)\mathscr{E}^\top \mathscr{P} \eta(t) + \eta^\top(t)\mathscr{P}^\top \mathscr{E} \dot{\eta}(t)$$

$$= (\mathscr{A}\eta(t) + \mathscr{B}\phi(t))^\top \mathscr{P} \eta(t) + \eta^\top(t)\mathscr{P}^\top (\mathscr{A}\eta(t) + \mathscr{B}\phi(t))$$

$$= \eta^\top(t)\mathscr{A}^\top \mathscr{P} \eta(t) + \eta^\top(t)\mathscr{P}^\top \mathscr{A} \eta(t) + \phi^\top(t)\mathscr{B}^\top \mathscr{P} \eta(t) + \eta^\top(t)\mathscr{P}^\top \mathscr{B}\phi(t)$$

$$= \eta^\top(t)\hat{A}^\top \mathscr{P} \eta(t) + \eta^\top(t)\hat{L}^\top \mathscr{P} \eta(t) + \eta^\top(t)\mathscr{P}^\top \hat{A}\eta(t) + \eta^\top(t)\mathscr{P}^\top \hat{L}\eta(t)$$

$$\quad + \phi^\top(t)\mathscr{B}^\top \mathscr{P} \eta(t) + \eta^\top(t)\mathscr{P}^\top \mathscr{B}\phi(t)$$

$$\overset{(4.3),(4.4)}{\leq} \eta^\top(t) \left(\hat{A}^\top \mathscr{P} + \mathscr{P}^\top \hat{A} + \mathscr{K} H + H^\top \mathscr{K}^\top \right) \eta(t)$$

$$+ \phi^\top(t)\mathscr{B}^\top \mathscr{P}\eta(t) + \eta^\top(t)\mathscr{P}^\top \mathscr{B}\phi(t) + \delta(\eta^\top(t)\mathscr{F}^\top \mathscr{F}\eta(t) - \phi_L^\top(t)\phi_L(t))$$

$$+ ([J,0]\eta(t))^\top \Theta \phi_M(t) + \phi_M^\top(t)\Theta^\top[J,0]\eta(t) - \mu([J,0]\eta(t))^\top([J,0]\eta(t))$$

$$= \eta^\top(t)\left(\hat{A}^\top \mathscr{P} + \mathscr{P}^\top \hat{A} + \mathscr{K}H + H^\top \mathscr{K}^\top + \delta \mathscr{F}^\top \mathscr{F} - \mu \mathscr{J}\right)\eta(t)$$

$$+ \phi^\top(t)\mathscr{B}^\top \mathscr{P}\eta(t) + \eta^\top(t)\mathscr{P}^\top \mathscr{B}\phi(t)$$

$$+ \eta^\top(t)\hat{\Theta}\phi(t) + \phi^\top(t)\hat{\Theta}^\top \eta(t) - \delta \phi^\top(t)\Lambda_{q_L}\phi(t)$$

$$= \begin{pmatrix} \eta(t) \\ \phi(t) \end{pmatrix}^\top \underbrace{\begin{bmatrix} \hat{A}^\top \mathscr{P} + \mathscr{P}^\top \hat{A} + H^\top \mathscr{K}^\top + \mathscr{K}H + \delta \mathscr{F}^\top \mathscr{F} - \mu \mathscr{J} & \mathscr{P}^\top \mathscr{B} + \hat{\Theta} \\ \mathscr{B}^\top \mathscr{P} + \hat{\Theta}^\top & -\delta \Lambda_{q_L} \end{bmatrix}}_{=\mathscr{Q}} \begin{pmatrix} \eta(t) \\ \phi(t) \end{pmatrix}.$$

$$(4.5)$$

Let $S \in \mathbb{R}^{(n+k+q_L+q_M)\times n_{\mathscr{V}}}$ with orthonormal columns be such that $\operatorname{im} S = \mathscr{V}^*_{[[\mathscr{E},0],[\mathscr{A},\mathscr{B}]]}$ and $\operatorname{rk}(S) = n_{\mathscr{V}}$. Then inequality (4.1) reads $\hat{\mathscr{Q}} := S^\top \mathscr{Q}S < 0$. Denote with $\lambda^-_{\hat{\mathscr{Q}}}$ the smallest eigenvalue of $-\hat{\mathscr{Q}}$, then $\lambda^-_{\hat{\mathscr{Q}}} > 0$. Since S has orthonormal columns we have $\|Sv\| = \|v\|$ for all $v \in \mathbb{R}^{n_{\mathscr{V}}}$.

By Lemma 3.1 we have $\begin{pmatrix} \eta(t) \\ \phi(t) \end{pmatrix} \in \mathscr{V}^*_{[[\mathscr{E},0],[\mathscr{A},\mathscr{B}]]}$ for all $t \geq t_0$, hence $\begin{pmatrix} \eta(t) \\ \phi(t) \end{pmatrix} = Sv(t)$ for some $v: [t_0, \infty) \to \mathbb{R}^{n_{\mathscr{V}}}$. Then (4.5) becomes

$$\forall t \geq t_0 : \frac{\mathrm{d}}{\mathrm{d}t}\tilde{V}(\eta(t)) \leq \begin{pmatrix} \eta(t) \\ \phi(t) \end{pmatrix}^\top \mathscr{Q} \begin{pmatrix} \eta(t) \\ \phi(t) \end{pmatrix} = v^\top(t)\hat{\mathscr{Q}}v(t)$$

$$\leq -\lambda^-_{\hat{\mathscr{Q}}}\|v(t)\|^2 = -\lambda^-_{\hat{\mathscr{Q}}}\left\|\begin{pmatrix} \eta(t) \\ \phi(t) \end{pmatrix}\right\|^2.$$

$$(4.6)$$

Let $\overline{S} \in \mathbb{R}^{(n+k)\times n_{\overline{\mathscr{V}}}}$ with orthonormal columns be such that $\operatorname{im}\overline{S} = \overline{\mathscr{V}}^*_{[[\mathscr{E},0],[\mathscr{A},\mathscr{B}]]}$ and $\operatorname{rk}(\overline{S}) = n_{\overline{\mathscr{V}}}$. Then condition (4.2) is equivalent to $\overline{S}^\top \mathscr{E}^\top \mathscr{P}\overline{S} > 0$. Since $\begin{pmatrix} \eta(t) \\ \phi(t) \end{pmatrix} \in \mathscr{V}^*_{[\mathscr{E},0],[\mathscr{A},\mathscr{B}]]}$ for all $t \geq t_0$ it is clear that $\eta(t) \in \overline{\mathscr{V}}^*_{[[\mathscr{E},0],[\mathscr{A},\mathscr{B}]]}$ for all $t \geq t_0$. If $\overline{\mathscr{V}}^*_{[[\mathscr{E},0],[\mathscr{A},\mathscr{B}]]} = \{0\}$ (which also holds when $\mathscr{V}^*_{[[\mathscr{E},0],[\mathscr{A},\mathscr{B}]]} = \{0\}$), then this implies $\eta(t) = 0$, thus $e(t) = 0$ for all $t \geq t_0$, which completes the proof. Otherwise, $n_{\overline{\mathscr{V}}} > 0$ and we set $\eta(t) = \overline{S}\bar{\eta}(t)$ for some $\bar{\eta}: [t_0, \infty) \to \mathbb{R}^{n_{\overline{\mathscr{V}}}}$ and denote with λ^+, λ^- the largest and smallest eigenvalue of $\overline{S}^\top \mathscr{E}^\top \mathscr{P}\overline{S}$, resp., where $\lambda^- > 0$ is a consequence of (4.2). Then we have

$$\tilde{V}(\eta(t)) = \eta^\top(t)\mathscr{E}^\top \mathscr{P}\eta(t) = \bar{\eta}^\top(t)\overline{S}^\top \mathscr{E}^\top \mathscr{P}\overline{S}\bar{\eta}(t) \leq \lambda^+\|\bar{\eta}(t)\|^2 = \lambda^+\|\eta(t)\|^2$$

$$(4.7)$$

and, analogously,

$$\forall t \geq t_0 : \lambda^- \|\eta(t)\|^2 \leq \bar{\eta}^\top(t) \overline{S}^\top \mathcal{E}^\top \mathscr{P} \overline{S} \bar{\eta}(t) = \tilde{V}(\eta(t)) \leq \lambda^+ \|\eta(t)\|^2. \tag{4.8}$$

Therefore,

$$\forall t \geq t_0 : \frac{\mathrm{d}}{\mathrm{d}t} \tilde{V}(\eta(t)) \overset{(4.6)}{\leq} -\lambda^-_{\hat{\mathcal{Q}}} \left\| \begin{pmatrix} \eta(t) \\ \phi(t) \end{pmatrix} \right\|^2 \leq -\lambda^-_{\hat{\mathcal{Q}}} \|\eta(t)\|^2 \overset{(4.7)}{\leq} -\frac{\lambda^-_{\hat{\mathcal{Q}}}}{\lambda^+} \tilde{V}(\eta(t)).$$

Now, abbreviate $\beta := \frac{\lambda^-_{\hat{\mathcal{Q}}}}{\lambda^+}$ and use Gronwall's Lemma to infer

$$\forall t \geq t_0 : \tilde{V}(\eta(t)) \leq \tilde{V}(\eta(0)) e^{-\beta t}. \tag{4.9}$$

Then we obtain

$$\forall t \geq t_0 : \|\eta(t)\|^2 \overset{(4.8)}{\leq} \frac{1}{\lambda^-} \tilde{V}(\eta(t)) \overset{(4.9)}{\leq} \frac{\tilde{V}(\eta(0))}{\lambda^-} e^{-\beta t},$$

and hence $\lim_{t \to \infty} e(t) = 0$, which completes the proof.   □

Remark 4.1

(i) Note that $\mathscr{A} = \hat{A} + \hat{L}$, where $\hat{L} = [0_{(l+p) \times n}, *]$ is a solution of $\mathscr{P}^\top \hat{L} = \mathscr{K} H$ and hence the space $\mathscr{V}^*_{[[\mathcal{E},0],[\mathscr{A},\mathscr{B}]]}$ on which (4.1) is considered depends on the sought solutions $\mathscr{P}$ and $\mathscr{K}$ as well; using $\mathscr{P}^\top \mathscr{A} = \mathscr{P}^\top \hat{A} + \mathscr{K} H$, this dependence is still linear. Furthermore, note that $\mathscr{K}$ only appears in union with the matrix $H = \begin{bmatrix} 0 & 0 \\ 0 & I_k \end{bmatrix}$, thus only the last k columns of $\mathscr{K}$ are of interest. In order to reduce the computational effort it is reasonable to fix the other entries beforehand, e.g. by setting them to zero.

(ii) We stress that the parameters in the description (3.1) of the system are not entirely fixed, especially regarding the linear parts. More precisely, an equation of the form $\frac{\mathrm{d}}{\mathrm{d}t} Ex(t) = Ax(t) + f(x(t), u(t))$, where f satisfies (3.2) can equivalently be written as $\frac{\mathrm{d}}{\mathrm{d}t} Ex(t) = f_L(x(t), u(t))$, where $f_L(x, u) = Ax + f(x, u)$ also satisfies (3.2), but with a different matrix F. However, this alternative (with $A = 0$) may not satisfy the necessary condition provided in Lemma 3.2, which hence should be checked in advance. Therefore, the system class (3.1) allows for a certain flexibility and different choices of the parameters may or may not satisfy the assumptions of Theorem 4.1.

(iii) In the special case $E = 0$, i.e., purely algebraic systems of the form $0 = Ax(t) + Bf(x(t), u(t), y(t))$, Theorem 4.1 may still be applicable. More precisely, condition (4.2) is satisfied in this case if, and only if,

$\overline{\mathcal{V}}^*_{[[\mathcal{E},0],[\mathcal{A},\mathcal{B}]]} = \{0\}$. This can be true, if for instance $\mathcal{B} = 0$ and $\mathcal{A}$ has full column rank, because then $\overline{\mathcal{V}}^*_{[[\mathcal{E},0],[\mathcal{A},\mathcal{B}]]} = [I_{n+k}, 0] \ker[\mathcal{A}, 0] = \{0\}$.

In the following theorem condition (4.2) is weakened to positive semi-definiteness. As a consequence, the system's matrices have to satisfy additional conditions, which are not present in Theorem 4.1. In particular, we require that $\mathcal{E}$ and $\mathcal{A}$ are square, which means that $k = l + p - n$. Furthermore, we require that JG_M is invertible for a certain matrix G_M and that the norms corresponding to F and J are compatible if both kinds of nonlinearities are present.

Theorem 4.2 *Use the notation from Theorem 4.1 and set $k = l+p-n$. In addition, denote with $\mathcal{V}^*_{[\mathcal{E},\mathcal{A}]}, \mathcal{W}^*_{[\mathcal{E},\mathcal{A}]} \subseteq \mathbb{R}^{n+k}$ the Wong limits of the pencil $s\mathcal{E} - \mathcal{A} \in \mathbb{R}[s]^{(l+p)\times(n+k)}$ and let $V \in \mathbb{R}^{(n+k)\times n_\mathcal{V}}$ and $W \in \mathbb{R}^{(n+k)\times n_\mathcal{W}}$ be basis matrices of $\mathcal{V}^*_{[\mathcal{E},\mathcal{A}]}$ and $\mathcal{W}^*_{[\mathcal{E},\mathcal{A}]}$, resp., where $n_\mathcal{V} = \dim(\mathcal{V}^*_{[\mathcal{E},\mathcal{A}]})$ and $n_\mathcal{W} = \dim(\mathcal{W}^*_{[\mathcal{E},\mathcal{A}]})$. Furthermore, denote with $\lambda_{max}(M)$ the largest eigenvalue of a matrix M.*

If there exist $\delta > 0$, $\mathcal{P} \in \mathbb{R}^{(l+p)\times(n+k)}$ invertible and $\mathcal{K} \in \mathbb{R}^{(n+k)\times(n+k)}$ such that (4.1) holds and

(a) $\mathcal{E}^\top \mathcal{P} = \mathcal{P}^\top \mathcal{E} \geq_{\overline{\mathcal{V}}^*_{[[\mathcal{E},0],[\mathcal{A},\mathcal{B}]]}} 0$,

(b) *the pencil $s\mathcal{E} - \mathcal{A} \in \mathbb{R}[s]^{(l+p)\times(n+k)}$ is regular and its index is at most one,*

(c) *F is such that $\|FG_L\| < 1$, where $G_L := -[I_n, 0]W[0, I_{n+k-r}][\mathcal{E}V, \mathcal{A}W]^{-1}\begin{bmatrix} B_L \\ 0 \end{bmatrix}$,*

(d) *JG_M is invertible and $\mu > \lambda_{max}(\Gamma)$, where $\Gamma := \tilde{\Theta} + \tilde{\Theta}^\top$, $\tilde{\Theta} := \Theta(JG_M)^{-1}$,*

$G_M := -[I_n, 0]W[0, I_{n+k-r}][\mathcal{E}V, \mathcal{A}W]^{-1}\begin{bmatrix} B_M \\ 0 \end{bmatrix}$,

(e) *there exists $\alpha > 0$ such that $\|Fx\| \leq \alpha\|Jx\|$ for all $x \in \mathbb{R}^n$ and, for*

$S := \tilde{\Theta}^\top(\Gamma - \mu I_{q_M})^{-1}\tilde{\Theta}$ *we have*

$$\frac{\alpha\|JG_L\|}{1 - \|FG_L\|}\left(\sqrt{\frac{\max\{0, \lambda_{max}(S)\}}{\mu - \lambda_{max}(\Gamma)}} + \|(\Gamma - \mu I_{q_M})^{-1}(\tilde{\Theta}^\top - \mu I_{q_M})\|\right) < 1,$$

(4.10)

then with $L_1 \in \mathbb{R}^{l\times k}$, $L_2 \in \mathbb{R}^{p\times k}$ such that $\begin{bmatrix} 0 & L_1 \\ 0 & L_2 \end{bmatrix} = \mathcal{P}^{-\top}\mathcal{K}H$ the system (3.5) is a state estimator for (3.1).

Proof Assume (4.1) and (4.10) (a)–(e) hold. Up to Eq. (4.9) the proof remains the same as for Theorem 4.1. By (4.10) (b) we may infer from [10, Thm. 2.6] that there exist invertible $\mathcal{M} = \begin{bmatrix} M_1^\top, M_2^\top \end{bmatrix}^\top \in \mathbb{R}^{(n+k)\times(l+p)}$ with $M_1 \in \mathbb{R}^{r\times(l+p)}$, $M_2 \in \mathbb{R}^{(n+k-r)\times(l+p)}$ and invertible $\mathcal{N} = [N_1, N_2] \in \mathbb{R}^{(n+k)\times(l+p)}$ with $N_1 \in \mathbb{R}^{(n+k)\times r}$,

$N_2 \in \mathbb{R}^{(n+k) \times (l+p-r)}$ such that

$$\mathcal{M}(\mathcal{E} - \mathcal{A})\mathcal{N} = \begin{bmatrix} I_r - A_r & 0 \\ 0 & -I_{n+k-r} \end{bmatrix}, \tag{4.11}$$

where $r = \mathrm{rk}(\mathcal{E})$ and $A_r \in \mathbb{R}^{r \times r}$, and that

$$\mathcal{N} = [V, W], \quad \mathcal{M} = [\mathcal{E}V, \mathcal{A}W]^{-1}. \tag{4.12}$$

Let

$$\mathscr{P} = \mathcal{M}^\top \begin{bmatrix} P_1 & P_2 \\ P_3 & P_4 \end{bmatrix} \mathcal{N}^{-1} \tag{4.13}$$

with $P_1 \in \mathbb{R}^{n_\mathcal{V} \times n_\mathcal{V}}$, $P_4 \in \mathbb{R}^{n_\mathcal{W} \times n_\mathcal{W}}$ and $P_2, P_3^\top \in \mathbb{R}^{n_\mathcal{V} \times n_\mathcal{W}}$. Then condition (4.10) (a) implies $P_1 > 0$ as follows. First, calculate

$$\mathcal{E}^\top \mathscr{P} = \mathcal{N}^{-T} \begin{bmatrix} I_r & 0 \\ 0 & 0 \end{bmatrix} \mathcal{M}^{-T} \mathcal{M}^\top \begin{bmatrix} P_1 & P_2 \\ P_3 & P_4 \end{bmatrix} \mathcal{N}^{-1} = \mathcal{N}^{-T} \begin{bmatrix} P_1 & P_2 \\ 0 & 0 \end{bmatrix} \mathcal{N}^{-1} \tag{4.14}$$

which gives $P_2 = 0$ as $\mathscr{P}^\top \mathcal{E} = \mathcal{E}^\top \mathscr{P}$. Note that therefore P_1 and P_4 in (4.13) are invertible since $\mathscr{P}$ is invertible by assumption. By (4.14) we have

$$\mathcal{E}^\top \mathscr{P} = \mathcal{N}^{-T} \begin{bmatrix} P_1 & 0 \\ 0 & 0 \end{bmatrix} \mathcal{N}^{-1} = [V, W]^{-T} \begin{bmatrix} P_1 & 0 \\ 0 & 0 \end{bmatrix} [V, W]^{-1}. \tag{4.15}$$

It remains to show $P_1 \geq 0$. Next, we prove the inclusion

$$\mathcal{V}_{[\mathcal{E},\mathcal{A}]}^* \subseteq \overline{\mathcal{V}}_{[[\mathcal{E},0],[\mathcal{A},\mathcal{B}]]}^* = [I_{n+k}, 0] \mathcal{V}_{[[\mathcal{E},0],[\mathcal{A},\mathcal{B}]]}^*. \tag{4.16}$$

To this end, we show $\mathcal{V}_{[\mathcal{E},\mathcal{A}]}^i \subseteq [I_{n+k},0] \mathcal{V}_{[[\mathcal{E},0],[\mathcal{A},\mathcal{B}]]}^i$ for all $i \in \mathbb{N}_0$. For $i = 0$ this is clear. Now assume it is true for some $i \in \mathbb{N}_0$. Then

$$[I_{n+k}, 0] \mathcal{V}_{[[\mathcal{E},0],[\mathcal{A},\mathcal{B}]]}^{i+1} = [I_{n+k}, 0][\mathcal{A}, \mathcal{B}]^{-1}([\mathcal{E}, 0] \mathcal{V}_{[[\mathcal{E},0],[\mathcal{A},\mathcal{B}]]}^i)$$

$$= [I_{n+k}, 0] \left\{ \begin{pmatrix} \eta(t) \\ \phi(t) \end{pmatrix} \in \mathbb{R}^{n+k+q} \,\middle|\, \mathcal{A}\eta \right.$$

$$\left. + \mathcal{B}\phi \in \mathcal{E}\left([I_{n+k}, 0] \mathcal{V}_{[[\mathcal{E},0],[\mathcal{A},\mathcal{B}]]}^i\right) \right\}$$

$$= \left\{ \eta \in \mathbb{R}^{n+k} \; \middle| \; \exists \phi \in \mathbb{R}^q : \; \mathscr{A}\eta + \mathscr{B}\phi \in \mathscr{E}\overline{\mathscr{V}}^i_{[[\mathscr{E},0],[\mathscr{A},\mathscr{B}]]} \right\}$$

$$\overset{\phi=0}{\supseteq} \left\{ \eta \in \mathbb{R}^{n+k} \; \middle| \; \mathscr{A}\eta \in \mathscr{E}\overline{\mathscr{V}}^i_{[[\mathscr{E},0],[\mathscr{A},\mathscr{B}]]} \right\}$$

$$= \mathscr{A}^{-1}\left(\mathscr{E}\overline{\mathscr{V}}^i_{[[\mathscr{E},0],[\mathscr{A},\mathscr{B}]]} \right)$$

$$\supseteq \mathscr{A}^{-1}\left(\mathscr{E}\mathscr{V}^i_{[\mathscr{E},\mathscr{A}]} \right) = \mathscr{V}^{i+1}_{[\mathscr{E},\mathscr{A}]},$$

which is the statement. Therefore it is clear that $\operatorname{im} V \subseteq \overline{\mathscr{V}}^*_{[[\mathscr{E},0],[\mathscr{A},\mathscr{B}]]} = \operatorname{im}\overline{V}$, with $\overline{V} \in \mathbb{R}^{(n+k)\times n_{\overline{\mathscr{V}}}}$ a basis matrix of $\overline{\mathscr{V}}^*_{[[\mathscr{E},0],[\mathscr{A},\mathscr{B}]]}$ and $n_{\overline{\mathscr{V}}} = \dim(\overline{\mathscr{V}}^*_{[[\mathscr{E},0],[\mathscr{A},\mathscr{B}]]})$. Thus there exists $R \in \mathbb{R}^{n_{\overline{\mathscr{V}}}\times n_{\mathscr{V}}}$ such that $V = \overline{V}R$. Now the inequality $\overline{V}^\top \mathscr{P}^\top \mathscr{E}\overline{V} \geq 0$ holds by condition (4.10) (a) and implies

$$0 \leq R^\top \overline{V}^\top \mathscr{P}^\top \mathscr{E}\overline{V}R = V^\top \mathscr{P}^\top \mathscr{E}V = \left([V,W]\begin{bmatrix} I_{n_{\mathscr{V}}} \\ 0 \end{bmatrix} \right)^\top \mathscr{P}^\top \mathscr{E}\left([V,W]\begin{bmatrix} I_{n_{\mathscr{V}}} \\ 0 \end{bmatrix} \right)$$

$$\overset{(4.15)}{=} [I_{n_{\mathscr{V}}},0]\begin{bmatrix} P_1 & 0 \\ 0 & 0 \end{bmatrix}\begin{bmatrix} I_{n_{\mathscr{V}}} \\ 0 \end{bmatrix} = P_1.$$

Now, let $\mathscr{N}^{-1}\eta(t) = \begin{pmatrix} \eta_1(t) \\ \eta_2(t) \end{pmatrix}$, with $\eta_1(t) \in \mathbb{R}^r$ and $\eta_2(t) \in \mathbb{R}^{n+k-r}$ and consider the Lyapunov function $\tilde{V}(\eta(t)) = \eta^\top(t)\mathscr{E}^\top \mathscr{P}\eta(t)$ in new coordinates:

$$\forall t \geq t_0 : \tilde{V}(\eta(t)) = \eta^\top(t)\mathscr{E}^\top \mathscr{P}\eta(t) \overset{(4.14)}{=} \begin{pmatrix} \eta_1(t) \\ \eta_2(t) \end{pmatrix}^\top \begin{bmatrix} P_1 & 0 \\ 0 & 0 \end{bmatrix}\begin{pmatrix} \eta_1(t) \\ \eta_2(t) \end{pmatrix} \qquad (4.17)$$

$$= \eta_1^\top(t)P_1\eta_1(t) \geq \lambda_{P_1}^- \|\eta_1(t)\|^2,$$

where $\lambda_{P_1}^- > 0$ denotes the smallest eigenvalue of P_1. Thus (4.17) implies

$$\forall t \geq t_0 : \|\eta_1(t)\|^2 \leq \frac{1}{\lambda_{P_1}^-}\eta^\top(t)\mathscr{E}^\top \mathscr{P}\eta(t) = \frac{1}{\lambda_{P_1}^-}\tilde{V}(\eta(t)) \overset{(4.9)}{\leq} \frac{\tilde{V}(\eta(0))}{\lambda_{P_1}^-}e^{-\beta t} \underset{t\to\infty}{\longrightarrow} 0.$$
$$(4.18)$$

Note that, if $\mathscr{V}^*_{[\mathscr{E},\mathscr{A}]} = \{0\}$, then $r = 0$ and $\mathscr{N}^{-1}\eta(t) = \eta_2(t)$, thus the above estimate (4.18) is superfluous (and, in fact, not feasible) in this case; it is straightforward to modify the remaining proof to this case. With the aid of transformation (4.11) we have:

$$\mathcal{M} \tfrac{d}{dt} \mathcal{E} \eta(t) = \mathcal{M} \mathcal{A} \eta(t) + \mathcal{M} \mathcal{B} \phi(t)$$

$$\Longleftrightarrow \mathcal{M} \mathcal{E} \mathcal{N} \tfrac{d}{dt} \begin{pmatrix} \eta_1(t) \\ \eta_2(t) \end{pmatrix} = \mathcal{M} \mathcal{A} \mathcal{N} \begin{pmatrix} \eta_1(t) \\ \eta_2(t) \end{pmatrix} + \mathcal{M} \mathcal{B} \phi(t) \tag{4.19}$$

$$\Longleftrightarrow \begin{bmatrix} I_r & 0 \\ 0 & 0 \end{bmatrix} \tfrac{d}{dt} \begin{pmatrix} \eta_1(t) \\ \eta_2(t) \end{pmatrix} = \begin{bmatrix} A_r & 0 \\ 0 & I_{n+k-r} \end{bmatrix} \begin{pmatrix} \eta_1(t) \\ \eta_2(t) \end{pmatrix} + \begin{bmatrix} M_1 \\ M_2 \end{bmatrix} \mathcal{B} \phi(t),$$

from which it is clear that $\eta_2(t) = -M_2 \mathcal{B} \phi(t)$. Observe

$$e(t) = [I_n, 0]\eta(t) = [I_n, 0]\mathcal{N}\begin{pmatrix} \eta_1(t) \\ \eta_2(t) \end{pmatrix} = [I_n, 0]V\eta_1(t) + [I_n, 0]W\eta_2(t) =: e_1(t) + e_2(t),$$

where $\lim_{t \to \infty} e_1(t) = 0$ by (4.18). We show $e_2(t) = -[I_n, 0]WM_2\mathcal{B}\phi(t) \to 0$ for $t \to \infty$. Set

$$e_2^L(t) := G_L \phi_L(t), \quad e_2^M(t) := G_M \phi_M(t)$$

so that $e_2(t) = e_2^L(t) + e_2^M(t)$. Next, we inspect the Lipschitz condition (3.2):

$$\|\phi_L(t)\| \le \|Fe(t)\| \le \|Fe_1(t)\| + \|Fe_2^L(t)\| + \|Fe_2^M(t)\|$$

$$\le \|Fe_1(t)\| + \|FG_L\|\|\phi_L(t)\| + \|Fe_2^M(t)\|,$$

which gives, invoking (4.10) (c),

$$\|\phi_L(t)\| \le \left(1 - \|FG_L\|\right)^{-1}\left(\|Fe_1(t)\| + \|Fe_2^M(t)\|\right). \tag{4.20}$$

Set $\hat{e}(t) := e_1(t) + e_2^L(t) = e_1(t) + G_L \phi_L(t)$ and $\kappa := \tfrac{\alpha\|JG_L\|}{1-\|FG_L\|}$ and observe that (4.20) together with (4.10) (e) implies

$$\|J\hat{e}(t)\| \le \|Je_1(t)\| + \|JG_L\|\|\phi_L(t)\| \overset{(4.20)}{\le} (1+\kappa)\|Je_1(t)\| + \kappa\|Je_2^M(t)\|. \tag{4.21}$$

Since JG_M is invertible by (4.10) (d) we find that

$$\phi_M(t) = (JG_M)^{-1}Je_2^M(t), \tag{4.22}$$

and hence the monotonicity condition (3.3) yields, for all $t \geq t_0$,

$$\mu \left\| Je(t) \right\|^2 \leq (Je(t))^\top \Theta \phi_M(t) + \phi_M^\top(t) \Theta^\top Je(t)$$

$$= (J\hat{e}(t) + Je_2^M(t))^\top \tilde{\Theta} Je_2^M(t) + (Je_2^M(t))^\top \tilde{\Theta}^\top (J\hat{e}(t) + Je_2^M(t))$$

$$= 2(J\hat{e}(t))^\top \tilde{\Theta} Je_2^M(t) + (Je_2^M(t))^\top \underbrace{\left(\tilde{\Theta} + \tilde{\Theta}^\top \right)}_{= \Gamma} Je_2^M(t)$$

and on the left-hand side

$$\mu \left\| Je(t) \right\|^2 = \mu \left(\left\| J\hat{e}(t) \right\|^2 + \left\| Je_2^M(t) \right\|^2 + 2(J\hat{e}(t))^\top (Je_2^M(t)) \right).$$

Therefore, we find that

$$0 \leq -\mu \left\| Je_2^M(t) \right\|^2 - \mu \left\| J\hat{e}(t) \right\|^2 - 2\mu(J\hat{e}(t))^\top (Je_2^M(t))$$

$$+ 2(J\hat{e}(t))^\top \tilde{\Theta}(Je_2^M(t)) + (Je_2^M(t))^\top \Gamma(Je_2^M(t))$$

$$= \begin{pmatrix} J\hat{e}(t) \\ Je_2^M(t) \end{pmatrix}^\top \begin{bmatrix} -\mu I_{q_M} & \tilde{\Theta} - \mu I_{q_M} \\ \tilde{\Theta}^\top - \mu I_{q_M} & \Gamma - \mu I_{q_M} \end{bmatrix} \begin{pmatrix} J\hat{e}(t) \\ Je_2^M(t) \end{pmatrix}.$$

Since $\Gamma - \mu I_{q_M}$ is invertible by (4.10) (d) we may set $\Xi := \tilde{\Theta}^\top - \mu I_{q_M}$ and $\tilde{e}_2^M(t) := (\Gamma - \mu I_{q_M})^{-1} \Xi J\hat{e}(t) + Je_2^M(t)$. Then

$$0 \leq \begin{pmatrix} J\hat{e}(t) \\ Je_2^M(t) \end{pmatrix}^\top \begin{bmatrix} -\mu I_{q_M} & \tilde{\Theta} - \mu I_{q_M} \\ \tilde{\Theta}^\top - \mu I_{q_M} & \Gamma - \mu I_{q_M} \end{bmatrix} \begin{pmatrix} J\hat{e}(t) \\ Je_2^M(t) \end{pmatrix}$$

$$= \begin{pmatrix} J\hat{e}(t) \\ \tilde{e}_2^M(t) \end{pmatrix}^\top \begin{bmatrix} -\mu I_{q_M} - \Xi^\top (\Gamma - \mu I_{q_M})^{-1} \Xi & 0 \\ 0 & \Gamma - \mu I_{q_M} \end{bmatrix} \begin{pmatrix} J\hat{e}(t) \\ \tilde{e}_2^M(t) \end{pmatrix}.$$

Therefore, using $\mu - \lambda_{\max}(\Gamma) > 0$ by (4.10) (d) and computing

$$-\mu I_{q_M} - \Xi^\top (\Gamma - \mu I_{q_M})^{-1} \Xi = \tilde{\Theta}^\top (\tilde{\Theta} + \tilde{\Theta}^\top - \mu I_{q_M})^{-1} \tilde{\Theta} = S,$$

we obtain

$$0 \leq \max\{0, \lambda_{\max}(S)\} \| J\hat{e}(t) \|^2 - (\mu - \lambda_{\max}(\Gamma)) \| \tilde{e}_2^M(t) \|^2,$$

which gives

$$\|Je_2^M(t)\| \le \|(\Gamma - \mu I_{r_M})^{-1}\Xi\|\|J\hat{e}(t)\| + \|\tilde{e}_2^M(t)\|$$

$$\le \left(\sqrt{\frac{\max\{0, \lambda_{\max}(S)\}}{\mu - \lambda_{\max}(\Gamma)}} + \|(\Gamma - \mu I_{r_M})^{-1}\Xi\| \right) \|J\hat{e}(t)\|$$

$$\overset{(4.21)}{\le} \left(\sqrt{\frac{\max\{0, \lambda_{\max}(S)\}}{\mu - \lambda_{\max}(\Gamma)}} + \|(\Gamma - \mu I_{r_M})^{-1}\Xi\| \right) \left((1+\kappa)\|Je_1(t)\| + \kappa\|Je_2^M(t)\| \right).$$

It then follows from (4.10) (e) that $\lim_{t\to\infty} Je_2^M(t) = 0$, and additionally invoking (4.20) and (4.22) gives $\lim_{t\to\infty} \phi_L(t) = 0$ and $\lim_{t\to\infty} \phi_M(t) = 0$, thus $\|e_2(t)\| \le \|G_L\phi_L(t)\| + \|G_M\phi_M(t)\| \xrightarrow[t\to\infty]{} 0$, and finally $\lim_{t\to\infty} e(t) = 0$. $\quad\square$

Remark 4.2

(i) If the nonlinearity f in (3.1) consists only of f_L satisfying the Lipschitz condition, then the conditions (4.10) (d) and (e) are not present. If it consists only of the monotone part f_M, then the conditions (4.10) (c) and (e) are not present. In fact, condition (4.10) (e) is a "mixed condition" in a certain sense which states additional requirements on the combination of both (classes of) nonlinearities.

(ii) The following observation is of practical interest. Whenever f_L satisfies (3.2) with a certain matrix F, it is obvious that f_L will satisfy (3.2) with any other $\tilde{F}$ such that $\|F\| \le \|\tilde{F}\|$. However, condition (4.10) (c) states an upper bound on the possible choices of F. Similarly, if f_M satisfies (3.3) with certain Θ and μ, then f_M satisfies (3.3) with any $\tilde{\mu} \le \mu$, for a fixed Θ. On the other hand, condition (4.10) (d) states lower bounds for μ (involving Θ as well). Additional bounds are provided by (4.1) and condition (4.10) (e). Analogous thoughts hold for the other parameters. Hence F, δ, J, Θ and μ can be utilized in solving the conditions of Theorems 4.1 and 4.2.

(iii) The condition $\|Fx\| \le \alpha\|Jx\|$ from (4.10) (e) is quite restrictive since it connects the Lipschitz estimation of f_L with the domain of f_M. This relation is far from natural and relaxing it is a topic of future research. The inequality would always be satisfied for $J = I_n$ by taking $\alpha = \|F\|$, however in view of the monotonicity condition (3.3), the inequality (4.1) and conditions (4.10) this would be even more restrictive.

(iv) In the case $\Xi = 0$ the assumptions of Theorem 4.2 simplify a lot. In fact, we may calculate that in this case we have $\mathcal{V}^*_{[[\mathcal{E},0],[\mathcal{A},\mathcal{B}]]} = \ker[\mathcal{A},\mathcal{B}]$ and hence the inequality (4.1) becomes

$$\mathcal{Q} = \begin{bmatrix} \mathscr{A}^\top \mathscr{P} + \mathscr{P}^\top \mathscr{A} + \delta \mathscr{F}^\top \mathscr{F} - \mu \mathscr{J} & \mathscr{P}^\top \mathscr{B} + \hat{\Theta} \\ \mathscr{B}^\top \mathscr{P} + \hat{\Theta}^\top & -\delta \Lambda_{qL} \end{bmatrix} <_{\ker[\mathscr{A},\mathscr{B}]} 0$$

$$\Longleftrightarrow \quad \forall \begin{pmatrix} \eta \\ \phi \end{pmatrix} \in \ker[\mathscr{A},\mathscr{B}]: \quad \begin{pmatrix} \eta \\ \phi \end{pmatrix}^\top \mathcal{Q} \begin{pmatrix} \eta \\ \phi \end{pmatrix} < 0$$

$$\Longleftrightarrow \quad \forall \begin{pmatrix} \eta \\ \phi \end{pmatrix} \in \ker[\mathscr{A},\mathscr{B}]: \quad \eta^\top \left(\mathscr{A}^\top \mathscr{P} + \mathscr{P}^\top \mathscr{A} + \delta \mathscr{F}^\top \mathscr{F} - \mu \mathscr{J} \right) \eta - \delta \left\| \Lambda_{qL} \phi \right\|^2$$

$$+ \phi^\top \left(\mathscr{B}^\top \mathscr{P} + \hat{\Theta} \right) \eta + \eta^\top \left(\mathscr{P}^\top \mathscr{B} + \hat{\Theta}^\top \right) \phi < 0$$

$$\Longleftrightarrow \quad \forall \begin{pmatrix} \eta \\ \phi \end{pmatrix} \in \ker[\mathscr{A},\mathscr{B}]: \quad \underbrace{(\eta^\top \mathscr{A}^\top + \phi^\top(t)\mathscr{B}^\top)}_{=0} \mathscr{P}\eta + \eta^\top \mathscr{P}^\top \underbrace{(\mathscr{A}\eta + \mathscr{B}\phi)}_{=0}$$

$$+ \delta \left(\|\mathscr{F}\eta\|^2 - \|\Lambda_{qL}\phi\|^2 \right) + \eta^\top \hat{\Theta}\phi + \phi^\top \hat{\Theta}^\top \eta - \mu \eta^\top \mathscr{J}\eta < 0.$$

Now, $\mathscr{A}$ is invertible by (4.10) (b) and hence $\eta = -\mathscr{A}^{-1}\mathscr{B}\phi$. Therefore, the inequality (4.1) is equivalent to

$$\delta \left((\mathscr{F}\mathscr{A}^{-1}\mathscr{B})^\top (\mathscr{F}\mathscr{A}^{-1}\mathscr{B}) - \Lambda_{qL} \right) - (\mathscr{A}^{-1}\mathscr{B})^\top \hat{\Theta} - \hat{\Theta}^\top \mathscr{A}^{-1}\mathscr{B}$$

$$- \mu(\mathscr{A}^{-1}\mathscr{B})^\top \mathscr{J}(\mathscr{A}^{-1}\mathscr{B}) < 0,$$

which is of a much simper shape.

(v) The conditions presented in Theorems 4.1 and 4.2 are sufficient conditions only. The following example does not satisfy the conditions in the theorems but a state estimator exists for it. Consider $\dot{x}(t) = -x(t)$, $y(t) = 0$, then the system $\dot{z}(t) = -z(t)$, $0 = d_1(t) - d_2(t)$ of the form (3.5) with $L_1 = [0, 0]$ and $L_2 = [1, -1]$ is obviously a state estimator, since the first equation is independent of the innovations d_1, d_2 and solutions satisfy $\lim_{t\to\infty} z(t) - x(t) = 0$. However, we have $n + k = 3 > 2 = l + p$ and therefore Theorem 4.2 is not applicable. Furthermore, the assumptions of Theorem 4.1 are not satisfied since

$$\overline{\mathscr{V}}^*_{[[\mathscr{E},0],[\mathscr{A},\mathscr{B}]]} = \mathscr{V}^*_{[\mathscr{E},\mathscr{A}]} = \operatorname{im} \begin{bmatrix} 1 & 0 \\ 0 & 1 \\ 0 & 1 \end{bmatrix} \quad \text{and} \quad \mathscr{E}\mathscr{V} = \begin{bmatrix} 1 & 0 \\ 0 & 0 \end{bmatrix},$$

by which $\ker \mathscr{E}\mathscr{V} \neq \{0\}$ and hence (4.2) cannot be true. We also like to stress that therefore, in virtue of Lemma 3.2, $n \leq l + p$ is a necessary condition for the existence of a state estimator of the form (3.5), but $n + k \leq l + p$ is not.

5 Sufficient Conditions for Asymptotic Observers

In the following theorem some additional conditions are asked to be satisfied in order to guarantee that the resulting observer candidate is in fact an asymptotic observer, i.e., it is a state estimator and additionally satisfies (2.3). To this end, we utilize an implicit function theorem from [21].

Theorem 5.1 *Use the notation from Theorem 4.2 and assume that $\mathscr{X} = \mathbb{R}^n$, $\mathscr{U} = \mathbb{R}^m$ and $\mathscr{Y} = \mathbb{R}^p$. Additionally, let $\mathscr{M}, \mathscr{N} \in \mathbb{R}^{(n+k)\times(l+p)}$ be as in (4.12), set $\bar{\mathscr{N}} := [I_n, 0]\mathscr{N}$ and $\begin{bmatrix} \hat{B}_1 \\ \hat{B}_2 \end{bmatrix} := \mathscr{M} \begin{bmatrix} B_L & B_M & 0 \\ 0 & 0 & I_p \end{bmatrix}$, where $\hat{B}_1 \in \mathbb{R}^{r\times(q_L+q_M+p)}$, $\hat{B}_2 \in \mathbb{R}^{(n+k-r)\times(q_L+q_M+p)}$. Let*

$$G \colon \mathbb{R}^r \times \mathbb{R}^{n+k-r} \times \mathbb{R}^m \times \mathbb{R}^p \to \mathbb{R}^{n+k-r}, \ (x_1, x_2, u, y) \mapsto x_2 + \hat{B}_2 \begin{pmatrix} f_L(\bar{\mathscr{N}}\binom{x_1}{x_2}, u, y) \\ f_M(J\bar{\mathscr{N}}\binom{x_1}{x_2}, u, y) \\ h(u) - y \end{pmatrix}$$

and $Z_0 := \left\{ (x_1, x_2, u, y) \in \mathbb{R}^r \times \mathbb{R}^{n+k-r} \times \mathbb{R}^m \times \mathbb{R}^p \ \middle| \ G(x_1, x_2, u, y) = 0 \right\}$.

If there exist $\delta > 0$, $\mathscr{P} \in \mathbb{R}^{(l+p)\times(n+k)}$ invertible and $\mathscr{K} \in \mathbb{R}^{(l+p)\times(n+k)}$ such that (4.1) and (4.10) hold and in addition

(a) $\frac{\partial}{\partial x_2}G(x_1, x_2, u, y)$ is invertible for all $(x_1, x_2, u, y) \in Z_0$,

(b) there exists $\omega \in \mathscr{C}([0,\infty) \to (0,\infty))$ nondecreasing with $\int_0^\infty \frac{dt}{\omega(t)} = \infty$ such that

$$\forall (x_1, x_2, u, y) \in Z_0 : \left\| \left(\tfrac{\partial}{\partial x_2}G(x_1, x_2, u, y)\right)^{-1} \right\| \left\| \tfrac{\partial}{\partial(x_1,u,y)}G(x_1, x_2, u, y) \right\| \le \omega(\|x_2\|),$$

(c) Z_0 is connected,

(d) f_M is locally Lipschitz continuous in the first variable,

(5.1)

then with $L_1 \in \mathbb{R}^{l\times k}$, $L_2 \in \mathbb{R}^{p\times k}$ such that $\begin{bmatrix} 0 & L_1 \\ 0 & L_2 \end{bmatrix} = \mathscr{P}^{-\top}\mathscr{K}H$ the system (3.5) is an asymptotic observer for (3.1).

Proof Since (3.5) is a state estimator for (3.1) by Theorem 4.2, it remains to show that (2.3) is satisfied. To this end, let $I \subseteq \mathbb{R}$ be an open interval, $t_0 \in I$, and $(x, u, y, z, d) \in \mathscr{C}(I \to \mathbb{R}^n \times \mathbb{R}^m \times \mathbb{R}^p \times \mathbb{R}^n \times \mathbb{R}^k)$ such that $(x, u, y) \in \mathfrak{B}_{(3.1)}$ and $\left(\binom{z}{d}, \binom{u}{y}, z\right) \in \mathfrak{B}_{(3.5)}$. Recall that $B = [B_L, B_M]$ and

$f(x, u, y) = \begin{pmatrix} f_L(x,u,y) \\ f_M(Jx,u,y) \end{pmatrix}$. Now assume $Ex(t_0) = Ez(t_0)$ and recall the equations

$$\tfrac{d}{dt}Ex(t) = Ax(t) + Bf(x(t), u(t), y(t)), \qquad \tfrac{d}{dt}Ez(t) = Az(t) + Bf(z(t), u(t), y(t)) + L_1 d(t),$$

$$y(t) = Cx(t) + h(u(t)), \qquad\qquad\qquad 0 = Cz(t) - y(t) + h(u(t)) + L_2 d(t).$$

This is equivalent to

$$\tfrac{d}{dt}\mathscr{E}\begin{pmatrix} x(t) \\ 0 \end{pmatrix} = \mathscr{A}\begin{pmatrix} x(t) \\ 0 \end{pmatrix} + \begin{bmatrix} B & 0 \\ 0 & I_p \end{bmatrix}\begin{pmatrix} f(x(t), u(t), y(t)) \\ h(u(t)) - y(t) \end{pmatrix} \tag{5.2}$$

and

$$\tfrac{d}{dt}\mathscr{E}\begin{pmatrix} z(t) \\ d(t) \end{pmatrix} = \mathscr{A}\begin{pmatrix} z(t) \\ d(t) \end{pmatrix} + \begin{bmatrix} B & 0 \\ 0 & I_p \end{bmatrix}\begin{pmatrix} f(x(t), u(t), y(t)) \\ h(u(t)) - y(t) \end{pmatrix}.$$

Let $\begin{pmatrix} x_1(t) \\ x_2(t) \end{pmatrix} = \mathscr{N}^{-1}\begin{pmatrix} x(t) \\ 0 \end{pmatrix}$ and $\begin{pmatrix} z_1(t) \\ z_2(t) \end{pmatrix} = \mathscr{N}^{-1}\begin{pmatrix} z(t) \\ d(t) \end{pmatrix}$. Application of transformations (4.11) to (5.2) gives

$$\begin{bmatrix} I_r & 0 \\ 0 & 0 \end{bmatrix}\begin{pmatrix} \dot{x}_1(t) \\ \dot{x}_2(t) \end{pmatrix} = \begin{bmatrix} A_r & 0 \\ 0 & I_{n+k-r} \end{bmatrix}\begin{pmatrix} x_1(t) \\ x_2(t) \end{pmatrix} + \begin{bmatrix} \hat{B}_1 \\ \hat{B}_2 \end{bmatrix}\begin{pmatrix} f(\mathscr{N}\begin{pmatrix} x_1(t) \\ x_2(t) \end{pmatrix}, u(t), y(t)) \\ h(u(t)) - y(t) \end{pmatrix}$$

or, equivalently,

$$\dot{x}_1(t) = A_r x_1(t) + \hat{B}_1\begin{pmatrix} f(\mathscr{N}\begin{pmatrix} x_1(t) \\ x_2(t) \end{pmatrix}, u(t), y(t)) \\ h(u(t)) - y(t) \end{pmatrix}$$

$$0 = \underbrace{x_2(t) + \hat{B}_2\begin{pmatrix} f(\mathscr{N}\begin{pmatrix} x_1(t) \\ x_2(t) \end{pmatrix}, u(t), y(t)) \\ h(u(t)) - y(t) \end{pmatrix}}_{=G(x_1(t),x_2(t),u(t),y(t))}$$

with $\bar{\mathscr{N}} := [I_n, 0]\mathscr{N}$ and $\mathscr{M}\begin{bmatrix} B & 0 \\ 0 & I_p \end{bmatrix} = \begin{bmatrix} \hat{B}_1 \\ \hat{B}_2 \end{bmatrix}$.

Since (5.1) (a)–(c) hold, the global implicit function theorem in [21, Cor. 5.3] ensures the existence of a unique continuous map $g: \mathbb{R}^r \times \mathbb{R}^m \times \mathbb{R}^p \to \mathbb{R}^{n+k-r}$ such that $G(x_1, g(x_1, u, y), u, y) = 0$ for all $(x_1, u, y) \in \mathbb{R}^r \times \mathbb{R}^m \times \mathbb{R}^p$, and hence $x_2(t) = g(x_1(t), u(t), y(t))$ for all $t \in I$. Thus x_1 solves the ordinary differential equation

$$\dot{x}_1(t) = A_r x_1(t) + \hat{B}_1\begin{pmatrix} f(\bar{\mathscr{N}}\begin{pmatrix} x_1(t) \\ g(x_1(t),u(t),y(t)) \end{pmatrix}, u(t), y(t)) \\ h(u(t)) - y(t) \end{pmatrix} \tag{5.3}$$

with initial value $x_1(t_0)$ for all $t \in I$; and $z_1(t)$ solves the same ODE with same initial value $z_1(t_0) = x_1(t_0)$. This can be seen as follows: $Ex(t_0) = Ez(t_0)$ implies $\mathscr{E}\left(\begin{smallmatrix} x(t_0) \\ 0 \end{smallmatrix}\right) = \mathscr{E}\left(\begin{smallmatrix} z(t_0) \\ d(t_0) \end{smallmatrix}\right)$, and the transformation (4.11) gives

$$\mathscr{E}\begin{pmatrix} x(t_0) \\ 0 \end{pmatrix} = \mathscr{M}^{-1}\begin{bmatrix} I_r & 0 \\ 0 & 0 \end{bmatrix}\mathscr{N}^{-1}\begin{pmatrix} x(t_0) \\ 0 \end{pmatrix} = \mathscr{M}^{-1}\begin{bmatrix} I_r & 0 \\ 0 & 0 \end{bmatrix}\begin{pmatrix} x_1(t_0) \\ x_2(t_0) \end{pmatrix} = \mathscr{M}^{-1}\begin{pmatrix} x_1(t_0) \\ 0 \end{pmatrix},$$

$$\mathscr{E}\begin{pmatrix} z(t_0) \\ d(t_0) \end{pmatrix} = \mathscr{M}^{-1}\begin{bmatrix} I_r & 0 \\ 0 & 0 \end{bmatrix}\mathscr{N}^{-1}\begin{pmatrix} z(t_0) \\ d(t_0) \end{pmatrix} = \mathscr{M}^{-1}\begin{bmatrix} I_r & 0 \\ 0 & 0 \end{bmatrix}\begin{pmatrix} z_1(t_0) \\ z_2(t_0) \end{pmatrix} = \mathscr{M}^{-1}\begin{pmatrix} z_1(t_0) \\ 0 \end{pmatrix},$$

which implies $x_1(t_0) = z_1(t_0)$.

Furthermore, $g(x_1, u, y)$ is differentiable, which follows from the properties of G: Let $v = (x_1, u, y)$ and write $G(x_1, g(v), u, y) = \tilde{G}(v, g(v))$, then taking the derivative yields

$$\frac{\mathrm{d}}{\mathrm{d}v}\tilde{G}(v, g(v)) = \frac{\partial}{\partial v}\tilde{G}(v, g(v)) + \frac{\partial}{\partial g}\tilde{G}(v, g(v))g'(v) = 0$$

$$\Rightarrow g'(v) = -\left(\frac{\partial \tilde{G}(v, g(v))}{\partial g}\right)^{-1}\left(\frac{\partial \tilde{G}(v, g(v))}{\partial v}\right),$$

which is well defined by assumption. Hence $g(x_1, u, y)$ is in particular locally Lipschitz. Since f_L is globally Lipschitz in the first variable by (3.2) and f_M is locally Lipschitz in the first variable by assumption (5.1) (d), $(x_1, u, y) \mapsto f\left(\bar{\mathscr{N}}\left(\begin{smallmatrix} x_1(t) \\ g(x_1(t), u(t), y(t)) \end{smallmatrix}\right), u(t), y(t)\right)$ is locally Lipschitz in the first variable and therefore the solution of (5.3) is unique by the Picard–Lindelöf theorem, see e.g. [28, Thm. 4.17]; this implies $z_1(t) = x_1(t)$ for all $t \in I$. Furthermore,

$$x_2(t) = g(x_1(t), u(t), y(t)) = g(z_1(t), u(t), y(t)) = z_2(t)$$

for all $t \in I$, and hence (3.5) is an observer for (3.1). Combining this with the fact that (3.5) is already a state estimator, (3.5) is an asymptotic observer for (3.1). □

6 Examples

We present some instructive examples to illustrate Theorems 4.1, 4.2 and 5.1. Note that the inequality (4.1) does not have unique solutions $\mathscr{P}$ and $\mathscr{K}$ and hence the resulting state estimator is just one possible choice. The first example illustrates Theorem 4.1.

Example 6.1 Consider the DAE

$$
\frac{d}{dt}
\begin{bmatrix} 1 & 0 \\ 1 & 1 \\ 0 & 0 \\ 0 & 1 \end{bmatrix}
\begin{pmatrix} x_1(t) \\ x_2(t) \end{pmatrix}
=
\begin{bmatrix} 0 & -3 \\ -2 & 0 \\ 1 & -2 \\ 0 & 0 \end{bmatrix}
\begin{pmatrix} x_1(t) \\ x_2(t) \end{pmatrix}
+
\begin{bmatrix} 0 & 2 \\ 1 & -1 \\ 0 & 1 \\ 1 & 0 \end{bmatrix}
\begin{pmatrix} \sin(x_1(t) - x_2(t)) \\ x_2(t) + \exp(x_2(t)) \end{pmatrix},
$$

$$
y(t) = \begin{bmatrix} 1 & -1 \end{bmatrix} \begin{pmatrix} x_1(t) \\ x_2(t) \end{pmatrix}.
$$

$$(6.1)$$

Choosing $F = [1, -1]$ the Lipschitz condition (3.2) is satisfied as

$$
\| f_L(x) - f_L(\hat{x}) \| = \| \sin(x_1 - x_2) - \sin(\hat{x}_1 - \hat{x}_2) \|
$$

$$
\leq \| (x_1 - x_2) - (\hat{x}_1 - \hat{x}_2) \| = \left\| \begin{bmatrix} 1 & -1 \end{bmatrix} \begin{pmatrix} x_1 - \hat{x}_1 \\ x_2 - \hat{x}_2 \end{pmatrix} \right\|
$$

for all $x, \hat{x} \in \mathscr{X} = \mathbb{R}^2$. The monotonicity condition (3.3) is satisfied with $\Theta = I_{q_M} = 1$, $\mu = 2$ and $J = [0, 1]$ since for all $x, z \in \hat{\mathscr{X}} = J\mathscr{X} = \mathbb{R}$ we have

$$
(z - x) \big(f_M(z) - f_M(x) \big) = (z - x) \big(z + \exp(z) - x - \exp(x) \big)
$$

$$
= (z - x)^2 + \underbrace{(z - x) \big(\exp(z) - \exp(x) \big)}_{\geq 0} \geq \frac{\mu}{2}(z - x)^2.
$$

To satisfy the conditions of Theorem 4.1 we choose $k = 2$. A straightforward computation yields that conditions (4.1) and (4.2) are satisfied with the following matrices $\mathscr{P} \in \mathbb{R}^{(4+1) \times (2+2)}$, $\mathscr{K} \in \mathbb{R}^{(2+2) \times (2+2)}$, $L_1 \in \mathbb{R}^{4 \times 2}$ and $L_2 \in \mathbb{R}^{1 \times 2}$ on the subspace $\mathscr{V}^*_{[[\mathscr{E}, 0], [\mathscr{A}, \mathscr{B}]]}$ with $\delta = 1$:

$$
\mathscr{P} = \frac{1}{10}
\begin{bmatrix}
2 & -2 & 0 & 0 \\
0 & 0 & 0 & 0 \\
0 & 0 & 0 & 0 \\
-2 & 3 & 0 & 0 \\
0 & 0 & 0 & 0
\end{bmatrix},
\quad
\mathscr{K} = \frac{1}{5}
\begin{bmatrix}
* & * & 4 & 10 \\
* & * & -4 & -10 \\
* & * & 0 & 0 \\
* & * & 0 & 0
\end{bmatrix},
$$

$$
\begin{bmatrix} L_1 \\ L_2 \end{bmatrix} =
\begin{bmatrix}
4 & 10 \\
1 & 9 \\
9 & 4 \\
0 & 0 \\
2 & 1
\end{bmatrix},
\quad
\mathscr{V}^*_{[[\mathscr{E}, 0], [\mathscr{A}, \mathscr{B}]]} = \text{im}
\begin{bmatrix}
1 & 0 & 0 \\
0 & 1 & 0 \\
5 & -4 & 0 \\
-11 & 9 & 0 \\
0 & 0 & 1 \\
-2 & 2 & 0
\end{bmatrix}.
$$

Then Theorem 4.1 implies that a state estimator for (6.1) is given by

$$
\frac{d}{dt}
\begin{bmatrix} 1 & 0 \\ 1 & 1 \\ 0 & 0 \\ 0 & 1 \end{bmatrix}
\begin{pmatrix} z_1(t) \\ z_2(t) \end{pmatrix}
=
\begin{bmatrix} 0 & -3 \\ -2 & 0 \\ 1 & -2 \\ 0 & 0 \end{bmatrix}
\begin{pmatrix} z_1(t) \\ z_2(t) \end{pmatrix}
+
\begin{bmatrix} 0 & 2 \\ 1 & -1 \\ 0 & 1 \\ 1 & 0 \end{bmatrix}
\begin{pmatrix} \sin(z_1(t) - z_2(t)) \\ z_2(t) + \exp(z_2(t)) \end{pmatrix}
+
\begin{bmatrix} 4 & 10 \\ 1 & 9 \\ 9 & 4 \\ 0 & 0 \end{bmatrix}
\begin{pmatrix} d_1(t) \\ d_2(t) \end{pmatrix}
$$

$$
0 = \begin{bmatrix} 1 & -1 \end{bmatrix}
\begin{pmatrix} z_1(t) \\ z_2(t) \end{pmatrix}
- y(t) + \begin{bmatrix} 2 & 1 \end{bmatrix}
\begin{pmatrix} d_1(t) \\ d_2(t) \end{pmatrix}
$$

(6.2)

Note, that L_2 is not invertible and thus the state estimator cannot be reformulated as a Luenberger type observer. Further, $n + k < l + p$ and therefore the pencil $s\mathscr{E} - \mathscr{A}$ is not square and hence in particular not regular; thus (4.10) (b) cannot be satisfied. In addition, for F and J in the present example, condition (4.10) (e) does not hold (and is independent of k), thus Theorem 4.2 is not applicable here. A closer investigation reveals that for $k = l + p - n$ inequality (4.2) cannot be satisfied. We like to emphasize that $\mathscr{Q} <_{\gamma^*_{[[\mathscr{E},0],[\mathscr{A},\mathscr{B}]]}} 0$ but $\mathscr{Q} < 0$ does not hold on $\mathbb{R}^{n+k+q_L+q_M} = \mathbb{R}^6$.

The next example illustrates Theorem 4.2.

Example 6.2 We consider the DAE

$$
\frac{d}{dt}
\begin{bmatrix} 1 & -1 \\ 0 & 0 \end{bmatrix}
\begin{pmatrix} x_1(t) \\ x_2(t) \end{pmatrix}
=
\begin{bmatrix} -1 & 0 \\ 0 & 1 \end{bmatrix}
\begin{pmatrix} x_1(t) \\ x_2(t) \end{pmatrix}
+
\begin{bmatrix} 2 & -1 \\ -1 & 1 \end{bmatrix}
\begin{pmatrix} \sin(x_1(t) + x_2(t)) \\ x_1(t) + x_2(t) + \exp(x_1(t) + x_2(t)) \end{pmatrix},
$$

$$
y(t) = \begin{bmatrix} 1 & 1 \end{bmatrix}
\begin{pmatrix} x_1(t) \\ x_2(t) \end{pmatrix}.
$$

(6.3)

Similar to Example 6.1 it can be shown that the monotonicity condition (3.3) is satisfied for $f_M(x) = x + \exp(x)$ with $J = [1, 1]$, $\Theta = 1$ and $\mu = 2$; the Lipschitz condition (3.2) is satisfied for $f_L(x_1, x_2) = \sin(x_1 + x_2)$ with $F = [1, 1]$.

Choosing $k = 1$ a straightforward computation yields that conditions (4.1) and (4.10) (a) are satisfied with $\delta = 1.5$, the following matrices $\mathscr{P}, \mathscr{K} \in \mathbb{R}^{(2+1)\times(2+1)}$, $L_1 \in \mathbb{R}^{2\times 1}$ and $L_2 \in \mathbb{R}^{1\times 1} = \mathbb{R}$ and subspaces $\mathscr{V}^*_{[[\mathscr{E},0],[\mathscr{A},\mathscr{B}]]}, \mathscr{V}^*_{[\mathscr{E},\mathscr{A}]}$ and $\mathscr{W}^*_{[\mathscr{E},\mathscr{A}]}$:

$$\mathscr{P} = \tfrac{1}{10} \begin{bmatrix} 1 & -1 & 0 \\ 1 & 17 & 0 \\ 0 & 0 & 17 \end{bmatrix}, \qquad \mathscr{K} = \tfrac{1}{10} \begin{bmatrix} * & * & 8 \\ * & * & -134 \\ * & * & 17 \end{bmatrix}, \quad \begin{bmatrix} L_1 \\ L_2 \end{bmatrix} = \begin{bmatrix} 15 \\ -7 \\ 1 \end{bmatrix},$$

$$\mathscr{V}^*_{[[\mathscr{E},0],[\mathscr{A},\mathscr{B}]]} = \mathrm{im} \begin{bmatrix} 1 & 0 & 0 \\ 0 & 1 & 0 \\ -1 & -1 & 0 \\ 0 & 0 & 1 \\ -7 & -8 & 1 \end{bmatrix}, \quad \mathscr{V}^*_{[\mathscr{E},\mathscr{A}]} = \mathrm{im} \begin{bmatrix} 8 \\ -7 \\ -1 \end{bmatrix}, \quad \mathscr{W}^*_{[\mathscr{E},\mathscr{A}]} = \mathrm{im} \begin{bmatrix} 1 & 0 \\ 1 & 0 \\ 0 & 1 \end{bmatrix}.$$

Conditions (4.10) (b)–(e) are satisfied as follows:

(b) $\det(s\mathscr{E} - \mathscr{A}) \neq 0$ and, using [2, Prop. 2.2.9], the index of $s\mathscr{E} - \mathscr{A}$ is $\nu = k^* = 1$, where k^* is from Def. 2.4;

(c) this holds since $G_L = [1/15, 1/15]^\top$ and thus $\|FG_L\| < 1$;

(d) JG_M is invertible since $G_M = -[1/15, 1/15]^\top$ and $\lambda_{max}(\Gamma) = -15 < 2 = \mu$;

(e) this condition is satisfied with e.g. $\alpha = 1$ since $F = J$, and

$$\frac{\alpha \|JG_L\|}{1 - \|FG_L\|} \left(\sqrt{\frac{\max\{0, \lambda_{max}(S)\}}{\mu - \lambda_{max}(\Gamma)}} + \|(\Gamma - \mu I_{q_M})^{-1}(\tilde{\Theta}^\top - \mu I_{q_M})\| \right) = \frac{19}{221} < 1.$$

Then Theorem 4.2 implies that a state estimator for system (6.3) is given by

$$\frac{d}{dt} \begin{bmatrix} 1 & -1 \\ 0 & 0 \end{bmatrix} \begin{pmatrix} z_1(t) \\ z_2(t) \end{pmatrix} = \begin{bmatrix} -1 & 0 \\ 0 & 1 \end{bmatrix} \begin{pmatrix} z_1(t) \\ z_2(t) \end{pmatrix} + \begin{bmatrix} 2 & -1 \\ -1 & 1 \end{bmatrix} \begin{pmatrix} \sin(z_1(t) + z_2(t)) \\ z_1(t) + z_2(t) + \exp(z_1(t) + z_2(t)) \end{pmatrix}$$

$$+ \begin{bmatrix} 15 \\ -7 \end{bmatrix} d(t),$$

$$0 = \begin{bmatrix} 1 & 1 \end{bmatrix} \begin{pmatrix} z_1(t) \\ z_2(t) \end{pmatrix} - y(t) + d(t).$$

(6.4)

Straightforward calculations show that conditions (4.10) (a)–(e) are satisfied, but condition (4.2) is violated; thus, Theorem 4.1 is not applicable for $k = l+p-n = 1$. The matrix L_2 is invertible and hence the state estimator (6.4) can be transformed as a standard Luenberger type observer. We emphasize that $\mathscr{Q} < 0$ does not hold on $\mathbb{R}^5$, i.e., the matrix inequality (4.1) on the subspace $\mathscr{V}^*_{[[\mathscr{E},0],[\mathscr{A},\mathscr{B}]]} \subseteq \mathbb{R}^5$ is a weaker condition.

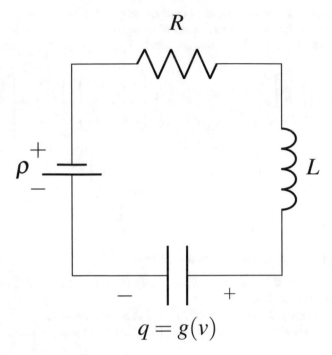

$$q = g(v)$$

Fig. 2 Nonlinear RLC circuit

The last example is an electric circuit where monotone nonlinearities occur, which is taken from [35].

Example 6.3 Consider the electric circuit depicted in Fig. 2, where a DC source with voltage ρ is connected in series to a linear resistor with resistance R, a linear inductor with inductance L and a nonlinear capacitor with the nonlinear characteristic

$$q = g(v) = (v - v_0)^3 - (v - v_0) + q_0, \tag{6.5}$$

where q is the electric charge and v is the voltage over the capacitor.

Using the magnetic flux ϕ in the inductor, the circuit admits the charge-flux description

$$\dot{q}(t) = \frac{1}{L}\phi(t),$$

$$\dot{\phi}(t) = -\frac{R}{L}\phi(t) - v(t) + \rho(t). \tag{6.6}$$

We scale the variables $q = C\,\tilde{q}$, $\phi = Vs\,\tilde{\phi}$, $v = V\,\tilde{v}$ (where s, V and C denote the SI units for seconds, Volt and Coulomb, resp.) in order to make them dimensionless. For simulation purposes we set $\rho = \rho_0 = 2$ V (i.e. ρ trivially satisfies condition (3.2)), $R = 1\,\Omega$ and $L = 0.5$ H, $\tilde{q}_0 = \tilde{v}_0 = 1$. Then with $(x_1, x_2, x_3)^\top = \left(\tilde{q} - \tilde{q}_0, \tilde{\phi}, \tilde{v} - \tilde{v}_0\right)^\top$ the circuit equations (6.5) and (6.6) can be written as the DAE

$$
\frac{d}{dt}
\begin{bmatrix} 1 & 0 & 0 \\ 0 & 1 & 0 \\ 0 & 0 & 0 \end{bmatrix}
\begin{pmatrix} x_1(t) \\ x_2(t) \\ x_3(t) \end{pmatrix}
=
\begin{bmatrix} 0 & 2 & 0 \\ 0 & -2 & -1 \\ -1 & 0 & -1 \end{bmatrix}
\begin{pmatrix} x_1(t) \\ x_2(t) \\ x_3(t) \end{pmatrix}
+
\begin{bmatrix} 0 & 0 \\ 1 & 0 \\ 0 & 1 \end{bmatrix}
\begin{pmatrix} 1 \\ x_3(t)^3 \end{pmatrix}
$$

$$
y(t) = \begin{bmatrix} 1 & 0 & -1 \end{bmatrix}
\begin{pmatrix} x_1(t) \\ x_2(t) \\ x_3(t) \end{pmatrix},
$$

(6.7)

where the output is taken as the difference $q(t) - v(t)$. Now, similar to the previous examples, a straightforward computation shows that Theorem 4.2 is applicable and yields parameters for a state estimator for (6.7), which has the form

$$
\frac{d}{dt}
\begin{bmatrix} 1 & 0 & 0 \\ 0 & 1 & 0 \\ 0 & 0 & 0 \end{bmatrix}
\begin{pmatrix} z_1(t) \\ z_2(t) \\ z_3(t) \end{pmatrix}
=
\begin{bmatrix} 0 & 2 & 0 \\ 0 & -2 & -1 \\ -1 & 0 & -1 \end{bmatrix}
\begin{pmatrix} z_1(t) \\ z_2(t) \\ z_3(t) \end{pmatrix}
+
\begin{bmatrix} 0 & 0 \\ 1 & 0 \\ 0 & 1 \end{bmatrix}
\begin{pmatrix} 1 \\ z_3^3(t) \end{pmatrix}
+
\begin{bmatrix} -1 \\ 5 \\ 5 \end{bmatrix}
d(t)
$$

$$
0 = \begin{bmatrix} 1 & 0 & -1 \end{bmatrix}
\begin{pmatrix} z_1(t) \\ z_2(t) \\ z_3(t) \end{pmatrix}
- y(t) + 4d(t).
$$

(6.8)

Note that since $L_2 = 4$ is invertible, the given state estimator can be reformulated as an observer of Luenberger type with gain matrix $L = L_1 L_2^{-1}$. As before we emphasize that $\mathcal{Q} < 0$ is not satisfied on $\mathbb{R}^6$.

Note that this example also satisfies the assumptions of Theorem 4.1 with $k = 0$, i.e., the system copy itself serves as a state estimator (no innovation terms d are present).

7   Comparison with the Literature

We compare the results found in [5, 15, 20, 29, 44] to the results in the present paper. In [29, Thm. 2.1] a way to construct an asymptotic observer of Luenberger type is presented. In the works [15, 20] reduced-order observer designs for non-square nonlinear DAEs are presented. An essential difference to Theorems 4.1, 4.2 and 5.1 is the space on which the LMIs are considered. While in [15, 20, 29] the LMI has to hold on the whole space $\mathbb{R}^{\bar{n}}$ for some $\bar{n} \in \mathbb{N}$, the inequalities stated in the present paper as well as the inequalities stated in [5, Thm. III.1] only have to be satisfied on a certain subspace where the solutions evolve in. While solving the

LMIs stated in [15, 20, 29] on the entire space $\mathbb{R}^{\bar{n}}$ is a much stronger condition, an advantage of this is that it can be solved numerically with little effort.

The LMI stated in [15] is similar to (4.1) and has to hold on $\mathbb{R}^{a+q_L}$, where $a \leq n$ is the observer's order ($a = n$ corresponds to a full-order observer comparable to the state estimator in the present work), and q_L is as in the present paper. Hence, the dimension of the space where the LMI has to be satisfied scales with the observer's order and the range of the Lipschitz nonlinearity. Similarly, the matrix inequality (4.1) in the present paper (in the case $q_M = 0$) is asked to hold on a subspace of $\mathbb{R}^{n+k+q_L}$ with dimension at most $n + k + q_L - \text{rk}[C, L_2]$. Therefore, the more independent information from the output is available, the smaller the dimension of the subspace $\mathcal{V}^*_{[[\mathcal{E},0],[\mathcal{A},\mathcal{B}]]}$ is. We stress that the detectability condition as identified in [15, Prop. 2] is implicitly encoded in the LMI (4.1) when $q_L = 0$ and $q_M = 0$, cf. also [5, Lem. III.2]. More precisely, a certain (behavioral) detectability of the linear part is a necessary condition for (4.1) to hold, since it is stated independent of the specific nonlinearities, which only need to satisfy (3.2) and (3.3), resp.

Another difference to [5, 15, 29] is that the nonlinearity has to satisfy a Lipschitz condition of the form (3.2), and the nonlinearity $f \in \mathcal{C}^1(\mathbb{R}^r \to \mathbb{R}^r)$ in [20] has to satisfy the generalized monotonicity condition $f'(s) + f'(s)^\top \geq \mu I_r$ for all $s \in \mathbb{R}^r$, which is less general than condition (3.3), cf. [26]. In the present paper we allow the function $f = \left(\begin{smallmatrix} f_L \\ f_M \end{smallmatrix}\right)$ to be a combination of a function f_L satisfying (3.2) and a function f_M satisfying (3.3). Therefore the presented theorems cover a larger class of systems. In the work [44], the Lipschitz condition on the nonlinearity is avoided. However, the results of this paper are restricted to a class of DAE systems, for which a certain transformation is feasible, that regularizes the system by introducing the derivative of the output in the differential equation for the state. Then classical Luenberger observer theory is applied to the resulting ODE system.

The work [29] considers square systems only, while in Theorems 4.1, 4.2 and 5.1 we allow for any rectangular systems with $n \neq l$. Therefore, the observer design presented in the present paper is a considerable generalization of the work [29].

Compared to [5, Thm. III.1], we may observe that in this work the invertibility of a matrix consisting of system parameters and the gain matrices L_2 and L_3 is required. This condition as well as the rank condition is comparable to the regularity condition (4.10) (b) in the present paper. However, in the present paper we do not state explicit conditions on the gains, which are unknown beforehand and constructed out of the solution of (4.1). Hence only the solution matrices $\mathcal{P}$ and $\mathcal{K}$ are required to meet certain conditions.

8   Computational Aspects

The sufficient conditions for the existence of a state estimator/asymptotic observer stated in Theorems 4.1, 4.2 and 5.1 need to be satisfied at the same time, in each of them. Hence it might be difficult to develop a computational procedure for the construction of a state estimator based on these results, in particular since the subspaces $\mathcal{V}^*_{[[\mathscr{E},0][\mathscr{A},\mathscr{B}]]}$, $\mathcal{V}^*_{[\mathscr{E},\mathscr{A}]}$ and $\mathcal{W}^*_{[\mathscr{E},\mathscr{A}]}$ depend on the solutions $\mathscr{P}$ and $\mathscr{K}$ of (4.1). The state estimators for the examples given in Sect. 6 are constructed using "trial and error" rather than a systematic numerical procedure. The development of such a numerical method will be the topic of future research.

Nevertheless, the theorems are helpful tools in examining if an alleged observer candidate is a state estimator for a given system. To this end, we may set $\mathscr{K}H = \mathscr{P}^\top \hat{L}$ with given $\hat{L}$. Then $\mathscr{A} = \hat{A} + \hat{L}$ and the subspace to which (4.1) is restricted is independent of its solutions and hence (4.1) can be rewritten as a LMI on the space $\mathbb{R}^{n^*}$, where $n^* = \dim \mathcal{V}^*_{[[\mathscr{E},0],[\mathscr{A},\mathscr{B}]]}$. This LMI can be solved numerically stable by standard MATLAB toolboxes like YALMIP [27] and PENLAB [17]. For other algorithmic approaches see e.g. the tutorial paper [38].

Acknowledgement This work was supported by the German Research Foundation (Deutsche Forschungsgemeinschaft) via the grant BE 6263/1-1.

References

1. Åslund, J., Frisk, E.: An observer for non-linear differential-algebraic systems. Automatica **42**(6), 959–965 (2006)
2. Berger, T.: On differential-algebraic control systems. Ph.D. Thesis, Institut für Mathematik, Technische Universität Ilmenau, Universitätsverlag Ilmenau (2014)
3. Berger, T.: Controlled invariance for nonlinear differential-algebraic systems. Automatica **64**, 226–233 (2016)
4. Berger, T.: The zero dynamics form for nonlinear differential-algebraic systems. IEEE Trans. Autom. Control **62**(8), 4131–4137 (2017)
5. Berger, T.: On observers for nonlinear differential-algebraic systems. IEEE Trans. Autom. Control **64**(5), 2150–2157 (2019)
6. Berger, T., Reis, T.: Observers and dynamic controllers for linear differential-algebraic systems. SIAM J. Control Optim. **55**(6), 3564–3591 (2017)
7. Berger, T., Reis, T.: ODE observers for DAE systems. IMA J. Math. Control Inf. **36**, 1375–1393 (2019)
8. Berger, T., Trenn, S.: The quasi-Kronecker form for matrix pencils. SIAM J. Matrix Anal. Appl. **33**(2), 336–368 (2012)
9. Berger, T., Trenn, S.: Addition to "The quasi-Kronecker form for matrix pencils". SIAM J. Matrix Anal. Appl. **34**(1), 94–101 (2013)
10. Berger, T., Ilchmann, A., Trenn, S.: The quasi-Weierstraß form for regular matrix pencils. Linear Algebra Appl. **436**(10), 4052–4069 (2012)
11. Boutayeb, M., Darouach, M., Rafaralahy, H.: Generalized state-space observers for chaotic synchronization and secure communication. IEEE Trans. Circuits Syst. I: Fundam. Theory Appl. **49**(3), 345–349 (2002)
12. Brenan, K.E., Campbell, S.L., Petzold, L.R.: Numerical Solution of Initial-Value Problems in Differential-Algebraic Equations. North-Holland, Amsterdam (1989)
13. Campbell, S.L.: Singular Systems of Differential Equations II. Pitman, New York (1982)
14. Corradini, M.L., Cristofaro, A., Pettinari, S.: Design of robust fault detection filters for linear descriptor systems using sliding-mode observers. IFAC Proc. Vol. **45**(13), 778–783 (2012)
15. Darouach, M., Boutat-Baddas, L.: Observers for a class of nonlinear singular systems. IEEE Trans. Autom. Control **53**(11), 2627–2633 (2008)

16. Eich-Soellner, E., Führer, C.: Numerical Methods in Multibody Dynamics. Teubner, Stuttgart (1998)
17. Fiala, J., Kočvara, M., Stingl, M.: PENLAB: a MATLAB solver for nonlinear semidefinite optimization (2013). https://arxiv.org/abs/1311.5240
18. Gantmacher, F.R.: The Theory of Matrices (Vol. I & II). Chelsea, New York (1959)
19. Gao, Z., Ho, D.W.: State/noise estimator for descriptor systems with application to sensor fault diagnosis. IEEE Trans. Signal Proc. **54**(4), 1316–1326 (2006)
20. Gupta, M.K., Tomar, N.K., Darouach, M.: Unknown inputs observer design for descriptor systems with monotone nonlinearities. Int. J. Robust Nonlinear Control **28**(17), 5481–5494 (2018)
21. Gutú, O., Jaramillo, J.A.: Global homeomorphisms and covering projections on metric spaces. Math. Ann. **338**, 75–95 (2007)
22. Ha, Q., Trinh, H.: State and input simultaneous estimation for a class of nonlinear systems. Automatica **40**, 1779–1785 (2004)
23. Kumar, A., Daoutidis, P.: Control of Nonlinear Differential Algebraic Equation Systems with Applications to Chemical Processes. Chapman and Hall/CRC Research Notes in Mathematics, vol. 397. Chapman and Hall, Boca Raton (1999)
24. Kunkel, P., Mehrmann, V.: Differential-Algebraic Equations. Analysis and Numerical Solution. EMS Publishing House, Zürich (2006)
25. Lamour, R., März, R., Tischendorf, C.: Differential Algebraic Equations: A Projector Based Analysis. Differential-Algebraic Equations Forum, vol. 1. Springer, Heidelberg (2013)
26. Liu, H.Y., Duan, Z.S.: Unknown input observer design for systems with monotone nonlinearities. IET Control Theory Appl. **6**(12), 1941–1947 (2012)
27. Löfberg, J.: YALMIP: a toolbox for modeling and optimization in MATLAB. In: Proceedings of the 2004 IEEE International Symposium on Computer Aided Control Systems Design, pp. 284–289 (2004)
28. Logemann, H., Ryan, E.P.: Ordinary Differential Equations. Springer, London (2014)
29. Lu, G., Ho, D.W.: Full-order and reduced-order observers for Lipschitz descriptor systems: the unified LMI approach. IEEE Trans. Circuits Syst. Express Briefs **53**(7), 563–567 (2006)
30. Luenberger, D.G.: Observing the state of a linear system. IEEE Trans. Mil. Electron. **MIL-8**, 74–80 (1964)
31. Luenberger, D.G.: An introduction to observers. IEEE Trans. Autom. Control **16**(6), 596–602 (1971)
32. Luenberger, D.G.: Nonlinear descriptor systems. J. Econ. Dyn. Control **1**, 219–242 (1979)
33. Luenberger, D.G., Arbel, A.: Singular dynamic Leontief systems. Econometrica **45**, 991–995 (1977)
34. Polderman, J.W., Willems, J.C.: Introduction to Mathematical Systems Theory. A Behavioral Approach. Springer, New York (1998)
35. Riaza, R.: Double SIB points in differential-algebraic systems. IEEE Trans. Autom. Control **48**(9), 1625–1629 (2003)
36. Riaza, R.: Differential-Algebraic Systems. Analytical Aspects and Circuit Applications. World Scientific, Basel (2008)
37. Valcher, M.E., Willems, J.C.: Observer synthesis in the behavioral approach. IEEE Trans. Autom. Control **44**(12), 2297–2307 (1999)
38. VanAntwerp, J.G., Braatz, R.D.: A tutorial on linear and bilinear matrix inequalities. J. Process Control **10**(4), 363–385 (2000)
39. Wong, K.T.: The eigenvalue problem $\lambda Tx + Sx$. J. Diff. Equ. **16**, 270–280 (1974)
40. Yang, C., Zhang, Q., Chai, T.: Nonlinear observers for a class of nonlinear descriptor systems. Optim. Control Appl. Methods **34**(3), 348–363 (2012)
41. Yang, C., Kong, Q., Zhang, Q.: Observer design for a class of nonlinear descriptor systems. J. Franklin Inst. **350**(5), 1284–1297 (2013)
42. Yeu, T.K., Kawaji, S.: Sliding mode observer based fault detection and isolation in descriptor systems. In: Proceedings of American Control Conference 2002, pp. 4543–4548. Anchorage (2002)

43. Zhang, J., Swain, A.K., Nguang, S.K.: Simultaneous estimation of actuator and sensor faults for descriptor systems. In: Robust Observer-Based Fault Diagnosis for Nonlinear Systems Using MATLAB®, Advances in Industrial Control, pp. 165–197. Springer, Berlin (2016)
44. Zheng, G., Boutat, D., Wang, H.: A nonlinear Luenberger-like observer for nonlinear singular systems. Automatica **86**, 11–17 (2017)
45. Zulfiqar, A., Rehan, M., Abid, M.: Observer design for one-sided Lipschitz descriptor systems. Appl. Math. Model. **40**(3), 2301–2311 (2016)

Error Analysis for the Implicit Euler Discretization of Linear-Quadratic Control Problems with Higher Index DAEs and Bang–Bang Solutions

Björn Martens and Matthias Gerdts

Abstract We investigate the implicit Euler discretization for linear-quadratic optimal control problems with index two DAEs. There is a discrepancy between the necessary conditions of problems with higher index DAEs and their discretizations, since the necessary conditions of the continuous problem coincide with the necessary conditions of the index reduced problem. This implicit index reduction does not occur for the discretized problem. Thus, the respective switching functions cannot be related. The discrepancy is overcome by reformulating the discretized problem, which yields an approximation of the index reduced problem with suitable necessary conditions. If the switching function has a certain structure, such that the optimal control is of bang–bang type, we can show that the controls converge with an order of $\frac{1}{2}$ in the L_1-norm. We then improve these error estimates with slightly stronger smoothness conditions of the problems data and switching function, which gives us a convergence order of one.

Keywords Optimal control · Linear-quadratic problems · Differential-algebraic equations · Discrete approximations · Implicit Euler · Bang–bang control

Mathematics Subject Classification (2010) 49K15, 49J15, 49N10, 34A09, 49M25, 49J30

B. Martens (✉) · M. Gerdts
Institute of Applied Mathematics and Scientific Computing, Department of Aerospace Engineering, Universität der Bundeswehr München, Neubiberg/Munich, Germany
e-mail: bjoern.martens@unibw.de; matthias.gerdts@unibw.de

291

T. Reis et al. (eds.), *Progress in Differential-Algebraic Equations II*,
Differential-Algebraic Equations Forum,
https://doi.org/10.1007/978-3-030-53905-4_10

1   Introduction

We consider the optimal control problem

$$(\textbf{OCP-DAE}) \qquad \text{Minimize} \quad f\left(x, y, u\right)$$

$$\text{subject to} \quad \dot{x}(t) = A(t)\, x(t) + B(t)\, y(t) + C(t)\, u(t), \qquad \text{a.e. in } \left[0, 1\right],$$
$$0 = D(t)\, x(t), \qquad\qquad\qquad\qquad \text{in } \left[0, 1\right],$$
$$\Xi\, x(0) = a,$$
$$u(t) \in U, \qquad\qquad\qquad\qquad\qquad\quad \text{a.e. in } \left[0, 1\right],$$

with the objective function

$$f\left(x, y, u\right) := \frac{1}{2} x(1)^\top Q\, x(1) + q^\top x(1)$$

$$+ \int_0^1 \frac{1}{2} x(t)^\top P(t)\, x(t) + p(t)^\top x(t) + r(t)^\top y(t) + g(t)^\top u(t)\, dt,$$

and the set of admissible controls $U := \left\{ u \in \mathbb{R}^{n_u} \mid b_l \le u \le b_u \right\}$. Using the techniques developed in [7, 8] and [22, 24] we aim to derive error estimates for the implicit Euler discretization of (**OCP-DAE**).

Convergence of approximations for non-linear problems has been analyzed in [9, 11–13, 17, 18, 20, 21, 23, 24, 29]. Herein, [17, 21] apply the Euler discretization for problems with mixed control-state constraints. Malanowski et al. [21] obtain convergence of order one for Lipschitz continuous optimal controls in the L_∞-norm, whereas [17] achieve convergence rate of $\frac{1}{p}$ in the L_p-norm for optimal controls of bounded variation. In [9, 11, 12], problems with pure state constraints of order one are discussed. Linear convergence in the L_2-norm and convergence of order $\frac{2}{3}$ in the L_∞-norm is obtained in [11, 12]. In [9], linear convergence in the L_∞-norm is achieved. [13, 18, 29] analyze Runge–Kutta methods for problems with set constraints on the control. A second order Runge–Kutta approximation is used in [13, 29] in order to prove convergence of order two. Convergence of arbitrary order is obtained in [18] with a Runge–Kutta scheme of appropriate order and a sufficiently smooth optimal control. In [20], convergence for the value of the objective function is achieved through a control parametrization enhancing technique. Convergence of approximations for optimal control problems with differential-algebraic equations (DAEs) has been studied in [22–24]. Linear convergence for problems with index one DAEs was proven in [23], using the general convergence theory provided in [28]. In [22, 24], problems with index two DAEs and mixed control-state constraints are considered, and linear convergence is achieved.

Optimal control problems with discontinuous controls have been considered in [3, 4, 6–8, 26, 27, 30]. Herein, [8, 26, 30] study linear problems. Convergence of order one in the L_1-norm, and of order $\frac{1}{2}$ in the L_2-norm for optimal controls of bang–bang type is proven in [8]. A controllability assumption is exploited in [26, 30] in order to obtain convergence of an order depending on the controllability index. In [3, 4, 7, 27], linear quadratic systems are analyzed. Herein, [7] achieve results similar to [8], whereas [4] augment a L_1 control cost depending on a parameter in order to obtain linear convergence for an optimal control of bang-zero-bang type. In [5, 6], discretizations for nonlinear problems with linearly appearing control and bang–bang solutions are examined.

The dynamic behavior of systems in process engineering, electric circuits, and mechanical multi-body systems is often modeled by DAEs, see [10, 16, Example 1.1.20], and the references therein. Moreover, DAEs can be generated by discretizing a $2D$ Stokes equation (cf. [16, Example 1.1.12, 3.1.14]). Linear systems can be obtained by linearizing a non-linear system around a reference solution.

Throughout the paper we use the following notation: By $\mathbb{R}^n$ we denote the n-dimensional Euclidean space with the norm $|\cdot|$. We equip the space of $n \times m$-matrices A with the spectral norm $\|A\|$ and denote the unit matrix by E. $L_p^n\left([0, 1]\right)$ is the Banach space of measurable vector functions $v : [0, 1] \to \mathbb{R}^n$ with

$$\|v\|_p = \left(\int_0^1 |v(t)|^p \, dt \right)^{\frac{1}{p}} < \infty$$

for $1 \le p < \infty$, and $L_\infty^n\left([0, 1]\right)$ is the Banach space of measurable, essentially bounded vector functions $v : [0, 1] \to \mathbb{R}^n$ with

$$\|v\|_\infty = \max_{1 \le j \le n} \operatorname{ess\,sup}_{t \in [0,1]} |v_j(t)| < \infty.$$

By $W_{1,p}^n\left([0, 1]\right) = \left\{ z \in L_p^n\left([0, 1]\right) \mid \dot{z} \in L_p^n\left([0, 1]\right) \right\}$ we denote the Sobolev spaces of absolutely continuous functions $z : [0, 1] \to \mathbb{R}^n$ equipped with the norm

$$\|z\|_{1,p} = \left(\|z\|_p^p + \|\dot{z}\|_p^p \right)^{\frac{1}{p}} \quad \text{for } 1 \le p < \infty, \quad \|z\|_{1,\infty} = \max\left\{ \|z\|_\infty, \|\dot{z}\|_\infty \right\}.$$

We denote by $BV^n\left([0, 1]\right)$ the space of vector functions $v : [0, 1] \to \mathbb{R}^n$ of bounded variation and by $\bigvee_{\tau_1}^{\tau_2} v$ the total variation of v on $[\tau_1, \tau_2] \subseteq [0, 1]$ for $\tau_1 < \tau_2$.

Definition 1.1 A feasible trajectory $(\hat{x}, \hat{y}, \hat{u})$ is called minimizer for **(OCP-DAE)**, if

$$f\left(\hat{x}, \hat{y}, \hat{u}\right) \le f\left(x, y, u\right)$$

for all admissible (x, y, u).

For $N \in \mathbb{N}$ we consider the equidistant mesh $\mathbb{G}_N := \{0 = t_0 < t_1 < \ldots < t_N = 1\}$ of the interval $[0, 1]$ with mesh size $h := \frac{1}{N}$ and $t_i := i\,h$ for $i = 0, 1, \ldots, N$. Using the backwards difference approximation

$$x_h'(t_i) := \frac{x_h(t_i) - x_h(t_{i-1})}{h} \tag{1.1}$$

we obtain the implicit Euler discretization of **(OCP-DAE)**

(DOCP-DAE)

$$\begin{aligned}
\text{Minimize} \quad & f_h\left(x_h, y_h, u_h\right)\\
\text{subject to} \quad & x_h'(t_i) = A(t_i)\,x_h(t_i) + B(t_i)\,y_h(t_i) + C(t_i)\,u_h(t_i), \quad i = 1, \ldots, N,\\
& 0 = D(t_i)\,x_h(t_i), \quad i = 0, 1, \ldots, N,\\
& \Xi\,x_h(t_0) = a,\\
& u_h(t_i) \in U, \quad i = 1, \ldots, N,
\end{aligned}$$

with the discrete objective function

$$f_h\left(x_h, y_h, u_h\right) := \frac{1}{2} x_h(t_N)^\top Q\,x_h(t_N) + q^\top x_h(t_N)$$

$$+ h \sum_{i=1}^{N} \frac{1}{2} x_h(t_i)^\top P(t_i)\,x_h(t_i) + p(t_i)^\top x_h(t_i) + r(t_i)^\top y_h(t_i) + g(t_i)^\top u_h(t_i).$$

We denote by $L_{p,h}^n\left([0,1]\right) \subset L_p^n\left([0,1]\right)$ the space of piecewise constant functions
$v_h : [0, 1] \to \mathbb{R}^n$ with

$$v_h \in L_p^n\left([0,1]\right), \quad v_h(t) = v_h(t_i), \quad t \in (t_{i-1}, t_i], \, i = 1, \ldots, N$$

equipped with the norm $\|\cdot\|_p$. $W_{1,p,h}^n\left([0,1]\right) \subset W_{1,p}^n\left([0,1]\right)$ equipped with the norm $\|\cdot\|_{1,p}$ is the space of continuous, piecewise linear functions $z_h : [0, 1] \to \mathbb{R}^n$ with

$$z_h \in W_{1,p}^n\left([0,1]\right), \quad z_h(t) = z_h'(t_i)\left(t - t_{i-1}\right) + z_h(t_{i-1}), \quad t \in (t_{i-1}, t_i], \, i = 1, \ldots, N.$$

If **(OCP-DAE)** has a solution $(\hat{x}, \hat{y}, \hat{u}) \in W_{1,\infty}^{n_x}([0,1]) \times L_{\infty}^{n_y}([0,1]) \times L_{\infty}^{n_u}([0,1])$ and certain smoothness and regularity assumptions hold (compare **(A1')** below), then there exist multipliers $\lambda \in W_{1,\infty}^{n_x}([0,1]), \mu \in L_{\infty}^{n_y}([0,1])$ satisfying the necessary conditions

$$\dot{\lambda}(t) = -A(t)^\top \lambda(t) - \left(\dot{D}(t) + D(t)A(t)\right)^\top \mu(t) - P(t)\hat{x}(t) - p(t) \qquad \text{a.e. in } [0,1]$$

$$0 = r(t) + B(t)^\top \lambda(t) + \left(D(t)B(t)\right)^\top \mu(t) \qquad \text{a.e. in } [0,1]$$

$$\lambda(1) = Q\hat{x}(1) + q$$

$$0 \le \left[g(t)^\top + \lambda(t)^\top C(t) + \mu(t)^\top D(t)C(t)\right](u - \hat{u}(t)), \quad \text{for all } u \in U, \quad \text{a.e. in } [0,1].$$

Furthermore, if there exists a solution $(\hat{x}_h, \hat{y}_h, \hat{u}_h) \in W_{1,\infty,h}^{n_x}([0,1]) \times L_{\infty,h}^{n_y}([0,1]) \times L_{\infty,h}^{n_u}([0,1])$ of **(DOCP-DAE)**, then the discrete necessary conditions read as

$$\tilde{\lambda}_h'(t_i) = -A(t_i)^\top \tilde{\lambda}_h(t_{i-1}) - D(t_i)^\top \tilde{\mu}_h(t_{i-1}) - P(t_i)\hat{x}_h(t_i) - p(t_i) \quad i = 1, \ldots, N$$

$$0 = r(t_i) + B(t_i)^\top \tilde{\lambda}_h(t_{i-1}) \qquad\qquad i = 1, \ldots, N$$

$$\tilde{\lambda}_h(t_N) = Q\hat{x}_h(t_N) + q$$

$$0 \le \left[g(t_i)^\top + \tilde{\lambda}_h(t_{i-1})^\top C(t_i)\right](u - \hat{u}_h(t_i)), \quad \text{for all } u \in U \quad i = 1, \ldots, N$$

for $\tilde{\lambda}_h \in W_{1,\infty,h}^{n_x}([0,1]), \tilde{\mu}_h \in L_{\infty,h}^{n_y}([0,1])$. Note that the continuous necessary conditions have the switching function

$$g(t)^\top + \lambda(t)^\top C(t) + \mu(t)^\top D(t)C(t), \quad t \in [0,1], \tag{1.2}$$

whereas the discrete necessary conditions have the switching function

$$g(t_i)^\top + \tilde{\lambda}_h(t_{i-1})^\top C(t_i), \quad i = 1, \ldots, N. \tag{1.3}$$

The switching functions are essential in our analysis, since they determine the structure of the optimal controls. However, we are not able to relate the continuous (1.2) and the discrete (1.3) switching functions, since there exists a discrepancy between the respective necessary conditions (cf. [22, 24]).

This paper is organized as follows: In Sect. 2, we consider an optimal control problem subject to an explicit differential equation and a general implicit approximation of that problem. We derive error estimates for optimal values of these problems (Theorem 3.1) in Sect. 3. In Sect. 4, we derive error estimates for optimal controls of bounded variation (Theorem 4.1) and in Sect. 5 we improve these estimates by assuming a structural stability property for the switching function (Theorem 5.1). In Sect. 6, we obtain the main result of this paper by reducing the index of the DAE in problem (OCP-DAE), which yields an optimal control problem subject to an explicit differential equation. Similarly, we transform the discretized problem and get an approximation of the reduced optimal control problem, which belongs to the general discrete problem class in Sect. 2. For the reduced problems we then obtain error estimates by applying the results of Sects. 4 and 5 (Theorem 6.1). An example is provided in Sect. 7.

2   Optimal Control Problem and Approximation

Throughout Sects. 2–5, we consider the optimal control problem

$$
\textbf{(OCP)} \qquad
\begin{aligned}
&\text{Minimize} \quad f\,(x, u) \\
&\text{subject to} \quad \dot{x}(t) = A(t)\,x(t) + B(t)\,u(t), \quad \text{a.e. in } [0, 1], \\
&\qquad\qquad\quad\; x(0) = a, \\
&\qquad\qquad\quad\; u(t) \in U, \qquad\qquad\qquad\qquad\;\; \text{a.e. in } [0, 1],
\end{aligned}
$$

with the linear-quadratic cost functional

$$
f\,(x, u) := \frac{1}{2} x(1)^\top Q\, x(1) + q^\top x(1) + \int_0^1 \frac{1}{2} x(t)^\top P(t)\, x(t) + p(t)^\top x(t) + r(t)^\top u(t)\, dt,
$$

and the bounded set $U := \{ u \in \mathbb{R}^{n_u} \mid b_l \leq u \leq b_u \}$. As an (implicit) approximation of (OCP) we consider

$$
\textbf{(DOCP)} \qquad
\begin{aligned}
&\text{Minimize} \quad f_h\,(x_h, u_h) \\
&\text{subject to} \quad x_h'(t_i) = A_h(t_i)\,x_h(t_i) + B_h(t_i)\,u_h(t_i), \quad i = 1, \ldots, N, \\
&\qquad\qquad\quad\; x_h(t_0) = a, \\
&\qquad\qquad\quad\; u_h(t_i) \in U, \qquad\qquad\qquad\qquad\qquad\qquad\; i = 1, \ldots, N,
\end{aligned}
$$

with the backwards difference approximation (1.1), and the objective function

$$
f_h\,(x_h, u_h) := \frac{1}{2} x_h(t_N)^\top Q_h\, x_h(t_N) + q_h^\top x_h(t_N)
$$

$$
+ h \sum_{i=1}^N \frac{1}{2} x_h(t_i)^\top P_h(t_i)\, x_h(t_i) + p_h(t_i)^\top x_h(t_i) + r_h(t_i)^\top u_h(t_i).
$$

Implicit discretization schemes have been investigated in, e.g., [3, 5]. We assume the following smoothness and approximation conditions:

(A1) Let $q \in \mathbb{R}^{n_x}$, let the matrix valued functions $A : [0, 1] \to \mathbb{R}^{n_x \times n_x}$, $B : [0, 1] \to \mathbb{R}^{n_x \times n_u}$, $P : [0, 1] \to \mathbb{R}^{n_x \times n_x}$ and the vector valued functions $p : [0, 1] \to \mathbb{R}^{n_x}$, $r : [0, 1] \to \mathbb{R}^{n_u}$ be Lipschitz continuous, and let $Q, P(t) \in \mathbb{R}^{n_x \times n_x}$, $t \in [0, 1]$, be symmetric and positive semidefinite.

(A2) Let $q_h \in \mathbb{R}^{n_x}$, $A_h \in W_{1,\infty,h}^{n_x \times n_x}([0, 1])$, $B_h \in W_{1,\infty,h}^{n_x \times n_u}([0, 1])$, $P_h \in W_{1,\infty,h}^{n_x \times n_x}([0, 1])$, $p_h \in W_{1,\infty,h}^{n_x}([0, 1])$, $r_h \in W_{1,\infty,h}^{n_u}([0, 1])$, and $Q_h, P_h(t_i) \in \mathbb{R}^{n_x \times n_x}$, $i = 1, \ldots, N$, be symmetric positive semidefinite. Furthermore, let the following approximation conditions hold

$$\|A - A_h\|_\infty \le L_A\, h, \quad \|B - B_h\|_\infty \le L_B\, h, \quad \|P - P_h\|_\infty \le L_P\, h, \quad \|Q - Q_h\| \le L_Q\, h,$$

$$|q - q_h| \le L_q\, h, \quad \|p - p_h\|_\infty \le L_p\, h, \quad \|r - r_h\|_\infty \le L_r\, h,$$

where the constants $L_A, L_B, L_P, L_Q, L_q, L_p, L_r \ge 0$ are independent of h.

We denote the set of admissible controls for **(OCP)** and **(DOCP)** by

$$\mathbf{U} := \left\{ u \in L_\infty^{n_u}([0, 1]) \mid u(t) \in U \text{ a.e. in } [0, 1] \right\},$$

$$\mathbf{U}_h := \left\{ u \in L_{\infty,h}^{n_u}([0, 1]) \mid u(t_i) \in U, \ i = 1, \ldots, N \right\},$$

and the feasible set of **(OCP)** and **(DOCP)** by

$$\mathbf{F} := \left\{ (x, u) \in X \;\middle|\; \begin{array}{l} \dot{x}(t) = A(t)\, x(t) + B(t)\, u(t), \text{ a.e. in } [0, 1], \\ x(0) = a, \\ u \in \mathbf{U} \end{array} \right\},$$

$$\mathbf{F}_h := \left\{ (x_h, u_h) \in X_h \;\middle|\; \begin{array}{l} x_h'(t_i) = A_h(t_i)\, x_h(t_i) + B_h(t_i)\, u_h(t_i), \ i = 1, \ldots, N, \\ x_h(t_0) = a, \\ u_h \in \mathbf{U}_h \end{array} \right\},$$

where $X := W_{1,\infty}^{n_x}([0, 1]) \times L_\infty^{n_u}([0, 1])$ and $X_h := W_{1,\infty,h}^{n_x}([0, 1]) \times L_{\infty,h}^{n_u}([0, 1])$. The set $\mathbf{F}$ is closed, convex, nonempty, and bounded, and the cost functional f is continuous and convex. Thus, there exists a minimizer $(\hat{x}, \hat{u}) \in W_{1,2}^{n_x}([0, 1]) \times L_2^{n_u}([0, 1])$ of **(OCP)**, according to [14, Chapter II, Proposition 1.2]. Moreover, since U is bounded, it holds $(\hat{x}, \hat{u}) \in X$. In addition, the compactness of $\mathbf{F}_h$ and the convexity and continuity of f_h imply that there exists a local minimizer $\left(\hat{x}_h, \hat{u}_h\right) \in X_h$ for (DOCP).

Now, we show that feasible trajectories and the objective functions of (OCP) and (DOCP) satisfy uniform Lipschitz conditions. To this end, let $(x, u) \in \mathbf{F}$ and $(x_h, u_h) \in \mathbf{F}_h$ be arbitrary. Since $\mathbf{U}$ and $\mathbf{U}_h$ are bounded, there exists a constant $c_u \geq 0$ such that $\|u\|_\infty \leq c_u$, $\|u_h\|_\infty \leq c_u$. Additionally, using

$$x(t) = \Phi_A(t) \left(a + \int_0^t \Phi_A(\tau)^{-1} B(\tau) u(\tau) \, d\tau \right), \qquad t \in [0, 1],$$

(2.1)

$$x_h(t_i) = \Psi_{A_h}(t_i) \left(a + h \sum_{k=1}^i \Psi_{A_h}(t_{k-1})^{-1} B_h(t_k) u_h(t_k) \right), \quad i = 1, \ldots, N,$$

(2.2)

where $\Phi_A(\cdot)$ and $\Psi_{A_h}(\cdot)$ denote the solutions of

$$\dot{\Phi}(t) = A(t)\,\Phi(t), \qquad \text{a.e. in } [0, 1], \qquad \Phi(0) = E, \qquad (2.3)$$

$$\Psi'(t_i) = A_h(t_i)\,\Psi(t_i), \qquad i = 1, \ldots, N, \qquad \Psi(t_0) = E, \qquad (2.4)$$

and $x_h(t) = x_h'(t_i)\left(t - t_{i-1}\right) + x_h(t_{i-1}) = \left(A_h(t_i)\,x_h(t_i) + B_h(t_i)\,u_h(t_i)\right)\left(t - t_{i-1}\right)$ $+ x_h(t_{i-1})$ for $t \in \left(t_{i-1}, t_i\right]$, $i = 1, \ldots, N$, we obtain a constant $c_x \geq 0$ such that $\|x\|_\infty \leq c_x$, $\|x_h\|_\infty \leq c_x$. Thus, exploiting the differential and difference equation yields a constant $L_x \geq 0$ with $\|\dot{x}\|_\infty \leq L_x$, $\left\|x_h'\right\|_\infty \leq L_x$. Summarizing, we have

$$\|u\|_\infty \leq c_u, \quad \|x\|_\infty \leq c_x, \quad \|\dot{x}\|_\infty \leq L_x, \text{ for all } (x, u) \in \mathbf{F},$$
$$\|u_h\|_\infty \leq c_u, \quad \|x_h\|_\infty \leq c_x, \quad \left\|x_h'\right\|_\infty \leq L_x, \text{ for all } (x_h, u_h) \in \mathbf{F}_h,$$

(2.5)

where c_u, c_x, L_x are independent of (x, u), (x_h, u_h), and h. Hence, the admissible trajectories for $\mathbf{F}$ and $\mathbf{F}_h$ are uniformly Lipschitz continuous with Lipschitz modulus L_x. Next, we derive Lipschitz conditions for the cost functionals f and f_h. To this end, let (x, u), $(z, v) \in \mathbf{F} \cup \mathbf{F}_h$ be arbitrary. Then,

$$\left| f(x, u) - f(z, v) \right| \leq \left| \frac{1}{2} x(1)^\top Q\, x(1) - \frac{1}{2} z(1)^\top Q\, z(1) \right| + \left| q^\top \left(x(1) - z(1) \right) \right|$$

$$+ \int_0^1 \left| \frac{1}{2} x(t)^\top P(t)\, x(t) - \frac{1}{2} z(t)^\top P(t)\, z(t) \right|$$

$$+ \left| p(t)^\top \left(x(t) - z(t) \right) \right| + \left| r(t)^\top \left(u(t) - v(t) \right) \right| \, dt$$

$$\leq \left(c_x \, \|Q\| + |q| + c_x \, \|P\|_\infty + \|p\|_\infty \right) \|x - z\|_\infty + \|r\|_\infty \, \|u - v\|_1 \, .$$

In the same way, we find a bound for f_h, hence we obtain

$$\left| f\left(x, u \right) - f\left(z, v \right) \right| \leq L_f \left(\|x - z\|_\infty + \|u - v\|_1 \right), \tag{2.6}$$

$$\left| f_h \left(x, u \right) - f_h \left(z, v \right) \right| \leq L_f \left(\|x - z\|_\infty + \|u - v\|_1 \right),$$

where $L_f \geq 0$ is independent of h.

Let $(\hat{x}, \hat{u}) \in W_{1,\infty}^{n_x} \left([0, 1] \right) \times L_\infty^{n_u} \left([0, 1] \right)$ and $(\hat{x}_h, \hat{u}_h) \in W_{1,\infty,h}^{n_x} \left([0, 1] \right) \times L_{\infty,h}^{n_u} \left([0, 1] \right)$ be minimizers of (OCP) and (DOCP), respectively. Then, there exists a multipliers $\lambda \in W_{1,\infty}^{n_x} \left([0, 1] \right)$ and $\lambda_h \in W_{1,\infty,h}^{n_x} \left([0, 1] \right)$ such that the following necessary conditions hold:

$$\dot{\lambda}(t) = -A(t)^\top \lambda(t) - P(t)\,\hat{x}(t) - p(t) \qquad\qquad \text{in } [0, 1]$$

$$\lambda(1) = Q\,\hat{x}(1) + q$$

$$0 \leq \left[r(t)^\top + \lambda(t)^\top B(t) \right] \left(u - \hat{u}(t) \right) \quad \text{for all } u \in U, \qquad \text{a.e. in } [0, 1]$$

and

$$\lambda_h(t_i)' = -A_h(t_i)^\top \lambda_h(t_{i-1}) - P_h(t_i)\,\hat{x}_h(t_i) - p_h(t_i) \qquad\qquad i = 1, \ldots, N$$

$$\lambda_h(t_N) = Q_h\,\hat{x}_h(t_N) + q_h$$

$$0 \leq \left[r_h(t_i)^\top + \lambda_h(t_{i-1})^\top B_h(t_i) \right] \left(u - \hat{u}_h(t_i) \right), \quad \text{for all } u \in U, \quad i = 1, \ldots, N.$$

We denote the respective switching functions by

$$\sigma(t) := r(t) + B(t)^\top \lambda(t), \qquad\qquad t \in [0, 1], \tag{2.7}$$

$$\sigma_h(t_i) := r_h(t_i) + B_h(t_i)^\top \lambda_h(t_{i-1}), \qquad i = 1, \ldots, N. \tag{2.8}$$

Since $r, r_h, B, B_h, \lambda, \lambda_h$ are Lipschitz continuous, the switching functions σ and σ_h are also Lipschitz continuous. In addition, using the local minimum principles

$$0 \leq \left[r(t)^\top + \lambda(t)^\top B(t) \right] \left(u - \hat{u}(t) \right), \qquad\qquad \text{a.e. in } [0, 1] \tag{2.9}$$

$$0 \leq \left[r_h(t_i)^\top + \lambda_h(t_{i-1})^\top B_h(t_i) \right] \left(u - \hat{u}_h(t) \right), \qquad i = 1, \ldots, N \tag{2.10}$$

for all $u \in U$, we obtain the well known structure of the optimal controls

$$
\hat{u}_j(t) = \begin{cases} b_{l,j}, & \text{if } \sigma_j(t) > 0 \\ b_{u,j}, & \text{if } \sigma_j(t) < 0 \\ \text{undetermined,} & \text{if } \sigma_j(t) = 0 \end{cases}, \quad \hat{u}_{h,j}(t_i) = \begin{cases} b_{l,j}, & \text{if } \sigma_{h,j}(t_i) > 0 \\ b_{u,j}, & \text{if } \sigma_{h,j}(t_i) < 0 \\ \text{undetermined,} & \text{if } \sigma_{h,j}(t_i) = 0 \end{cases}.
$$

$$\tag{2.11}$$

3 Error Analysis for Optimal Values

In this section, we aim to derive error estimates of order one for optimal values. To this end, we assume the control u to have bounded variation, and obtain the following:

Theorem 3.1 *Let (A1), (A2) be satisfied and let $(\hat{x}, \hat{u}) \in F$ be a minimizer of (OCP) with $\hat{u} \in BV^{n_u}\left([0, 1]\right)$. Then, for all solutions $(\hat{x}_h, \hat{u}_h) \in F_h$ of (DOCP) it holds*

$$
\left| f_h\left(\hat{x}_h, \hat{u}_h\right) - f(\hat{x}, \hat{u}) \right| \le c h,
$$

where $c \ge 0$ is independent of h and the choice of $(\hat{x}_h, \hat{u}_h) \in F_h$.

In order to prove Theorem 3.1, we require some intermediary results. First, we show the following estimation result for feasible $(x, u) \in F$ with $u \in BV^{n_u}\left([0, 1]\right)$:

Lemma 3.2 *Let (A1), (A2) be satisfied and let $(x, u) \in F$ with $u \in BV^{n_u}\left([0, 1]\right)$. Then, there exists $(x_h, u_h) \in F_h$ such that*

$$
\|u - u_h\|_1 \le h \bigvee_0^1 u, \quad \|u - u_h\|_2 \le \sqrt{h} \bigvee_0^1 u, \quad \|x - x_h\|_\infty \le c h,
$$

where $c \ge 0$ is independent of h.

Proof For arbitrary $(x, u) \in F$, where $u \in BV^{n_u}\left([0, 1]\right)$, let the piecewise constant function u_h be defined by $u_h(t) := u(t_i)$ for $t \in \left(t_{i-1}, t_i\right]$ and $i = 1, \ldots, N$. This implies $u_h \in U_h$. For $i = 1, \ldots, N$ and $t \in \left(t_{i-1}, t_i\right]$ it holds $\left|u(t) - u(t_i)\right| \le \left|u(t) - u(t_{i-1})\right| + \left|u(t_i) - u(t_{i-1})\right| \le \bigvee_{t_{i-1}}^{t_i} u$. Thus, we have

$$
\|u - u_h\|_1 = \sum_{i=1}^N \int_{t_{i-1}}^{t_i} \left|u(t) - u(t_i)\right| dt \le \sum_{i=1}^N \int_{t_{i-1}}^{t_i} \bigvee_{t_{i-1}}^{t_i} u \, dt \le h \bigvee_0^1 u,
$$

and for the L_2-norm we obtain

$$\|u - u_h\|_2^2 = \sum_{i=1}^{N} \int_{t_{i-1}}^{t_i} |u(t) - u(t_i)|^2 \, dt \le \sum_{i=1}^{N} \int_{t_{i-1}}^{t_i} \left(\bigvee_{t_{i-1}}^{t_i} u \right)^2 dt \le h \left(\bigvee_{0}^{1} u \right)^2.$$

For the chosen u_h, let x_h satisfy the difference equation

$$x_h'(t_i) = A_h(t_i) x_h(t_i) + B_h(t_i) u_h(t_i), \quad i = 1, \ldots, N, \quad x_h(t_0) = a,$$

hence $(x_h, u_h) \in F_h$. Furthermore, let $\Phi_A : [0, 1] \to \mathbb{R}^{n_x \times n_x}$ and $\Psi_{A_h} : [0, 1] \to \mathbb{R}^{n_x \times n_x}$ solve (2.3) and (2.4), respectively. According to [22, Lemma 2.4.7], assumption **(A2)** implies

$$\left\| \Phi_A - \Psi_{A_h} \right\|_\infty \le L_\Phi h$$

for some $L_\Phi \ge 0$ independent of h. Moreover, (2.1) and (2.2) hold, and therefore for $t \in (t_{i-1}, t_i], i = 1, \ldots, N$

$$|x(t) - x_h(t)| \le |x(t) - x_h(t_i)| + |x_h(t_i) - x_h(t)| \le \left(L_\Phi |a| + c_1 + c_2 \bigvee_{0}^{1} u + L_x \right) h,$$

which proves the assertion. $\qquad \square$

Note that in many applications the optimal control $\hat{u}$ is piecewise Lipschitz continuous, which implies $\hat{u} \in BV^{n_u}\left([0, 1] \right)$.

Lemma 3.3 *Let (A1), (A2) be satisfied and let $(x_h, u_h) \in F_h$. Then, there exists a function z such that $(z, u_h) \in F$ and*

$$\|z - x_h\|_\infty \le c h,$$

where $c \ge 0$ is independent of h and the choice of $(x_h, u_h) \in F_h$.

Proof Since $(x_h, u_h) \in F_h$, we have $u_h \in U_h \subset U$. Let z satisfy the initial value problem

$$\dot{z}(t) = A(t) z(t) + B(t) u_h(t), \quad \text{a.e. in } [0, 1], \quad z(0) = a,$$

where u_h is piecewise constant. Thus, $(z, u_h) \in F$ and furthermore, x_h solves the perturbed differential equation

$$\dot{x}_h(t) = A(t) x_h(t) + B(t) u_h(t) + \omega(t), \quad \text{a.e. in } [0, 1], \quad x_h(0) = a,$$

with perturbation $\omega(t) := A_h(t_i) x_h(t_i) - A(t) x_h(t) + \big(B_h(t_i) - B(t)\big) u_h(t)$ for $t \in (t_{i-1}, t_i]$ and $i = 1, \ldots, N$. Since x_h is uniformly Lipschitz continuous, u_h is uniformly bounded, and **(A2)** is satisfied, there exists $c_1 \geq 0$ independent of h and (x_h, u_h) such that $|\omega(t)| \leq c_1 h$ for almost every $t \in [0, 1]$. Hence, using (2.1) we find a constant $c_2 \geq 0$ such that

$$\|z - x_h\|_\infty \leq c_2 \|\omega\|_\infty \leq c_1 c_2 h,$$

which completes the proof.   □

Next, we show that the functions f and f_h satisfy a linear error estimate for all $(x_h, u_h) \in \mathbf{F}_h$.

Lemma 3.4 *Let (A1), (A2) be satisfied and let $(x_h, u_h) \in \mathbf{F}_h$. Then, we have*

$$\big| f(x_h, u_h) - f_h(x_h, u_h) \big| \leq c h,$$

where $c \geq 0$ is independent of h and the choice of $(x_h, u_h) \in \mathbf{F}_h$.

Proof We recall (compare (2.5)) there exist constants $c_x, c_u, L_x \geq 0$ independent of h such that $\|x_h\|_\infty \leq c_x$, $\|u_h\|_\infty \leq c_u$, and $\|\dot{x}_h\|_\infty \leq L_x$ for all $(x_h, u_h) \in \mathbf{F}_h$. Moreover, it holds

$$f(x_h, u_h) - f_h(x_h, u_h) = \frac{1}{2} x_h(t_N)^\top (Q - Q_h) x_h(t_N) + (q - q_h)^\top x_h(t_N)$$

$$+ \sum_{i=1}^{N} \int_{t_{i-1}}^{t_i} \frac{1}{2} x_h(t)^\top P(t) x_h(t) - \frac{1}{2} x_h(t_i)^\top P_h(t_i) x_h(t_i)$$

$$+ p(t)^\top x_h(t) - p(t_i)^\top x_h(t_i) + \big(r(t) - r_h(t_i)\big)^\top u_h(t_i) \, dt.$$

Since we have

$$\xi(t) := x_h(t)^\top P(t) x_h(t) - x_h(t_i)^\top P_h(t_i) x_h(t_i) = \big(x_h(t) + x_h(t_i)\big)^\top P(t) \big(x_h(t) - x_h(t_i)\big)$$

$$+ x_h(t_i)^\top \big(P(t) - P_h(t_i)\big) x_h(t_i),$$

using (2.5) and **(A2)** we obtain $|\xi(t)| \leq 2 c_x \|P\|_\infty L_x h + c_x^2 L_P h$. Similarly we find bounds for the remaining terms, which proves the assertion.   □

Now, using the results of Lemma 3.2, 3.3, and 3.4, we are able to prove Theorem 3.1:

Proof of Theorem 3.1 By Lemma 3.2, there exists $(x_h, u_h) \in \mathbf{F}_h$ such that $\|x_h - \hat{x}\|_\infty \leq c_1 h$, $\|u_h - \hat{u}\|_1 \leq c_2 h$, where $c_1, c_2 \geq 0$ are independent of h.

Then, for an arbitrary solution $(\hat{x}_h, \hat{u}_h) \in \mathbf{F}_h$ of **(DOCP)** it holds $f_h(\hat{x}_h, \hat{u}_h) \leq f_h(x_h, u_h)$. This implies

$$0 \leq f_h(x_h, u_h) - f_h(\hat{x}_h, \hat{u}_h) = f_h(x_h, u_h) - f(\hat{x}, \hat{u}) + f(\hat{x}, \hat{u}) - f_h(\hat{x}_h, \hat{u}_h),$$

and thus

$$f_h(\hat{x}_h, \hat{u}_h) - f(\hat{x}, \hat{u}) \leq f_h(x_h, u_h) - f(\hat{x}, \hat{u})$$
$$= f_h(x_h, u_h) - f(x_h, u_h) + f(x_h, u_h) - f(\hat{x}, \hat{u}).$$

By Lemma 3.4 and (2.6) this yields $f_h(\hat{x}_h, \hat{u}_h) - f(\hat{x}, \hat{u}) \leq c_3 h + L_f c_4 h$. Furthermore, Lemma 3.3 implies the existence of $\hat{z}$ such that $(\hat{z}, \hat{u}_h) \in \mathbf{F}$ and $\|\hat{z} - \hat{x}_h\|_\infty \leq c_5 h$. Additionally, we have $f(\hat{x}, \hat{u}) \leq f(\hat{z}, \hat{u}_h)$, and therefore

$$0 \leq f(\hat{z}, \hat{u}_h) - f(\hat{x}, \hat{u}) = f(\hat{z}, \hat{u}_h) - f_h(\hat{x}_h, \hat{u}_h) + f_h(\hat{x}_h, \hat{u}_h) - f(\hat{x}, \hat{u}).$$

This gives us

$$f(\hat{x}, \hat{u}) - f_h(\hat{x}_h, \hat{u}_h) \leq f(\hat{z}, \hat{u}_h) - f_h(\hat{x}_h, \hat{u}_h)$$
$$= f(\hat{z}, \hat{u}_h) - f(\hat{x}_h, \hat{u}_h) + f(\hat{x}_h, \hat{u}_h) - f_h(\hat{x}_h, \hat{u}_h).$$

Using (2.6) and Lemma 3.4 we obtain $f(\hat{x}, \hat{u}) - f_h(\hat{x}_h, \hat{u}_h) \leq L_f c_5 h + c_3 h$. Hence,

$$\left| f(\hat{x}, \hat{u}) - f_h(\hat{x}_h, \hat{u}_h) \right| \leq \left(L_f (c_4 + c_5) + c_3 \right) h,$$

which completes the proof. $\qquad\square$

Note that the constant c in Theorem 3.1 is independent of h, but depends on the total variation of $\hat{u}$ (compare Lemma 3.2).

4 Error Analysis for Problems with Bang–Bang Solutions

In the sequel, we introduce conditions which guarantee that the optimal control is of bang–bang type (cf. [7, 8]). The control u is of bang–bang type, if its values are on the boundary of the control set U for almost every $t \in [0, 1]$, that is, $u_j(t) \in \{b_{l,j}, b_{u,j}\}$ for all $j = 1, \ldots, n_u$ and almost every $t \in [0, 1]$. The isolated time points where some component of u switches from the upper bound to the lower bound or vice versa are called switching times. The structure of the optimal control is determined by the switching function (compare (2.11)). Additionally, the following conditions imply that an optimality condition is satisfied for **(OCP)** (see Theorem 4.2).

(A3) There exists a solution $(\hat{x}, \hat{u}) \in \mathbf{F}$ of **(OCP)** such that the set Σ of zeros of the components σ_j, $j = 1, \ldots, n_u$, of the switching function is finite and $0, 1 \notin \Sigma$, i.e., we have $\Sigma = \{s_1, \ldots, s_\ell\}$ with $0 < s_1 < \ldots < s_\ell < 1$.

Note that, if $0, 1 \notin \Sigma$ holds, for sufficiently small h we have $t_2 < s_1$ and $s_\ell < t_{N-1}$. Let $J(s_i) := \{1 \le j \le n_u \mid \sigma_j(s_i) = 0\}$ denote the set of active indices of the switching function. In order to obtain a bang–bang type optimal control, we require the switching function to satisfy a certain stability condition around its zeros:

(A4) There exist $\varsigma > 0$, $\varrho > 0$ such that $\left|\sigma_j(\tau)\right| \ge \varsigma \left|\tau - s_i\right|$ for all $i = 1, \ldots, \ell$, $j \in J(s_i)$, and every $\tau \in \left[s_i - \varrho, s_i + \varrho\right]$, and σ_j changes sign in s_i, i.e., $\sigma_j(s_i - \varrho)\,\sigma_j(s_i + \varrho) < 0$.

With these assumptions we are able to obtain the following error estimates:

Theorem 4.1 *Let $(\hat{x}, \hat{u})$ be a minimizer for* **(OCP)** *and let* **(A1)–(A4)** *be satisfied. Then, for sufficiently small h, any solution $(\hat{x}_h, \hat{u}_h) \in \mathbf{F}_h$ of* **(DOCP)***, the associated multiplier, and the switching function can be estimated by*

$$\left\|\hat{u}_h - \hat{u}\right\|_1 \le c\sqrt{h}, \qquad\qquad \left\|\hat{x}_h - \hat{x}\right\|_\infty \le c\sqrt{h},$$

$$\left\|\lambda_h - \lambda\right\|_\infty \le c\sqrt{h}, \qquad \max_{t \in [t_1, t_N]}\left|\sigma_h(t) - \sigma(t)\right| \le c\sqrt{h}, \tag{4.1}$$

where $c \ge 0$ is independent of h. Moreover, there exists a constant κ independent of h such that for sufficiently small h any discrete optimal control $\hat{u}_h$ coincides with $\hat{u}$ except on a set of measure smaller than $\kappa\sqrt{h}$.

The rest of Sect. 4 is dedicated to the proof of this result. To this end, we first introduce the following optimality result for solutions of **(OCP)**. In [7, Theorem 4.2]) it was shown, that assumptions **(A3)** and **(A4)** are sufficient for a quadratic minorant for optimal values of **(OCP)** in a sufficiently small L_1-neighborhood of the optimal control. Additionally, for trajectories outside that neighborhood a linear minorant is satisfied:

Theorem 4.2 ([7, Theorem 4.2]) *Let $(\hat{x}, \hat{u})$ be a minimizer for* **(OCP)***. If* **(A1)***,* **(A3)***, and* **(A4)** *are satisfied, then there exist constants $\alpha, \beta, \tilde{\varepsilon} > 0$ such that for every $(x, u) \in \mathbf{F}$ it holds,*

$$f(x, u) - f(\hat{x}, \hat{u}) \ge \alpha\left(\left\|u - \hat{u}\right\|_1^2 + \left\|x - \hat{x}\right\|_{1,1}^2\right) \ge \alpha\left\|u - \hat{u}\right\|_1^2 \quad if \ \left\|u - \hat{u}\right\|_1 \le 2\beta\tilde{\varepsilon},$$

$$f(x, u) - f(\hat{x}, \hat{u}) \ge \alpha\left(\left\|u - \hat{u}\right\|_1 + \left\|x - \hat{x}\right\|_{1,1}\right) \ge \alpha\left\|u - \hat{u}\right\|_1 \quad if \ \left\|u - \hat{u}\right\|_1 > 2\beta\tilde{\varepsilon}.$$

Theorem 4.2 implies uniqueness of $(\hat{x}, \hat{u})$ for **(OCP)** (cf. [15, Theorem 2.2]), since for any solution $(x, u) \in \mathbf{F}$ we have $f(x, u) = f(\hat{x}, \hat{u})$. Hence, $0 = \|u - \hat{u}\|_1 + \|x - \hat{x}\|_{1,1}$, which implies $(x, u) = (\hat{x}, \hat{u})$.

Next, we prove that, if the switching function satisfies an error estimate of order h^ν for a constant $0 < \nu \le 1$, then the optimal controls coincide (are of bang–bang type) except on a set of measure of order h^ν. If the error estimate is satisfied for $\nu > 1$, then the set has measure of order h (cf. [7, Theorem 4.5]). To that end, for $0 < \varepsilon \le \varrho$ we define

$$I(\varepsilon) := \bigcup_{1 \le i \le \ell} [s_i - \varepsilon, s_i + \varepsilon].$$

For $j = 1, \ldots, n_u$ we denote the set of zeros of the component σ_j by $\Sigma_j := \{\tau_1, \ldots, \tau_{\ell_j}\} \subset \Sigma$ with $0 < \tau_1 < \ldots < \tau_{\ell_j} < 1$, and

$$I_j^-(\varepsilon) := \bigcup_{i=1,\ldots,\ell_j} [\tau_i - \varepsilon, \tau_i + \varepsilon], \quad I_j^+(\varepsilon) := [0, 1] \setminus I_j^-(\varepsilon) \tag{4.2}$$

Since σ_j is Lipschitz continuous, **(A4)** implies

$$0 < \sigma_{j,\min} = \min_{t \in \overline{I_j^+(\varrho)}} |\sigma_j(t)|, \tag{4.3}$$

where $\overline{I_j^+(\varrho)}$ denotes the closure of $I_j^+(\varrho)$.

Theorem 4.3 *Let assumptions (A1)–(A4) be satisfied. Suppose for sufficiently small h it holds*

$$\max_{t \in [t_1, t_N]} |\sigma_h(t) - \sigma(t)| \le c_\sigma h^\nu \tag{4.4}$$

with constant $c_\sigma \ge 0$ independent of h and $\nu > 0$. Then, there exists a constant $\kappa \ge 0$ independent of h such that for sufficiently small h any discrete optimal control $\hat{u}_h$ coincides with $\hat{u}$ except on a set of measure smaller than κh^ν for $0 < \nu \le 1$ and κh for $\nu > 1$.

Proof Let $j \in \{1, \ldots, n_u\}$ be arbitrary, and let $\sigma_{j,\min}$ be defined by (4.3). Then, for sufficiently small h, (4.4) implies

$$|\sigma_{h,j}(t)| \ge |\sigma_j(t)| - c_\sigma h^\nu \ge \sigma_{j,\min} - c_\sigma h^\nu \ge \frac{1}{2} \sigma_{j,\min} > 0 \quad \text{for all } t \in I_j^+(\varrho).$$

Moreover, for $i = 1, \ldots, \ell_j$, $\tau_i \in \Sigma_j$, and $\tau \in [\tau_i - \varrho, \tau_i + \varrho]$, according to **(A4)** and (4.4), it holds

$$\left|\sigma_{h,j}(\tau)\right| \geq \left|\sigma_j(\tau)\right| - c_\sigma \, h^\nu \geq \varsigma \, |\tau - \tau_i| - c_\sigma \, h^\nu. \tag{4.5}$$

Hence, we have $\sigma_{h,j}(\tau) \neq 0$, if $|\tau - \tau_i| > \frac{c_\sigma}{\varsigma} h^\nu$. For each zero $\tau_i \in \Sigma_j$ we aim to construct natural numbers $k_i^-, k_i^+ \in \{1, \ldots, N\}$, $k_i^- < k_i^+$ such that

$$\tau_i \in \left[t_{k_i^-}, t_{k_i^+}\right] \subset [\tau_i - \varrho, \tau_i + \varrho], \quad |\tau - \tau_i| > \frac{c_\sigma}{\varsigma} h^\nu \text{ for all } \tau \in \left[\tau_i - \varrho, t_{k_i^-}\right] \cup \left[t_{k_i^+}, \tau_i + \varrho\right].$$

To this end, for sufficiently small h we choose numbers $\iota \in \{2, \ldots, N-2\}$ and $k \in \mathbb{N}$ such that $\iota - k, \iota + k + 1 \in \{1, \ldots, N\}$, $\tau_i \in [t_\iota, t_{\iota+1}]$, and

$$\frac{c_\sigma}{\varsigma} h^{\nu-1} < k \leq \frac{c_\sigma}{\varsigma} h^{\nu-1} + 1. \tag{4.6}$$

We denote $k_i^- := \iota - k$, $k_i^+ := \iota + k + 1$. Then, it holds

$$t_{k_i^+} - t_{k_i^-} = (2k+1) \, h \leq \left(2\left(\frac{c_\sigma}{\varsigma} h^{\nu-1} + 1\right) + 1\right) h = \left(2\frac{c_\sigma}{\varsigma} h^{\nu-1} + 3\right) h. \tag{4.7}$$

For $0 < \nu \leq 1$ we conclude

$$\left(2\frac{c_\sigma}{\varsigma} h^{\nu-1} + 3\right) h = \left(2\frac{c_\sigma}{\varsigma} + 3 h^{1-\nu}\right) h^\nu \leq \left(2\frac{c_\sigma}{\varsigma} + 3\right) h^\nu,$$

and for $\nu > 1$

$$\left(2\frac{c_\sigma}{\varsigma} h^{\nu-1} + 3\right) h \leq \left(2\frac{c_\sigma}{\varsigma} + 3\right) h.$$

For sufficiently small h this implies $\tau_i \in \left[t_{k_i^-}, t_{k_i^+}\right] \subset [\tau_i - \varrho, \tau_i + \varrho]$. In addition, with (4.6) we obtain

$$t_{k_i^+} - \tau_i = t_{\iota+k+1} - \tau_i \geq t_{\iota+k+1} - t_{\iota+1} = k \, h > \frac{c_\sigma}{\varsigma} h^{\nu-1} h = \frac{c_\sigma}{\varsigma} h^\nu,$$

$$\tau_i - t_{k_i^-} = \tau_i - t_{\iota-k} \geq t_\iota - t_{\iota-k} = k \, h > \frac{c_\sigma}{\varsigma} h^{\nu-1} h = \frac{c_\sigma}{\varsigma} h^\nu.$$

Thus, $|\tau - \tau_i| > \frac{c_\sigma}{\varsigma} h^\nu$ for all $\tau \in \left[\tau_i - \varrho, t_{k_i^-}\right] \cup \left[t_{k_i^+}, \tau_i + \varrho\right]$. Using this and (4.5) we conclude $\left|\sigma_{h,j}(t)\right| > 0$ on $\left[\tau_i - \varrho, t_{k_i^-}\right] \cup \left[t_{k_i^+}, \tau_i + \varrho\right]$. We denote

$$I_j^- := \bigcup_{i=1,\dots,\ell_j} \left[t_{k_i^-}, t_{k_i^+}\right] \subset I_j^-(\varrho), \quad I_j^+ := [0, 1] \setminus I_j^- \supset I_j^+(\varrho). \tag{4.8}$$

We have shown that

$$\left|\sigma_{h,j}(t)\right| > 0 \quad \text{on } I_j^+ \supset I_j^+(\varrho), \tag{4.9}$$

which implies $\hat{u}_{h,j}(t) = \hat{u}_j(t)$ for all $t \in I_j^+$ for any discrete optimal control $\hat{u}_h$. Hence, the components of the optimal controls coincide except on I_j^-, which has measure smaller than $\ell_j \left(2\frac{c_\sigma}{\varsigma} + 3\right) h^\nu$ for $0 < \nu \leq 1$ and $\ell_j \left(2\frac{c_\sigma}{\varsigma} + 3\right) h$ for $\nu > 1$. This proves the assertion for $\kappa := \sum_{j=1}^{n_u} \ell_j \left(2\frac{c_\sigma}{\varsigma} + 3\right)$. $\qquad\square$

Now, we are able to prove the main result of this section:

Proof of Theorem 4.1 First, we derive Hölder type estimates for the optimal controls using the minorants in Theorem 4.2 (cf. [1, 25]). Let $(\hat{x}, \hat{u})$ be a solution of **(OCP)** and let $(\hat{x}_h, \hat{u}_h)$ be a solution of **(DOCP)**. According to Lemma 3.3, there exists $\hat{z}$ with $(\hat{z}, \hat{u}_h) \in \mathbf{F}$ and

$$\|\hat{z} - \hat{x}_h\|_\infty \leq c_1 h \tag{4.10}$$

for $c_1 \geq 0$ independent of h. Then, by Theorem 4.2, for $\alpha > 0$ independent of h we have,

$$f(\hat{z}, \hat{u}_h) - f(\hat{x}, \hat{u}) \geq \alpha \left(\|\hat{u}_h - \hat{u}\|_1^2 + \|\hat{z} - \hat{x}\|_{1,1}^2\right) \quad \text{if } \|\hat{u}_h - \hat{u}\|_1 \leq 2\beta\tilde{\varepsilon}, \tag{4.11}$$

$$f(\hat{z}, \hat{u}_h) - f(\hat{x}, \hat{u}) \geq \alpha \left(\|\hat{u}_h - \hat{u}\|_1 + \|\hat{z} - \hat{x}\|_{1,1}\right) \quad \text{if } \|\hat{u}_h - \hat{u}\|_1 > 2\beta\tilde{\varepsilon}. \tag{4.12}$$

Using (2.6), Lemma 3.4, and Theorem 3.1 we obtain

$$\begin{aligned}
f(\hat{z}, \hat{u}_h) - f(\hat{x}, \hat{u}) &= f(\hat{z}, \hat{u}_h) - f(\hat{x}_h, \hat{u}_h) + f(\hat{x}_h, \hat{u}_h) \\
&\quad - f_h(\hat{x}_h, \hat{u}_h) + f_h(\hat{x}_h, \hat{u}_h) - f(\hat{x}, \hat{u}) \\
&\leq L_f \|\hat{z} - \hat{x}_h\|_\infty + c_2 h + c_3 h \leq c_4 h.
\end{aligned}$$

Then, (4.11) and (4.12) imply

$$\left\| \hat{u}_h - \hat{u} \right\|_1 \le c_5 \max \left\{ h, \sqrt{h} \right\}, \tag{4.13}$$

and using the Sobolev inequality $\|v\|_\infty \le c_6 \|v\|_{1,1}$ for all $v \in W_{1,1}^n \left([0,1] \right)$ we obtain

$$\left\| \hat{z} - \hat{x} \right\|_\infty \le c_6 \left\| \hat{z} - \hat{x} \right\|_{1,1} \le c_7 \max \left\{ h, \sqrt{h} \right\}. \tag{4.14}$$

Thus, for sufficiently small h, it holds $\left\| \hat{u}_h - \hat{u} \right\|_1 \le 2 \beta \tilde{\varepsilon}$. Moreover, (4.10) and (4.14) yield

$$\left\| \hat{x}_h - \hat{x} \right\|_\infty \le \left\| \hat{x}_h - \hat{z} \right\|_\infty + \left\| \hat{z} - \hat{x} \right\|_\infty \le c_1 h + c_7 \sqrt{h} \le c_8 \sqrt{h} \tag{4.15}$$

for sufficiently small h. In conclusion, by (4.13) and (4.15), we obtain the following error estimates

$$\left\| \hat{u}_h - \hat{u} \right\|_1 \le c_1 \sqrt{h}, \quad \left\| \hat{x}_h - \hat{x} \right\|_\infty \le c_2 \sqrt{h} \tag{4.16}$$

for sufficiently small h. Next, we show the bound for the multipliers. To that end, we consider the adjoint equations

$$\dot{\lambda}(t) = -A(t)^\top \lambda(t) - b(t), \qquad \text{in } [0,1] \qquad \lambda(1) = \chi,$$

$$\lambda_h'(t_i) = -A_h(t_i)^\top \lambda_h(t_{i-1}) - b_h(t_i), \qquad i = 1, \ldots, N \qquad \lambda_h(t_N) = \chi_h,$$

where

$$b(t) = P(t)\,\hat{x}(t) + p(t), \qquad t \in [0,1] \qquad \chi = Q\,\hat{x}(1) + q,$$

$$b_h(t_i) = P_h(t_i)\,\hat{x}_h(t_i) + p_h(t_i), \qquad i = 1, \ldots, N \qquad \chi_h = Q_h\,\hat{x}_h(t_N) + q_h.$$

These equations have the solutions

$$\lambda(t) = \left(\Phi_A(t)^\top \right)^{-1} \left(\Phi_A(1)^\top \chi + \int_t^1 \Phi_A(\tau)^\top b(\tau)\,d\tau \right), \qquad t \in [0,1],$$

$$\lambda_h(t_i) = \left(\Psi_{A_h}(t_i)^\top \right)^{-1} \left(\Psi_{A_h}(t_N)^\top \chi_h + h \sum_{k=i+1}^N \Psi_{A_h}(t_k)^\top b_h(t_k) \right), \qquad i = 1, \ldots, N.$$

Then, using [22, Lemma 2.4.7], **(A2)**, and (4.16) we find a constant $c_3 \geq 0$ independent of h such that

$$\|\lambda_h - \lambda\|_\infty \leq c_3 \sqrt{h}.$$

This and **(A2)** immediately imply

$$\max_{t \in [t_1, t_N]} |\sigma_h(t) - \sigma(t)| \leq c \sqrt{h}$$

for the switching functions defined in (2.7), (2.8) (cf. [8, Theorem 2.3]). Finally, applying Theorem 4.3 for $\nu = \frac{1}{2}$ proves the assertion. □

5 Improved Error Estimates

According to (4.9), for $j = 1, \ldots, n_u$ the component $\sigma_{h,j}$ has no zero in I_j^+ (see (4.2) and (4.8)), and by (4.4), at least one zero in each interval $\left[t_{k_i^-}, t_{k_i^+}\right]$. By slightly strengthening the condition **(A4)**, we can show that the zero in $\left[t_{k_i^-}, t_{k_i^+}\right]$ is unique. Thus, the optimal controls $\hat{u}$ and $\hat{u}_h$ have the same structure.

(A5) Let B, r be differentiable, $\dot{B}, \dot{r}$ be Lipschitz continuous, and for sufficiently small h let

$$\left\| \dot{B}(t_i) - B_h'(t_i) \right\| \leq L_{\dot{B}} h, \quad \left| \dot{r}(t_i) - r_h'(t_i) \right| \leq L_{\dot{r}} h$$

be satisfied for $i = 2, \ldots, N$. Furthermore, there exists $\varsigma > 0$ such that

$$\min_{1 \leq i \leq \ell} \min_{j \in J(s_i)} \left\{ \left| \dot{\sigma}_j(s_i) \right| \right\} \geq 2\varsigma.$$

Assumption **(A5)** allows us to improve the error estimates in Theorem 4.1 to linear order. In addition, the switching functions have the same structure and the zeros satisfy a linear error estimate.

Theorem 5.1 *Let $(\hat{x}, \hat{u})$ be a minimizer for* **(OCP)** *and let* **(A1)–(A3)**, **(A5)** *be satisfied. Then, for sufficiently small h, any solution $(\hat{x}_h, \hat{u}_h) \in F_h$ of* **(DOCP)**, *the associated multiplier, and the switching function can be estimated by*

$$\left\| \hat{u}_h - \hat{u} \right\|_1 \leq c\,h, \quad \left\| \hat{x}_h - \hat{x} \right\|_\infty \leq c\,h, \quad \left\| \lambda_h - \lambda \right\|_\infty \leq c\,h, \quad \left\| \sigma_h - \sigma \right\|_\infty \leq c\,h, \tag{5.1}$$

where $c \geq 0$ is independent of h. Moreover, there exist constants $\kappa, \vartheta \geq 0$ independent of h such that for sufficiently small h any discrete optimal control $\hat{u}_h$

*coincides with û except on a set of measure smaller than κ h, and the switching
functions σ and σ$_h$ have the same structure, i.e., the components of σ$_h$ have ℓ zeros
$0 < s_{h,1} < \ldots < s_{h,\ell} < 1$, which satisfy the estimate*

$$|s_i - s_{h,i}| \leq \vartheta \, h, \quad i = 1, \ldots, \ell.$$

Before we prove Theorem 5.1, we first show some results for the switching
functions, which follow from condition **(A5)**.

The derivative $\dot{\lambda}$ is Lipschitz continuous, since λ satisfies the adjoint equation

$$\dot{\lambda}(t) = -A(t)^\top \lambda(t) - P(t) \hat{x}(t) - p(t) \quad \text{in } \big[0, 1\big].$$

Therefore, **(A5)** and (2.7) imply that $\dot{\sigma}$ is also Lipschitz continuous. Thus, there
exists $\varrho > 0$ such that for all $i = 1, \ldots, \ell$ and $j \in J(s_i)$ it holds

$$|\dot{\sigma}_j(t)| \geq \varsigma \quad \text{on } \big[s_i - \varrho, s_i + \varrho\big]. \tag{5.2}$$

Moreover, using the mean value theorem we obtain for all $\tau \in \big[s_i - \varrho, s_i + \varrho\big]$

$$|\sigma_j(\tau)| = |\sigma_j(\tau) - \sigma_j(s_i)| = |\dot{\sigma}_j(\theta)(\tau - s_i)| \geq \varsigma \, |\tau - s_i|,$$

where $\theta \in (\tau, s_i)$, if $\tau \leq s_i$ and $\theta \in (s_i, \tau)$, if $\tau \geq s_i$. Hence, **(A5)** implies condition
(A4). Additionally, since it holds $\sigma_j(s_i - \varrho) \sigma_j(s_i + \varrho) < 0$, i.e., σ_j changes sign in
s_i, according to Theorem 4.1, for sufficiently small h it holds $\sigma_{h,j}(s_i - \varrho) \sigma_{h,j}(s_i +
\varrho) < 0$. The discrete switching function σ_h in (2.8) is differentiable on each interval
$\big(t_{i-1}, t_i\big]$ for $i = 2, \ldots, N$ with the derivative $\dot{\sigma}_h(t) := \frac{\sigma_h(t_i) - \sigma_h(t_{i-1})}{h}$. In addition,
if **(A1)**–**(A3)**, and **(A5)** are satisfied, for sufficiently small h we have the following
error estimate (cf. [8, Theorem 2.6]):

$$\max_{t \in [t_1, t_N]} |\dot{\sigma}_h(t) - \dot{\sigma}(t)| \leq c \sqrt{h}, \tag{5.3}$$

where $c \geq 0$ is independent of h. Then, by (5.2) and (5.3), for arbitrary $i = 1, \ldots, \ell$,
$j \in J(s_i)$, and sufficiently small h it holds

$$|\dot{\sigma}_{h,j}(\tau)| \geq \frac{1}{2} \varsigma \quad \text{on } \big[s_i - \varrho, s_i + \varrho\big]. \tag{5.4}$$

Therefore, $\sigma_{h,j}$ is strictly decreasing or increasing on $\big[s_i - \varrho, s_i + \varrho\big]$. Furthermore,
since $\sigma_{h,j}(s_i - \varrho) \sigma_{h,j}(s_i + \varrho) < 0$, we conclude that $\sigma_{h,j}$ has exactly one zero
$s_{h,i}$ in $\big[s_i - \varrho, s_i + \varrho\big]$. Additionally, (4.9) implies $s_{h,i} \in \big[t_{k_i^-}, t_{k_i^+}\big]$ and by (4.7) for
$\nu = \frac{1}{2}$ we have

$$|s_i - s_{h,i}| \leq \vartheta \sqrt{h}, \quad j = 1, \ldots, \ell \tag{5.5}$$

for a constant $\vartheta \geq 0$ independent of h. Thus, the discrete switching function σ_h has the same structure as σ. This allows us to prove Theorem 5.1.

Proof of Theorem 5.1 Since assumptions **(A1)**–**(A3)**, and **(A5)** hold, the discrete switching function σ_h has the same structure as σ. Suppose $(\hat{x}, \hat{u})$ is the optimal solution for **(OCP)** and $(\hat{x}_h, \hat{u}_h)$ is the optimal solution for **(DOCP)**. For $0 < \varepsilon \leq \varrho$ we denote

$$I_h(\varepsilon) := \bigcup_{1 \leq i \leq \ell} \left[s_{h,i} - \varepsilon, s_{h,i} + \varepsilon \right],$$

and for $j = 1, \ldots, n_u$ we define the set of zeros of $\sigma_{h,j}$ by $\Sigma_{h,j} := \left\{ \tau_{h,1}, \ldots, \tau_{h,\ell_j} \right\}$ with $0 < \tau_{h,1} < \ldots < \tau_{h,\ell_j} < 1$ and

$$I_{h,j}^-(\varepsilon) := \bigcup_{i=1,\ldots,\ell_j} \left[\tau_{h,i} - \varepsilon, \tau_{h,i} + \varepsilon \right], \quad I_{h,j}^+(\varepsilon) := [0,1] \setminus I_{h,j}^-(\varepsilon)$$

Since $\sigma_{h,j}$ is Lipschitz continuous, **(A5)** implies

$$0 < \sigma_{h,j,\min} = \min_{t \in \overline{I_{h,j}^+(\varrho)}} \left| \sigma_{h,j}(t) \right|, \tag{5.6}$$

where $\overline{I_{h,j}^+(\varrho)}$ denotes the closure of $I_{h,j}^+(\varrho)$. Using (4.3) and Theorem 4.1 for sufficiently small h we obtain

$$\left| \sigma_{h,j}(t) \right| \geq \sigma_{h,j,\min} \geq \frac{1}{2} \sigma_{j,\min} > 0 \quad \text{on } t \in I_{h,j}^+(\varrho).$$

Additionally, exploiting (5.4) and (5.5) yields

$$\left| \dot{\sigma}_{h,j}(\tau) \right| \geq \frac{1}{4} \varsigma \quad \text{on } \left[s_{h,i} - \varrho, s_{h,i} + \varrho \right] \tag{5.7}$$

for sufficiently small h. Thus, for $0 < \varepsilon \leq \varrho$ we get

$$\left| \sigma_{h,j}(t) \right| \geq \frac{1}{4} \varsigma \varepsilon > 0 \quad \text{on } \left[s_{h,i} - \varrho, s_{h,i} + \varrho \right] \setminus \left[s_{h,i} - \varepsilon, s_{h,i} + \varepsilon \right].$$

Choose $0 < \bar{\varepsilon} \leq \varrho$ such that

$$\bar{\varepsilon} \varsigma \leq \min_{1 \leq j \leq n_u} \sigma_{j,\min}.$$

Then, for all $j = 1, \ldots, n_u$ and every $0 < \varepsilon \leq \bar{\varepsilon}$ we conclude

$$\frac{1}{4}\varsigma\,\varepsilon \leq \left|\sigma_{h,j}(t)\right| \quad \text{on } [0,1] \setminus I_h(\varepsilon). \tag{5.8}$$

For $\tau_{h,i} \in \Sigma_{h,j}$ we denote

$$k_{h,i}^-(\varepsilon) := \max\left\{\iota \in \mathbb{N} \mid t_\iota \leq \tau_{h,i} - \varepsilon\right\}, \quad k_{h,i}^+(\varepsilon) := \min\left\{\iota \in \mathbb{N} \mid t_\iota \geq \tau_{h,i} + \varepsilon\right\},$$

and define the sets

$$\Upsilon_{h,j}(\varepsilon) := \bigcup_{1 \leq i \leq \ell_j} \left\{0 \leq \iota \leq N \mid k_{h,i}^-(\varepsilon) \leq \iota \leq k_{h,i}^+(\varepsilon)\right\}, \quad \Delta_{h,j}(\varepsilon) := \bigcup_{1 \leq i \leq \ell_j} \left[t_{k_{h,i}^-(\varepsilon)}, t_{k_{h,i}^+(\varepsilon)}\right].$$

Then, it holds $t_{k_{h,i}^+(\varepsilon)} - t_{k_{h,i}^-(\varepsilon)} \leq 2\,(\varepsilon + h)$ and the measure of the set $\Delta_{h,j}(\varepsilon)$ is bounded by $2\,\ell_j\,(\varepsilon + h)$. Moreover, since $\left[t_{k_{h,i}^-(\varepsilon)}, t_{k_{h,i}^+(\varepsilon)}\right] \supset \left[s_{h,i} - \varepsilon, s_{h,i} + \varepsilon\right]$, for every $0 < \varepsilon \leq \bar{\varepsilon}$ we obtain

$$\frac{1}{4}\varsigma\,\varepsilon \leq \left|\sigma_{h,j}(t)\right| \quad \text{on } [0,1] \setminus \Delta_{h,j}(\varepsilon).$$

Let $(x_h, u_h) \in \mathbf{F}_h$ be arbitrary. Then, for all $j = 1, \ldots, n_u$ and $i = 1, \ldots, N$ the signs of $\sigma_{h,j}(t_i)$ and $u_{h,j}(t_i) - \hat{u}_{h,j}(t_i)$ coincide, according to the discrete minimum principle (2.10). Hence, $\sigma_{h,j}(t_i)\left(u_{h,j}(t_i) - \hat{u}_{h,j}(t_i)\right) \geq 0$ and therefore by (5.8) it holds

$$\Omega_{h,j} := h\sum_{i=1}^N \sigma_{h,j}(t_i)\left(u_{h,j}(t_i) - \hat{u}_{h,j}(t_i)\right) \geq h\sum_{i \notin \Upsilon_{h,j}(\varepsilon)} \sigma_{h,j}(t_i)\left(u_{h,j}(t_i) - \hat{u}_{h,j}(t_i)\right)$$

$$\geq \frac{1}{4}\varsigma\,\varepsilon\,h\sum_{i \notin \Upsilon_{h,j}(\varepsilon)} \left|u_{h,j}(t_i) - \hat{u}_{h,j}(t_i)\right|.$$

Furthermore, we have

$$h\sum_{i \in \Upsilon_{h,j}} \left|u_{h,j}(t_i) - \hat{u}_{h,j}(t_i)\right| \leq \max_{1 \leq j \leq n_u}\left(b_{u,j} - b_{l,j}\right)\sum_{i \in \Upsilon_{h,j}} h$$

$$\leq \max_{1 \leq j \leq n_u}\left(b_{u,j} - b_{l,j}\right)\operatorname{meas}\left(\Delta_{h,j}(\varepsilon)\right) \leq \gamma_j\,(\varepsilon + h),$$

where $\gamma_j := 2\ell_j \max\limits_{1 \le j \le n_u} \left(b_{u,j} - b_{l,j}\right)$ and $\mathrm{meas}\left(\Delta_{h,j}(\varepsilon)\right)$ is the measure of $\Delta_{h,j}(\varepsilon)$. Thus, it holds

$$\Omega_{h,j} \ge \frac{1}{4}\varsigma\varepsilon h \sum_{i=1}^{N} \left|u_{h,j}(t_i) - \hat{u}_{h,j}(t_i)\right| - \frac{1}{4}\varsigma\varepsilon\gamma_j\,(\varepsilon + h), \quad j = 1, \dots, n_u,$$

$$\Omega_h := \sum_{j=1}^{n_u} \Omega_{h,j} = h \sum_{j=1}^{n_u} \sum_{i=1}^{N} \sigma_{h,j}(t_i)\left(u_{h,j}(t_i) - \hat{u}_{h,j}(t_i)\right)$$

$$\ge \frac{1}{4}\varsigma\varepsilon h \sum_{j=1}^{n_u} \sum_{i=1}^{N} \left|u_{h,j}(t_i) - \hat{u}_{h,j}(t_i)\right| - \frac{1}{4}\varsigma\varepsilon \sum_{j=1}^{n_u} \gamma_j\,(\varepsilon + h).$$

For $\gamma := \sum_{j=1}^{n_u} \gamma_j$ we obtain the lower bound $\Omega_h \ge \frac{1}{4}\varsigma\varepsilon\left(\left\|u_h - \hat{u}_h\right\|_1 - \gamma\,(\varepsilon + h)\right)$.

Now, we choose $u_h = v_h$, where $v_h(t) := \hat{u}(t_i)$ for $t \in \left(t_{i-1}, t_i\right]$, $i = 1, \dots, N$. Hence, $v_h \in U$ and by Lemma 3.2 it holds $\left\|\hat{u} - v_h\right\|_1 \le h \bigvee\limits_0^1 \hat{u}$. According to Theorem 4.1, this implies

$$\left\|\hat{u}_h - v_h\right\|_1 \le \left\|\hat{u}_h - \hat{u}\right\|_1 + \left\|\hat{u} - v_h\right\|_1 \le c\sqrt{h} + h\bigvee_0^1 \hat{u} \le 2\gamma\bar{\varepsilon}$$

for sufficiently small h. For $\varepsilon := \frac{1}{2\gamma}\left\|\hat{u}_h - v_h\right\|_1 \le \bar{\varepsilon}$ we get

$$\Omega_h \ge \frac{1}{4}\varsigma\varepsilon\left(\left\|v_h - \hat{u}_h\right\|_1 - \gamma\,(\varepsilon + h)\right)$$

$$= \frac{1}{4}\varsigma\frac{1}{2\gamma}\left\|v_h - \hat{u}_h\right\|_1\left(\left\|v_h - \hat{u}_h\right\|_1 - \gamma\left(\frac{1}{2\gamma}\left\|v_h - \hat{u}_h\right\|_1 + h\right)\right)$$

$$= \frac{\varsigma}{16\gamma}\left\|v_h - \hat{u}_h\right\|_1\left(\left\|v_h - \hat{u}_h\right\|_1 - 2\gamma h\right).$$

Consider the two cases: If $\left\|v_h - \hat{u}_h\right\|_1 \le 4\gamma h$, then we have a discrete error estimate of order one. Otherwise, it holds $\frac{1}{2}\left\|v_h - \hat{u}_h\right\|_1 > 2\gamma h$. Thus, we get

$$\Omega_h \ge \frac{\varsigma}{32\gamma}\left\|v_h - \hat{u}_h\right\|_1^2.$$

We aim to derive an upper bound for Ω_h (cf. [2, 19]) of the form

$$\Omega_h \le c_\Omega h\left\|v_h - \hat{u}_h\right\|_1,$$

where the constant $c_\Omega \geq 0$ is independent of h. This would imply

$$\frac{\varsigma}{32\,\gamma}\,\left\|v_h - \hat{u}_h\right\|_1^2 \leq \Omega_h \leq c_\Omega\,h\,\left\|v_h - \hat{u}_h\right\|_1,$$

and therefore $\left\|v_h - \hat{u}_h\right\|_1 \leq c\,h$ for a constant $c \geq 0$ independent of h. Hence, we would also have a discrete error estimate of order one.

We exploit the continuous minimum principle (2.9) for $u = \hat{u}_h$ to obtain

$$\sigma(t_i)^\top\,\left(\hat{u}_h(t_i) - \hat{u}(t_i)\right) = \sigma(t_i)^\top\,\left(\hat{u}_h(t_i) - v_h(t_i)\right) \geq 0, \quad i = 1, \ldots, N.$$

This yields

$$\Omega_h = h\,\sum_{i=1}^N \sigma_h(t_i)^\top\,\left(v_h(t_i) - \hat{u}_h(t_i)\right)$$

$$\leq h\,\sum_{i=1}^N \sigma_h(t_i)^\top\,\left(v_h(t_i) - \hat{u}_h(t_i)\right) + h\,\sum_{i=1}^N \sigma(t_i)^\top\,\left(\hat{u}_h(t_i) - v_h(t_i)\right)$$

$$= h\,\sum_{i=1}^N \left(r_h(t_i) + B_h(t_i)^\top\,\lambda_h(t_{i-1}) - r(t_i) - B(t_i)^\top\,\lambda(t_i)\right)^\top\,\left(v_h(t_i) - \hat{u}_h(t_i)\right)$$

$$\leq \left(L_r + L_B\,\|\lambda\|_\infty\right)\,h\,\left\|v_h - \hat{u}_h\right\|_1 + h\,\sum_{i=1}^N \left(\lambda_h(t_{i-1}) - \lambda(t_i)\right)^\top\,B_h(t_i)\,\left(v_h(t_i) - \hat{u}_h(t_i)\right)$$

$$= \left(L_r + L_B\,\|\lambda\|_\infty\right)\,h\,\left\|v_h - \hat{u}_h\right\|_1 + h\,\sum_{i=1}^N \left(\lambda(t_{i-1}) - \lambda(t_i)\right)^\top\,B_h(t_i)\,\left(v_h(t_i) - \hat{u}_h(t_i)\right)$$

$$+ h\,\sum_{i=1}^N \left(\lambda_h(t_{i-1}) - \lambda(t_{i-1})\right)^\top\,B_h(t_i)\,\left(v_h(t_i) - \hat{u}_h(t_i)\right)$$

$$\leq \left(L_r + L_B\,\|\lambda\|_\infty + L_\lambda\,\|B_h\|_\infty\right)\,h\,\left\|v_h - \hat{u}_h\right\|_1 + \tilde{\Omega}_h,$$

where

$$\tilde{\Omega}_h := h\,\sum_{i=1}^N \left(\lambda_h(t_{i-1}) - \lambda(t_{i-1})\right)^\top\,B_h(t_i)\,\left(v_h(t_i) - \hat{u}_h(t_i)\right).$$

It remains to find an upper bound for $\tilde{\Omega}_h$. Let z_h be such that $(z_h, v_h) \in \mathbf{F}_h$, and let η_h be the associated multiplier, i.e.,

$$z_h'(t_i) = A_h(t_i)\,z_h(t_i) + B_h(t_i)\,v_h(t_i), \qquad\qquad i = 1, \ldots, N,$$

$$z_h(t_0) = a,$$

$$\eta_h'(t_i) = -A_h(t_i)^\top \eta_h(t_{i-1}) - P_h(t_i) z_h(t_i) - p_h(t_i), \qquad i = 1, \ldots, N,$$

$$\eta_h(t_N) = Q_h z_h(t_N) + q_h.$$

We write $\tilde{\Omega}_h = \tilde{\Omega}_{1,h} + \tilde{\Omega}_{2,h}$ with

$$\tilde{\Omega}_{1,h} := h \sum_{i=1}^{N} \left(\lambda_h(t_{i-1}) - \eta_h(t_{i-1}) \right)^\top B_h(t_i) \left(v_h(t_i) - \hat{u}_h(t_i) \right),$$

$$\tilde{\Omega}_{2,h} := h \sum_{i=1}^{N} \left(\eta_h(t_{i-1}) - \lambda(t_{i-1}) \right)^\top B_h(t_i) \left(v_h(t_i) - \hat{u}_h(t_i) \right).$$

Analog to the proof of Theorem 4.1, we find a constant $c \geq 0$ independent of h such that $\left\| \eta_h - \lambda \right\|_\infty \leq c h$. Hence, we obtain $\tilde{\Omega}_{2,h} \leq c h \left\| B_h \right\|_\infty \left\| v_h - \hat{u}_h \right\|_1$. Furthermore, the difference equations and the adjoint equations yield for $i = 1, \ldots, N$

$$h B_h(t_i) \left(v_h(t_i) - \hat{u}_h(t_i) \right) = \left(z_h(t_i) - z_h(t_{i-1}) \right) - \left(\hat{x}_h(t_i) - \hat{x}_h(t_{i-1}) \right)$$
$$- h A_h(t_i) \left(z_h(t_i) - \hat{x}_h(t_i) \right),$$

$$-h A_h(t_i)^\top \left(\lambda_h(t_{i-1}) - \eta_h(t_{i-1}) \right) = \left(\lambda_h(t_i) - \lambda_h(t_{i-1}) \right) - \left(\eta_h(t_i) - \eta_h(t_{i-1}) \right)$$
$$- h P_h(t_i) \left(z_h(t_i) - \hat{x}_h(t_i) \right).$$

This allows us to write $\tilde{\Omega}_{1,h}$ as

$$\tilde{\Omega}_{1,h} = h \sum_{i=1}^{N} \left(\lambda_h(t_{i-1}) - \eta_h(t_{i-1}) \right)^\top B_h(t_i) \left(v_h(t_i) - \hat{u}_h(t_i) \right),$$

$$= \sum_{i=1}^{N} \left(\lambda_h(t_{i-1}) - \eta_h(t_{i-1}) \right)^\top \left[\left(z_h(t_i) - z_h(t_{i-1}) \right) - \left(\hat{x}_h(t_i) - \hat{x}_h(t_{i-1}) \right) \right]$$

$$- h \sum_{i=1}^{N} \left(\lambda_h(t_{i-1}) - \eta_h(t_{i-1}) \right)^\top A_h(t_i) \left(z_h(t_i) - \hat{x}_h(t_i) \right)$$

$$= \sum_{i=1}^{N} \left(\lambda_h(t_{i-1}) - \eta_h(t_{i-1}) \right)^\top \left[\left(z_h(t_i) - z_h(t_{i-1}) \right) - \left(\hat{x}_h(t_i) - \hat{x}_h(t_{i-1}) \right) \right]$$

$$+ \sum_{i=1}^{N} \left[\left(\lambda_h(t_i) - \lambda_h(t_{i-1}) \right) - \left(\eta_h(t_i) - \eta_h(t_{i-1}) \right) \right]^\top \left(z_h(t_i) - \hat{x}_h(t_i) \right)$$

$$- h \sum_{i=1}^{N} \left(z_h(t_i) - \hat{x}_h(t_i) \right)^\top P_h(t_i) \left(z_h(t_i) - \hat{x}_h(t_i) \right)$$

$$= -\left(\lambda_h(t_0) - \eta_h(t_0)\right)^\top \left(z_h(t_0) - \hat{x}_h(t_0)\right) + \left(\lambda_h(t_N) - \eta_h(t_N)\right)^\top \left(z_h(t_N) - \hat{x}_h(t_N)\right)$$

$$- h \sum_{i=1}^{N} \left(z_h(t_i) - \hat{x}_h(t_i)\right)^\top P_h(t_i) \left(z_h(t_i) - \hat{x}_h(t_i)\right)$$

$$= -\left(z_h(t_N) - \hat{x}_h(t_N)\right)^\top Q_h \left(z_h(t_N) - \hat{x}_h(t_N)\right)$$

$$- h \sum_{i=1}^{N} \left(z_h(t_i) - \hat{x}_h(t_i)\right)^\top P_h(t_i) \left(z_h(t_i) - \hat{x}_h(t_i)\right).$$

Exploiting the positive definiteness of Q_h and $P_h(t_i)$ for $i = 1, \ldots, N$ gives us $\tilde{\Omega}_{1,h} \leq 0$. Thus, we obtain

$$\tilde{\Omega}_h = \tilde{\Omega}_{1,h} + \tilde{\Omega}_{2,h} \leq \tilde{\Omega}_{2,h} \leq c\, h \, \|B_h\|_\infty \, \|v_h - \hat{u}_h\|_1 .$$

Finally, we have $\Omega_h \leq c_\Omega\, h \, \|v_h - \hat{u}_h\|_1$ with $c_\Omega \geq 0$ independent of h. This implies

$$\frac{\varsigma}{32\,\gamma} \, \|v_h - \hat{u}_h\|_1^2 \leq \Omega_h \leq c_\Omega\, h \, \|v_h - \hat{u}_h\|_1 ,$$

and therefore dividing by $\frac{\varsigma}{32\gamma} \, \|v_h - \hat{u}_h\|_1$ yields $\|v_h - \hat{u}_h\|_1 \leq \frac{32\,\gamma\,c_\Omega}{\varsigma}\, h$. Hence, we have a linear error estimate for $\hat{u}_h$. Analog to the proof of Theorem 4.1, we also obtain improved error estimates of linear order for $\hat{x}_h$ and the associated multiplier λ_h. Then, **(A5)** immediately yields a linear error estimate for the switching function σ_h. Finally, by (4.7) for $\nu = 1$, we obtain

$$\left|s_i - s_{h,i}\right| \leq \vartheta\, h, \quad j = 1, \ldots, \ell,$$

which completes the proof.   □

6   Index Reduction and Error Analysis

In this section, we consider the DAE optimal control problem

(OCP-DAE)   Minimize $f\,(x, y, u)$
               subject to $\dot{x}(t) = A(t)\,x(t) + B(t)\,y(t) + C(t)\,u(t),$ a.e. in $[0, 1]$,
                            $0 = D(t)\,x(t),$ in $[0, 1]$
                      $\Xi\, x(0) = a,$
                            $u(t) \in U,$ a.e. in $[0, 1]$.

with the objective function

$$f(x, y, u) := \frac{1}{2} x(1)^\top Q x(1) + q^\top x(1)$$

$$+ \int_0^1 \frac{1}{2} x(t)^\top P(t) x(t) + p(t)^\top x(t) + r(t)^\top y(t) + g(t)^\top u(t) \, dt,$$

and the set of admissible controls $U := \{ u \in \mathbb{R}^{n_u} \mid b_l \leq u \leq b_u \}$, and assume the following smoothness and regularity properties:

(A1') Let $q \in \mathbb{R}^{n_x}$, $\mathcal{E} \in \mathbb{R}^{n_y \times n_x}$, let the functions $A : [0, 1] \to \mathbb{R}^{n_x \times n_x}$, $B : [0, 1] \to \mathbb{R}^{n_x \times n_y}$, $C : [0, 1] \to \mathbb{R}^{n_x \times n_u}$, $P : [0, 1] \to \mathbb{R}^{n_x \times n_x}$, $p : [0, 1] \to \mathbb{R}^{n_x}$, $r : [0, 1] \to \mathbb{R}^{n_y}$, $g : [0, 1] \to \mathbb{R}^{n_u}$ be Lipschitz continuous, and $D : [0, 1] \to \mathbb{R}^{n_y \times n_x}$ be differentiable with Lipschitz continuous derivative $\dot{D} : [0, 1] \to \mathbb{R}^{n_y \times n_x}$, and let the matrices $Q, P(t) \in \mathbb{R}^{n_x \times n_x}$, $t \in [0, 1]$, be symmetric and positive semidefinite. Additionally, let $D(t) B(t)$ be non-singular for all $t \in [0, 1]$ with continuous and uniformly bounded inverse, and let $\begin{pmatrix} D(0) \\ \mathcal{E} \end{pmatrix}$ be non-singular.

The admissible set of **(OCP-DAE)** is closed, convex, nonempty, and bounded, and the cost functional f is continuous and convex. Thus, by [14, Chapter II, Proposition 1.2], there exists a minimizer $(\hat{x}, \hat{y}, \hat{u}) \in W_{1,2}^{n_x} ([0, 1]) \times L_2^{n_y} ([0, 1]) \times L_2^{n_u} ([0, 1])$ of **(OCP-DAE)**. Moreover, since U is bounded, it holds $(\hat{x}, \hat{y}, \hat{u}) \in W_{1,\infty}^{n_x} ([0, 1]) \times L_\infty^{n_y} ([0, 1]) \times L_\infty^{n_u} ([0, 1])$.

With assumption **(A1')** we are able to reduce the index of the algebraic equation, since it holds

$$0 = D(t) x(t) \quad \text{in } [0, 1] \quad \Leftrightarrow \quad 0 = \frac{d}{dt} (D(t) x(t)) \quad \text{a.e. in } [0, 1], \quad 0 = D(0) x(0)$$

Thus, we obtain the reduced DAE

$$\dot{x}(t) = A(t) x(t) + B(t) y(t) + C(t) u(t), \qquad \text{a.e. in } [0, 1],$$
$$0 = \left(\dot{D}(t) + D(t) A(t) \right) x(t) + D(t) B(t) y(t) + D(t) C(t) u(t), \qquad \text{a.e. in } [0, 1]$$
$$D(0) x(0) = 0, \quad \mathcal{E} x(0) = a.$$

and the associated necessary (and sufficient) conditions

$$\dot{\lambda}(t) = -P(t)\,\hat{x}(t) - p(t) - A(t)^\top \lambda(t) - \left(\dot{D}(t) + D(t)\,A(t)\right)^\top \mu(t) \qquad \text{in } [0, 1]$$

$$0 = r(t) + B(t)^\top \lambda(t) + \left(D(t)\,B(t)\right)^\top \mu(t) \qquad \text{in } [0, 1]$$

$$\lambda(1) = Q\,\hat{x}(1) + q$$

$$0 \leq \left[g(t)^\top + \lambda(t)^\top C(t) + \mu(t)^\top D(t)\,C(t)\right]\left(u - \hat{u}(t)\right), \quad \text{for all } u \in U, \quad \text{a.e. in } [0, 1],$$

which are the same as for **(OCP-DAE)**. Now, we consider the discretized problem

(DOCP-DAE)
> Minimize $\quad f_h\left(x_h, y_h, u_h\right)$
> subject to $\quad x_h'(t_i) = A(t_i)\,x_h(t_i) + B(t_i)\,y_h(t_i) + C(t_i)\,u_h(t_i),\; i = 1, \ldots, N,$
> $\qquad\qquad\qquad 0 = D(t_i)\,x_h(t_i), \qquad\qquad\qquad\qquad\qquad i = 0, 1, \ldots, N,$
> $\qquad\quad \varXi\, x_h(t_0) = a,$
> $\qquad\qquad u_h(t_i) \in U, \qquad\qquad\qquad\qquad\qquad\qquad i = 1, \ldots, N,$

with the discrete objective function

$$f_h\left(x_h, y_h, u_h\right) := \frac{1}{2}\,x_h(t_N)^\top Q\,x_h(t_N) + q^\top x_h(t_N)$$

$$+ h \sum_{i=1}^{N} \frac{1}{2}\,x_h(t_i)^\top P(t_i)\,x_h(t_i) + p(t_i)^\top x_h(t_i) + r(t_i)^\top y_h(t_i) + g(t_i)^\top u_h(t_i).$$

Similarly to the continuous case we are able to reduce the index of the discretized DAE (cf. [22, 24]), since it holds

$$0 = D(t_i)\,x_h(t_i) \quad i = 0, 1, \ldots, N$$

$$\Leftrightarrow\quad 0 = \frac{D(t_i)\,x_h(t_i) - D(t_{i-1})\,x_h(t_{i-1})}{h} \quad i = 1, \ldots, N, \quad 0 = D(t_0)\,x_h(t_0).$$

Using the difference equation we obtain

$$x_h(t_{i-1}) = x_h(t_i) - h\left(A(t_i)\,x_h(t_i) + B(t_i)\,y_h(t_i) + C(t_i)\,u_h(t_i)\right), \quad i = 1, \ldots, N,$$

and therefore for $i = 1, \ldots, N$

$$0 = \left(\frac{D(t_i) - D(t_{i-1})}{h} + D(t_{i-1})\,A(t_i)\right)x_h(t_i) + D(t_{i-1})\,B(t_i)\,y_h(t_i) + D(t_{i-1})\,C(t_i)\,u_h(t_i).$$

This gives us the reduced system

$$x'_h(t_i) = A(t_i)\, x_h(t_i) + B(t_i)\, y_h(t_i) + C(t_i)\, u_h(t_i), \qquad i = 1, \ldots, N,$$

$$0 = \left(\frac{D(t_i) - D(t_{i-1})}{h} + D(t_{i-1})\, A(t_i) \right) x_h(t_i)$$

$$+ D(t_{i-1})\, B(t_i)\, y_h(t_i) + D(t_{i-1})\, C(t_i)\, u_h(t_i), \qquad i = 1, \ldots, N,$$

$$D(t_0)\, x_h(t_0) = 0, \qquad \mathcal{E}\, x_h(t_0) = a,$$

and the associated necessary conditions for $i = 1, \ldots, N$

$$\lambda'_h(t_i) = -P(t_i)\, \hat{x}_h(t_i) - p(t_i) - A(t_i)^\top \lambda_h(t_{i-1})$$

$$- \left(\frac{D(t_i) - D(t_{i-1})}{h} + D(t_{i-1})\, A(t_i) \right)^\top \mu_h(t_{i-1}),$$

$$0 = r(t_i) + B(t_i)^\top \lambda_h(t_{i-1}) + \left(D(t_{i-1})\, B(t_i) \right)^\top \mu_h(t_{i-1}),$$

$$\lambda_h(t_N) = Q\, \hat{x}_h(t_N) + q,$$

$$0 \leq \left[g(t_i)^\top + \lambda_h(t_{i-1})^\top C(t_i) + \mu_h(t_{i-1})^\top D(t_{i-1})\, C(t_i) \right] \left(u_h - \hat{u}_h(t_i) \right), \qquad \text{for all } u_h \in U.$$

The continuous and discrete local minimum principles yield the consistent switching functions (compare (1.2) and (1.3))

$$\sigma(t)^\top := g(t)^\top + \lambda(t)^\top C(t) + \mu(t)^\top D(t)\, C(t), \qquad t \in [0, 1] \tag{6.1}$$

$$\sigma_h(t_i)^\top := g(t_i)^\top + \lambda_h(t_{i-1})^\top C(t_i) + \mu_h(t_{i-1})^\top D(t_{i-1})\, C(t_i), \qquad i = 1, \ldots, N.$$

In order to apply the above theorems, we require additional assumptions regarding the continuous switching function in (6.1):

(A3') There exists a solution $(\hat{x}, \hat{y}, \hat{u})$ of **(OCP-DAE)** such that the set Σ of zeros of the components σ_j, $j = 1, \ldots, n_u$ of the switching function is finite and $0, 1 \notin \Sigma$, i.e., we have $\Sigma = \{s_1, \ldots, s_\ell\}$ with $0 < s_1 < \ldots < s_\ell < 1$.

(A4') Let $J(s_i) := \left\{ 1 \leq j \leq n_u \mid \sigma_j(s_i) = 0 \right\}$ denote the set of active indices. Let there exist constants $\varsigma > 0$, $\varrho > 0$ such that $|\sigma_j(\tau)| \geq \varsigma\, |\tau - s_i|$ for all $i = 1, \ldots, \ell$, $j \in J(s_i)$, and every $\tau \in [s_i - \varrho, s_i + \varrho]$, and σ_j changes sign in s_i, i.e., $\sigma_j(s_i - \varrho)\, \sigma_j(s_i + \varrho) < 0$.

(A5') Let the derivatives $\dot{B} : [0, 1] \to \mathbb{R}^{n_x \times n_y}$, $\dot{C} : [0, 1] \to \mathbb{R}^{n_x \times n_u}$, $\dot{r} : [0, 1] \to \mathbb{R}^{n_y}$, $\dot{g} : [0, 1] \to \mathbb{R}^{n_u}$ be Lipschitz continuous. Furthermore, let there exist $\varsigma > 0$ such that

$$\min_{1 \leq i \leq \ell} \min_{j \in J(s_i)} \left\{ |\dot{\sigma}_j(s_i)| \right\} \geq 2\varsigma.$$

With these assumptions we obtain the main results of this paper

Theorem 6.1 *Let $(\hat{x}, \hat{y}, \hat{u})$ be a minimizer for (OCP-DAE) and let (A1'), (A3')
hold. If (A4') is satisfied, then for sufficiently small h, any solution $(\hat{x}_h, \hat{y}_h, \hat{u}_h)$
of (DOCP-DAE), the associated multipliers, and the switching function can be
estimated by*

$$\|\hat{x}_h - \hat{x}\|_\infty \le c\sqrt{h}, \quad \|\hat{y}_h - \hat{y}\|_1 \le c\sqrt{h}, \quad \|\hat{u}_h - \hat{u}\|_1 \le c\sqrt{h},$$

$$\|\lambda_h - \lambda\|_\infty \le c\sqrt{h}, \quad \|\mu_h - \mu\|_\infty \le c\sqrt{h}, \quad \|\sigma_h - \sigma\|_\infty \le c\sqrt{h},$$

*where $c \ge 0$ is independent of h. Moreover, there exists a constant κ independent of
h such that for sufficiently small h any discrete optimal control $\hat{u}_h$ coincides with $\hat{u}$
except on a set of measure smaller than $\kappa\sqrt{h}$.*

*If (A5') is satisfied, then for sufficiently small h, any solution $(\hat{x}_h, \hat{y}_h, \hat{u}_h)$
of (DOCP-DAE), the associated multipliers, and the switching function can be
estimated by*

$$\|\hat{x}_h - \hat{x}\|_\infty \le ch, \quad \|\hat{y}_h - \hat{y}\|_1 \le ch, \quad \|\hat{u}_h - \hat{u}\|_1 \le ch,$$

$$\|\lambda_h - \lambda\|_\infty \le ch, \quad \|\mu_h - \mu\|_\infty \le ch, \quad \|\sigma_h - \sigma\|_\infty \le ch,$$

*where $c \ge 0$ is independent of h. Moreover, there exist constants $\kappa, \vartheta \ge 0$
independent of h such that for sufficiently small h any discrete optimal control $\hat{u}_h$
coincides with $\hat{u}$ except on a set of measure smaller than κh, and the switching
functions σ and σ_h have the same structure, i.e., the components of σ_h have ℓ zeros
$0 < s_{h,1} < \ldots < s_{h,\ell} < 1$, which satisfy the estimate*

$$|s_i - s_{h,i}| \le \vartheta h, \quad i = 1, \ldots, \ell.$$

Proof In order to apply Theorem 4.1 and 5.1, we need to verify the conditions
(A1)–(A5) using the above assumptions (A1'), (A3'), (A4'), and (A5').

Clearly, (A1') implies (A1). Next, we transform (OCP-DAE) and (DOCP-DAE)
such that they are contained in the problem classes (OCP) and (DOCP) in Sect. 2.

Since $D(t) B(t)$ is non-singular for all $t \in [0, 1]$, we are able to solve the
respective algebraic equations for y and μ in order to get

$$y(\cdot) = -\left(D(\cdot) B(\cdot)\right)^{-1}\left[\left(\dot{D}(\cdot) + D(\cdot) A(\cdot)\right) x(\cdot) + D(\cdot) C(\cdot) u(\cdot)\right], \qquad (6.2)$$

$$\mu(\cdot) = -\left(\left(D(\cdot) B(\cdot)\right)^{-1}\right)^\top \left[r(\cdot) + B(\cdot)^\top \lambda(\cdot)\right].$$

Note that since B, D, r, λ are Lipschitz continuous, μ is also Lipschitz continuous. Inserting (6.2) into the differential equation, adjoint equation, and local minimum principles yields

$$
\begin{aligned}
\dot{x}(t) &= \tilde{A}(t)\, x(t) + \tilde{B}(t)\, u(t), & &\text{a.e. in } [0, 1], \\
x(0) &= \tilde{a}, \\
\dot{\lambda}(t) &= -P(t)\,\hat{x}(t) - \tilde{p}(t) - \tilde{A}(t)^\top \lambda(t), & &\text{in } [0, 1], \\
\lambda(1) &= Q\,\hat{x}(1) + q \\
0 &\le \left[\tilde{r}(t)^\top + \lambda(t)^\top \tilde{B}(t) \right] (u - \hat{u}(t)), & &\text{for all } u \in U, \ \text{a.e. in } [0, 1],
\end{aligned}
$$

with

$$
\tilde{A}(t) := \left(I - B(t)\, \big(D(t)\, B(t)\big)^{-1} D(t) \right) A(t) - B(t)\, \big(D(t)\, B(t)\big)^{-1} \dot{D}(t)
$$

$$
\tilde{B}(t) := \left(I - B(t)\, \big(D(t)\, B(t)\big)^{-1} D(t) \right) C(t)
$$

$$
\tilde{a} := \begin{pmatrix} D(0) \\ \Xi \end{pmatrix}^{-1} \begin{pmatrix} 0 \\ a \end{pmatrix}
$$

$$
\tilde{p}(t)^\top := p(t)^\top - r(t)^\top \big(D(t)\, B(t)\big)^{-1} \big(\dot{D}(t) + D(t)\, A(t) \big),
$$

$$
\tilde{r}(t)^\top := g(t)^\top - r(t)^\top \big(D(t)\, B(t)\big)^{-1} D(t)\, C(t).
$$

In addition, we obtain the new objective function

$$
\tilde{f}(x, u) := \frac{1}{2} x(1)^\top Q\, x(1) + q^\top x(1) + \int_0^1 \frac{1}{2} x(t)^\top P(t)\, x(t) + \tilde{p}(t)^\top x(t) + \tilde{r}(t)^\top u(t)\, dt.
$$

Since $\big(D(\cdot)\, B(\cdot)\big)^{-1}$ is uniformly bounded, there exists $\beta > 0$ such that for all $t \in [0, 1]$ we have $\left\| D(t)\, B(t)\, d \right\| \ge \beta \|d\|$ for all $d \in \mathbb{R}^{n_y}$. Then, it holds for all $d \in \mathbb{R}^{n_y}$ and $i = 1, \ldots, N$

$$
\left\| D(t_{i-1})\, B(t_i)\, d \right\| \ge \left\| D(t_i)\, B(t_i)\, d \right\| - \left\| \big(D(t_i)\, B(t_i) - D(t_{i-1})\, B(t_i)\big)\, d \right\| \ge \frac{\beta}{2} \|d\|,
$$

if $0 < h \le \frac{\beta}{2 L_D \|B\|}$. Hence, for sufficiently small h the inverse $\big(D(t_{i-1})\, B(t_i)\big)^{-1}$ exists for $i = 1, \ldots, N$. This allows us to solve the respective algebraic equations for y_h and μ_h

$$y_h(t_i) = -\left(D(t_{i-1})\,B(t_i)\right)^{-1}\left(\frac{D(t_i) - D(t_{i-1})}{h} + D(t_{i-1})\,A(t_i)\right) x_h(t_i)$$

$$- \left(D(t_{i-1})\,B(t_i)\right)^{-1} D(t_{i-1})\,C(t_i)\,u_h(t_i), \tag{6.3}$$

$$\mu_h(t_{i-1}) = -\left(\left(D(t_{i-1})\,B(t_i)\right)^{-1}\right)^{\top}\left[r(t_i) + B(t_i)^{\top}\,\lambda_h(t_{i-1})\right]$$

for $i = 1, \ldots, N$, in order to obtain

$$
\begin{aligned}
x_h'(t_i) &= \tilde{A}_h(t_i)\,x_h(t_i) + \tilde{B}_h(t_i)\,u_h(t_i), && i = 1, \ldots, N,\\
x_h(t_0) &= \tilde{a},\\
\lambda_h'(t_i) &= -P(t_i)\,\hat{x}_h(t_i) - \tilde{p}_h(t_i) - \tilde{A}_h(t_i)^{\top}\,\lambda_h(t_{i-1}), && i = 1, \ldots, N\\
\lambda_h(t_N) &= Q\,\hat{x}_h(t_N) + q\\
0 &\le \left[\tilde{r}_h(t_i)^{\top} + \lambda_h(t_{i-1})^{\top}\,\tilde{B}_h(t_i)\right]\left(u_h - \hat{u}_h(t_i)\right), && \text{for all } u_h \in U,\, i = 1, \ldots, N
\end{aligned}
$$

with

$$\tilde{A}_h(t_i) := \left(I - B(t_i)\left(D(t_{i-1})\,B(t_i)\right)^{-1} D(t_{i-1})\right) A(t_i)$$

$$- B(t_i)\left(D(t_{i-1})\,B(t_i)\right)^{-1}\frac{D(t_i) - D(t_{i-1})}{h},$$

$$\tilde{B}_h(t_i) := \left(I - B(t_i)\left(D(t_{i-1})\,B(t_i)\right)^{-1} D(t_{i-1})\right) C(t_i),$$

$$\tilde{p}_h(t_i)^{\top} := p(t_i)^{\top} - r(t_i)^{\top}\left(D(t_{i-1})\,B(t_i)\right)^{-1}\left(\frac{D(t_i) - D(t_{i-1})}{h} + D(t_{i-1})\,A(t_i)\right),$$

$$\tilde{r}_h(t_i)^{\top} := g(t_i)^{\top} - r(t_i)^{\top}\left(D(t_{i-1})\,B(t_i)\right)^{-1} D(t_{i-1})\,C(t_i)$$

for $i = 1, \ldots, N$, and the discrete objective function

$$\tilde{f}_h\,(x_h, u_h) := \frac{1}{2}\,x_h(t_N)^{\top}\,Q\,x_h(t_N) + q^{\top}\,x_h(t_N)$$

$$+ h\sum_{i=1}^{N}\frac{1}{2}\,x_h(t_i)^{\top}\,P(t_i)\,x_h(t_i) + \tilde{p}_h(t_i)^{\top}\,x_h(t_i) + \tilde{r}_h(t_i)^{\top}\,u_h(t_i).$$

Summarizing, we have the reduced optimal control problem
(ROCP)

$$
\begin{aligned}
\text{Minimize} \quad & \tilde{f}\,(x, u)\\
\text{subject to} \quad & \dot{x}(t) = \tilde{A}(t)\,x(t) + \tilde{B}(t)\,u(t), && \text{a.e. in } \left[0, 1\right],\\
& x(0) = \tilde{a},\\
& u(t) \in U, && \text{a.e. in } \left[0, 1\right],
\end{aligned}
$$

and the reduced approximation
(RDOCP)

$$
\begin{aligned}
\text{Minimize} \quad & \tilde{f}_h\,(x_h, u_h) \\
\text{subject to} \quad & x_h'(t_i) = \tilde{A}_h(t_i)\, x_h(t_i) + \tilde{B}_h(t_i)\, u_h(t_i), \quad i = 1, \dots, N, \\
& x_h(t_0) = \tilde{a}, \\
& u_h(t_i) \in U, \qquad\qquad\qquad\qquad\qquad i = 1, \dots, N,
\end{aligned}
$$

which are contained in the problem classes **(OCP)** and **(DOCP)**, respectively (compare Sect. 2). Moreover, these problems have the switching functions

$$
\tilde{\sigma}(t)^\top := \tilde{r}(t)^\top + \lambda(t)^\top \tilde{B}(t), \quad \tilde{\sigma}_h(t_i)^\top := \tilde{r}_h(t_i)^\top + \lambda_h(t_{i-1})^\top \tilde{B}_h(t_i).
$$

Note that by (6.1), (6.2), and (6.3) we have $\tilde{\sigma} = \sigma$ and $\tilde{\sigma}_h = \sigma_h$. Hence, **(A3')**–**(A5')** imply the conditions for the switching function in **(A3)**–**(A5)**.

It remains to verify the smoothness and approximation conditions in **(A2)** and **(A5)** for the system functions in **(ROCP)** and **(RDOCP)**. Since D and $\dot{D}$ are Lipschitz continuous by **(A1')**, there exists a constant $L_D \geq 0$ such that

$$
\left\| D(t_i) - D(t_{i-1}) \right\| \leq L_D\, h, \quad \left\| \dot{D}(t_i) - \frac{D(t_i) - D(t_{i-1})}{h} \right\| \leq L_D\, h, \quad i = 1, \dots, N.
$$

Furthermore, by [22, Lemma A.2], $\left(D(\cdot)\, B(\cdot) \right)^{-1}$ and $\left(D(t_{i-1})\, B(t_i) \right)^{-1}$ are Lipschitz continuous. Thus, there exist $L_{\tilde{A}}, L_{\tilde{B}}, L_{\tilde{p}}, L_{\tilde{r}} \geq 0$ independent of h such that

$$
\left\| \tilde{A} - \tilde{A}_h \right\|_\infty \leq L_{\tilde{A}}\, h, \quad \left\| \tilde{B} - \tilde{B}_h \right\|_\infty \leq L_{\tilde{B}}\, h, \quad \left\| \tilde{p} - \tilde{p}_h \right\|_\infty \leq L_{\tilde{p}}\, h, \quad \left\| \tilde{r} - \tilde{r}_h \right\|_\infty \leq L_{\tilde{r}}\, h,
$$

which verifies **(A2)**. For the conditions in **(A5)**, we first verify that the time derivative of $\tilde{B}(t) = \left(I - B(t)\, \left(D(t)\, B(t) \right)^{-1} D(t) \right) C(t)$ exists and is Lipschitz continuous, if **(A1')** and **(A5')** hold. It remains to show that $\frac{d}{dt}\left(\left(D(t)\, B(t) \right)^{-1} \right)$ exists and is Lipschitz continuous. To this end, we consider for $t, t + \epsilon \in [0, 1]$, $\epsilon \neq 0$

$$
\begin{aligned}
0 &= \left(D(t+\epsilon)\, B(t+\epsilon) \right) \left(D(t+\epsilon)\, B(t+\epsilon) \right)^{-1} - \left(D(t)\, B(t) \right) \left(D(t)\, B(t) \right)^{-1} \\
&= \left[\left(D(t+\epsilon)\, B(t+\epsilon) \right) - \left(D(t)\, B(t) \right) \right] \left(D(t+\epsilon)\, B(t+\epsilon) \right)^{-1} \\
&\quad + \left(D(t)\, B(t) \right) \left[\left(D(t+\epsilon)\, B(t+\epsilon) \right)^{-1} - \left(D(t)\, B(t) \right)^{-1} \right],
\end{aligned}
$$

which yields

$$\frac{1}{\epsilon} \left[\left(D(t+\epsilon) B(t+\epsilon) \right)^{-1} - \left(D(t) B(t) \right)^{-1} \right] \tag{6.4}$$

$$= - \left(D(t) B(t) \right)^{-1} \frac{1}{\epsilon} \left[\left(D(t+\epsilon) B(t+\epsilon) \right) - \left(D(t) B(t) \right) \right] \left(D(t+\epsilon) B(t+\epsilon) \right)^{-1}$$

$$= - \left(D(t) B(t) \right)^{-1} \int_0^1 \frac{d}{dt} \left(D(t+\theta\,\epsilon) B(t+\theta\,\epsilon) \right) d\theta \left(D(t+\epsilon) B(t+\epsilon) \right)^{-1}$$

Since $\dot{B}, \dot{D}$ exist and are Lipschitz continuous, and $\left(D(\cdot) B(\cdot) \right)^{-1}$ is Lipschitz continuous, the limit for $\epsilon \to 0$ on the right hand side exists. Therefore, $\frac{d}{dt} \left(\left(D(t) B(t) \right)^{-1} \right)$ exists and is Lipschitz continuous. Thus, $\ddot{B}$ exists and is Lipschitz continuous, and analog we conclude that $\ddot{r}$ exists and is Lipschitz continuous. Next, according to (6.4), for $i \in \{2, \ldots, N\}$ we get

$$\frac{1}{h} \left[\left(D(t_{i-1}) B(t_i) \right)^{-1} - \left(D(t_{i-2}) B(t_{i-1}) \right)^{-1} \right]$$

$$= - \left(D(t_{i-2}) B(t_{i-1}) \right)^{-1} \frac{1}{h} \left[\left(D(t_{i-1}) B(t_i) \right) - \left(D(t_{i-2}) B(t_{i-1}) \right) \right] \left(D(t_{i-1}) B(t_i) \right)^{-1}.$$

Hence, for $i = 2, \ldots, N$ it holds

$$\left\| \frac{d}{dt} \left(\left(D(t_i) B(t_i) \right)^{-1} \right) - \frac{1}{h} \left[\left(D(t_{i-1}) B(t_i) \right)^{-1} - \left(D(t_{i-2}) B(t_{i-1}) \right)^{-1} \right] \right\| \leq c\,h$$

for a constant $c \geq 0$ independent of h. Moreover, we have

$$\ddot{B}(t) = \dot{C}(t) - \dot{B}(t) \left(D(t) B(t) \right)^{-1} D(t) C(t) - B(t) \frac{d}{dt} \left(\left(D(t) B(t) \right)^{-1} \right) D(t) C(t)$$

$$\qquad - B(t) \left(D(t) B(t) \right)^{-1} \dot{D}(t) C(t) - B(t) \left(D(t) B(t) \right)^{-1} D(t) \dot{C}(t)$$

and

$$\tilde{B}_h'(t_i) = C'(t_i) - \frac{1}{h} \left[B(t_i) \left(D(t_{i-1}) B(t_i) \right)^{-1} D(t_{i-1}) C(t_i) \right.$$

$$\qquad \left. - B(t_{i-1}) \left(D(t_{i-2}) B(t_{i-1}) \right)^{-1} D(t_{i-2}) C(t_{i-1}) \right]$$

$$= C'(t_i) - B'(t_i) \left(D(t_{i-1}) B(t_i) \right)^{-1} D(t_{i-1}) C(t_i)$$

$$\qquad - B(t_{i-1}) \frac{1}{h} \left[\left(D(t_{i-1}) B(t_i) \right)^{-1} D(t_{i-1}) C(t_i) - \left(D(t_{i-2}) B(t_{i-1}) \right)^{-1} D(t_{i-2}) C(t_{i-1}) \right]$$

$$= C'(t_i) - B'(t_i) \left(D(t_{i-1}) B(t_i) \right)^{-1} D(t_{i-1}) C(t_i)$$

$$- B(t_{i-1}) \frac{1}{h} \left[\left(D(t_{i-1}) B(t_i) \right)^{-1} - \left(D(t_{i-2}) B(t_{i-1}) \right)^{-1} \right] D(t_{i-1}) C(t_i)$$

$$- B(t_{i-1}) \left(D(t_{i-2}) B(t_{i-1}) \right)^{-1} D'(t_{i-1}) C(t_i)$$

$$- B(t_{i-1}) \left(D(t_{i-2}) B(t_{i-1}) \right)^{-1} D(t_{i-2}) C'(t_{i-1}).$$

Thus, we conclude there exists $L_{\tilde{B}} \geq 0$ independent of h such that $\left\| \dot{\tilde{B}}(t_i) - \tilde{B}'(t_i) \right\| \leq L_{\tilde{B}} h$ for $i = 2, \ldots, N$. Analog we find $L_{\tilde{r}} \geq 0$ independent of h such that $\left\| \dot{\tilde{r}}(t_i) - \tilde{r}'(t_i) \right\| \leq L_{\tilde{r}} h$ for $i = 2, \ldots, N$. Hence, the smoothness and approximation conditions in (A5) are verified.

Applying Theorem 4.1 and 5.1, respectively, yields error estimates for x_h, u_h, λ_h. According to the relations (6.1), (6.2), and (6.3) we automatically get error estimates for y_h, μ_h, and for the switching function σ_h and its zeros. This completes the proof.

$$\square$$

7 Numerical Example

We consider the optimal control problem

$$\text{Minimize} \quad \frac{\alpha}{2} \left(x_1(T) - 1 \right)^2 + \frac{1}{2} \int\limits_0^T x_2(t)^2 \, dt$$

subject to
$$\dot{x}_1(t) = u(t), \qquad \qquad \text{a.e. in } [0, T],$$
$$\dot{x}_2(t) = y(t), \qquad \qquad \text{a.e. in } [0, T],$$
$$0 = x_1(t) - x_2(t), \qquad \text{in } [0, T]$$
$$x_1(0) = 1,$$
$$u(t) \in [-1, 1], \qquad \qquad \text{a.e. in } [0, T].$$

with final time $T > 0$ and parameter $\alpha > 0$. Differentiating the algebraic constraint with respect to t yields $0 = u(t) - y(t)$, for almost every $t \in [0, T]$, which we can (explicitly) solve for y. Thus, the DAE has index two. The KKT-conditions of the problem read as

$$\dot{\lambda}_1(t) = 0, \qquad \qquad \qquad \qquad \qquad \text{in } [0, T],$$

$$\dot{\lambda}_2(t) = -\hat{x}_2(t), \qquad \qquad \qquad \qquad \text{in } [0, T],$$

$$0 = \lambda_2(t) - \mu(t), \qquad \qquad \qquad \qquad \text{in } [0, T],$$

$$\lambda_1(T) = \alpha \left(\hat{x}_1(T) - 1 \right),$$

$$\lambda_2(T) = 0,$$

$$0 \leq \left(\lambda_1(t) + \mu(t) \right) \left(u - \hat{u}(t) \right), \quad \text{for all } u \in [-1, 1], \quad \text{a.e. in } [0, T].$$

By solving the algebraic equation for μ we obtain $\mu(t) = \lambda_2(t)$ for $t \in [0, 1]$, and therefore we get the switching function

$$\sigma(t) := \lambda_1(t) + \lambda_2(t), \quad t \in [0, 1].$$

Depending on the final time T and the parameter α, the cost functional implies that the optimal control has a bang–bang or a bang-singular-bang switching structure. The structure changes with the critical final time

$$T_{crit}(\alpha) = 1 - \alpha + \sqrt{\alpha^2 + 2\alpha},$$

which satisfies $\lim_{\alpha \to \infty} T_{crit}(\alpha) = 2$. For final time $T > T_{crit}(\alpha)$ the switching function has the zeros 1 and $s(\alpha, T) = T + \alpha - \sqrt{\alpha^2 + 2\alpha} = T - T_{crit}(\alpha) + 1$, and we obtain the solution

$$y(t) = u(t) = \begin{cases} -1, & t \in [0, 1) \\ 0, & t \in [1, s(\alpha, T)) \\ 1, & t \in [s(\alpha, T), T] \end{cases}, \quad x_1(t) = x_2(t) = \begin{cases} 1 - t, & t \in [0, 1) \\ 0, & t \in [1, s(\alpha, T)) \\ t - s(\alpha, T), & t \in [s(\alpha, T), T] \end{cases},$$

$$\lambda_1(t) = \alpha \left(T - s(\alpha, T) - 1 \right),$$

$$\lambda_2(t) = \mu(t) = \begin{cases} \frac{1}{2} \left(T - s(\alpha, T) \right)^2 + \frac{1}{2} (1 - t)^2, & t \in [0, 1) \\ \frac{1}{2} \left(T - s(\alpha, T) \right)^2, & t \in [1, s(\alpha, T)) \\ \frac{1}{2} \left(T - s(\alpha, T) \right)^2 - \frac{1}{2} \left(t - s(\alpha, T) \right)^2, & t \in [s(\alpha, T), T] \end{cases}.$$

Note that the switching function is zero on the interval $[1, s(\alpha, T))$. Hence, the conditions **(A3')–(A5')** are not satisfied and Theorem 6.1 cannot be applied.

For final time $T \leq T_{crit}(\alpha)$ the switching function has one zero at

$$s(\alpha, T) = \frac{1}{3} \left(2\alpha + 2T + 1 - \sqrt{(2\alpha + 1)^2 + 2(\alpha - 1)T + T^2} \right),$$

which satisfies $s(\alpha, T) < 1$, $\lim_{\alpha \to \infty} s(\alpha, T) = \frac{T}{2}$ for $T < T_{crit}(\alpha)$ and $s(\alpha, T_{crit}(\alpha)) = 1$. This yields the solution and switching function

$$y(t) = u(t) = \begin{cases} -1, & t \in [0, s(\alpha, T)) \\ 1, & t \in [s(\alpha, T), T] \end{cases}, \quad x_1(t) = x_2(t) = \begin{cases} 1 - t, & t \in [0, s(\alpha, T)) \\ t + 1 - 2s(\alpha, T), & t \in [s(\alpha, T), T] \end{cases},$$

$$\lambda_1(t) = \alpha \left(T - 2s(\alpha, T) \right),$$

$$\lambda_2(t) = \mu(t) = \begin{cases} \frac{1}{2} \left(T + 1 - 2s(\alpha, T) \right)^2 - \left(1 - s(\alpha, T) \right)^2 + \frac{1}{2} (1 - t)^2, & t \in [0, s(\alpha, T)) \\ \frac{1}{2} \left(T + 1 - 2s(\alpha, T) \right)^2 - \frac{1}{2} \left(t + 1 - 2s(\alpha, T) \right)^2, & t \in [s(\alpha, T), T] \end{cases}.$$

$$\sigma(t) = \begin{cases} -\frac{1}{2}\left(1 - s(\alpha, T)\right)^2 + \frac{1}{2}\left(1 - t\right)^2, & t \in [0, s(\alpha, T)) \\ \frac{1}{2}\left(1 - s(\alpha, T)\right)^2 - \frac{1}{2}\left(t + 1 - 2s(\alpha, T)\right)^2, & t \in [s(\alpha, T), T] \end{cases},$$

which satisfies $\dot{\sigma}(s(\alpha, T)) = 1 - s(\alpha, T)$. Thus, for $T < T_{crit}(\alpha)$ there exists $\varsigma > 0$ such that

$$\left|\dot{\sigma}(s(\alpha, T))\right| = \left|1 - s(\alpha, T)\right| \geq \varsigma > 0,$$

i.e., condition (A5') holds. For $T = T_{crit}(\alpha)$ we have the switching function

$$\sigma(t) = \begin{cases} \frac{1}{2}\left(1 - t\right)^2, & t \in [0, 1) \\ -\frac{1}{2}\left(t - 1\right)^2, & t \in [1, T_{crit}(\alpha)], \end{cases}$$

which satisfies neither condition (A5') nor (A4').

For $\alpha = 1000$ we have $T_{crit}(1000) = 1.9995004993759267$ and for final times $T = 1.5$ and $T = 2.0$ we get the zeros

$$s(1000, 1.5) = 0.7502343000000000, \quad s(1000, 2.0) = 1.0002497502965753.$$

For final time $T = 1.5$ the discrete optimal control and the zero of the discrete switching function converge with order one (compare Table 1). For final time $T = 2.0 \approx T_{crit}(1000)$ conditions (A4') and (A5') are not satisfied, but the discrete optimal control and the zero of the discrete switching function still have linear convergence order (compare Table 2). Cases like this and problems, where the switching function remains zero for a finite length of time, will be studied in the future. In Figs. 1 and 2, we see the optimal control and the switching function for $T = 1.5$ and $T = 2.0$, respectively.

Table 1 Implicit Euler discretization for $\alpha = 1000$, $T = 1.5$

N	$\|\hat{u}_h - \hat{u}\|_1$	Convergence order	$|s - s_h|$	Convergence order
10	0.1984000288809703	1.011406920082053	0.15023430	0.9977517775392430
20	0.0984187648297279	1.023088031191810	0.07523430	0.9955140382199442
40	0.0484281328011752	1.047314992770759	0.03773430	0.9910697073672796
80	0.0234328167871758	1.099556217643134	0.01898430	0.9823035590362839
160	0.0109351587820418	1.222443783190412	0.00960930	0.9652453057076787
320	0.0046863297783591	1.585142270993060	0.00492180	0.9329057899039909
640	0.0015619152880134	–	0.00257806	–

Table 2 Implicit Euler discretization for $\alpha = 1000$, $T = 2.0$

N	$\|\hat{u}_h - \hat{u}\|_1$	Convergence order	$|s - s_h|$	Convergence order
10	0.1983875509228102	1.011407638673553	0.20024975	0.9982018025970495
20	0.0984125259674461	1.023089503364656	0.10024975	0.9964103151109476
40	0.0484250134797348	1.047318084448415	0.05024975	0.9928473149729322
80	0.0234312572348496	1.099563063645673	0.02524975	0.9858001452851699
160	0.0109343791144259	1.222460900264533	0.01274975	0.9720128245428712
320	0.0046859400496147	1.585203576451960	0.00649975	0.9456035968731975
640	0.0015617190299793	–	0.00337475	–

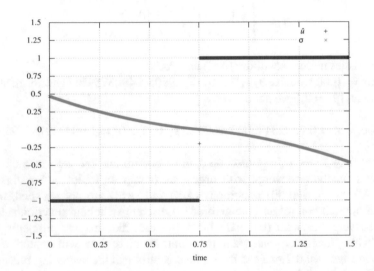

Fig. 1 Optimal control and switching function for $\alpha = 1000$, $T = 1.5$

8 Conclusion

For the implicit Euler discretization of optimal control problems with DAEs and bang–bang optimal controls we derived error estimates (Theorem 6.1). This was achieved by reducing the discretized algebraic constraints (cf. [22, 24]) such that the resulting discrete switching function could be related to the continuous switching function. We therefore extended the results of [7, 8] by including algebraic constraints. A byproduct of our analysis are error estimates for a general class of implicit approximations of linear quadratic optimal control problems (Theorem 4.1 and 5.1).

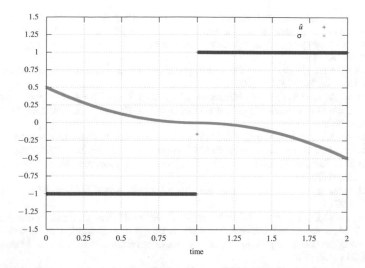

Fig. 2 Optimal control and switching function for $\alpha = 1000$, $T = 2.0$

Acknowledgement This work was supported by the German Research Foundation DFG under the contract GE 1163/8-2

References

1. Alt, W.: Local stability of solutions to differentiable optimization problems in Banach spaces. J. Optim. Theory Appl. **70**, 443–466 (1991).
2. Alt, W., Bräutigam, N.: Finite-difference discretizations of quadratic control problems governed by ordinary elliptic differential equations. Comp. Optim. Appl. **43**, 133–150 (2009)
3. Alt, W., Seydenschwanz, M.: An implicit discretization scheme for linear-quadratic control problems with bang–bang solutions. Optim. Methods Softw. **29**(3), 535–560 (2014)
4. Alt, W., Schneider, C.: Linear-quadratic control problems with L^1-control costs. Optimal Control Appl. Methods **36**(4), 512–534 (2015)
5. Alt, W., Schneider, C., Seydenschwanz, M.: Regularization and implicit Euler discretization of linear-quadratic optimal control problems with bang–bang solutions. Appl. Math. Comput. **287–288**, 104–124 (2016)
6. Alt, W., Felgenhauer, U., Seydenschwanz, M.: Euler discretization for a class of nonlinear optimal control problems with control appearing linearly. Comput. Optim. Appl. **69**(3), 825–856 (2018)
7. Alt, W., Baier, R., Gerdts, M., Lempio, F.: Error bounds for Euler approximation of linear-quadratic control problems with bang–bang solutions. Numer. Algebra Control Optim. **2**(3), 547–570 (2012)
8. Alt, W., Baier, R., Lempio, F., Gerdts, M.: Approximations of linear control problems with bang–bang solutions. Optimization **62**(1), 9–32 (2013)
9. Bonnans, F., Festa, A.: Error estimates for the Euler discretization of an optimal control problem with first-order state constraints. SIAM J. Numer. Anal. **55**(2), 445–471 (2017)

10. Burger, M., Gerdts, M.: DAE aspects in vehicle dynamics and mobile robotics. In: Campbell, S., Ilchmann, A., Mehrmann, V., Reis, T. (eds.) Applications of Differential-Algebraic Equations: Examples and Benchmarks. Differential-Algebraic Equations Forum. Springer, Cham (2018)
11. Dontchev, A.L., Hager, W.W.: The Euler approximation in state constrained optimal control. Math. Comput. **70**, 173–203 (2001)
12. Dontchev, A.L., Hager, W.W., Malanowski, K.: Error bounds for Euler approximation of a state and control constrained optimal control problem. Numer. Funct. Anal. Optim. **21**(5–6), 653–682 (2000)
13. Dontchev, A.L., Hager, W.W., Veliov, V.M.: Second-order Runge–Kutta approximations in control constrained optimal control. SIAM J. Numer. Anal. **38**(1), 202–226 (2000)
14. Ekeland, I., Temam, R.: Convex Analysis and Variational Problems. North Holland, Amsterdam, Oxford (1976)
15. U. Felgenhauer, On stability of bang–bang type controls. SIAM J. Control Optim. **41**, 1843–1867 (2003)
16. Gerdts, M.: Optimal control of ODEs and DAEs. Walter de Gruyter, Berlin/Boston (2012)
17. Gerdts, M., Kunkel, M.: Convergence analysis of Euler discretization of control-state constrained optimal control problems with controls of bounded variation. J. Ind. Manag. Optim. **10**(1), 311–336 (2014)
18. Hager, W.W.: Runge–Kutta methods in optimal control and the transformed adjoint system. Numer. Math. **87**, 247–282 (2000)
19. M. Hinze, A variational discretization concept in control constrained optimization: the linear-quadratic case. Comp. Optim. Appl. **30**, 45–61 (2005)
20. Loxton, R., Lin, Q., Rehbock, V., Teo, K.L.: Control parametrization for optimal control problems with continuous inequality constraints: new convergence results. Numer. Algebra Control Optim. **2**, 571–599 (2012)
21. Malanowski, K., Büskens, C., Maurer, H.: Convergence of approximations to nonlinear optimal control problems. In: Fiacco, A. (eds.) Mathematical Programming with Data Perturbations, vol. 195, pp. 253–284. Lecture Notes in Pure and Applied Mathematics, Dekker (1997)
22. Martens, B.: Necessary conditions, sufficient conditions, and convergence analysis for optimal control problems with differential-algebraic equations. Ph.D. Thesis, Fakultät für Luft- und Raumfahrttechnik, Universität der Bundeswehr München (2019). https://athene-forschung. unibw.de/doc/130232/130232.pdf
23. Martens, B., Gerdts, M.: Convergence analysis of the implicit Euler-discretization and sufficient conditions for optimal control problems subject to index-one differential-algebraic equations. Set-Valued Var. Anal. **27**, 405–431 (2019). https://doi.org/10.1007/s11228-018-0471-x
24. Martens, B., Gerdts, M.: Convergence analysis for approximations of optimal control problems subject to higher index differential-algebraic equations and mixed control-state constraints. SIAM J. Control Optim. **58**(1), 1–33 (2020)
25. P. Merino, F. Tröltzsch, B. Vexler, Error estimates for the finite element approximation of a semilinear elliptic control problem with state constraints and finite dimensional control space. ESAIM Math. Model. Numer. Anal. **44**, 167–188 (2010)
26. Pietrus, A., Scarinci, T., Veliov, V.: High order discrete approximations to Mayer's problems for linear systems. SIAM J. Control Optim. **56**(1), 102–119 (2018)
27. Scarinci, T., Veliov, V.: Higher-order numerical scheme for linear-quadratic problems with bang–bang controls. Comput. Optim. Appl. **69**(2), 403–422 (2018)
28. Stetter, H.J.: Analysis of Discretization Methods for Ordinary Differential Equations. Springer Tracts in Natural Philosophy, vol.23. Springer, Berlin (1973)
29. Veliov, V.: On the time-discretization of control systems. SIAM J. Control Optim. **35**(5), 1470–1486 (1997)
30. Veliov, V.M.: Error analysis of discrete approximations to bang–bang optimal control problems: the linear case. Control. Cybern. **34**, 967–982 (2005)

Part IV
Applications

Port-Hamiltonian Modeling of District Heating Networks

Sarah-Alexa Hauschild, Nicole Marheineke, Volker Mehrmann, Jan Mohring, Arbi Moses Badlyan, Markus Rein, and Martin Schmidt

Abstract This paper provides a first contribution to port-Hamiltonian modeling of district heating networks. By introducing a model hierarchy of flow equations on the network, this work aims at a thermodynamically consistent port-Hamiltonian embedding of the partial differential-algebraic systems. We show that a spatially discretized network model describing the advection of the internal energy density with respect to an underlying incompressible stationary Euler-type hydrodynamics can be considered as a parameter-dependent finite-dimensional port-Hamiltonian system. Moreover, we present an infinite-dimensional port-Hamiltonian formulation for a compressible instationary thermodynamic fluid flow in a pipe. Based on these first promising results, we raise open questions and point out research perspectives concerning structure-preserving discretization, model reduction, and optimization.

Keywords Partial differential equations on networks · Port-Hamiltonian model framework · Energy-based formulation · District heating network · Thermodynamic fluid flow · Turbulent pipe flow · Euler-like equations

Mathematics Subject Classification (2010) 93A30, 35Q31, 37D35, 76-XX

S.-A. Hauschild (✉) · N. Marheineke · M. Schmidt
Trier University, Department of Mathematics, Trier, Germany
e-mail: hauschild@uni-trier.de; marheineke@uni-trier.de; martin.schmidt@uni-trier.de

V. Mehrmann · A. Moses Badlyan
Technische Universität Berlin, Institut für Mathematik, Berlin, Germany
e-mail: mehrmann@math.tu-berlin.de; badlyan@math.tu-berlin.de

J. Mohring · M. Rein
Fraunhofer-ITWM, Kaiserslautern, Germany
e-mail: jan.mohring@itwm.fraunhofer.de; markus.rein@itwm.fraunhofer.de

1 Introduction

A very important part of a successful energy transition is an increasing supply of renewable energies. However, the power supply through such energies is highly volatile. That is why a balancing of this volatility and more energy efficiency is needed. An important player in this context are district heating networks. They show a high potential to balance the fluctuating supply of renewable energies due to their ability to absorb more or less excess power while keeping the heat supply unchanged. A long-term objective is to strongly increase energy efficiency through the intelligent control of district heating networks. The basis for achieving this goal is the dynamic modeling of the district heating network itself, which is not available in the optimization tools currently used in industry. Such a dynamic modeling would allow for optimization of the fluctuating operating resources, e.g., waste incineration, electric power, or gas. However, as power and heating networks act on different time scales and since their descriptions lead to mathematical problems of high spatial dimension, their coupling for a dynamic simulation that is efficiently realizable involves various mathematical challenges. One possible remedy is a port-Hamiltonian modeling framework: Such an energy-based formulation brings the different scales on a single level, the port-Hamiltonian character is inherited during the coupling of individual systems, and in a port-Hamiltonian system the physical principles (stability, passivity, conservation of energy and momentum) are ideally encoded in the algebraic and geometric structures of the model. Deriving model hierarchies by using adequate Galerkin projection-based techniques for structure-preserving discretization as well as model reduction, and combining them with efficient adaptive optimization strategies opens up a new promising approach to complex application issues.

Against the background of this vision, this paper provides a first contribution to port-Hamiltonian modeling of district heating networks, illustrating the potential for optimization in a case study, and raising open research questions and challenges. Port-Hamiltonian (pH) systems have been elaborately studied in literature lately; see, e.g., [3, 23, 40, 43] and the references therein. The standard form of the finite-dimensional dissipative pH-system appears as

$$\frac{dz}{dt} = (J - R)\nabla_z \mathcal{H}(z) + (B - P)u, \quad y = (B + P)^T \nabla_z \mathcal{H}(z) + (S + N)u$$
(1.1a)

with

$$W = W^T \geq 0, \quad W = \begin{bmatrix} R & P \\ P^T & S \end{bmatrix},$$
(1.1b)

cf., e.g., [40]. The Hamiltonian $\mathcal{H}$ is an energy storage function, $J = -J^T$ is the structure matrix describing energy flux among energy storage elements, $R = R^T$

is the dissipation matrix, $B \pm P$ are port matrices for energy in- and output, and $S = S^T$, $N = -N^T$ are matrices associated with the direct feed-through from input u to output y. The system satisfies a dissipation inequality, which is an immediate consequence of the positive (semi-)definiteness of the passivity matrix W. This also holds even when the coefficient matrices depend on the state z, [8], or explicitly on time t, [24], or when they are defined as linear operators on infinite-dimensional spaces [16, 20, 41]. Including time-varying state constraints yields a pH-descriptor system of differential-algebraic equations [3, 23, 39]. Port-Hamiltonian systems on graphs have been studied in [42]. Port-Hamiltonian partial differential equations on networks (port-Hamiltonian PDAE) are topic in, e.g., [10] for linear damped wave equations or in [22] for nonlinear isothermal Euler equations. The adequate handling of thermal effects is a novelty of this work. Thermodynamical aspects have been investigated in the port-Hamiltonian formalism on top of infinite-dimensional pH-models with entropy balance in different fields (such as thermo-magneto-hydrodynamics of plasmas; anisotropic heterogeneous heat equation), see, e.g., [31, 35, 44–46]. In non-equilibrium thermodynamics the GENERIC-framework (GENERIC—General Equation for Non-Equilibrium Reversible-Irreversible Coupling) handles systems with both reversible and irreversible dynamics generated by energy and entropy, respectively, [14, 29, 30]. This framework has been brought together with the port-Hamiltonian formalism in [26, 27]. In this paper we extend the work of [26, 27]. We make use of a thermodynamically consistent generalization of the port-Hamiltonian framework in which the Hamiltonian is combined with an entropy function. The resulting dynamic system consists of a (reversible) Hamiltonian system and a generalized (dissipative) gradient system. Degeneracy conditions ensure that the flows of the two parts do not overlap. Respective pH-models in operator form can be found, e.g., for the Vlasov–Maxwell system in plasma physics in [18, 19], for the Navier–Stokes equations for reactive flows in [1] or for finite strain thermoelastodynamics in [4].

The paper is structured as follows. Starting with the description of a district heating network as a connected and directed graph in Sect. 2, we present models associated to the arcs for the pipelines, consumers, and the depot of the network operator that are coupled with respect to conservation of mass and energy as well as continuity of pressure at the network's nodes. We especially introduce a hierarchy of pipe models ranging from the compressible instationary Navier–Stokes equations for a thermodynamic fluid flow to an advection equation for the internal energy density coupled with incompressible stationary Euler-like equations for the hydrodynamics. Focusing on the latter, we show that the associated spatially discretized network model can be embedded into a family of parameter-dependent standard port-Hamiltonian systems in Sect. 3 and numerically explore the network's behavior in Sect. 4. In a study on operating the heating network with respect to the avoidance of power peaks in the feed-in, we particularly reveal the potential for optimization. In view of the other pipe models, a generalization of the port-Hamiltonian framework to cover the dissipative thermal effects is necessary. In Sect. 5 we develop an infinite-dimensional thermodynamically consistent port-Hamiltonian formulation for the one-dimensional partial differential equations of a compressible instationary turbulent pipe flow. From this, we raise open research questions and perspectives concerning structure-preserving discretization, model reduction, and optimization in Sect. 6.

2 Network Modeling

The district heating network is modeled by a connected and directed graph $G = (N, A)$ with node set N and arc set A. This graph consists of (1) a foreflow part, which provides the consumers with hot water; (2) consumers, that obtain power via heat exchangers; (3) a backflow part, which transports the cooled water back to the depot; and (4) the depot, where the heating of the cooled water takes place; see Fig. 1 for a schematic illustration. The nodes $N = N_{ff} \cup N_{bf}$ are the disjoint union of nodes N_{ff} of the foreflow part and nodes N_{bf} of the backflow part of the network. The arcs $A = A_{ff} \cup A_{bf} \cup A_c \cup \{a_d\}$ are divided into foreflow arcs A_{ff}, backflow arcs A_{bf}, consumer arcs A_c, and the depot arc a_d of the district heating network operator. The set of pipelines is thus given by $A_p = A_{ff} \cup A_{bf}$.

In the following we introduce a model hierarchy for the flow in a single pipe (cf. Fig. 2) and afterward discuss the nodal coupling conditions for the network. Models for consumers (households) and the depot yield the closure conditions for the modeling of the network.

2.1 Model Hierarchy for Pipe Flow

Let $a \in A_p$ be a pipe. Starting point for the modeling of the flow in a pipe are the cross-sectionally averaged one-dimensional instationary compressible Navier–Stokes equations for a thermodynamic fluid flow [34]. We assume that the pipe is cylindrically shaped, that it has constant circular cross-sections, and that the flow quantities are only varying along the cylinder axis. Consider $(x, t) \in (0, \ell) \times (t_0, t_{end}] \subseteq \mathbb{R}^2$ with pipe length ℓ as well as start and end time $t_0, t_{end} > 0$. Mass density, velocity, and internal energy density, i.e., $\rho, v, e : (0, \ell) \times (t_0, t_{end}] \to \mathbb{R}$,

Fig. 1 A schematic district heating network: foreflow arcs are plotted in solid black, backflow arcs in dashed black, consumers (households) in dotted blue, and the depot in dash-dotted red

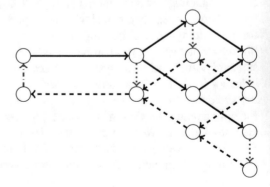

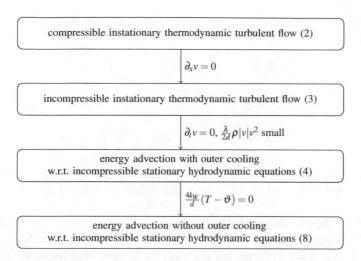

Fig. 2 Hierarchy of pipe flow models

are then described by the balance equations

$$0 = \partial_t \rho + \partial_x (\rho v),$$

$$0 = \partial_t (\rho v) + \partial_x (\rho v^2) + \partial_x p + \frac{\lambda}{2d} \rho |v| v + \rho g \partial_x h, \qquad (2.1)$$

$$0 = \partial_t e + \partial_x (ev) + p \partial_x v - \frac{\lambda}{2d} \rho |v| v^2 + \frac{4k_w}{d} (T - \vartheta).$$

Pressure and temperature, i.e., $p, T : (0, \ell) \times (t_0, t_{end}] \to \mathbb{R}$, are determined by respective state equations. In the momentum balance the frictional forces with friction factor λ and pipe diameter d come from the three-dimensional surface conditions on the pipe walls, the outer forces arise from gravity with gravitational acceleration g and pipe level h (with constant pipe slope $\partial_x h$). The energy exchange with the outer surrounding is modeled by Newton's cooling law in terms of the pipe's heat transmission coefficient k_w and the outer ground temperature ϑ. System (2.1) are (Euler-like) non-linear hyperbolic partial differential equations of first order for a turbulent pipe flow.

The hot water in the pipe is under such a high pressure that it does not turn into steam. Thus, the transition to the incompressible limit of (2.1) makes sense, yielding the following partial differential-algebraic system for velocity v and internal energy density e, where the pressure p acts as a Lagrange multiplier to the incompressibility constraint:

$$0 = \partial_x v,$$

$$0 = \partial_t v + \frac{1}{\rho}\partial_x p + \frac{\lambda}{2d}|v|v + g\partial_x h, \qquad (2.2)$$

$$0 = \partial_t e + v\partial_x e - \frac{\lambda}{2d}\rho|v|v^2 + \frac{4k_w}{d}(T - \vartheta).$$

The system is supplemented with state equations for density ρ and temperature T. Note that the energy term due to friction is negligibly small in this case and can be omitted.

Since the hydrodynamic and thermal effects act on different time scales, System (2.2) may be simplified even further by setting $\partial_t v = 0$. This can be understood as a balancing of the frictional and gravitational forces by the pressure term, while the acceleration is negligibly small, i.e.,

$$0 = \partial_x v,$$

$$0 = \partial_x p + \frac{\lambda}{2d}\rho|v|v + \rho g\partial_x h, \qquad (2.3)$$

$$0 = \partial_t e + v\partial_x e + \frac{4k_w}{d}(T - \vartheta),$$

again supplemented with state equations for ρ, T. System (2.3) describes the heat transport in the pipe where flow velocity and pressure act as Lagrange multipliers to the stationary hydrodynamic equations. However, the flow field is not stationary at all because of the time-dependent closure (boundary) conditions (at households and the depot). In the presented model hierarchy one might even go a step further and ignore the term concerning the heat transition with the outer surrounding of the pipe, i.e., $4k_w(T - \vartheta)/d = 0$, when studying the overall network behavior caused by different operation of the depot; see Sects. 3 and 4.

State Equations and Material Models In the pressure and temperature regime being relevant for operating district heating networks, we model the material properties of water by polynomials depending exclusively on the internal energy density, and not on the pressure. The relations for temperature T, mass density ρ, and kinematic viscosity $\bar{v}$ summarized in Table 1 are based on a fitting of data taken from the NIST Chemistry WebBook [28]. The relative error of the approximation is of order $O(10^{-3})$, which is slightly higher than the error $O(5 \times 10^{-4})$ we observe due to neglecting the pressure dependence. The quadratic state equation for the temperature allows a simple conversion between e and T, which is necessary since closure conditions (households, depot) are usually stated in terms of T; cf. Sect. 2.3. Obviously, $e_\star(T_\star) = 0.5\,T_2^{-1}(-T_1 + (T_1^2 - 4T_2(T_0 - T_\star))^{1/2})$ holds

Table 1 Material properties of water as functions of the internal energy density $z(e) = z_0 \, z_\star(e/e_0)$, $z \in \{T, \rho, \bar{\nu}\}$, where $z_\star$ denotes the dimensionless quantity scaled with the reference value z_0; in particular $e_\star = e/e_0$ with $e_0 = 10^9 \, \mathrm{J \, m^{-3}}$

Reference	Material model	Rel. error
$T_0 = 1\,^{\circ}\mathrm{C}$	$T_\star(e_\star) = 59.2453\,e_\star^2 + 220.536\,e_\star + 1.93729$	1.2×10^{-3}
$\rho_0 = 10^3 \, \mathrm{kg \, m^{-3}}$	$\rho_\star(e_\star) = -0.208084\,e_\star^2 - 0.025576\,e_\star + 1.00280$	6.0×10^{-4}
$\bar{\nu}_0 = 10^{-6} \, \mathrm{m^2 \, s^{-1}}$	$\bar{\nu}_\star(e_\star) = 11.9285\,e_\star^4 - 22.8079\,e_\star^3 + 17.6559\,e_\star^2 - 7.00355\,e_\star + 1.42624$	9.9×10^{-4}

The stated relative errors of the underlying polynomial approximation hold in the regime $e \in [0.2, 0.5] \, \mathrm{GJ \, m^{-3}}$ and $p \in [5, 25] \, \mathrm{bar}$, implying $T \in [50, 130]\,^{\circ}\mathrm{C}$

for $T_\star(e_\star) = \sum_{i=0}^{2} T_i e_\star^i$, $e_\star \geq 0$, where the subscript $\star$ indicates the associated dimensionless quantities.

Remark 2.1 Alternatively to the specific data-driven approach, the state equations can be certainly also deduced more rigorously from thermodynamic laws. A thermodynamic fluid flow described by (2.1) satisfies the entropy balance for $s : (0, \ell) \times (t_0, t_{\text{end}}] \to \mathbb{R}$, i.e.,

$$0 = \partial_t s + \partial_x (sv) - \frac{\lambda}{2d} \frac{1}{T} \rho |v| v^2 + \frac{4k_{\text{w}}}{d} \frac{1}{T} (T - \vartheta).$$

Considering the entropy as a function of mass density and internal energy density, $s = s(\rho, e)$, yields the Gibbs identities which can be used as state equations for pressure p and temperature T, i.e.,

$$\partial_\rho s = -(\rho T)^{-1}(e + p - Ts), \quad \partial_e s = T^{-1}.$$

Pipe-Related Models The pipe flow is mainly driven in a turbulent regime, i.e., with Reynolds number Re $> 10^3$. Thus, the pipe friction factor λ can be described by the Colebrook–White equation in terms of the Reynolds number Re and the ratio of pipe roughness and diameter k_r/d,

$$\frac{1}{\sqrt{\lambda}}(v, e) = -2 \log_{10} \left(\frac{2.52}{\text{Re}(v, e) \sqrt{\lambda}(v, e)} + \frac{1}{3.71} \frac{k_r}{d} \right), \quad \text{Re}(v, e) = \frac{|v| d}{\bar{\nu}(e)}.$$

The model is used for technically rough pipes. Its limit behavior corresponds to the relation by Prandtl and Karman for a hydraulically smooth pipe, i.e., $1/\sqrt{\lambda} = 2 \log_{10}(\text{Re}\sqrt{\lambda}) - 0.8$ for $k_r/d \to 0$, and to the relation by Prandtl, Karman, and Nikuradse for a completely rough pipe, i.e., $1/\sqrt{\lambda} = 1.14 - 2 \log_{10}(k_r/d)$ for Re $\to \infty$, [38]. The underlying root finding problem for λ can be solved using the Lambert W-function; see [7]. However, in view of the computational effort it can also be reasonable to consider a fixed constant Reynolds number for the pipe as further simplification.

The pipe quantities—length ℓ, diameter d, slope $\partial_x h$, roughness k_r, and heat transmission coefficient k_{w}—are assumed to be constant in the pipe model. Moreover, note that in this work we also consider the outer ground temperature ϑ as constant, which will play a role for our port-Hamiltonian formulation of (2.1) in Sect. 5.

2.2 Nodal Coupling Conditions

For the network modeling it is convenient to use the following standard notation. Quantities related to an arc $a = (m, n) \in A$, $m, n \in N$, are marked with the subscript a, quantities associated to a node $n \in N$ with the subscript n. For a node $n \in N$, let $\delta_n^{\text{in}}, \delta_n^{\text{out}}$ be the sets of all topological ingoing and outgoing arcs, i.e.,

$$\delta_n^{\text{in}} = \{a \in A : \exists m \text{ with } a = (m, n)\}, \quad \delta_n^{\text{out}} = \{a \in A : \exists m \text{ with } a = (n, m)\},$$

and let $\mathscr{I}_n(t)$, $\mathscr{O}_n(t)$, $t \in [t_0, t_{\text{end}}]$, be the sets of all flow-specific ingoing and outgoing arcs,

$$\mathscr{I}_n(t) = \{a \in \delta_n^{\text{in}} : q_a(\ell_a, t) \geq 0\} \cup \{a \in \delta_n^{\text{out}} : q_a(0, t) \leq 0\},$$

$$\mathscr{O}_n(t) = \{a \in \delta_n^{\text{in}} : q_a(\ell_a, t) < 0\} \cup \{a \in \delta_n^{\text{out}} : q_a(0, t) > 0\};$$

see, e.g., [12, 13, 15] where a similar notation is used in the context of gas networks. Note that $\mathscr{I}_n(t) \cup \mathscr{O}_n(t) = \delta_n^{\text{in}} \cup \delta_n^{\text{out}}$ holds for all t and that the sets $\mathscr{I}_n(t)$, $\mathscr{O}_n(t)$ depend on the flow q_a, $a \in A$, in the network, which is not known a priori.

The coupling conditions we require for the network ensure the conservation of mass and energy as well as the continuity of pressure at every node $n \in N$ and for all time $t \in [t_0, t_{\text{end}}]$, i.e.,

$$\sum_{a \in \delta_n^{\text{in}}} q_a(\ell_a, t) = \sum_{a \in \delta_n^{\text{out}}} q_a(0, t), \tag{2.4a}$$

$$\sum_{a \in \delta_n^{\text{in}}} \hat{q}_a(\ell_a, t) e_a(\ell_a, t) = \sum_{a \in \delta_n^{\text{out}}} \hat{q}_a(0, t) e_a(0, t), \quad e_a(0, t) = e_n(t), \quad a \in \mathscr{O}_n(t), \tag{2.4b}$$

$$p_a(\ell_a, t) = p_n(t), \quad a \in \delta_n^{\text{in}}, \qquad p_a(0, t) = p_n(t), \quad a \in \delta_n^{\text{out}}. \tag{2.4c}$$

Here, q_a and $\hat{q}_a$ denote the mass flow and the volumetric flow in pipe a, respectively. They scale with the mass density, i.e., $q_a = \rho_a v_a \varsigma_a$ and $\hat{q}_a = q_a / \rho_a$, where $\varsigma_a = d_a^2 \pi / 4$ is the cross-sectional area of the pipe. In case of incompressibility, it holds that $\hat{q}_a(x, t) = \hat{q}_a(t)$ is constant along the pipe. The functions e_n and p_n are auxiliary variables describing internal energy density and pressure at node n. Note that the second condition in (2.4b), namely that the out-flowing energy densities are identical in all (flow-specific outgoing) pipes, rests upon the assumption of instant mixing of the in-flowing energy densities.

2.3 Households, Depot, and Operational Constraints

The network modeling is closed by models for the consumers (households) and the depot of the network operator. Quantities associated to the arc a at node n are indicated by the subscript $a : n$.

For the consumer at $a = (m, n) \in A_c$, where the nodes m and n belong to the foreflow and backflow part of the network, respectively (cf. Fig. 1), the following conditions are posed for $t \in [t_0, t_{end}]$,

$$P_a(t) = \hat{q}_a(t)\Delta e_a(t), \qquad v_a(t) \geq 0, \qquad\qquad \Delta e_a(t) = e_{a:m}(t) - e_{a:n}(t), \tag{2.5a}$$

$$T_{a:n}(t) = T^{bf}, \qquad T_{a:m}(t) \in [T_-^{ff}, T_+^{ff}], \quad T_{a:m}(t) - T_{a:n}(t) \leq \Delta T^c, \tag{2.5b}$$

$$p_{a:n}(t) \in [p_-^{bf}, p_+^{bf}], \qquad p_{a:m}(t) \in [p_-^{ff}, p_+^{ff}], \quad p_{a:m}(t) - p_{a:n}(t) \in [\Delta p_-^c, \Delta p_+^c]. \tag{2.5c}$$

The prescribed power consumption P_a of the household is realized by the product of the energy density difference at the arc and the volumetric flow in (2.5a). Moreover, the underlying flow velocity has a pre-specified direction. The consumer's outflow temperature is set to be equal to the contractually agreed temperature T^{bf}. Moreover, the operational constraints ensure a certain temperature range at each consumption point and define a maximal temperature difference between foreflow and backflow part of the consumers. In addition, minimal and maximal values for the pressure level at both backflow and foreflow part of the consumer arcs are prescribed. Finally, the pressure difference between foreflow and backflow part is bounded.

The depot $a_d = (m, n)$ for operating the district heating network is modeled by the following conditions for $t \in [t_0, t_{end}]$:

$$e_{a_d:n}(t) = u^e(t), \quad T_{a_d:n}(t) \leq T^{net}, \qquad\qquad v_{a_d}(t) \geq 0, \tag{2.6a}$$

$$p_{a_d:m}(t) = u^p(t), \quad p_{a_d:n}(t) = p_{a_d:m}(t) + u^{\Delta p}(t). \tag{2.6b}$$

Here, u^p prescribes the so-called stagnation pressure of the network and $u^{\Delta p}$ is the realized pressure increase at the depot. The energy density injected at the depot to the foreflow part of the network is denoted by u^e. The resulting temperature is bounded above by T^{net}, which also acts as temperature limit for all network nodes.

In addition to the operational constraints in (2.5) and (2.6), the pressure in all network nodes is bounded, i.e., $p_n(t) \leq p^{net}$ for $n \in N$ and $t \in [t_0, t_{end}]$.

3 Port-Hamiltonian Formulation of a Semi-Discrete Network Model

In this section we present a spatially semi-discrete model variant for the district heating network and discuss its formulation in the port-Hamiltonian context. Making use of the different hydrodynamic and thermal time scales, a finite volume upwind discretization yields a port-Hamiltonian descriptor system for the internal energy density, in which the solenoidal flow field acts as a time-varying parameter.

We describe the network by means of the following partial differential-algebraic system for $t \in [t_0, t_{end}]$,

$$\partial_t e_a = -v_a \partial_x e_a, \quad a \in A_p, \tag{3.1a}$$

$$e_a(0, t) = e_n(t), \quad a \in \mathcal{O}_n(t), \quad \sum_{a \in \delta_n^{in}} \hat{q}_a e_a(\ell_a, t) = \sum_{a \in \delta_n^{out}} \hat{q}_a e_a(0, t), \quad n \in N, \tag{3.1b}$$

$$e_{a:n}(t) = e^{bf}, \quad a \in A_c, \tag{3.1c}$$

$$e_{a:n}(t) = u^e(t), \quad a = a_d, \tag{3.1d}$$

$$g(e, v, p) = 0. \tag{3.1e}$$

This system results from the incompressible pipe model in (2.3) and neglecting the cooling term in the energy balance (i.e., $k_w = 0$). Here, the condition on the backflow temperature for the consumers is expressed in terms of the internal energy density, cf., $e^{bf} = e(T^{bf})$ in (3.1c). In the formulation we use the separation of thermal and hydrodynamic effects and state the temporal advection of the internal energy density with respect to the algebraic equations covering the hydrodynamics. So, $g(e, v, p) = 0$ in (3.1e) contains the hydrodynamic pipe equations, the pressure continuity at the nodes (2.4c), the condition on the households' power consumption (2.5a), the pressure conditions at the depot (2.6b), and the conservation of volume

$$\sum_{b \in \delta_n^{in}} \hat{q}_b(t) = \sum_{a \in \delta_n^{out}} \hat{q}_a(t), \quad n \in N. \tag{3.2}$$

Considering the volume balance (3.2) instead of the mass balance (2.4a) is very convenient in the incompressible setting, since the velocity field and hence the induced volumetric flow are constant along a pipe. Moreover, this description naturally fits the numerical method of finite volumes.

For the spatial discretization of the hyperbolic-like system (3.1) we apply a classical finite volume upwind scheme [21]. Let $\alpha \in A_p$, $\alpha \in \mathcal{O}_n(t_0)$, $n \in N$, and consider an equidistant mesh of cell size Δx_α, then

$$\frac{d}{dt} e_{\alpha,\beta} = -\frac{v_\alpha}{\Delta x_\alpha}(e_{\alpha,\beta} - e_{\alpha,\beta-1}), \quad \beta \in V_\alpha,$$

$$e_{\alpha,0} = e_n, \quad e_n = \frac{\sum_{b \in \mathcal{I}_n} \hat{q}_b \, e_{b,|V_b|}}{\sum_{a \in \mathcal{O}_n} \hat{q}_a},$$

where $e_{\alpha,\beta}$ denotes the internal energy density with respect to the finite volume cell β of pipe α with cell index set V_α. For the first cell ($\beta = 1$) we make use of the quantity at the node that results from (3.1b). We summarize the unknown energy densities in a vector $e = (e_1, \dots, e_\kappa)^T$, $e_{f(\alpha,\beta)} = e_{\alpha,\beta}$ by ordering pipe- and cell-wise according to the mapping $f(\alpha, \beta) = \beta + \sum_{k=1}^{\alpha-1} |V_k|$, $\alpha \in A_p$, $\beta \in V_\alpha$, in particular $\kappa = \sum_{\alpha \in A_p} |V_\alpha|$. Then, a semi-discrete version of the network model (3.1) is given by the following descriptor system

$$\frac{d}{dt} e = A(v)\, e + B(v)\, u, \quad y = Ce, \tag{3.3}$$

subject to $v = G(e)$.

The system matrices $A(w) \in \mathbb{R}^{\kappa \times \kappa}$ and $B(w) \in \mathbb{R}^{\kappa \times 2}$ can be interpreted as parameter-dependent quantities, where the (vector-valued) parameter w represents a spatially discretized solenoidal volume-preserving velocity field. So,

$$A_{f(\alpha,\beta), f(\mu,\sigma)}(w) = \partial \frac{d}{dt} e_{\alpha,\beta}(w)/\partial e_{\mu,\sigma}$$

holds. The special velocity field belonging to the hydrodynamic network equations (3.1e) is formally stated as $v = G(e)$. We assume a setting in which v is time-continuous. In (3.3) the input u consists of the energy densities u^e injected at the depot into the foreflow part and e^{bf} returning from the consumers into the backflow part of the network, $u = (u^e, e^{bf})^T \in \mathbb{R}^2$. The output y typically refers to energy densities in pipes supplying the consumers, implying $C \in \mathbb{R}^{c \times \kappa}$.

Theorem 3.1 *Let w be a (spatially discretized) solenoidal volume-preserving time-continuous velocity field. Then, the semi-discrete network model (3.3) can be embedded into a family of parameter-dependent port-Hamiltonian systems*

$$\frac{d}{dt} e = (J(w) - R(w))Qe + \tilde{B}(w)\tilde{u}, \quad \tilde{y} = \tilde{B}^T(w)Qe, \tag{3.4}$$

with $\tilde{u} = (u^T, 0, \dots, 0)^T \in \mathbb{R}^{2+c}$ which contains the original outputs as subset.

Remark 3.1 Theorem 3.1 implies that there exists an energy matrix Q such that

$$QA(w) + A^T(w)Q \leq 0 \tag{3.5}$$

for all solenoidal volume-preserving velocity fields w. Thus, the Hamiltonian $\mathcal{H}(e) = e^T Q e$ is a Lyapunov function for the parameter-dependent system [2]. The energy matrix Q can be particularly constructed as a diagonal matrix with positive entries, i.e., $Q_{f(\alpha,\beta),f(\alpha,\beta)} = \varsigma_\alpha \Delta x_\alpha$ for $\alpha \in A_p$, $\beta \in V_\alpha$, where $\varsigma_\alpha \Delta x_\alpha$ is the volume of each discretization cell in pipe α.

Note that a change of the flow direction, which might occur in case of cycles, yields a structural modification of the system matrix $A(w)$, but does not affect the stability of the system. However, it might cause a discontinuity in the velocity field such that (3.3), or (3.4) respectively, only allows for a weak solution.

Proof of Theorem 3.1 Let the positive definite diagonal matrix $Q \in \mathbb{R}^{\kappa \times \kappa}$ with $Q_{f(\alpha,\beta),f(\alpha,\beta)} = \varsigma_\alpha \Delta x_\alpha > 0$ be given. Then, we define the matrices J and R by

$$J(w) = \frac{1}{2}(A(w)Q^{-1} - (A(w)Q^{-1})^T), \quad R(w) = -\frac{1}{2}(A(w)Q^{-1} + (A(w)Q^{-1})^T).$$

Obviously, $A(w) = (J(w) - R(w))Q$ holds. The properties $J = -J^T$ and $R = R^T$ of port-Hamiltonian system matrices are satisfied by construction for any parameter w. The positive semi-definiteness of R follows from the Lyapunov inequality (3.5). Considering

$$L(w) = QA(w) + A^T(w)Q, \quad L_{f(\alpha,\beta),f(\alpha,\beta)}(w) = -2Q_{f(\alpha,\beta),f(\alpha,\beta)}\frac{w_\alpha}{\Delta x_\alpha} = -2\hat{q}_\alpha \leq 0,$$

the volume-preservation of w ensures that the symmetric matrix $L(w)$ is weakly diagonal dominant. Hence, $L(w)$ is negative semi-definite, yielding

$$x^T R(w)x = -\frac{1}{2}(Q^{-1}x)^T L(w)(Q^{-1}x) \geq 0 \quad \text{for all} \quad x \in \mathbb{R}^\kappa.$$

Here, $R(w)$ acts as the passivity matrix since the system has no feed-through term. The port matrix $\tilde{B}(w) \in \mathbb{R}^{\kappa \times 2 + c}$ defined by

$$\tilde{B}(w) = [B(w), \ (CQ^{-1})^T]$$

ensures that the outputs of the network model are contained in the output set of the port-Hamiltonian system, i.e., $\tilde{B}^T(w)Q = [B^T(w)Q, \ C]^T$. Finally note that the parameter-dependent port-Hamiltonian system matrices $J(w)$, $R(w)$, and $\tilde{B}(w)$ are continuous in time due to the given time-regularity of the parameter w.

Remark 3.2 We point out that applying the stated framework to the other pipe models presented in Sect. 2.1 is non-trivial. Already the consideration of the cooling term in the energy balance, cf. pipe model (2.3), which acts dissipative requires a generalization of the port-Hamiltonian description. We refer to Sect. 5 for an infinite-dimensional port-Hamiltonian formulation of the compressible thermodynamic pipe flow (2.1).

4 Numerical Study on Network Operation

In this section we demonstrate the potential for optimization of district heating networks. Operating the network according to certain exogenously given temporal profiles for the internal energy densities injected at the depot may lead to high amplitudes in the feed-in power. The avoidance of such power peaks in the feed-in prevents that using additional energy sources, such as gas storages, is required for covering the heating demand of the consumers. This is environmental friendly, while saving resources and operational costs.

In the numerical case study we employ a real-world district heating network supplying different streets by means of the port-Hamiltonian semi-discrete network model (3.3). The model describes the advection-driven internal energy density with respect to an underlying incompressible flow field with negligible acceleration. Thermal losses/outer cooling effects are neglected. For the time integration we use an implicit midpoint rule with constant time step Δt. The topology of the network and the data of the pipelines come from the Technische Werke Ludwigshafen AG; see Fig. 3 and Table 2. Mass density and friction factor are taken as constant in time for every pipeline. For the presented simulation, a time horizon of 50 h is studied. The consumption behavior of the households is modeled by standardized profiles used in the operation of district heating networks [5] for a mean environmental temperature of 3 °C. The total consumption of all households is 108 kW on temporal average and rises up to a maximum of 160 kW. Given the internal energy density u^e injected at the depot as input, the feed-in power can be considered as the response of the network system, i.e.,

$$P_{\text{in}} = (u^e - e_{a_d:m}) \sum_{a \in A_c} \hat{q}_a.$$

Note that due to the neglect of cooling in (3.1), $e_{a_d:m} = e(T^{bf})$ holds, where the backflow temperature at the consumers is fixed here to $T^{bf} = 60\,°\text{C}$.

The traveling time of the heated water from the depot to the consumers (households) allows to choose from different injection profiles, when covering the aggregated heating demand in the network. Figure 4 shows the injected temperature $T(u^e)$ and the corresponding feed-in power for two different input profiles. Supplying an almost constant energy density u^e over time yields pronounced power peaks

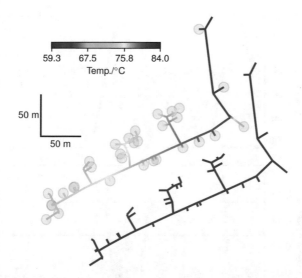

59.3   67.5   75.8   84.0
Temp./°C

50 m

50 m

Fig. 3 Real-world heating network supplying several streets. The network consists of the foreflow part (top) and the backflow part (bottom), where the households are indicated by circles. The topology has been provided by Technische Werke Ludwigshafen AG, Germany. The color plot visualizes a simulated temperature distribution for a certain time t^*, where $T(u^e(t^*)) = 84\,°C$. The backflow temperature is constant at $T^{bf} = 60\,°C$ due to the use of the network model (3.1) where cooling effects are neglected

Table 2 Graph-associated outline data for the street network in Fig. 3

Pipes $|A_p|$	Consumers $|A_c|$	Depot	Arcs $|A|$	Nodes $|N|$	Loops
162	32	1	195	162	2

The total pipe length of the foreflow part is 835.5 m and of the backflow part 837.0 m

(dashed-dotted red curves). These undesired peaks can be avoided when using an input that is varying in time with respect to the expected consumer demands. The improved input conducts a preheating strategy. By anticipating typical maxima in demand patterns, the injected input energy density is increased in times of small power demands. If the dynamically changing transport time from the power plant to the households is reflected correctly, the additionally injected thermal energy is then available to the consumers in times of high consumption. This strategy allows to successfully bound the feed-in power, here, as illustrated, by $\bar{P}_{in} = 134\,kW$ (dashed green curves). This promising result asks for a rigorous optimal control of the network in further studies.

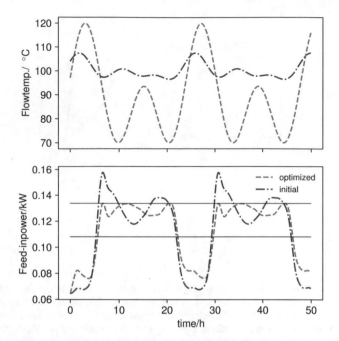

Fig. 4 Flow temperature at depot $T(u^e)$ (top) and corresponding feed-in power (bottom) over time for two different injection profiles marked in dashed-dotted red and dashed green, $\Delta t = 5$ min. The upper solid, black line indicates the power threshold $\bar{P}$, the lower one the mean feed-in power over time

5 Port-Hamiltonian Formulation of Compressible Thermodynamic Pipe Flow

The adequate handling of thermal effects requires the generalization of the port-Hamiltonian framework by combining the Hamiltonian with an entropy function. In this section we embed the partial differential model (2.1) for a compressible thermodynamic turbulent pipe flow into the GENERIC-formalism, which has lately been studied in [26, 27], and present an infinite-dimensional thermodynamically consistent port-Hamiltonian description. The following state space model encodes (2.1) in a weak form. Assuming the existence of a smooth solution and nicely behaving boundary terms, the partial differential model can be obtained through integration by parts as shown for a simplified example in [27].

The thermodynamic pipe flow model (2.1) can be reformulated as a generalized (non-linear) port-Hamiltonian system in operator form for $z = (\rho, M, e)^T$, $M = (\rho v)$,

$$\frac{dz}{dt} = \left(\mathscr{J}(z) - \mathscr{R}(z)\right)\frac{\delta\mathscr{E}(z)}{\delta z} + \mathscr{B}(z)u(z) \qquad \text{in } \mathscr{D}_z^*,$$

$$y(z) = \mathscr{B}^*(z)\frac{\delta\mathscr{E}(z)}{\delta z} \qquad \text{in } \mathscr{D}_u^*,$$

(5.1)

where $\mathscr{Z} = \{z \in \mathscr{D}_z \mid \rho \geq \delta \text{ with } \delta > 0 \text{ almost everywhere}\} \subset \mathscr{D}_z$ denotes the state space with the Sobolev space $\mathscr{D}_z = W^{1,3}((0, \ell); \mathbb{R}^3)$ being a reflexive Banach space. For $z \in \mathscr{Z}$ the operators $\mathscr{J}(z)[\cdot]$, $\mathscr{R}(z)[\cdot] : \mathscr{D}_z \to \mathscr{D}_z^*$ are linear and continuous, moreover $\mathscr{J}(z)$ is skew-adjoint and $\mathscr{R}(z)$ is self-adjoint semi-elliptic, i.e., $\langle \varphi, \mathscr{J}(z)\psi \rangle = -\langle \psi, \mathscr{J}(z)\varphi \rangle$ and $\langle \varphi, \mathscr{R}(z)\psi \rangle = \langle \psi, \mathscr{R}(z)\varphi \rangle \geq 0$ for all $\varphi, \psi \in \mathscr{D}_z$. The system theoretic input is given by $u(z) \in \mathscr{D}_u = L^q(\{0, \ell\})$ with linear continuous operator $\mathscr{B}(z)[\cdot] : \mathscr{D}_u \to \mathscr{D}_z^*$ and dual space $\mathscr{D}_u^* = L^p(\{0, \ell\})$, $1/q + 1/p = 1$. The system theoretic output is denoted by $y(z)$. The form of the energy functional $\mathscr{E}$ and the port-Hamiltonian operators $\mathscr{J}(z)[\cdot]$, $\mathscr{R}(z)[\cdot]$ and $\mathscr{B}(z)[\cdot]$ are derived as follows.

Remark 5.1 We assume that all relevant mathematical statements hold for an arbitrary but fixed time parameter $t \in (t_0, t_{\text{end}}]$. The function spaces $\mathscr{D}_z$ and $\mathscr{D}_u$ associated with the spatial evolution are chosen in an ad-hoc manner, i.e., we assume that the considered fields and functions satisfy certain regularity requirements. A mathematically rigorous justification requires an analytical consideration of the generalized port-Hamiltonian system. The corresponding functional analytical and structural questions are the focus of ongoing work.

Accounting for the thermodynamic behavior of the pipe flow, (5.1) is composed of a Hamiltonian and a generalized gradient system. This is reflected in the energy functional that is an exergy-like functional consisting of a Hamiltonian and an entropy part, i.e.,

$$\mathscr{E}(z) = \mathscr{H}(z) - \vartheta\mathscr{S}(z), \qquad \mathscr{H}(z) = \int_0^\ell \left(\frac{|M|^2}{2\rho} + e + \rho g h\right) dx, \qquad \mathscr{S}(z) = \int_0^\ell s(\rho, e)\, dx.$$

where the outer ground temperature ϑ is assumed to be constant. Introducing the ballistic free energy $H(\rho, e) = e - \vartheta s(\rho, e)$ [11], the functional $\mathscr{E}$ and its variational derivatives become

$$\mathscr{E}(z) = \int_0^\ell \left(\frac{|M|^2}{2\rho} + H(\rho, e) + \rho g h\right) dx$$

$$\frac{\delta\mathscr{E}(z)}{\delta z} = \left(\frac{\delta\mathscr{E}(z)}{\delta\rho}, \frac{\delta\mathscr{E}(z)}{\delta M}, \frac{\delta\mathscr{E}(z)}{\delta e}\right)^T = \left(\left(-\frac{|M|^2}{2\rho^2} + \frac{\partial H}{\partial\rho} + gh\right), \frac{M}{\rho}, \frac{\partial H}{\partial e}\right)^T.$$

The port-Hamiltonian operators in (5.1) are assembled with respect to the (block-) structure of the state z. Let $\varphi, \psi \in \mathcal{D}_z$ be two block-structured test functions, i.e., $\varphi = (\varphi_\rho, \varphi_M, \varphi_e)^T$. Then the skew-adjoint operator $\mathcal{J}(z)$ is given by

$$
\mathcal{J}(z) = \begin{bmatrix} 0 & \mathcal{J}_{\rho,M}(z) & 0 \\ \mathcal{J}_{M,\rho}(z) & \mathcal{J}_{M,M}(z) & \mathcal{J}_{M,e}(z) \\ 0 & \mathcal{J}_{e,M}(z) & 0 \end{bmatrix}, \tag{5.2a}
$$

associated with the bilinear form

$$
\langle \varphi, \mathcal{J}(z)\psi \rangle = \langle \varphi_\rho, \mathcal{J}_{\rho,M}(z)\psi_M \rangle + \langle \varphi_M, \mathcal{J}_{M,\rho}(z)\psi_\rho \rangle + \langle \varphi_M, \mathcal{J}_{M,M}(z)\psi_M \rangle
$$
$$
+ \langle \varphi_M, \mathcal{J}_{M,e}(z)\psi_e \rangle + \langle \varphi_e, \mathcal{J}_{e,M}(z)\psi_M \rangle.
$$

Its entries are particularly defined by the following relations,

$$
\langle \varphi_\rho, \mathcal{J}_{\rho,M}(z)\psi_M \rangle = -\langle \psi_M, \mathcal{J}_{M,\rho}(z)\varphi_\rho \rangle = \int_0^\ell \rho(\psi_M \partial_x)\varphi_\rho \, dx, \tag{5.2b}
$$

$$
\langle \varphi_M, \mathcal{J}_{M,M}(z)\psi_M \rangle = -\langle \psi_M, \mathcal{J}_{M,M}(z)\varphi_M \rangle = \int_0^\ell M((\psi_M \partial_x)\varphi_M - (\varphi_M \partial_x)\psi_M) \, dx, \tag{5.2c}
$$

$$
\langle \varphi_e, \mathcal{J}_{e,M}(z)\psi_M \rangle = -\langle \psi_M, \mathcal{J}_{M,e}(z)\varphi_e \rangle = \int_0^\ell e(\psi_M \partial_x)\varphi_e + (\psi_M \partial_x)(\varphi_e p) \, dx \tag{5.2d}
$$

that result from the partial derivatives in (2.1). The self-adjoint semi-elliptic operator $\mathcal{R}(z)$ is composed of two operators that correspond to the friction in the pipe $\mathcal{R}^\lambda(z)$ and the temperature loss through the pipe walls $\mathcal{R}^{kw}(z)$. It is given by

$$
\mathcal{R}(z) = \mathcal{R}^\lambda(z) + \mathcal{R}^{kw}(z) = \begin{bmatrix} 0 & 0 & 0 \\ 0 & \mathcal{R}^\lambda_{M,M}(z) & \mathcal{R}^\lambda_{M,e}(z) \\ 0 & \mathcal{R}^\lambda_{e,M}(z) & \mathcal{R}^\lambda_{e,e}(z) + \mathcal{R}^{kw}_{e,e}(z) \end{bmatrix}, \tag{5.3a}
$$

associated with the bilinear form,

$$
\langle \varphi, \mathcal{R}(z)\psi \rangle = \langle \varphi_M, \mathcal{R}^\lambda_{M,M}(z)\psi_M \rangle + \langle \varphi_M, \mathcal{R}^\lambda_{M,e}(z)\psi_e \rangle + \langle \varphi_e, \mathcal{R}^\lambda_{e,M}(z)\psi_M \rangle
$$
$$
+ \langle \varphi_e, (\mathcal{R}^\lambda_{e,e}(z) + \mathcal{R}^{kw}_{e,e}(z))\psi_e \rangle.
$$

Its entries are

$$\langle \varphi_M, \mathcal{R}^{\lambda}_{M,M}(z)\psi_M \rangle = \int_0^{\ell} \varphi_M \left(\frac{\lambda}{2d} \frac{T}{\vartheta} \rho|v| \right) \psi_M \, dx, \qquad (5.3b)$$

$$\langle \varphi_M, \mathcal{R}^{\lambda}_{M,e}(z)\psi_e \rangle = \langle \psi_e, \mathcal{R}^{\lambda}_{e,M}(z)\varphi_M \rangle = \int_0^{\ell} -\varphi_M \left(\frac{\lambda}{2d} \frac{T}{\vartheta} \rho|v|v \right) \psi_e \, dx,$$

$$(5.3c)$$

$$\langle \varphi_e, (\mathcal{R}^{\lambda}_{e,e}(z) + \mathcal{R}^{k_w}_{e,e}(z))\psi_e \rangle = \int_0^{\ell} \varphi_e \left(\frac{\lambda}{2d} \frac{T}{\vartheta} \rho|v|v^2 + \frac{4k_w}{d} T \right) \psi_e \, dx.$$

$$(5.3d)$$

Note that the state dependencies of pressure $p = p(\rho, e)$ and temperature $T = T(\rho, e)$ occurring in (5.2d) and (5.3b)–(5.3d) are prescribed by the state equations, cf. Remark 2.1. Moreover, $v = M/\rho$ and $\lambda = \lambda(v, e)$ hold for the velocity and the friction factor, respectively. Assuming consistent state equations, e.g., ideal gas law, cf. Remark 5.2, the operators in (5.2) and (5.3) fulfill the non-interacting conditions

$$\mathcal{J}(z) \frac{\delta \mathcal{S}(z)}{\delta z} = 0, \qquad \mathcal{R}^{\lambda}(z) \frac{\delta \mathcal{H}(z)}{\delta z} = 0,$$

which arise in the GENERIC context [26, 27] and ensure that the flows of the Hamiltonian and the gradient system do not overlap. Finally, concerning the system theoretic input and output, the state dependent input is given as $u(z) \in \mathcal{D}_u$ by $u(z) = [M/\rho]|_0^{\ell}$. Then, the port operator $\mathcal{B}(z)[\cdot] : \mathcal{D}_u \to \mathcal{D}_z^*$ is specified through the pairing

$$\langle \varphi, \mathcal{B}(z)u(z) \rangle = - \left[(\varphi_{\rho}\rho + \varphi_M M + \varphi_e(e + p)) u(z) \right]\Big|_0^{\ell},$$

which originates from the boundary terms, when applying partial integration to parts of (2.1). With the adjoint operator $\mathcal{B}^*(z)[\cdot] : \mathcal{D}_z \to \mathcal{D}_u^*$, i.e., $\langle \varphi, \mathcal{B}(z)u(z) \rangle = \langle \mathcal{B}^*(z)\varphi, u(z) \rangle$, the system theoretic output reads

$$y(z) = \mathcal{B}^*(z) \frac{\delta \mathcal{E}(z)}{\partial z} = - \left[\frac{|M|^2}{2\rho} + p + H(\rho, e) + \rho g h \right]\Bigg|_0^{\ell}.$$

Remark 5.2 In the port-Hamiltonian framework the choice of the state variables in the interplay with the energy functional is crucial for encoding the physical properties in the system operators. Hence, asymptotic simplifications as, e.g., the limit to incompressibility in the hydrodynamics (2.2), are not straightforward, since they change the underlying equation structure. However, system (5.1) is well suited when, e.g., dealing with gas networks. Then, it can be closed by using, e.g., the ideal gas law, implying

$$s(\rho, e) = \frac{R}{2}\rho \ln\left(c_p \frac{e^3}{\rho^5}\right), \quad T(\rho, e) = \frac{2}{3R}\frac{e}{\rho}, \quad p(\rho, e) = \frac{2}{3}e,$$

with specific gas constant R and heat capacity c_p.

6 Research Perspectives

An energy-based port-Hamiltonian framework is very suitable for optimization and control when dealing with subsystems coming from various different physical domains, such as hydraulic, electrical, or mechanical ones, as it occurs when coupling a district heating network with a power grid, a waste incineration plant, or a gas turbine. The formulation is advantageous as it brings different scales on a single level, the port-Hamiltonian character is inherited by the coupling, and the physical properties are directly encoded in the structure of the equations. However, to come up with efficient adaptive optimization strategies based on port-Hamiltonian model hierarchies for complex application issues on district heating networks, there are still many mathematical challenges to be handled.

In this paper we contributed with an infinite-dimensional and thermodynamically consistent formulation for a compressible turbulent pipe flow, which required to set up a (reversible) Hamiltonian system and a generalized (dissipative) gradient system with suitable degeneracy conditions. In particular, the choice of an appropriate energy function was demanding. The asymptotic transition to an incompressible pipe flow is non-trivial in this framework, since it changes the differential-algebraic structure of the equations and hence requires the reconsideration of the variables and the modification of the energy function. In view of structure-preserving discretization and model reduction the use of Galerkin projection-based techniques seems to be promising. Lately, partitioned finite element methods for structure-preserving discretization have been developed in [36, 37]. However, the choice of the variables and the formulation of the system matrices crucially determine the complexity of the numerics as, e.g., the works [6, 9, 22] show. Especially, the handling of the nonlinearities requires adequate complexity-reduction strategies. Interesting to explore are certainly also structure-preserving time-integration schemes, see, e.g., [17, 24]. The port-Hamiltonian formulation of the complete network is topic of current research.

In the special case of the presented semi-discrete district heating network model that makes use of the different hydrodynamic and thermal time scales and a suitable finite volume upwind discretization we came up with a finite-dimensional port-Hamiltonian system for the internal energy density where the solenoidal flow field acts a time-varying parameter. This system is employed for model reduction (moment matching) in [32] and for optimal control in [33].

The application of the port-Hamiltonian modeling framework for coupled systems leads to many promising ideas for the optimization of these systems. Due to the complexity and size of the respective optimization models, a subsystem-specific port-Hamiltonian modeling together with suitable model reduction techniques allows for setting up a coupled model hierarchy for optimization, which paves the way for highly efficient adaptive optimization methods; cf., e.g., [25], where a related approach has shown to be useful for the related field of gas network optimization.

Acknowledgments The authors acknowledge the support by the German BMBF, Project *EiFer—Energy efficiency via intelligent heating networks* and are very grateful for the provision of the data by their industrial partner Technische Werke Ludwigshafen AG. Moreover, the support of the DFG within the CRC TRR 154, subprojects A05, B03, and B08, as well as within the RTG 2126 *Algorithmic Optimization* is acknowledged.

References

1. Altmann, R., Schulze, P.: A port-Hamiltonian formulation of the Navier-Stokes equations for reactive flows. Syst. Control Lett. **100**, 51–55 (2017). https://doi.org/10.10.16/j.sysconle.2016.12.005
2. Antoulas, A.: Approximation of Large-Scale Dynamical Systems. SIAM, Philadelphia (2005). https://doi.org/10.1137/1.9780898718713
3. Beattie, C., Mehrmann, V., Xu, H., Zwart, H.: Linear port-Hamiltonian descriptor systems. Math. Control Signals Syst. **30**, 17 (2018). https://doi.org/10.1007/s00498-018-0223-3
4. Betsch, P., Schiebl, M.: Energy-momentum-entropy consistent numerical methods for large-strain thermoelasticity relying on the GENERIC formalism. Int. J. Numer. Methods Eng. **119**(12), 1216–1244 (2019). https://doi.org/10.1002/nme.6089
5. Bundesverband der deutschen Gas- und Wasserwirtschaft (BGW): Anwendung von Standardlastprofilen zur Belieferung nichtleistungsgemessener Kunden (2006). http://www.gwb-netz.de/wa_files/05_bgw_leitfaden_lastprofile_56550.pdf. Accessed July 23 2019
6. Chaturantabut, S., Beattie, C., Gugercin, S.: Structure-preserving model reduction for nonlinear port-Hamiltonian systems. SIAM J. Sci. Comput. **38**(5), B837–B865 (2016). https://doi.org/10.1137/15M1055085
7. Clamond, D.: Efficient resolution of the Colebrook equation. Ind. Eng. Chem. Res. **48**(7), 3665–3671 (2009). https://doi.org/10.1021/ie801626g
8. Dalsmo, M., van der Schaft, A.: On representations and integrability of mathematical structures in energy-conserving physical systems. SIAM J. Control Optim. **37**(1), 54–91 (1999). https://doi.org/10.1137/S0363012996312039
9. Egger, H.: Energy stable Galerkin approximation of Hamiltonian and gradient systems (2018). arXiv:1812.04253
10. Egger, H., Kugler, T., Liljegren-Sailer, B., Marheineke, N., Mehrmann, V.: On structure-preserving model reduction for damped wave propagation in transport networks. SIAM J. Sci. Comput. **40**(1), A331–A365 (2018). https://doi.org/10.1137/17M1125303
11. Feireisl, E.: Relative entropies in thermodynamics of complete fluid systems. Discrete Continuous Dynam. Syst. **32**(9), 3059–3080 (2012). https://doi.org/10.3934/dcds.2012.32.3059
12. Geißler, B., Morsi, A., Schewe, L., Schmidt, M.: Solving power-constrained gas transportation problems using an MIP-based alternating direction method. Comput. Chem. Eng. **82**, 303–317 (2015). https://doi.org/10.1016/j.compchemeng.2015.07.005

13. Geißler, B., Morsi, A., Schewe, L., Schmidt, M.: Solving highly detailed gas transport MINLPs: block separability and penalty alternating direction methods. INFORMS J. Comput. **30**(2), 309–323 (2018). https://doi.org/10.1287/ijoc.2017.0780

14. Grmela, M., Öttinger, H.C.: Dynamics and thermodynamics of complex fluids. I. Development of a general formalism. Phys. Rev. E **56**(6), 6620–6632 (1997). https://doi.org/10.1103/PhysRevE.56.6620

15. Hante, F.M., Schmidt, M.: Complementarity-based nonlinear programming techniques for optimal mixing in gas networks. EURO J. Comput. Optim. **7**, 299–323, (2019). https://doi.org/10.1007/s13675-019-00112-w

16. Jacob, B., Zwart, H.: Linear Port-Hamiltonian Systems on Infinite-Dimensional Spaces. Birkhäuser/Springer, Basel (2012). https://doi.org/10.1007/978-3-0348-0399-1

17. Kotyczka, P., Lefèvre, L.: Discrete-time port-Hamiltonian systems based on Gauss-Legendre collocation. IFAC-Papers OnLine **51**(3), 125–130 (2018). https://doi.org/10.1016/j.ifacol.2018.06.035

18. Kraus, M., Hirvijoki, E.: Metriplectic integrators for the Landau collision operator. Phys. Plasmas **24**(10), 102311 (2017). https://doi.org/10.1063/1.4998610

19. Kraus, M., Kormann, K., Morrison, P.J., Sonnendrücker, E.: GEMPIC: Geometric electromagnetic particle-in-cell methods. J. Plasma Phys. **83**(4), 905830401 (2017). https://doi.org/10.1017/S002237781700040X

20. Le Gorrec, Y., Zwart, H., Maschke, B.: Dirac structures and boundary control systems associated with skew-symmetric differential operators. SIAM J. Control Optim. **44**(5), 1864–1892 (2005). https://doi.org/10.1137/040611677

21. LeVeque, R.J.: Numerical Methods for Conservation Laws, 2 edn. Birkhäuser, Basel (2008). https://doi.org/10.1007/978-3-0348-8629-1

22. Liljegren-Sailer, B., Marheineke, N.: Structure-preserving Galerkin approximation for a class of nonlinear port-Hamiltonian partial differential equations on networks. Proc. Appl. Math. Mech. **19**(1), e201900399 (2019). https://doi.org/10.1002/pamm.201900399

23. Mehl, C., Mehrmann, V., Wojtylak, M.: Linear algebra properties of dissipative port-Hamiltonian descriptor systems. SIAM J. Matrix Anal. Appl. **39**(3), 1489–1519 (2018). https://doi.org/10.1137/18M1164275

24. Mehrmann, V., Morandin, R.: Structure-preserving discretization for port-Hamiltonian descriptor systems (2019). arXiv:1903.10451

25. Mehrmann, V., Schmidt, M., Stolwijk, J.J.: Model and discretization error adaptivity within stationary gas transport optimization. Vietnam J. Math. **46**(4), 779–801 (2018). https://doi.org/10.1007/s10013-018-0303-1

26. Moses Badlyan, A., Zimmer, C.: Operator-GENERIC formulation of thermodynamics of irreversible processes (2018). arXiv:1807.09822

27. Moses Badlyan, A., Maschke, B., Beattie, C., Mehrmann, V.: Open physical systems: from GENERIC to port-Hamiltonian systems. In: Proceedings of the 23rd International Symposium on Mathematical Theory of Systems and Networks, pp. 204–211 (2018)

28. National Institute of Standards and Technology: Thermophysical Properties of Fluid Systems (2016). http://webbook.nist.gov/chemistry/fluid

29. Öttinger, H.C.: Nonequilibrium thermodynamics for open systems. Phys. Rev. E **73**(3), 036126 (2006). https://doi.org/10.1103/PhysRevE.73.036126

30. Öttinger, H.C., Grmela, M.: Dynamics and thermodynamics of complex fluids. II. Illustrations of a general formalism. Phys. Rev. E **56**(6), 6633–6655 (1997). https://doi.org/10.1103/PhysRevE.56.6633

31. Polner, M., van der Vegt, J.: A Hamiltonian vorticity-dilatation formulation of the compressible Euler equations. Nonlinear Anal. **109**, 113–135 (2014). https://doi.org/10.1016/j.na.2014.07.005

32. Rein, M., Mohring, J., Damm, T., Klar, A.: Model order reduction of hyperbolic systems at the example of district heating networks (2019). arXiv:1903.03342

33. Rein, M., Mohring, J., Damm, T., Klar, A.: Optimal control of district heating networks using a reduced order model (2019). arXiv:1907.05255

34. Schlichting, H., Gersten, K.: Grenzschicht-Theorie, 10th edn. Springer, Berlin (2006). https://doi.org/10.1007/3-540-32985-4
35. Serhani, A., Haine, G., Matignon, D.: Anisotropic heterogeneous n-D heat equation with boundary control and observation: I. Modeling as port-Hamiltonian system. IFAC-PapersOnLine **52**(7), 51–56 (2019). https://doi.org/10.1016/j.ifacol.2019.07.009
36. Serhani, A., Haine, G., Matignon, D.: Anisotropic heterogeneous n-D heat equation with boundary control and observation: II. Structure-preserving discretization. IFAC-PapersOnLine **52**(7), 57–62 (2019). https://doi.org/10.1016/j.ifacol.2019.07.010
37. Serhani, A., Matignon, D., Haine, G.: A partitioned finite element method for the structure-preserving discretization of damped infinite-dimensional port-Hamiltonian systems with boundary control. In: F. Nielsen, F. Barbaresco (eds.) Geometric Science of Information, pp. 549–558. Springer (2019)
38. Shashi Menon, E.: Transmission Pipeline Calculations and Simulations Manual. Elsevier, Amsterdam (2015). https://doi.org/10.1016/C2009-0-60912-0
39. van der Schaft, A.: Port-Hamiltonian differential-algebraic systems. In: Surveys in Differential-Algebraic Equations I, pp. 173–226. Springer, Berlin (2013). https://doi.org/10.1007/978-3-642-34928-7_5
40. van der Schaft, A., Jeltsema, D.: Port-Hamiltonian systems theory: an introductory overview. Found. Trends Syst. Control **1**(2–3), 173–378 (2014). https://doi.org/10.1561/2600000002
41. van der Schaft, A., Maschke, B.: Hamiltonian formulation of distributed-parameter systems with boundary energy flow. J. Geom. Phys. **42**(1–2), 166–194 (2002). https://doi.org/10.1016/S0393-0440(01)00083-3
42. van der Schaft, A., Maschke, B.: Port-Hamiltonian systems on graphs. SIAM J. Control Optim. **51**(2), 906–937 (2013). https://doi.org/10.1137/110840091
43. van der Schaft, A., Maschke, B.: Generalized port-Hamiltonian DAE systems. Syst. Control Lett. **121**, 31–37 (2018). https://doi.org/10.1016/j.sysconle.2018.09.008
44. van der Schaft, A., Maschke, B.: Geometry of thermodynamic processes. Entropy **20**(12), 92523 (2018). https://doi.org/10.3390/e20120925
45. Vu, N., Lefèvre, L., Maschke, B.: A structured control model for the thermo-magneto-hydrodynamics of plasmas in tokamaks. Math. Comput. Model. Dyn. Syst. **22**(3), 181–206 (2016). https://doi.org/10.1080/13873954.2016.1154874
46. Zhou, W., Hamroun, B., Couenne, F., Le Gorrec, Y.: Distributed port-Hamiltonian modelling for irreversible processes. Math. Comput. Model. Dyn. Syst. **23**(1), 3–22 (2017). https://doi.org/10.1080/13873954.2016.1237970

Coupled Systems of Linear Differential-Algebraic and Kinetic Equations with Application to the Mathematical Modelling of Muscle Tissue

Steffen Plunder and Bernd Simeon

Abstract We consider a coupled system composed of a linear differential-algebraic equation (DAE) and a linear large-scale system of ordinary differential equations where the latter stands for the dynamics of numerous identical particles. Replacing the discrete particles by a kinetic equation for a particle density, we obtain in the mean-field limit the new class of partially kinetic systems.

We investigate the influence of constraints on the kinetic theory of those systems and present necessary adjustments. We adapt the mean-field limit to the DAE model and show that index reduction and the mean-field limit commute. As a main result, we prove Dobrushin's stability estimate for linear systems. The estimate implies convergence of the mean-field limit and provides a rigorous link between the particle dynamics and their kinetic description.

Our research is inspired by mathematical models for muscle tissue where the macroscopic behaviour is governed by the equations of continuum mechanics, often discretised by the finite element method, and the microscopic muscle contraction process is described by Huxley's sliding filament theory. The latter represents a kinetic equation that characterises the state of the actin-myosin bindings in the muscle filaments. Linear partially kinetic systems are a simplified version of such models, with focus on the constraints.

Keywords Kinetic theory · Statistical physics · Differential-algebraic equations · Mathematical modelling · Skeletal muscle tissue

S. Plunder (✉)
Faculty of Mathematics, University of Vienna, Vienna, Austria
e-mail: steffen.plunder@univie.ac.at

B. Simeon
Felix Klein Zentrum für Mathematik, TU Kaiserslautern, Kaiserslautern, Germany
e-mail: simeon@mathematik.uni-kl.de

© The Editor(s) (if applicable) and The Author(s), under exclusive licence   357
to Springer Nature Switzerland AG 2020
T. Reis et al. (eds.), *Progress in Differential-Algebraic Equations II*,
Differential-Algebraic Equations Forum,
https://doi.org/10.1007/978-3-030-53905-4_12

1   Introduction

Differential-algebraic equations (DAEs) and kinetic equations are usually considered as separate and quite independent topics. While DAEs stem from models that are in some sense constrained, kinetic theory deals with identical particles such as atoms or molecules and their mutual interaction. In this work, we introduce a problem class that combines these two mathematical structures. More precisely, we demonstrate how to couple a macroscopic component with a kinetic system using algebraic constraints.

For given microscopic laws, the kinetic description of a particle system is obtained by the mean-field limit which replaces the discrete particle states by a particle distribution. Application of the mean-field limit comes with a loss of information: Individual particle positions are lost and only their statistical distribution is available. This gives rise to a challenge in the coupling process since we cannot impose an algebraic constraint on the individual particle positions of a kinetic system. However, if we know the microscopic laws governing the particles of a kinetic system, we can impose the algebraic-constraint on the microscopic level for each particle and then apply the mean-field limit to obtain new kinetic equations for the particles. We refer to the resulting system as the partially kinetic system.

To the best of our knowledge, there is no kinetic theory for systems where DAEs describe the microscopic law. In order to provide a rigorous theory for partially kinetic systems, we extend ideas from classical kinetic theory [5, 17, 26, 27]. To streamline the presentation, we restrict ourself to linear systems.

Important examples for partially kinetic system are mathematical models for muscle contraction. Muscle tissue, with all its supporting tissue (macroscopic component), contracts due to the accumulated force of numerous actin-myosin cross-bridges (particles). In this specific case, the kinetic theory of cross-bridges without the coupling is already well-studied and led to the famous Huxley model [15, 18, 30]. On the other hand, models from continuum mechanics are today in use to simulate the muscle contraction at the macro-scale in combination with the finite element method. For the coupling of both scales, simplifications and ad-hoc procedures are used so far [1, 2, 10, 11] that call for a theoretical foundation.

This article is organized as follows: Sect. 2 presents a strongly simplified DAE model for muscle cells with attached actin-myosin cross-bridges and derives an equivalent ODE formulation for the DAE model. Next, Sect. 3 derives formally partially kinetic equations for the DAE model. Basics of kinetic theory are outlined during the application of the mean-field limit onto the ODE formulation. For the DAE formulation, the mean-field limit requires modification. To justify the formal computations rigorously, Sect. 4 adapts and proves Dobrushin's stability estimate for linear partially kinetic systems. With regard to the application fields, Sect. 5 sketches possible generalisations of linear partially kinetic systems, while Sect. 6 provides details about the numerical implementation of the simulations presented in this article. The numerical challenge of partially kinetic systems is demonstrated by an example in which energy conservation is violated by the discretisation.

2 A Differential-Algebraic Model for Muscle Cells with Attached Cross-Bridges

The emergence of macroscopic effects from microscopic properties is a central theme in kinetic theory. In laymen terms, emergence describes how the big picture arises from the laws that govern the model at a smaller scale. Understanding this transition is essential in many biological applications [24]. Muscle tissue consists of millions of small contractible molecules called actin-myosin cross-bridges. Kinetic theory allows the up-scaling of these microscopic units to the organ level and provides a means to derive macroscopic models for muscle tissue. Most macroscopic models focus on the emergence of a contraction force as the result of the synchronization between muscle cells [12, 18, 21]. However, there are applications where more than just the macroscopic contraction force is of interest.

One example is vibrational medicine, in particular, the medical therapy concept called *Matrix-Rhythm-Therapy* [23] that treats diseased muscle tissue by vibrational stimulation in a specific frequency range. In order to understand this therapy approach, it is crucial to study how the mechanical stimulation influences the physiological health of cells. In laymen terms: How does the big picture influence the small scale? A first mathematical model for the interplay between mechanics and the physiology of muscle cells was proposed in [25].

We extend this work in the direction of more detailed physiological models for muscle cells that are based on the sliding filament theory for cross-bridges. In mathematical terms, this requires an understanding of muscles at both, the micro and the macro scale. To study the influence of mechanics on the physiology of muscle cells, the coupling between mechanical properties and physiological models is essential. In the following, we will study a prototype of a system which couples a physiological model for cross-bridges with a prototypical mechanical system. We have to mention that Ma and Zahalak [21, 31] already studied cross-bridge dynamics with kinetic and thermodynamic methods and also extended their cross-bridge models for refined coupling with continuum models. In contrast, we study a simpler model and focus on mathematical details of the coupling. In Sect. 5.3, we relate our mathematical analysis to established models.

Fig. 1 Sketch of the parallel actin filaments (purple, outside) and myosin filaments (orange, central). The myosin heads (red) can attach to binding sides at the actin filament, which forms a so-called cross-bridge. Skeletal muscle fibers are a large array of parallel actin-myosin filaments

2.1 Sliding Filament Theory for Cross-Bridges

Compared to many other biological phenomena, the contraction of muscles cells is a relatively well-studied field [12, 14, 18]. For a mathematical introduction to muscle models, we refer to [14, 18]. *Sliding filament theory* [12, 15] is the mainstream theory to explain muscle contraction.

In its simplest form, sliding filament theory suggests that muscle cells consist of parallel myosin and actin filaments, as visualised in Fig. 1. On each actin filament, small binding sides allow myosin heads to attach and to form a bridge between both filaments, a so-called *cross-bridge*. Due to the molecular configuration of newly attached cross-bridges, they pull the two filaments such that they slide alongside each other, which causes a shortening of the muscle cell. This pulling step is called a power stroke. After each power stroke, the myosin head can unbind from the binding side, release the ADP (adenosine diphosphate) molecule and obtain new free energy by hydrolyzing another ATP (adenosine triphosphate) molecule.

The cycling of binding, power stroke, unbinding and resetting of myosin heads is called the cross-bridge cycle. Since numerous muscle cells contract due to this mechanism, the whole muscle tissue contracts on the macroscopic scale. The rate at which the cross-bridge cycling takes place controls the contraction strength. The contraction process varies for different muscle types. However, blocking and unblocking the binding sides at the actin filaments is always part of the control mechanism. In skeletal muscle tissue, the periodic release of calcium ions unblocks the binding sides. A higher frequency of calcium ion bursts leads to a stronger contraction.

From the variety of available mathematical models, we extract the common core, which is given by the sliding filament theory with cross-bridges modelled as springs [12, 15, 18, 30]. To obtain a linear model, we assume that the springs are linear, which is valid for some models [15, 30], but not the case for more detailed models [16], [18, Section 15.6]. We also simplify the model radically by considering only the attached cross-bridges. Hence, the actual cross-bridge cycling does not take place in the system we present. However, Sect. 5.3 outlines possible extensions,

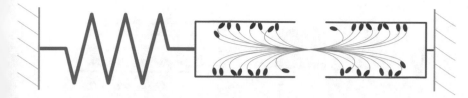

Fig. 2 Model for the coupling between a macroscopic linear spring (blue) and microscopic myosin filaments (orange) with their corresponding pair of actin filaments (purple)

which we neglect for most of the exposition to avoid distraction from the main mathematical ideas.

2.2 A Differential-Algebraic Model for Attached Cross-Bridges

Without further ado, we present the mathematical model for attached cross-bridges in the presence of constraints. Our goal is to model a muscle cell which is coupled to a macroscopic linear spring, as displayed in Fig. 2. It is sufficient to model only one half of the actin-myosin filaments from Fig. 2, since their arrangement is mirror-symmetrical.

We define the dimensions $n_r := 1, n_q := 1$, where n_r denotes the dimension of the macroscopic spring and n_q denotes the degrees of freedom of a single cross-bridge. However, throughout this article, we will continue to distinguish between $\mathbb{R}$, $\mathbb{R}^{n_r}$ and $\mathbb{R}^{n_q}$ to indicate real numbers and position variables in the according spaces. The reader is welcome to read this section with $n_q, n_r \geq 1$ in mind. While the coupled cross-bridge model leads to the one-dimensional case, we simultaneously include an abstract model where the macroscopic system has n_r degrees of freedom and each particle has n_q degrees of freedom.

To model the macroscopic spring, we use $r \in \mathbb{R}^{n_r}$ as the extension of a linear spring with mass M_r and force $F_r(r) = -\gamma_r r$.

For the microscopic model, we label the attached cross-bridges with $j = 1, \ldots, N$, where N is the total number of cross-bridges. The extension of a single cross-bridge is denoted by $Q_j \in \mathbb{R}^{n_q}$ and each cross-bridge is a linear spring with mass M_q and force $F_q(Q_j) = -\gamma_q Q_j$. Because a single cross-bridge is very light compared to the macroscopic spring, the dynamics of cross-bridges are typically fast compared to the macroscopic spring. The small individual mass is compensated by the large number of cross-bridges N, which is a crucial parameter of the system. The situation is sketched in Fig. 3.

For the abstract model $n_r, n_q \geq 1$, we define the function

$$g(r, Q_j) := Q_j + G_r r,$$

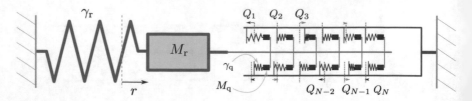

Fig. 3 Simplified model for attached cross-bridges (red) between parallelly sliding actin and myosin filaments coupled with a macroscopic linear spring (blue)

where $G_r \in \mathbb{R}^{n_q \times n_r}$ is an arbitrary, possibly singular matrix. In the cross-bridge model, the macroscopic spring and the actin filament are considered to be fixed to the walls at both sides, as displayed in Fig. 3. Therefore, we require the total length to remain constant and pick $G_r := -1$ in the one-dimensional case $n_r = n_q = 1$. For each cross-bridge, we define the linear constraint as

$$g(r, Q_j) = g(r^{\mathrm{in}}, Q_j^{\mathrm{in}}) \quad \text{for } j = 1, \ldots, N,$$

where $r^{\mathrm{in}} \in \mathbb{R}^{n_r}$ and $Q_j^{\mathrm{in}} \in \mathbb{R}^{n_q}$ denote the initial states of the macroscopic system and the cross-bridges. The corresponding Lagrangian multipliers are denoted by $\lambda_1, \ldots, \lambda_N \in \mathbb{R}^{n_q}$. Overall, we arrive at the following linear differential-algebraic system that models a linear spring coupled to a sliding actin-myosin filament pair with N cross-bridges

$$M_r \ddot{r} = -\gamma_r r - \sum_{i=1}^{N} G_r^T \lambda_i, \tag{2.1}$$

$$M_q \ddot{Q}_j = -\gamma_q Q_j - \lambda_j \quad \text{for } j = 1, \ldots, N, \tag{2.2}$$

$$Q_j + G_r r = Q_j^{\mathrm{in}} + G_r r^{\mathrm{in}} \quad \text{for } j = 1, \ldots, N \tag{2.3}$$

with initial conditions

$$r(0) = r^{\mathrm{in}} \in \mathbb{R}^{n_r}, \quad \dot{r}(0) = s^{\mathrm{in}} \in \mathbb{R}^{n_r} \quad \text{and} \quad Q_j(0) = Q_j^{\mathrm{in}} \in \mathbb{R}^{n_q} \quad \text{for } j = 1, \ldots, N.$$

In the following, we will refer to (2.1)–(2.3) as the *DAE formulation*. There is no initial condition for the velocities $\dot{Q}_j$, since the constraint implies the compatibility condition

$$\dot{Q}_j(0) = -G_r s^{\mathrm{in}} \quad \text{for } j = 1, \ldots, N.$$

We require the mass matrices $M_r \in \mathbb{R}^{n_r \times n_r}$, $M_q \in \mathbb{R}^{n_q \times n_q}$ to be positive definite. We remark that $\frac{\partial g}{\partial Q_j}(r, Q_j) = \mathbb{1} \in \mathbb{R}^{n_q \times n_q}$, which implies the full-rank condition required for local existence and uniqueness of solutions [6, Section VII.I, Eq. (1.16)].

2.3 Derivation of an Effective Balance Law via Index Reduction and Elimination of Multipliers

The system (2.1)–(2.3) has differential index 3 [2], [6, Section VII.I]. Due to the particular structure, it is possible to eliminate the Lagrange multipliers and derive an ODE formulation. Differentiating (2.3) with respect to time yields

$$\dot{Q}_j = -G_r \dot{r} \tag{2.4}$$

and

$$\ddot{Q}_j = -G_r \ddot{r}. \tag{2.5}$$

Using (2.5), we solve (2.2) for λ_j and insert the result into (2.1), which leads to

$$M_r \ddot{r} = -\gamma_r r - \sum_{i=1}^{N} G_r^T \left(-\gamma_q Q_i - M_q \ddot{Q}_i \right)$$

$$= -\gamma_r r - \sum_{i=1}^{N} G_r^T \left(-\gamma_q Q_i + M_q G_r \ddot{r} \right).$$

After collecting the acceleration terms on the left-hand side, one obtains

$$\underbrace{\left(M_r + \sum_{i=1}^{N} G_r^T M_q G_r \right)}_{=:M_{\text{eff}}{}^{(N)}} \ddot{r} = \underbrace{-\gamma_r r + \sum_{i=1}^{N} G_r^T \gamma_q Q_i}_{=:F_{\text{eff}}{}^{(N)}(r, Q_1, \dots, Q_N)}. \tag{2.6}$$

This system of ordinary differential equations describes the *effective* balance of forces after elimination of the constraint equation, and thus we use the subscript $()_{\text{eff}}$.

In Eq. (2.6), the Lagrangian multipliers are eliminated, but the equation is not closed since Q_i is needed to compute $F_{\text{eff}}{}^{(N)}$. We employ (2.4) to generate a first order differential equation for all Q_j, i.e.

$$\dot{Q}_j = -G_r \dot{r} \quad \text{for } j = 1, \dots, N. \tag{2.7}$$

Now, the Eqs. (2.6) and (2.7) form a closed linear ordinary differential equation (ODE) which we will call the *ODE formulation*. There are other ODE formulations, but we prefer (2.6) and (2.7) since this form leads to a direct derivation of the mean-field PDE in Sect. 3.3.

A numerical simulation of (2.6) and (2.7) is presented in Fig. 4. For the simulation, the initial conditions of the cross-bridge extensions Q_i are samples

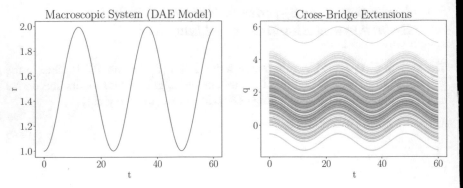

Fig. 4 The trajectory of the macroscopic system (left) and the cross-bridge extensions (right)

of a normal distribution. The cross-bridges influence the effective mass and the effective force of the macroscopic system. This influence leads to a shift of the macroscopic system's equilibrium to $r_0 \approx 1.5$ instead of $r_0 = 0$. Since the constraint is $r - Q_j = r^{\text{in}} - Q_j^{\text{in}}$, the trajectories of the cross-bridge extensions just differ by constant shifts. For details on the numerical method and the used parameters, we refer to Sect. 6.

2.4 Explicit Solutions

The linear constraint (2.3) can be solved for Q_i, which yields

$$Q_i = -G_r r + G_r r^{\text{in}} + Q_i^{\text{in}}.$$

With this formula, we can reformulate (2.6) as

$$M_{\text{eff}}^{(N)} \ddot{r} = -\gamma_{\text{eff}}^{(N)}(r - r_0^{(N)}) \tag{2.8}$$

with the effective stiffness

$$\gamma_{\text{eff}}^{(N)} := \gamma_r + N G_r^T \gamma_q$$

and the new equilibrium

$$r_0^{(N)} = G_r^T \gamma_q \sum_{i=1}^{N} (G_r r^{\text{in}} + Q_i^{\text{in}}).$$

Equation (2.8) has well-known explicit solutions. The system (2.1)–(2.3) is therefore a benchmark case for partially kinetic systems.

In the setting of more realistic muscle models, explicit solutions are not available any more, since attachment and detachment of cross-bridges lead to a switched system where the number of cross-bridges N changes over time, as outlined in Sect. 5.3. Moreover, if the constraint function $g(r, Q_j)$ is nonlinear with respect to Q_j, then explicit solutions are not known in general. These extensions are discussed in Sect. 5.

Despite the existence of explicit solutions, we will continue without using (2.3) to solve for Q_j, since this benchmark-setting is convenient to demonstrate the influence of constraints on kinetic theory. The calculations in Sects. 3 and 4 generalise well to relevant extensions as sketched in Sect. 5. Only in proofs of this article, we will use (2.3) explicitly.

3 Partially Kinetic Model for Muscle Cells with Attached Cross-Bridges

The kinetic theory for cross-bridges was investigated already in the early eighties [15]. The first approaches suggested to model attached cross-bridges as linear springs, while many refinements have been introduced later on and are still today subject of current research [13]. To compute the contraction force, all models known to us [1, 10–12, 18] assume implicitly that the kinetic equations remain valid without modification in the presence of constraints, cf. [12]. Moreover, the masses of the cross-bridges are assumed to add no kinetic energy to the macroscopic system. These two assumptions are very reasonable and lead to successful models. In the following, we want to compute explicitly how the kinetic equations look like in the presence of constraints and give a mathematical quantification for common modelling assumptions.

We remark that Sect. 3.1 comprises a short outline of the fundamentals of kinetic theory. For this purpose, it is preferable to use the ODE formulation (2.6) and (2.7) instead of the DAE formulation (2.1)–(2.3). Afterwards, in Sect. 3.4, similar steps are applied to the DAE formulation.

3.1 Partially Kinetic Equations for the ODE Formulation

This section outlines the derivation of the partially kinetic equations for the ODE formulation (2.6) and (2.7). We obtain the kinetic equations by the formal mean-field limit. In Sect. 4, this approach is justified by a rigorous estimate which proves that the solutions of the ODE formulation convergence under the scaling assumptions of the mean-field limit to the solutions of the partially kinetic equations. There are different techniques to compute the mean-field limit for ODE models. The dual pairing between measures and smooth test functions is a powerful formalism to

derive the mean-field PDE in its weak formulation [17]. However, we will motivate
the mean-field limit as a generalisation of the strong law of large numbers, which is
a less abstract approach. The derivation follows these steps:

Step 1:   Introduction of a scaling factor,
Step 2:   Introduction of a measure μ^t to describe the statistical distribution of the
              cross-bridge extensions,
Step 3:   Derivation of the mean-field characteristic flow equations, which govern
              the evolution of the cross-bridge distribution μ^t,
Step 4:   Derivation of a kinetic balance law for the macroscopic system.

Step 1 As the number of cross-bridges is large, we want to study the limit of
infinitely many cross-bridges $N \to \infty$. A naive limit $N \to \infty$ leads to a system
with infinitely many identical linear springs. Such a system is either in equilibrium
or entails an infinite force. The force term in (2.6)

$$F_{\text{eff}}^{(N)}(r, Q_1, \ldots, Q_N) = -\gamma_r r - G_r^T \sum_{j=1}^{N} \gamma_q Q_j$$

will either be divergent or the cross-bridge extensions form a zero sequence.
Therefore, the naive limit is mathematically and physically unreasonable since it
describes either states close to equilibrium or states with infinite energy.

 The scaling assumption of the mean-field limit is that the force $\sum_{j=1}^{N} F_q(Q_j)$
is replaced by the mean-field force $\frac{1}{N} \sum_{j=1}^{N} F_q(Q_j)$. Therefore, while increasing
the number of cross-bridges, we scale the mass and force of each cross-bridge,
such that the total energy remains constant. To maintain the right ratio between
the macroscopic system and the cross-bridges, we add another factor N_{real}, which
denotes the realistic numbers of cross-bridges. Hence, we apply the following
scaling to the ODE formulation (2.6) and (2.7)

$$\tilde{M}_q := \frac{N_{\text{real}}}{N} M_q \quad \text{and} \quad \tilde{F}_q(Q_j) := \frac{N_{\text{real}}}{N} F_q(Q_j) = -\frac{N_{\text{real}}}{N} \gamma_q Q_j. \tag{3.1}$$

After this modification, (2.6) and (2.7) take the form

$$\Big(M_r + \overbrace{\frac{N_{\text{real}}}{N} \sum_{i=1}^{N} G_r^T M_q G_r}^{=:M_{\text{mean}}^{(N)}} \Big) \ddot{r} = -\gamma_r r + \overbrace{\frac{N_{\text{real}}}{N} \sum_{i=1}^{N} G_r^T \gamma_q Q_i}^{=:F_{\text{mean}}^{(N)}(r, Q_1, \ldots, Q_N)}, \tag{3.2}$$

$$\dot{Q}_j = -G_r \dot{r}. \tag{3.3}$$

For the mathematical discussion, we might assume without loss of generality
$N_{\text{real}} = 1$, which is a typical simplification in kinetic theory [5, 17]. However,
for partially kinetic systems, the correct ratio between masses and forces of both

systems is relevant. Therefore, in contrast to the classical case, different values of N_{real} change the properties of the kinetic equations.

Step 2 A key observation in Eq. (3.2) is that only the mean value of the cross-bridge masses $M_{\text{mean}}^{(N)}$ and the mean value of the cross-bridge forces $F_{\text{mean}}^{(N)}$ are relevant. In other words, we are just interested in the statistics of $(Q_1, \ldots, Q_N)$ but not in particular states of single cross-bridges. This observation motivates the use of a probability measure to quantify the distribution of the cross-bridges.

We will use the following notations from measure theory with notation as in [5]: The Borel σ-algebra on $\mathbb{R}^{n_q}$ is denoted by $\mathscr{B}(\mathbb{R}^{n_q})$ and the corresponding space of probability measures on $\mathbb{R}^{n_q}$ is $\mathscr{P}(\mathbb{R}^{n_q})$. The space of probability measures $\mu \in \mathscr{P}(\mathbb{R}^{n_q})$ with finite first moments $\int_{\mathbb{R}^{n_q}} \|q\| \, d\mu(q) < \infty$ is denoted by $\mathscr{P}^1(\mathbb{R}^{n_q})$.

We assume that for each fixed time t, there is a probability measure $\mu^t \in \mathscr{P}^1(\mathbb{R}^{n_q})$, such that the cross-bridge extensions $Q_j(t)$ are independent and identically distributed random variables with probability law μ^t. We use the notation

$$Q_j(t) \sim \mu^t \quad :\Leftrightarrow \quad \mathbb{P}(Q_j(t) \in A) = \mu^t(A), \quad \text{for all } A \in \mathscr{B}(\mathbb{R}^{n_q}).$$

We call μ^t the cross-bridge distribution.[1]

Step 3 To characterise the evolution of μ^t, we will now define the characteristic flow. We assume that just the initial cross-bridge distribution $\mu^{\text{in}} \in \mathscr{P}^1(\mathbb{R}^{n_q})$ is known, i.e.

$$Q_j^{\text{in}} \sim \mu^{\text{in}}, \quad \text{for all } j = 1, 2, \ldots.$$

We interpret the velocity constraint $\dot{Q}_j = -G_r \dot{r}$ (2.7) as a first order differential equation and denote its flow by $Q(t, q^{\text{in}})$. Hence, $Q(t, q^{\text{in}})$ satisfies for all $q^{\text{in}} \in \mathbb{R}^{n_q}$

$$\dot{Q}(t, q^{\text{in}}) = -G_r \dot{r}(t), \tag{3.4}$$

$$Q(0, q^{\text{in}}) = q^{\text{in}}. \tag{3.5}$$

In the setting of (3.2) and (3.3), the discrete cross-bridge states satisfy

$$Q_j(t) = Q(t, Q_j^{\text{in}}). \tag{3.6}$$

Since the cross-bridge extensions follow the characteristic flow of $Q(t, \cdot)$, the distribution of cross-bridges is also determined by the characteristic flow. More precisely, the cross-bridge distribution μ^t is the transformation of the initial cross-bridge distribution μ^{in} under the flow $Q(t, \cdot) : \mathbb{R}^{n_q} \to \mathbb{R}^{n_q}$. This transformation

---

[1]It is not trivial to argue why *all* cross-bridges Q_j are well described by *one* common probability measure μ^t. This property is related to the concept of *propagation of chaos* [17]. The mean-field limit, which generalises the strong law of large numbers, is one possibility to overcome this issue.

Fig. 5 The flow of the
particles does also induce a
transformation of the initial
particle measure μ^{in}

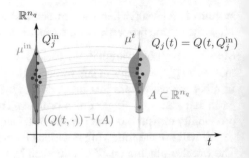

is visualized in Fig. 5. To measure how many cross-bridges have extensions in $A \in \mathscr{B}(\mathbb{R}^{n_q})$, we can count how many cross-bridges have an initial extension in $\left(Q(t, \cdot)\right)^{-1}(A)$, i.e.

$$Q_j(t) \in A \quad \Leftrightarrow \quad Q_j^{\text{in}} \in \left(Q(t, \cdot)\right)^{-1}(A).$$

This relation characterises the pushforward of a measure [5, 17]. For a map $\varphi : \mathbb{R}^{n_q} \to \mathbb{R}^{n_q}$, the pushforward of μ^{in} under φ is defined as

$$\varphi \# \mu^{\text{in}}(A) := \mu^{\text{in}}(\varphi^{-1}(A)), \quad \text{for all } A \in \mathscr{B}(\mathbb{R}^{n_q}).$$

Applied to our situation, with $\varphi = Q(t, \cdot)$, the cross-bridge distribution at time t is the pushforward of μ^{in} under $Q(t, \cdot)$. Therefore, the evolution of μ^t is characterised by

$$\mu^t := Q(t, \cdot) \# \mu^{\text{in}}. \tag{3.7}$$

Step 4 Our goal is to approximate the limit $N \to \infty$ of (3.2) by an expression depending on μ^t. Consequently, we continue with computing $M_{\text{mean}}^{(N)}$ and $F_{\text{mean}}^{(N)}$ (3.2) in the limit $N \to \infty$. Now, we make use of the assumption that all cross-bridges are independent and identically distributed with law $Q_j(t) \sim \mu^t$. Application of the *strong law of large numbers* [19, Section 5.3] yields

$$\lim_{N \to \infty} F_{\text{mean}}^{(N)}(t) = N_{\text{real}} \lim_{N \to \infty} G_r^T \gamma_q \frac{1}{N} \sum_{i=1}^{N} Q_i(t)$$

$$= N_{\text{real}} G_r^T \gamma_q \mathbb{E}\left[Q_1(t)\right] \quad \text{almost surely.} \tag{3.8}$$

The sum converges almost surely (a.s.), i.e. with probability 1 the last equality holds.

Using $\mathbb{E}[Q_1(t)] = \int_{\mathbb{R}^{nq}} q \, d\mu^t(q)$, we define the mean-field force as

$$f_{\text{mean}}(\mu^t) := N_{\text{real}} G_r^T \gamma_q \int_{\mathbb{R}^{nq}} q \, d\mu^t(q). \tag{3.9}$$

Due to (3.8), the mean-field force satisfies

$$f_{\text{mean}}(\mu^t) = \lim_{N \to \infty} F_{\text{mean}}^{(N)}(t) \quad a.s.$$

Similarly, the mean-field mass is

$$m_{\text{mean}}(\mu^t) := N_{\text{real}} \int_{\mathbb{R}^{nq}} G_r^T M_q G_r \, d\mu^t(q) \tag{3.10}$$

$$= N_{\text{real}} G_r^T M_q G_r \int_{\mathbb{R}^{nq}} d\mu^t(q) = N_{\text{real}} G_r^T M_q G_r.$$

In the setting of our model, the mean-field mass is constant, since we only consider linear constraints. We remark that for nonlinear constraints, the term G_r would depend on q in general, which calls for a more profound analysis.

With (3.9) and (3.10), we obtain the kinetic formulation of the effective balance law (3.2) as

$$\underbrace{\left(M_r + N_{\text{real}} G_r^T M_q G_r \right)}_{=:m_{\text{eff}}} \ddot{r} = \underbrace{-\gamma_r r + N_{\text{real}} G_r^T \gamma_q \int_{\mathbb{R}^{nq}} q \, d\mu^t(q)}_{=:f_{\text{eff}}(r,\mu^t)}. \tag{3.11}$$

Finally, the kinetic description of the ODE formulation (2.5) and (2.6) is given by the effective balance law (3.11), the characteristic flow equations (3.4) and the pushforward relation (3.7).

The *partially kinetic model* for muscle cells with attached cross-bridges is given by (3.4), (3.5), (3.7), (3.11), summarised as

$$\left(M_r + N_{\text{real}} \int_{\mathbb{R}^{nq}} G_r^T M_q G_r \, d\mu^t(q) \right) \ddot{r} = -\gamma_r r + N_{\text{real}} G_r^T \gamma_q \int_{\mathbb{R}^{nq}} q \, d\mu^t(q), \tag{3.12}$$

$$\dot{Q}(\cdot, q^{\text{in}}) = -G_r \dot{r}, \quad \text{for all } q^{\text{in}} \in \mathbb{R}^{nq}, \tag{3.13}$$

$$\mu^t := Q(t, \cdot) \# \mu^{\text{in}} \tag{3.14}$$

with initial conditions

$$r(0) = r^{\text{in}} \in \mathbb{R}^{nr}, \quad \dot{r}(0) = s^{\text{in}} \in \mathbb{R}^{nr} \quad \text{and} \quad Q(0, q^{\text{in}}) = q^{\text{in}}$$

and initial cross-bridge distribution $\mu^{in} \in \mathscr{P}^1(\mathbb{R}^{nq})$. In kinetic theory, systems of the form (3.13) and (3.14) are called the *mean-field characteristic flow equations* [5, Section 1.3] since $Q(t, \cdot)$ describes the flow of the cross-bridge distribution in the presence of the mean-field forces. For that reason, we call the model (3.12)–(3.14) *partially kinetic*, since it combines an effective balance law for the macroscopic system (3.12) and the *mean-field characteristic flow equations* (3.13) and (3.14).

3.2 Consistency Between the ODE Formulation and the Partially Kinetic Equations

The cross-bridge distribution μ^t is usually modelled as a continuous measure. However, inserting an empirical measure for the initial state μ^{in} yields a consistency check of (3.12)–(3.14). We define the empirical measure as

$$\mu^{(emp)}_{Q_1,\ldots,Q_N} := \frac{1}{N} \sum_{j=1}^N \delta_{Q_j} \in \mathscr{P}^1(\mathbb{R}^{nq}),$$

where δ_{Q_j} denotes the Dirac measure which assigns unit mass to the position $Q_j \in \mathbb{R}^{nq}$. This measure allows us to treat the discrete system (2.6) and (2.7) as a special case of (3.12)–(3.14).

Lemma 3.1 (Consistency with the ODE Formulation) *For a solution* $(r(t), Q_1(t), \ldots, Q_N(t))$ *of (2.6) and (2.7), we define*

$$\mu^t = \mu^{(emp)}_{Q_1(t),\ldots,Q_N(t)}.$$

Then $(r(t), \mu^t)$ *is a solution of (3.12)–(3.14) with* $N_{real} := N$ *and initial conditions* $r(0) = r^{in}, \dot{r}(0) = s^{in}$ *and* $\mu^{in} = \mu^{(emp)}_{Q_1^{in},\ldots,Q_N^{in}}.$

Proof Let $(r(t), Q_1(t), \ldots, Q_N(t))$ be a solution of (2.6) and (2.7) with the corresponding initial conditions $r(0) = r^{in}, \dot{r}(0) = s^{in}$ and $Q_j(0) = q_j^{in}$ for all $j = 1, \ldots, N$.

1. We define the characteristic flow as

$$Q(t, q^{in}) := -G_r(r(t) - r^{in}) + q^{in},$$

which is the integral of (3.13) and hence a solution of (3.13) for all $q^{in} \in \mathbb{R}^{nq}$.
2. Since $Q_1(t), \ldots, Q_N(t)$ satisfy (2.3)

$$Q_j(t) = -G_r(r(t) - r^{in}) + Q_j^{in},$$

we obtain $Q(t, Q_j^{in}) = Q_j(t)$, which yields

$$
\begin{aligned}
Q(t, \cdot) \# \mu^{in} &= Q(t, \cdot) \# \mu_{Q_1^{in}, \dots, Q_N^{in}}^{(emp)} \\
&= \mu_{Q(t, Q_1^{in}), \dots, Q(t, Q_N^{in})}^{(emp)} \\
&= \mu_{Q_1(t), \dots, Q_N(t)}^{(emp)} = \mu^t.
\end{aligned}
$$

As a result, μ^t satisfies (3.14).

3. Finally, we insert $\mu^t := \mu_{Q_1(t), \dots, Q_N(t)}^{(emp)}$ into (3.12) and compute

$$
\left(M_r + N_{\text{real}} \int_{\mathbb{R}^{nq}} G_r^T M_q G_r \, d\mu_{Q_1(t), \dots, Q_N(t)}^{(emp)}(q) \right) \ddot{r}(t) = -\gamma_r r(t)
$$

$$
+ N_{\text{real}} G_r^T \gamma_q \int_{\mathbb{R}^{nq}} q \, d\mu_{Q_1(t), \dots, Q_N(t)}^{(emp)}(q)
$$

$$
\Leftrightarrow \left(M_r + N_{\text{real}} \frac{1}{N} \sum_{j=1}^{N} G_r^T M_q G_r \right) \ddot{r}(t) = -\gamma_r r(t) + N_{\text{real}} \frac{1}{N} \sum_{j=1}^{N} G_r^T \gamma_q Q_j(t)
$$

$$
\Leftrightarrow \left(M_r + N_{\text{real}} G_r^T M_q G_r \right) \ddot{r}(t) = -\gamma_r r(t) + \frac{N_{\text{real}}}{N} \sum_{j=1}^{N} G_r^T \gamma_q Q_j(t).
$$

$$(3.15)$$

For $N_{\text{real}} := N$, the last line (3.15) is exactly (2.6). Therefore, $(r(t), \mu^t)$ solve (3.12), which concludes the proof. □

Lemma 3.1 shows that the ODE formulation (2.5) and (2.6) is a special case of the partially kinetic system (3.12)–(3.14) with an empirical initial measure $\mu^{in} = \mu_{Q_1^{in}, \dots, Q_N^{in}}^{(emp)}$. The consistency check from Lemma 3.1 does not prove anything for the limit $N \to \infty$, but it relates the limit $N \to \infty$ to the stability of the partially kinetic system with respect to initial data. We will prove the stability of linear partially kinetic systems with respect to perturbation in the initial data in Sect. 4.

3.3 Partially Kinetic Mean-Field PDE

The mean-field characteristic flow equations (3.13) and (3.14) are too complex for direct numerical simulation, as they contain an infinite family of differential equations. The method of characteristics allows us to relate the family of ODEs (3.13) and (3.14) to a first order partial differential equation (PDE). The resulting PDE is called the *mean-field PDE*.

To derive the mean-field PDE, we assume that μ^{in} has a probability density, i.e. there exists a function $u(t, q)$ such that

$$u(t, q')\,dq' = d\mu^t(q') \quad \text{for all } t \in [0, \infty), \ q' \in \mathbb{R}^{n_q}$$

where dq' denotes the Lebesgue measure on $\mathbb{R}^{n_q}$ with variable q'. Then, the pushforward relation (3.14) implies that $u(t, q')$ is constant along the characteristic curves

$$t \mapsto Q(t, q^{in}).$$

Using this invariance, we can compute

$$0 = \frac{d\,u(t, Q(t, q^{in}))}{dt} = \frac{\partial u}{\partial t} + \frac{\partial u}{\partial q}\dot{Q}$$

$$\Leftrightarrow \quad 0 = \frac{\partial u}{\partial t} - \frac{\partial u}{\partial q}G_r \dot{r}. \tag{3.16}$$

The transport equation (3.16) is the mean-field PDE for the ODE formulation (2.6) and (2.7) and the Eqs. (3.12) and (3.16) are another kinetic description for attached cross-bridges. The possibility to derive a mean-field PDE is the main advantage of the ODE formulation. In the literature [14, 18], the foundational model for cross-bridge dynamics is the transport equation as in (3.16) with additional source terms. In Sect. 5.3, the relation between the simplified model of attached cross-bridges and more realistic models is outlined.

A numerical simulation of (3.12) and (3.16) is presented in Fig. 6.

In contrast to the ODE/DAE simulation, the computational complexity of the transport equation does not increase in complexity for different values of N_{real}. For more details on the used numerical methods, we refer to Sect. 6. The simulation

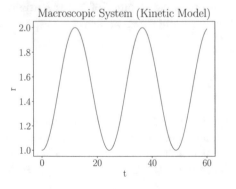
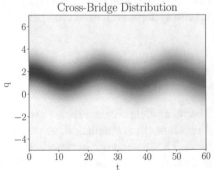

Fig. 6 Evolution of the macroscopic system (left) and the cross-bridge distribution (right). The colour intensity represents the cross-bridge density $u(t, q)$

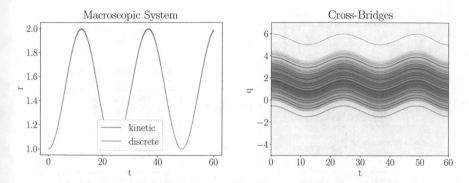

Fig. 7 Comparison of the discrete simulation Fig. 4 and the corresponding kinetic trajectory from Fig. 6. The trajectory of the macroscopic system with 250 cross-bridges is well approximated by the corresponding mean-field equation

results are not surprising, and the results fit well to the simulation of the ODE/DAE formulation, as visualised in Fig. 7.

However, Fig. 7 can be misleading: If we denote the solution of the partially kinetic system as $r^{\text{kin}}(t; \mu^{\text{in}})$ and the solution of the DAE formulation as $r^{\text{DAE}}(t; Q_1^{\text{in}}, \ldots, Q_N^{\text{in}})$, then the relation is given by

$$\lim_{N \to \infty} \mathbb{E}[r^{\text{DAE}}(t; Q_1^{\text{in}}, \ldots, Q_N^{\text{in}})] = r^{\text{kin}}(t; \mu^{\text{in}}), \qquad (3.17)$$

where $Q_j^{\text{in}} \sim \mu^{\text{in}}$ are independent and identically distributed random variables. Hence, instead of comparing single trajectories $r^{\text{DAE}}(t)$ and $r^{\text{kin}}(t)$, we need to compare the mean-trajectory $\mathbb{E}[r^{\text{DAE}}(t)]$ with $r^{\text{kin}}(t)$. Section 4.1 gives a numerical validation of (3.17) and presents numerical evidence for the mean-field limit in Figs. 9 and 10.

3.4 Partially Kinetic Equations for the DAE Formulation

To analyse the influence of index reduction, we demonstrate how the mean-field limit applies directly to the DAE (2.1)–(2.3) without prior index reduction. Here, we need to generalise the Lagrangian multipliers to fit into the kinetic framework. The resulting characteristic flow has to satisfy the algebraic constraint (2.3), which leads to the new concept of constrained characteristic flows. Since the constraint (2.3) is uniform for all $j \in \{1, \ldots, N\}$, the notation of a differentiability index [2, 7] carries over to constrained characteristic flows. After two index reduction steps and elimination of multipliers, as in Sect. 2.3, the constrained characteristic flow equations transform into the (unconstrained) characteristic flow equations (3.13).

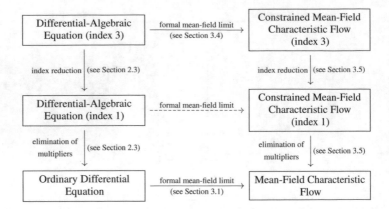

Fig. 8 Different paths to derive the mean-field characteristic flow equations for (2.1)–(2.3)

In abstract terms, the index reduction and the mean-field limit commute, as summarised in Fig. 8.

In order to formally derive the kinetic equations for the system of DAEs (2.1)–(2.3), we follow similar steps as in Sect. 3.1.

Step 1: Introduction of a scaling factor,
Step 2: Introduction of a measure μ^t to describe the statistical distribution of the cross-bridge extensions,
Step 3: Derivation of the *constrained* mean-field characteristic flow equations, which govern the evolution of the cross-bridge distribution μ^t,
Step 4: Derivation of a kinetic balance law for the macroscopic system.

Step 1 Exactly as in (3.1), we scale the mass and force of each cross-bridge via

$$\tilde{M}_q := \frac{N_{\text{real}}}{N} M_q \quad \text{and} \quad \tilde{F}_q(Q_j) := \frac{N_{\text{real}}}{N} F_q(Q_j) = -\frac{N_{\text{real}}}{N} \gamma_q Q_j, \qquad (3.18)$$

where N_{real} is the fixed number of cross-bridges that a realistic system would have, and N is the number of cross-bridges of the systems for the limit process $N \to \infty$. After this rescaling, the system of DAEs (2.1)–(2.3) is

$$M_r \ddot{r} = -\gamma_r r - \sum_{i=1}^{N} G_r^T \tilde{\lambda}_i, \qquad (3.19)$$

$$\frac{N_{\text{real}}}{N} M_q \ddot{Q}_j = -\frac{N_{\text{real}}}{N} \gamma_q Q_j - \tilde{\lambda}_j \quad \text{for } j = 1, \ldots, N, \qquad (3.20)$$

$$Q_j + G_r r = Q_j^{\text{in}} + G_r r^{\text{in}} \quad \text{for } j = 1, \ldots, N \qquad (3.21)$$

where $\tilde{\lambda}_i \in \mathbb{R}^{n_q}$ are the Lagrangian multipliers of the scaled system (3.19)–(3.21).

Step 2 Precisely as in Sect. 3.1, we assume that the cross-bridge distribution $\mu^t \in \mathscr{P}^1(\mathbb{R}^{n_q})$ exists. In the following, we want to derive a law for the evolution of μ^t. The initial cross-bridge distribution is given by $\mu^{in} \in \mathscr{P}^1(\mathbb{R}^{n_q})$.

Step 3 In contrast to the effective balance law (2.6), the balance law of the macroscopic system (3.19) contains Lagrangian multipliers. The value of the multiplier $\tilde{\lambda}_j$ in (3.20) represents a force acting on the jth cross-bridge. In a statistical description, there are no individually labelled cross-bridges any more. Instead, the characteristic flow $Q(t, q^{in})$ tracks the dynamics of cross-bridges with initial condition $q^{in} \in \mathbb{R}^{n_q}$. Hence, instead of one multiplier $\tilde{\lambda}_j(t)$ per cross-bridge, the kinetic description requires a multiplier such as $\tilde{\lambda}(t, Q(t, q^{in}))$ for each initial state.

The characteristic flow $Q(t, q^{in})$ and the generalised Lagrangian multipliers $\tilde{\lambda}(t, q)$ are solutions of the following family of DAEs

$$\frac{N_{\text{real}}}{N} M_q \ddot{Q}(t, q^{in}) = -\frac{N_{\text{real}}}{N} \gamma_q Q(t, q^{in}) - \tilde{\lambda}(t, Q(t, q^{in})) \qquad \text{for all } q^{in} \in \mathbb{R}^{n_q},$$

(3.22)

$$Q(t, q^{in}) + G_r r = q^{in} + G_r r^{in} \qquad \text{for all } q^{in} \in \mathbb{R}^{n_q}.$$

(3.23)

However, this formulation is not satisfying since we can formally compute

$$\tilde{\lambda}(t, Q(t, q^{in})) = -\frac{N_{\text{real}}}{N} \left(M_q \ddot{Q}(t, q^{in}) - \frac{N_{\text{real}}}{N} \gamma_q Q(t, q^{in}) \right) \qquad (3.24)$$

$$\to 0 \quad \text{(formally) for } N \to \infty. \qquad (3.25)$$

As a result, the quantities of interest in the mean-field limit are not the Lagrangian multipliers $\tilde{\lambda}(t, Q(t, q^{in}))$ but the mean-field Lagrangian multipliers

$$\lambda_{\text{mf}}(t, Q(t, q^{in})) := \frac{N}{N_{\text{real}}} \tilde{\lambda}(t, Q(t, q^{in})) \quad \text{for all } q^{in} \in \mathbb{R}^{n_q}.$$

With this definition, the system (3.22) and (3.23) reads

$$M_q \ddot{Q}(t, q^{in}) = -\gamma_q Q(t, q^{in}) - \lambda_{\text{mf}}(t, Q(t, q^{in})) \qquad \text{for all } q^{in} \in \mathbb{R}^{n_q},$$

(3.26)

$$Q(t, q^{in}) + G_r r = q^{in} + G_r r^{in} \qquad \text{for all } q^{in} \in \mathbb{R}^{n_q}.$$

(3.27)

We refer to (3.26) and (3.27) as the *constrained mean-field characteristic flow equations*. For the same arguments as in Sect. 3.1 and Fig. 5, the evolution of the cross-bridge distribution μ^t is the pushforward of μ^{in} under the constrained characteristic flow

$$\mu^t := Q(t, \cdot) \# \mu^{in}. \tag{3.28}$$

Step 4 To replace (3.19) by its kinetic counterpart, we apply the scaling $\lambda_j = \frac{N}{N_{real}} \tilde{\lambda}_j$, to obtain formally

$$M_r \ddot{r} = -\gamma_r r - \sum_{i=1}^{N} G_r^T \tilde{\lambda}_i$$

$$= -\gamma_r r - \frac{N_{real}}{N} \sum_{i=1}^{N} G_r^T \lambda_i$$

$$= -\gamma_r r - N_{real} \left(\frac{1}{N} \sum_{i=1}^{N} G_r^T \lambda_{mf}(t, Q(t, Q_i^{in})) \right)$$

$$\rightarrow -\gamma_r r - N_{real} \int_{\mathbb{R}^{nq}} G_r^T \lambda_{mf}(t, Q(t, q^{in})) \mu^{in} \quad \text{(formally) for } N \rightarrow \infty \tag{3.29}$$

$$= -\gamma_r r - N_{real} \int_{\mathbb{R}^{nq}} G_r^T \lambda_{mf}(t, q) \mu^t$$

This formal argument yields the kinetic balance law

$$M_r \ddot{r} = -\gamma_r r - N_{real} \int_{\mathbb{R}^{nq}} G_r^T \lambda_{mf}(t, q) \, d\mu^t(q), \tag{3.30}$$

We remark that, in contrast to the derivation in Sect. 3.1, the strong law of large numbers is not applicable in (3.29).

Finally, we arrive at the *partially kinetic index-3 formulation* of (2.1)–(2.3) which is given by Eqs. (3.26)–(3.28) and (3.30). We summarise these equations

$$M_r \ddot{r} = -\gamma_r r - N_{real} \int_{\mathbb{R}^{nq}} G_r^T \lambda_{mf}(t, q) \, d\mu^t(q), \tag{3.31}$$

$$M_q \ddot{Q}(t, q^{in}) = -\gamma_q Q(t, q^{in}) - \lambda_{mf}(t, Q(t, q^{in})) \qquad \text{for all } q^{in} \in \mathbb{R}^{nq}, \tag{3.32}$$

$$Q(t, q^{in}) + G_r r = q^{in} + G_r r^{in} \qquad \text{for all } q^{in} \in \mathbb{R}^{nq}, \tag{3.33}$$

$$\mu^t := Q(t, \cdot) \# \mu^{in} \tag{3.34}$$

with initial conditions

$$r(0) = r^{\text{in}} \in \mathbb{R}^{n_r}, \quad \dot{r}(0) = s^{\text{in}} \in \mathbb{R}^{n_r} \quad \text{and} \quad Q(0, q^{\text{in}}) = q^{\text{in}} \quad \text{for all } q^{\text{in}} \in \mathbb{R}^{n_q}. \tag{3.35}$$

Since the derivation is formal, regularity aspects are not considered. Indeed, the integral in (3.31) might not exist for $\mu^{\text{in}} \in \mathscr{P}^1(\mathbb{R}^{n_q})$. It could be, that higher regularity of the initial data is necessary to ensure the existence of solutions of the partially kinetic index-3 formulation. Hence, we do not claim that the system has solutions.

3.5 Index Reduction and Elimination of the Multipliers for Partially Kinetic Systems

In Sect. 2.3, for every cross-bridge index $j \in \{1, \ldots, N\}$, the same algebraic transformations recast the DAE formulation into the index-1 formulation. Despite (3.32) and (3.33) being infinite-dimensional, the index reduction is possible with the same algebraic transformations as in Sect. 2.3 but applied for each $q^{\text{in}} \in \mathbb{R}^{n_q}$. The key assumption here is the uniformity of the constraints, since for each $q^{\text{in}} \in \mathbb{R}^{n_q}$ the algebraic constraint in (3.33) just differs by a shift.

The first two time derivatives of (3.33) are

$$\dot{Q}(t, q^{\text{in}}) = -G_r \dot{r}(t), \tag{3.36}$$

$$\ddot{Q}(t, q^{\text{in}}) = -G_r \ddot{r}(t). \tag{3.37}$$

Now, we solve (3.32) for the mean-field Lagrangian multipliers

$$\lambda_{\text{mf}}(t, Q(t, q^{\text{in}})) = -\gamma_q Q(t, q^{\text{in}}) - M_q \ddot{Q}(t, q^{\text{in}}),$$

and with (3.37) we obtain

$$\lambda_{\text{mf}}(t, Q(t, q^{\text{in}})) = -\gamma_q Q(t, q^{\text{in}}) + M_q G_r \ddot{r}(t). \tag{3.38}$$

Finally, we substitute (3.38) into the kinetic balance law (3.31)

$$
\begin{aligned}
M_r \ddot{r} &= -\gamma_r r - N_{\text{real}} \int_{\mathbb{R}^{n_q}} G_r^T \lambda_{\text{mf}}(t, q) \, d\mu^t(q) \\
&= -\gamma_r r - N_{\text{real}} \int_{\mathbb{R}^{n_q}} G_r^T (-\gamma_q q + M_q G_r \ddot{r}(t)) \, d\mu^t(q) \\
&= -\gamma_r r + N_{\text{real}} \int_{\mathbb{R}^{n_q}} G_r^T \gamma_q q \, d\mu^t(q) - \left(\int_{\mathbb{R}^{n_q}} N_{\text{real}} G_r^T M_q G_r \, d\mu^t(q) \right) \ddot{r}(t).
\end{aligned}
\tag{3.39}
$$

The resulting equation (3.39) is exactly the previously derived effective kinetic balance law for the macroscopic system in (3.12). The equations (3.34), (3.36) and (3.39) are exactly the previously derived partially kinetic equations (3.12)–(3.14). This concludes the claim from the beginning of Sect. 3.4: The formal mean-field limit and index-reduction commute for the DAE formulation (2.1)–(2.3).

4 The Mean-Field Limit for Partially Kinetic Systems

Until now, all derivations of partially kinetic equations in this article have been formal. In this section, we will give a rigorous link between the discrete cross-bridge dynamics and their kinetic description. In Eq. (3.8), the strong law of large numbers motivates the use of a mean-field force. Nevertheless, the strong law of large numbers is not sufficient to show that the kinetic description is a good approximation for systems with many cross-bridges. The law requires all cross-bridges to be stochastically independent, which is not easy to prove in general. Moreover, the argument in (3.8) only applies for a fixed time t. The mean-field limit is a generalization of the strong law of large numbers and yields the required convergence result.

Section 4.1 motivates the mean-field limit with a numerical evidence. Section 4.2 shows that the mean-field limit follows from Dobrushin's stability estimate for partially kinetic systems. Thereafter, Sect. 4.3 proofs the stability estimate.

4.1 Numerical Evidence of the Mean-Field Convergence

A consequence of the mean-field limit is that the mean-trajectories of the macroscopic system in the discrete formulation tend towards the solution of the partially kinetic description, as explained in Sect. 3.3. We will now perform a numerical test for this claim.

In Fig. 9, we perform numerical simulations of the ODE formulation (2.6) and (2.7) with fixed initial conditions $r^{in} = 1$ and $s^{in} = 0$. The statistics of the initial cross-bridge extensions are given by a normal distribution

$$d\mu^{in}(q) = \frac{1}{\sqrt{2\pi\sigma^2}} \exp\left(-\frac{(q-m)^2}{2\sigma}\right) dq,$$

with mean $m = -2$ and variance $\sigma^2 = 1$. As initial data for the ODE formulation, we sample $n_{samples} = 100$ many initial cross-bridge extensions $Q^{in}_{i,k}, \ldots, Q^{in}_{N,m} \in \mathbb{R}^{n_q}$, with law

$$Q^{in}_{i,k} \sim \mu^{in} \quad \text{for } i \in \{1, \ldots, N\}, k \in \{1, \ldots, n_{samples}\}.$$

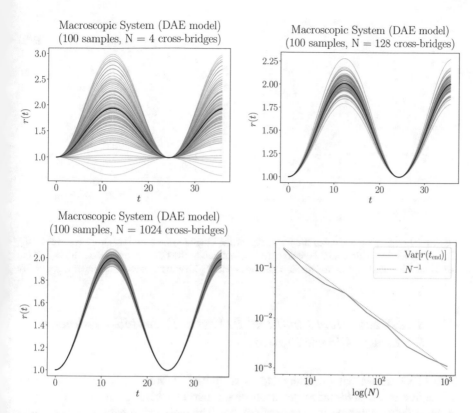

Fig. 9 Samples of trajectories $r(t)$ of the macroscopic system for different numbers of cross-bridges $N \in \{4, 128, 2048\}$ (top and bottom left). For an increasing number of particles n, the estimated variance of $r(t)$ decreases as $\frac{1}{n}$. (bottom right)

The sampling yields n_{samples} different initial conditions and therefore, n_{samples} trajectories of the macroscopic system and the cross-bridges. The trajectories of the macroscopic system are plotted in Fig. 9.

To quantify the distance of single trajectories from the mean-trajectory, we estimate the variance of $r(t)$ with respect to the randomly sampled initial conditions. It turns out that the variance reduces asymptotically as fast as $\frac{1}{N}$, which is displayed in Fig. 9. Therefore, even single trajectories are close to the mean-trajectory.

The mean-trajectories in Fig. 9 indeed converge towards the trajectory of the partially kinetic systems, as visualised in Fig. 10. We remark that this convergence also depends on the number of samples n_{samples}. Increasing the number of samples leads to faster convergence in Fig. 10. In a nonlinear setting, this convergence behaviour might change radically.

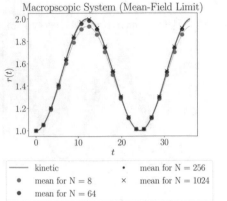

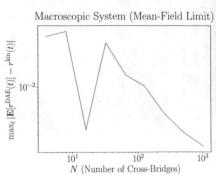

Fig. 10 The left plot compares the trajectory of the kinetic system and mean-trajectories of the discrete dynamics. This visual convergence is quantified in the right figure, which shows that the maximum distance between the mean-trajectories and the kinetic trajectory converges numerically

4.2 The Mean-Field Limit and Dobrushin's Stability Estimate for Linear Kinetic Systems

In this section, we will state an adapted version of Dobrushiun's stability estimate and define mean-field convergence for linear partially kinetic systems. The proof of the stability estimate is given in Sect. 4.3. The classical version of Dobrushin's stability estimate for the Vlasov equation can be found in [5, theorem 1.4.3].

The numerical experiments in Sect. 4.1 indicate that the mean-trajectories of the macroscopic systems converge to the mean-field trajectory. Now, to compare the distance between the discrete cross-bridge extensions $(Q_1(t), \ldots, Q_N(t))$ and the kinetic cross-bridge distribution μ^t, we need to choose a metric. The empirical measure $\mu^{(emp)}_{Q_1(t),\ldots,Q_N(t)} \in \mathscr{P}^1(\mathbb{R}^{n_q})$ is a possibility to represent the discrete cross-bridge extensions in the space of probability measures, which leads to the new goal of finding a metric for the space $\mathscr{P}^1(\mathbb{R}^{n_q})$. Moreover, such a metric should be compatible with the discrete distance between cross-bridge states. Therefore, we require

$$\text{dist}\left(\mu^{(emp)}_{Q_1,\ldots,Q_N}, \mu^{(emp)}_{\tilde{Q}_1,\ldots,\tilde{Q}_N}\right) \approx \frac{1}{N}\sum_{i=1}^{N}\left\|Q_i - \tilde{Q}_i\right\|. \qquad (4.1)$$

If $Q_i \approx \tilde{Q}_i$ holds, then the Monge-Kantorovich distance (also called Wasserstein distance) satisfies the geometric requirement (4.1). For a detailed study of the Monge-Kantorovich distance, we refer to [29, Chapter 6]. For our purpose, the duality formula for the Monge-Kantorovich distance with exponent 1 is most useful and hence serves as the definition here.

Definition 4.1 (Monge-Kantorovich Distance [5, Proposition 1.4.2])
The Monge-Kantorovich distance with exponent 1 is given by the formula

$$W_1(\nu, \mu) = \sup_{\substack{\phi \in \text{Lip}(\mathbb{R}^{n_q}) \\ \text{Lip}(\phi) \leq 1}} \left| \int_{\mathbb{R}^{n_q}} \phi(q) \, d\nu(q) - \int_{\mathbb{R}^{n_q}} \phi(q) \, d\mu(q) \right|, \tag{4.2}$$

with the notation

$$\text{Lip}(\phi) := \sup_{\substack{x, y \in \mathbb{R}^{n_q} \\ x \neq y}} \frac{|\phi(x) - \phi(y)|}{\|x - y\|}$$

and

$$\text{Lip}(\mathbb{R}^{n_q}) := \{\phi : \mathbb{R}^{n_q} \to \mathbb{R} \mid \text{Lip}(\phi) < \infty\}.$$

The Monge-Kantorovich distance is a complete metric on $\mathscr{P}^1(\mathbb{R}^{n_q})$ [29, Lemma 6.14].

For the attached cross-bridge model with $n_q = 1$, the duality formula (4.2) with $\phi(q) = q$ is directly applicable to estimate the difference of the mean-field forces

$$\|f_{\text{mean}}(\nu) - f_{\text{mean}}(\mu)\| \leq \left\| G_r^T \gamma_q \right\| \left\| \int_{\mathbb{R}^{n_q}} q \, d\nu(q) - \int_{\mathbb{R}^{n_q}} q \, d\mu(q) \right\| \tag{4.3}$$

$$\leq \left\| G_r^T \gamma_q \right\| W_1(\nu, \mu). \tag{4.4}$$

This estimate is at the core of the relation between partially kinetic systems and the Monge-Kantorovich distance. Together with a classical stability estimate for ODEs, this yields the following theorem.

Theorem 4.1 (Dobrushin's Stability Estimate for Linear Partially Kinetic Systems) *Suppose that for $i \in \{1, 2\}$ the tuples $(r_i(t), \mu_i(t)) \in \mathbb{R}^{n_r} \times \mathscr{P}^1(\mathbb{R}^{n_q})$ are solutions of (3.12)–(3.14) with initial conditions*

$$r_i(0) = r_i^{\text{in}}, \quad \dot{r}_i(0) = s_i^{\text{in}} \quad \text{and} \quad \mu_i(0) = \mu_i^{\text{in}}.$$

Then

$$\|r_1(t) - r_2(t)\| + \|\dot{r}_1(t) - \dot{r}_2(t)\| + W_1(\mu_1(t), \mu_2(t))$$

$$\leq C e^{Lt} \left(\left\| r_1^{\text{in}} - r_2^{\text{in}} \right\| + \left\| s_1^{\text{in}} - s_2^{\text{in}} \right\| + W_1(\mu_1^{\text{in}}, \mu_2^{\text{in}}) \right), \tag{4.5}$$

for some constants $L, C > 0$ which are independent of the initial conditions.

The Proof of Theorem 4.1 is content of Sect. 4.3.

Dobrushin's stability estimate for linear partially kinetic systems (4.5) provides a concrete answer to the approximation quality of the kinetic description. Lemma 3.1 shows that solutions $(r(t), Q_1(t), \ldots, Q_N(t))$ of the discrete system (2.5) and (2.6) yield a solution of the kinetic formulation with initial data $\mu^{\text{in}} = \mu^{(emp)}_{Q_1^{\text{in}}, \ldots, Q_N^{\text{in}}}$. If we increase the number of cross-bridges such that the empirical measures converge to a probability distribution $\mu^{\text{in}} \in \mathscr{P}^1(\mathbb{R}^{n_q})$, i.e.

$$W_1(\mu^{(emp)}_{Q_1^{\text{in}}, \ldots, Q_N^{\text{in}}}, \mu^{\text{in}}) \to 0, \quad \text{for } N \to \infty,$$

then (4.5) provides a bound for the approximation error

$$\|r_N(t) - r(t)\| + \|s_N(t) - s(t)\| + W_1(\mu^{(emp)}_{Q_1(t), \ldots, Q_N(t)}, \mu^t) \leq Ce^{tL} W_1(\mu^{(emp)}_{Q_1^{\text{in}}, \ldots, Q_N^{\text{in}}}, \mu^{\text{in}})$$

$$\to 0, \quad \text{for } N \to \infty. \tag{4.6}$$

This estimate provides a rigorous argument for the use of kinetic models. Moreover, if the initial distribution of N cross-bridges $(Q_j)_{j=1,\ldots,N}$ is very close to a continuous distribution μ^{in}, then approximation error of the kinetic description is bounded.

Using tools from probability theory and functional analysis, the topology for the convergence in (4.6) can be refined to the topology of weak convergence of measures (also called convergence in distribution for probability measures) [5, Lemma 1.4.6], [29, theorem 6.9]. This leads to the precise definition of mean-field convergence [17]. We omit these details here as they are out of scope for the present paper.

4.3 Proof of Dobrushin's Stability Estimate for Linear Partially Kinetic Systems

This section provides a proof of Theorem 4.1. Compared to classical proofs of Dobrushin's stability estimate for the Vlasov equation [5, 17], the particular structure of linear partially kinetic system allows for a more elementary proof, which essentially reuses the stability estimate for ODEs with respect to initial conditions and parameters, as stated in Theorem 4.3.

In contrast to many other metrics for probability measures, the Monge-Kantorovich distance is motivated by geometry. In particular, shifting two measures increases their distance at most by the length of the shift, see Lemma 4.2. Due to this property, the proof of Theorem 4.1 boils down to the ODE estimate from Theorem 4.3.

Lemma 4.2 (Shift of the Monge-Kantorovich Distance) *Let $\mu, \nu \in \mathscr{P}^1(\mathbb{R}^{n_q})$ be two probability measures with finite first moment. For $w \in \mathbb{R}^{n_q}$ we define the shift mapping as $T_w : \mathbb{R}^{n_q} \to \mathbb{R}^{n_q} : q \mapsto q + w$. Then for $w_1, w_2 \in \mathbb{R}^{n_q}$ it holds*

$$W_1(T_{w_1}\#\mu, T_{w_2}\#\nu) \leq W_1(\mu, \nu) + \|w_2 - w_1\|.$$

Lemma 4.2 is usually a special case of more general theorems on the relation between geodesic flows and the Monge-Kantorovich metric [29, Chapter 8]. For completeness, we give an elementary proof.

Proof The Monge-Kantorovich distance is invariant with respect to shifts $W_1(T_w\#\nu, T_w\#\mu) = W_1(\nu, \mu)$, hence we can assume $w_1 = 0$. For an arbitrary vector $w \in \mathbb{R}^{n_q}$, we use the duality formula (4.2) and compute

$$W_1(\nu, T_w\#\mu) = \sup_{\mathrm{Lip}(\phi) \leq 1} \left| \int_{\mathbb{R}^{n_q}} \phi(q)\, d\nu(q) - \int_{\mathbb{R}^{n_q}} \phi(q)\, d(T_w\#\mu)(q) \right|$$

$$= \sup_{\mathrm{Lip}(\phi) \leq 1} \left| \int_{\mathbb{R}^{n_q}} \phi(q)\, d\nu(q) - \int_{\mathbb{R}^{n_q}} \phi(T_w(q))\, d\mu(q) \right|$$

$$= \sup_{\mathrm{Lip}(\phi) \leq 1} \left| \int_{\mathbb{R}^{n_q}} \phi(q)\, d\nu(q) - \int_{\mathbb{R}^{n_q}} \phi(q)\, d\mu(q) + \int_{\mathbb{R}^{n_q}} \phi(q) - \phi(q + w)\, d\mu(q) \right|$$

$$\leq W_1(\nu, \mu) + \sup_{\mathrm{Lip}(\phi) \leq 1} \left| \int_{\mathbb{R}^{n_q}} \phi(q) - \phi(q + w)\, d\mu(q) \right| \tag{4.7}$$

where we have applied the transformation formula to the integral of the pushforward operator. Next, we employ $\mathrm{Lip}(\phi) \leq 1$ to obtain the upper bound

$$\left| \int_{\mathbb{R}^{n_q}} \phi(q) - \phi(q + w)\, d\mu(q) \right| \leq \int_{\mathbb{R}^{n_q}} |\phi(q) - \phi(q + w)|\, d\mu(q)$$

$$\leq \int_{\mathbb{R}^{n_q}} \|w\|\, d\mu(q) = \|w\|.$$

We conclude

$$W_1(\nu, T_w\#\mu) \leq W_1(\nu, \mu) + \|w\|. \tag{4.8}$$

$\square$

Theorem 4.3 (The "Fundamental Lemma", [9, theorem 10.2]) *Suppose that for $i \in \{1, 2\}$ the functions $x_i(t)$ solve*

$$\dot{x}_i(t) = f_i(x_i(t)), \tag{4.9}$$

$$x_i(0) = x_i^{\mathrm{in}}. \tag{4.10}$$

If for some constants ϱ, ε, $L > 0$ the following bounds hold

(i) $\left\| x_1^{\text{in}} - x_2^{\text{in}} \right\| \leq \varrho$,

(ii) $\left\| f_1(x) - f_2(x) \right\| \leq \varepsilon$ *for all $x \in \mathbb{R}^n$,*

(iii) $\left\| f_1(a) - f_1(b) \right\| \leq L \, \|a - b\|$ *for all $a, b \in \mathbb{R}^n$,*

then

$$\left\| x_1(t) - x_2(t) \right\| \leq \varrho e^{Lt} + \frac{\varepsilon}{L} \left(e^{Lt} - 1 \right). \tag{4.11}$$

Theorem 4.1 can be considered as a generalisation of Theorem 4.3.

Proof of Theorem 4.1 First, we will reformulate the ODE formulation (3.12)–(3.14) such that Theorem 4.3 is applicable. The characteristic flow is explicitly given by

$$Q(t, q^{\text{in}}) := -G_r(r(t) - r^{\text{in}}) + q^{\text{in}}, \tag{4.12}$$

which is the unique solution of (3.13) and the initial condition. With (4.12), we can compute the first moment of μ^t as

$$\int_{\mathbb{R}^{nq}} q \, \mathrm{d}\mu^t(q) = \int_{\mathbb{R}^{nq}} Q(t, q') \, \mathrm{d}\mu^{\text{in}}(q')$$

$$= G_r(r(t) - r^{\text{in}}) + \int_{\mathbb{R}^{nq}} q' \, \mathrm{d}\mu^{\text{in}}(q'). \tag{4.13}$$

As a result, the effective force can be written as

$$f_{\text{eff}}(r(t), \mu^t) = -\gamma_r r(t) + N_{\text{real}} G_r^T \gamma_q \int_{\mathbb{R}^{nq}} q \, \mathrm{d}\mu^t(q)$$

$$= -\gamma_r r(t) + N_{\text{real}} G_r^T \gamma_q \left(G_r(r(t) - r^{\text{in}}) + \int_{\mathbb{R}^{nq}} q \, \mathrm{d}\mu^{\text{in}}(q) \right)$$

$$=: \tilde{f}_{\text{eff}}(r(t); \mu^{\text{in}}).$$

Now, we consider two different initial conditions $(r_1^{\text{in}}, s_1^{\text{in}}, \mu_1^{\text{in}}) \in \mathbb{R}^{n_r} \times \mathbb{R}^{n_r} \times \mathscr{P}^1(\mathbb{R}^{nq})$ and $(r_2^{\text{in}}, s_2^{\text{in}}, \mu_2^{\text{in}}) \in \mathbb{R}^{n_r} \times \mathbb{R}^{n_r} \times \mathscr{P}^1(\mathbb{R}^{nq})$. To prepare the application of Theorem 4.3, we transform $m_{\text{eff}} \ddot{r}_i = f_{\text{eff}}$ (3.12) into a first order ODE, with $x_i(t) = (r_i(t), \dot{r}_i(t))^T \in \mathbb{R}^{2n_r}$ and

$$\dot{x}_i(t) = \begin{pmatrix} \dot{r}_i(t) \\ m_{\text{eff}}^{-1} \tilde{f}_{\text{eff}}(r(t); \mu_i^{\text{in}}) \end{pmatrix} =: f_i(x_i(t)) \qquad \text{for } i \in \{1, 2\}, \tag{4.14}$$

$$x_i(0) = (r_i^{\text{in}}, s_i^{\text{in}})^T \qquad\qquad\qquad \text{for } i \in \{1, 2\}.$$

We remark that $m_{\mathrm{eff}} = M_r + N_{\mathrm{real}} G_r^T M_q G_r$ is invertible, since M_r and M_q are defined to be positive definite, see Sect. 2.

The difference between the right-hand sides f_i at a fixed state $x = (r, s) \in \mathbb{R}^{2n_r}$ is

$$\|f_1(r) - f_2(r)\| = \left\| m_{\mathrm{eff}}^{-1} \left(\tilde{f}_{\mathrm{eff}}(r; \mu_1^{\mathrm{in}}) - \tilde{f}_{\mathrm{eff}}(r; \mu_2^{\mathrm{in}}) \right) \right\|$$

$$= \left\| m_{\mathrm{eff}}^{-1} N_{\mathrm{real}} G_r^T \gamma_q \left(G_r (r_1^{\mathrm{in}} - r_2^{\mathrm{in}}) + \int_{\mathbb{R}^{n_q}} q \, d\mu_1^{\mathrm{in}}(q) - \int_{\mathbb{R}^{n_q}} q \, d\mu_2^{\mathrm{in}}(q) \right) \right\|$$

$$=: \varepsilon. \tag{4.15}$$

Next, we compute the Lipschitz constant of f_1. The partial derivatives are

$$\frac{\partial f_1}{\partial \dot{r}_1} = 1 \quad \text{and} \quad \frac{\partial f_1}{\partial r_1} = m_{\mathrm{eff}}^{-1}(\gamma_r + N_{\mathrm{real}} G_r^T \gamma_q G_r),$$

which implies that

$$L := \left\| m_{\mathrm{eff}}^{-1} \right\| \left(\|\gamma_r\| + N_{\mathrm{real}} \left\| G_r^T \gamma_q G_r \right\| \right) + 1$$

is a Lipschitz constant for f_1.

Then Theorem 4.3 yields

$$\sqrt{\|r_1(t) - r_2(t)\|^2 + \|s_1(t) - s_2(t)\|^2} \leq \varrho e^{Lt} + \frac{\varepsilon}{L} \left(e^{Lt} - 1 \right) \tag{4.16}$$

with

$$\varrho := \|x_1(0) - x_2(0)\| = \sqrt{\left\| r_1^{\mathrm{in}} - r_2^{\mathrm{in}} \right\|^2 + \left\| s_1^{\mathrm{in}} - s_2^{\mathrm{in}} \right\|^2}. \tag{4.17}$$

Next, we apply the duality formula for the Monge-Kantorovich distance (4.2) to each component of $q \in \mathbb{R}^{n_q}$. Hence, we use $\phi(q) := q_l \in \mathbb{R}$ in (4.2), which yields

$$\left\| \int_{\mathbb{R}^{n_q}} q \, d\mu_1^{\mathrm{in}}(q) - \int_{\mathbb{R}^{n_q}} q \, d\mu_2^{\mathrm{in}}(q) \right\|$$

$$\leq \sum_{l=1}^{n_q} \left| \int_{\mathbb{R}^{n_q}} q_l \, d\mu_1^{\mathrm{in}}(q) - \int_{\mathbb{R}^{n_q}} q_l \, d\mu_2^{\mathrm{in}}(q) \right|$$

$$\leq n_q W_1(\mu_1^{\mathrm{in}}, \mu_2^{\mathrm{in}}). \tag{4.18}$$

The estimate (4.18) provides the upper bound

$$
\varepsilon \leq \left\| m_{\mathrm{eff}}^{-1} N_{\mathrm{real}} G_r^T \gamma_q \right\| \left(\|G_r\| \left\| r_1^{\mathrm{in}} - r_2^{\mathrm{in}} \right\| + \left\| \int_{\mathbb{R}^{nq}} q \, \mathrm{d}\mu_1^{\mathrm{in}}(q) - \int_{\mathbb{R}^{nq}} q \, \mathrm{d}\mu_2^{\mathrm{in}}(q) \right\| \right)
$$

$$
\leq C_1(\varrho + W_1(\mu_1^{\mathrm{in}}, \mu_2^{\mathrm{in}}))
$$

with $C_1 := \| m_{\mathrm{eff}}^{-1} N_{\mathrm{real}} G_r^T \gamma_q \| (\|G_r\| + n_q)$.

Now, we use $\frac{1}{2}(|a| + |b|) \leq \sqrt{a^2 + b^2}$ to transform (4.16) into

$$
\left\| r_1(t) - r_2(t) \right\| + \left\| s_1(t) - s_2(t) \right\| \leq 2 \sqrt{ \left\| r_1(t) - r_2(t) \right\|^2 + \left\| s_1(t) - s_2(t) \right\|^2 }
$$

$$
\leq 2\rho e^{Lt} + 2 \frac{C_1(\rho + W_1(\mu_1^{\mathrm{in}}, \mu_2^{\mathrm{in}}))}{L} (e^{Lt} - 1)
$$

$$
\leq 2(1 + \frac{C_1}{L})\rho e^{Lt} + 2 \frac{C_1}{L} W_1(\mu_1^{\mathrm{in}}, \mu_2^{\mathrm{in}}))e^{Lt}
$$

and we bound ϱ with $\sqrt{a^2 + b^2} \leq |a| + |b|$ to obtain

$$
\left\| r_1(t) - r_2(t) \right\| + \left\| s_1(t) - s_2(t) \right\|
$$

$$
\leq C_2 \left(\left\| r_1^{\mathrm{in}} - r_2^{\mathrm{in}} \right\| + \left\| s_1^{\mathrm{in}} - s_2^{\mathrm{in}} \right\| + W_1(\mu_1^{\mathrm{in}}, \mu_2^{\mathrm{in}}) \right) e^{Lt}
$$

$$
\tag{4.19}
$$

with the constant $C_2 := 2(1 + \frac{C_1}{L})$.

The estimate (4.19) already looks similar to the claim. It only misses an estimate for the difference of the cross-bridge distributions $\mu_1(t)$ and $\mu_2(t)$. We notice that

$$
\mu_i(t) = Q_i(t, \cdot) \# \mu_i^{\mathrm{in}} = T_{w_i} \mu_i^{\mathrm{in}}
$$

with

$$
w_i(t) = -G_r(r_i(t) - r_i^{\mathrm{in}}).
$$

Therefore, Lemma 4.2 gives

$$
W_1(\mu_1(t), \mu_2(t)) = W_1(T_{w_1(t)}\mu_1^{\mathrm{in}}, T_{w_2(t)}\mu_2^{\mathrm{in}})
$$

$$
= W_1(\mu_1(t), T_{w_2(t) - w_1(t)}\mu_2(t))
$$

$$
\leq W_1(\mu_1^{\mathrm{in}}, \mu_2^{\mathrm{in}}) + \left\| w_2(t) - w_1(t) \right\|
$$

$$\leq W_1(\mu_1^{\text{in}}, \mu_2^{\text{in}}) + \|G_r\| \left(\|r_1(t) - r_2(t)\| + \left\|r_1^{\text{in}} - r_2^{\text{in}}\right\| \right).$$

$$\leq C_2(\|G_r\| + 1) \left(\left\|r_1^{\text{in}} - r_2^{\text{in}}\right\| + \left\|s_1^{\text{in}} - s_2^{\text{in}}\right\| + W_1(\mu_1^{\text{in}}, \mu_2^{\text{in}}) \right) e^{Lt} \qquad (4.20)$$

where we use $e^{Lt} \geq 1$ and (4.19) in the last equation.

Combining (4.19) and (4.20), we obtain (4.5) with the constants

$$C := C_2(2 + \|G_r\|), \quad L := \left\|m_{\text{eff}}^{-1}\right\| \left(\|\gamma_r\| + N_{\text{real}} \left\|G_r^T \gamma_q G_r\right\| \right) + 1.$$

□

5 Generalisations and Relations to Established Models

The model for attached cross-bridges demands for extensions in two directions: For applications in biology, the cross-bridge model should incorporate more biological effects, for example, cross-bridge cycling. For further mathematical investigations, partially kinetic systems can be studied in more generality, most notably with nonlinear constraints or stochastic jumps. In this section, we give a brief outlook on these extensions.

5.1 An Abstract Class of Nonlinear Partially Kinetic Systems

The underlying system (2.1)–(2.3) is a prototype for linear partially kinetic systems. A possible extension would be to consider systems with nonlinear forces $F_r(r)$, $F_q(Q_j)$ and with uniform but nonlinear constraints $g(r, Q_j) = g(r^{\text{in}}, Q_j^{\text{in}})$, for example

$$M_r \ddot{r} = F_r(r) - \sum_{i=1}^{N} \frac{\partial g^T}{\partial r} \lambda_i, \qquad (5.1)$$

$$M_q \ddot{Q}_j = F_q(Q_j) - \frac{\partial g^T}{\partial Q_j} \lambda_j \quad \text{for } j = 1, \ldots, N, \qquad (5.2)$$

$$g(r, Q_j) = g(r^{\text{in}}, Q_j^{\text{in}}) \quad \text{for } j = 1, \ldots, N. \qquad (5.3)$$

Despite the nonlinear terms, the formal derivation of kinetic equations for (5.1)–(5.3) follows similar algebraic steps as in Sects. 3.1 and 3.4.

Remark 5.1 We conjecture that Theorem 4.1 generalises to nonlinear partially kinetic systems (5.1)–(5.3), if certain Lipschitz conditions and technical bounds are satisfied. In contrast to the linear case, the effective mass matrix m_{eff} depends on the cross-bridge distribution μ^t. Therefore, it is more challenging to obtain the Lipschitz constant L and the defect ε as in (4.15). Moreover, the mean-field characteristic flow $Q(t, \cdot)$ will be nonlinear and there is no explicit formula as in (4.12). Therefore Lemma 4.2 is not applicable in the nonlinear case.

5.2   Coupling with Nonlinear Elasticity

The muscle model in Sect. 2 allows only linear constraints and linear forces. This is not sufficient for realistic multi-scale models. On a large scale, muscles can be modelled as nonlinear, quasi-incompressible, hyperelastic solids [25].

Using the framework of partially kinetic systems, a nonlinear constraint can link a hyperelastic model at the large scale for the muscle tissue and the cross-bridge model at the physiological scale. The resulting constraint is linear with respect to the extension of cross-bridges but nonlinear with respect to the deformation of the tissue. In this sense, the situation is similar to Sect. 5.1 Additional complexity arises since the macroscopic system in this setting is described by the PDEs of elasticity, which results in an infinite-dimensional system already in the discrete case. We can assume that most material points of the muscle are occupied parallel actin and myosin filaments. This assumption leads mathematically to an infinite family of cross-bridge models, one at each spatial point of the muscle. A formal derivation is possible and not very different from the theory presented in this article. Analytical results, however, are far more challenging in this setting. Even more, the cross-bridges of sarcomeres at neighbouring spacial points can be in very different states. Therefore, a rigorous mathematical approach requires a justification for spatial averaging over the cross-bridge states.

5.3   Comparison with Established Cross-Bridge Models

The model for attached cross-bridges neglects a fundamental part of cross-bridge dynamics: Cross-bridges can attach and detach dynamically. The repeated attachment and detachment is called cross-bridge cycling. Only with this mechanism, muscle cells can contract far beyond the working range of a single cross-bridge. Non-kinetic models, like the popular Hill model fail to capture some phenomena which depend on cross-bridge cycling [18, Section 15.3.1]. This motivates the use of kinetic models in muscle simulations.

The most common models for cross-bridge cycling are probabilistic. One example is the *two-state model* [30]. The two-state model is usually only formulated on the kinetic level as a source term in the transport equation (3.16). A function

$h_+(t, q, u)$ determines the creation rate of new cross-bridges with extension $q \in \mathbb{R}^{n_q}$ at time t. The counterpart is a function $h_-(t, q, u)$, which gives the annihilation rate of cross-bridges with extension $q \in \mathbb{R}^{n_q}$. The two-state model leads to a kinetic transport equation with source terms

$$\frac{\partial u}{\partial t}(t, q) + v_{\text{eff}} \frac{\partial u}{\partial q}(t, q) = h_+(t, q, u(t, q)) - h_-(t, q, u(t, q)). \qquad (5.4)$$

Here, the transport velocity v_{eff} is the contraction speed of the muscle. In the setting of Sect. 3.3, contraction speed of the muscle is the velocity of the macroscopic system, hence $v_{\text{eff}} = -G_r \dot{r}(t)$. The rate functions $h_\pm$ allow controlling the contraction speed in the muscle model. For a contracting muscle, they are such that many cross-bridges with large positive extension are created, and cross-bridges with negative extension are annihilated. A discussion of the two-state model is not subject of this publication. Instead, we refer to [12, 14, 18]. Models with more than just two-states are also studied and applied, for example in [10, 11, 13].

All these models for cross-bridge cycling have in common, that their underlying discrete microscopic model for cross-bridge cycling is, to the best of our knowledge, not specified. Instead, these models are built directly from the kinetic perspective, to avoid unnecessary complexity. The rate functions h_+ and h_- are heuristic and usually fitted to experimental data [30]. We are not aware of a derivation for the rates h_+ and h_- using kinetic theory. As far as we know, the most rigorous approach to derive the rate functions is based on thermodynamics principles [21].

Nonetheless, a rigorous connection between the microscopic world and macroscopic simulations requires a microscopic law for cross-bridge cycling. One possibility could be to model the microscopic law as a pieces-wise deterministic Markov process [4] where a Markov process models the creation and annihilation of cross-bridges, and the deterministic model for attached cross-bridges (2.2), (2.3) and (2.6) governs the system at all other times. The corresponding mean-field limit is formally similar to the mean-field limit for chemical reactions, as outlined in [3, Section 3.3].

Finally, we want to point out a detail which differs between established models and the kinetic model developed in this article. Since we developed a kinetic theory which includes the constraints from the beginning, the influence of the cross-bridges onto the macroscopic system is exactly represented. The effective balance law which governs the macroscopic system is (3.12)

$$m_{\text{eff}} \ddot{r} = f_{\text{eff}},$$

where f_{eff} includes the influence of the cross-bridges onto the macroscopic system, i.e. the muscle tissue. Moreover, m_{eff} integrates the inertia of the cross-bridges to the effective balance law (3.12). In contrast, established muscle models [1, 11, 21] just compute the force f_{eff} and use

$$M_r \ddot{r} = f_{\text{eff}}.$$

For applications, this approximation is very reasonable, since the momentum of cross-bridges M_q is considered to be very small, compared to the mass of the muscle tissue, i.e. the underlying assumption is $N_{\text{real}} \|M_q\| \ll \|M_r\|$, which implies $M_r \approx m_{\text{eff}}$.

6 Numerical Simulation of Partially Kinetic Systems

Partially kinetic systems are mixed systems involving Newton's equations of motion for the macroscopic components (3.12) and a non-linear transport equation for the particle density (3.16). Therefore, a perfect numerical scheme for such a system should not only nearly conserve energy and momentum, but also the mass of the particle density. It is an open issue if such a scheme exists.

In the literature on sliding filament theory, a popular method is the distributed moment method (DM method) [30]. By assuming a specific shape for the cross-bridge distribution μ^t, it is possible to derive a closed set of differential equations for the first moments of μ^t and thus approximate the solution of the transport equation for cross-bridges (5.4) by a three-dimensional ODE. If the particle measure is close to a Gaussian distribution, then the DM method works best. The DM method is successful and has been used in many multi-scale simulations [1, 11]. However, the state of cross-bridges is often very different from a normal distribution. In these cases, the DM method does not yield a numerically convergent discretisation of (3.12) and (3.16). This drawback is acceptable for most applications, but precise information about the physical state of the cross-bridges is lost.

6.1 Implementation Details

The numerical simulations in this article are performed straightforwardly. More advanced and adapted methods are out of scope for this article.

For numerical time integration, we used the RADAU [6, Section IV.8.] and the LSODA [22] methods from the python package scipy.integrate [28], both with numerical parameters atol $= 10^{-8}$, rtol $= 10^{-8}$ for the adaptive time-stepping scheme. For all examples in this article, the time integration was successful without any indicators for numerical instabilities.

For space discretisation of the transport equation (3.16), the standard upwind discretisation was used [20, Section 10.4]. For simplicity, we assumed zero boundary conditions for the numerical spatial domain. The grid was chosen sufficiently large such that the boundary conditions do not influence the simulation results. In detail, we solved the transport equation (3.16) restricted onto the spacial domain $[-5, 7]$

Table 1 All simulations in the article are for these parameters

Model parameters		
Description	Symbol	Value
Degrees of freedom (macroscopic system)	n_r	1
Degrees of freedom (single cross-bridge)	n_q	1
Number of cross-bridges	N_{real}	250
Actually simulated cross-bridges	N	250
Mass (macroscopic system)	M_r	20
Mass (single cross-bridge)	M_q	$\frac{10}{N_{real}} = 0.04$
Stiffness (macroscopic system)	γ_r	1
Stiffness (single cross-bridge)	γ_q	$\frac{1}{N_{real}} = 0.004$
Initial position (macroscopic system)	r^{in}	1
Initial velocity (macroscopic system)	s^{in}	0
Initial extensions (cross-bridges)	Q_j	$\sim \mathcal{N}(\mu = 2, \sigma^2 = 1)$
Initial distribution (cross-bridges)	$u^{in}(q)$	$= \frac{1}{\sqrt{2\pi}}\exp(-\frac{(q-2)^2}{2})$
Time interval		$[0, 60]$.

The parameters are chosen to demonstrate the mathematical structure, not to represent a realistic biological setting

on an equidistant grid with 101 grid-points, with

$$\dot{u}(t, y_i) = \begin{cases} v_{\text{eff}}(t)\frac{1}{\Delta x}(u(t, x_i) - u(t, x_{i-1})) & v_{\text{eff}}(t) \geq 0, \\ v_{\text{eff}}(t)\frac{1}{\Delta x}(u(t, x_{i+1}) - u(t, x_i)) & v_{\text{eff}}(t) < 0, \end{cases}$$

where $v_{\text{eff}}(t) = -G_r\dot{r}(t)$ denotes the velocity according to (3.16), and $\Delta x = \frac{12}{100}$ is the space between two grid-points $y_i = -5 + i\Delta x$.

For all simulations in this article, we used the model parameters from Table 1. The parameters are not related to real cross-bridges; the numerical simulations should merely demonstrate the mathematical model, not its application to reality.

For systems with linear constraints, the use of the upwind method might appear exaggerated. Instead, it would be sufficient to approximate the shift between μ^t and the initial measure μ^{in}, which is given by $w = G_r(r(t) - r^{in})$. Since the transport equation (3.16) has no source terms, such a numerical scheme is fast and stable. In the presence of source terms, as in (5.4), the resulting numerical scheme contains stiff differential equations. Since source terms are essential for realistic muscle models, we neglect this specialised method to focus on the upwind discretisation instead.

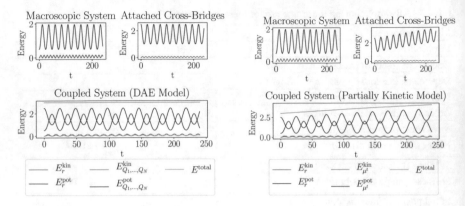

Fig. 11 Energies of the discrete system (left) and the kinetic system (right). The kinetic energy is denoted by T_r for the macroscopic system and T_q for the particles, the potential energy is denoted by U_r and U_q respectively. The total energy E_{total} is well preserved in the discrete case, but not in the mesoscopic simulation

6.2 Loss of Numerical Energy Conservation in the Partially Kinetic Description

In Fig. 11, we compare the energies of the ODE formulation and the partially kinetic formulation with the same initial data as in Sect. 4.1. The results demonstrate an essential drawback of the numerical scheme for the kinetic description. There are numerical methods for DAE formulation (2.1), (2.2) and (2.5) and the ODE formulation (2.5) and (2.6) which conserve the energy asymptotically [8]. In this particular linear example, the numerical conservation of energy is not very difficult even for stiff integrators. However, numerical diffusion in the upwind scheme leads to an increase in the total energy, as shown in Fig. 11.

To explain the energy increase, we recall that the potential energy of the cross-bridges is

$$E_{Q_1,\dots,Q_N}^{\text{pot}} = \frac{\gamma_q}{2} \sum_{i=1}^{N} Q_i^2 \tag{6.1}$$

which becomes

$$E_{\mu^t}^{\text{pot}} = N_{\text{real}} \frac{\gamma_q}{2} \int_{\mathbb{R}^{n_q}} q^2 \, d\mu^t(q) \tag{6.2}$$

in the kinetic setting. The upwind scheme increases the potential energy artificially since numerical diffusion leads to a more widespread cross-bridge distribution, as demonstrated in Fig. 12. We are not aware of a method which conserves the total energy numerically and at the same time works in the presence of source terms.

Fig. 12 Numerical diffusion of the upwind scheme leads to a diffusion of the cross-bridge distribution. As a result, the potential energy of the cross-bridges increases

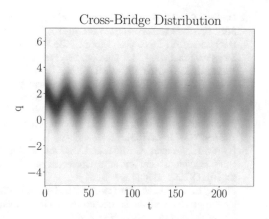

Cross-Bridge Distribution

7 Conclusion

In this article, we have presented the new framework of linear partially kinetic systems. We have motivated this abstract class as a kinetic model for cross-bridge dynamics in skeletal muscles, in a manner which allows us to add constraints. The linear setting represents a toy example, for which ideas from kinetic theory are applicable. It can be argued that the analysis so far is restricted to a rather simple model scenario. Thus, there is a need to generalise these results to a wider class of models, which then would yield a rigorous link between existing physiological models at different scales.

Finally, the numerics of partially kinetic systems is still in its infancy. The investigation of the stability estimate was motivated by the lack of numerical analysis for linear partially kinetic systems. We have presented an example in which the conservation of energy is violated, which already indicates the limitations of a naive discretisation.

Acknowledgments We thank Claudia Totzeck and Sara Merino-Aceituno for the fruitful discussions and hints regarding the mean-field limit. The work is motivated by the aim to develop a mathematical foundation for muscle tissue with a two-way coupling between cells and tissue. We thank Ulrich Randoll for his advice and numerous discussion on the physiology of skeletal muscle tissue. This research is supported by the German Federal Ministry of Education and Research (BMBF) under grant no. 05M16UKD (project DYMARA).

References

1. Böl, M., Reese, S.: Micromechanical modelling of skeletal muscles based on the finite element method. Comput. Method Biomec. Biomed. Eng. **11**(5), 489–504 (2008). https://doi.org/10. 1080/10255840701771750
2. Brenan, K.E., Campbell, S.L., Petzold, L.R.: The Numerical Solution of Initial Value Problems in Ordinary Differential-Algebraic Equations. SIAM, Philadelphia (1996)

3. Darling, R.W.R., Norris, J.R.: Differential equation approximations for Markov chains. Probab. Surv. **5**, 37–79 (2008). https://doi.org/10.1214/07-PS121
4. Davis, M.H.A.: Piecewise-deterministic Markov processes: a general class of non-diffusion Stochastic models. J. R. Stat. Soc. Ser. B (Methodol.) **46**(3), 353–376 (1984). https://doi.org/10.1111/j.2517-6161.1984.tb01308.x
5. Golse, F.: On the dynamics of large particle systems in the mean field limit. In: Muntean, A., Rademacher, J., Zagaris, A. (eds.) Macroscopic and Large Scale Phenomena: Coarse Graining, Mean Field Limits and Ergodicity. Lecture Notes in Applied Mathematics and Mechanics, pp. 1–144. Springer International Publishing, Cham (2016). https://doi.org/10.1007/978-3-319-26883-5_1
6. Hairer, E., Wanner, G.: Stiff and Differential-Algebraic Problems. Solving Ordinary Differential Equations, 2nd rev. edn., Corrected printing, 1. softcover printing edn. Springer, Berlin (2010). OCLC: 837885597
7. Hairer, E., Roche, M., Lubich, C.: The Numerical Solution of Differential-Algebraic Systems by Runge-Kutta Methods. Lecture Notes in Mathematics, vol. 1409. Springer, Berlin (1989). https://doi.org/10.1007/BFb0093947
8. Hairer, E., Lubich, C., Wanner, G.: Geometric Numerical Integration: Structure-Preserving Algorithms for Ordinary Differential Equations. Springer Series in Computational Mathematics, no. 31, 2nd edn. Springer, Berlin (2006). OCLC: ocm69223213
9. Hairer, E., Nørsett, S.P., Wanner, G.: Solving Ordinary Differential Equations I: Nonstiff Problems. Springer Series in Computational Mathematics, No. 8, 2nd rev. edn. Springer, Heidelberg (2009). OCLC: ocn620251790
10. Heidlauf, T., Klotz, T., Rode, C., Altan, E., Bleiler, C., Siebert, T., Röhrle, O.: A multiscale continuum model of skeletal muscle mechanics predicting force enhancement based on actin–titin interaction. Biomech. Model. Mechanobiol. **15**(6), 1423–1437 (2016). https://doi.org/10.1007/s10237-016-0772-7
11. Heidlauf, T., Klotz, T., Rode, C., Siebert, T., Röhrle, O.: A continuum-mechanical skeletal muscle model including actin-titin interaction predicts stable contractions on the descending limb of the force-length relation. PLoS Comput. Biol. **13**(10), e1005773 (2017). https://doi.org/10.1371/journal.pcbi.1005773
12. Herzog, W.: Skeletal Muscle Mechanics: From Mechanisms to Function. Wiley, Hoboken (2000)
13. Herzog, W.: Skeletal muscle mechanics: questions, problems and possible solutions. J. NeuroEng. Rehabil. **14**(1), 98 (2017). https://doi.org/10.1186/s12984-017-0310-6
14. Howard, J.: Mechanics of Motor Proteins and the Cytoskeleton, nachdr. edn. Sinauer, Sunderland (2001). OCLC: 247917499
15. Huxley, A.F.: Muscle structure and theories of contraction. Prog. Biophys. Biophys. Chem. **7**, 255–318 (1957)
16. Huxley, A.F., Simmons, R.M.: Proposed mechanism of force generation in striated muscle. Nature **233**(5321), 533–538 (1971). https://doi.org/10.1038/233533a0
17. Jabin, P.E.: A review of the mean field limits for Vlasov equations. Kinet. Relat. Models **7**(4), 661–711 (2014). https://doi.org/10.3934/krm.2014.7.661
18. Keener, J.P., Sneyd, J.: Mathematical Physiology. Interdisciplinary Applied Mathematics, no. 8, 2nd edn. Springer, New York (2009). OCLC: ocn298595247
19. Klenke, A.: Probability Theory: A Comprehensive Course. Universitext. Springer, London (2008)
20. LeVeque, R.J.: Finite Difference Methods for Ordinary and Partial Differential Equations. Other Titles in Applied Mathematics. Society for Industrial and Applied Mathematics, Philadelphia (2007). https://doi.org/10.1137/1.9780898717839
21. Ma, S.P., Zahalak, G.I.: A distribution-moment model of energetics in skeletal muscle. J. Biomech. **24**(1), 21–35 (1991)
22. Petzold, L.: Automatic Selection of Methods for Solving Stiff and Nonstiff Systems of Ordinary Differential Equations. SIAM J. Sci. Stat. Comput. **4**(1), 136–148 (1983). https://doi.org/10.1137/0904010

23. Randoll, U.: Matrix-rhythm-therapy of dynamic illnesses. In: Heine, H., Rimpler M. (eds.) Extracellular Matrix and Groundregulation System in Health and Disease, pp. 57–70. G. Fischer (1997)
24. Resat, H., Petzold, L., Pettigrew, M.F.: Kinetic Modeling of biological systems. Method Mol.Biol. **541**, 311–335 (2009). https://doi.org/10.1007/978-1-59745-243-4_14
25. Simeon, B., Serban, R., Petzold, L.R.: A model of macroscale deformation and microvibration in skeletal muscle tissue. ESAIM Math. Modell. Numer. Anal. **43**(4), 805–823 (2009). https://doi.org/10.1051/m2an/2009030
26. Spohn, H.: Kinetic equations from hamiltonian dynamics: Markovian limits. Rev. Modern Phys. **52**(3), 569 (1980)
27. Spohn, H.: Large Scale Dynamics of Interacting Particles. Springer Science & Business Media, Berlin (2012)
28. van der Walt, S., Colbert, S.C., Varoquaux, G.: The NumPy array: a structure for efficient numerical computation. Comput. Sci. Eng. **13**(2), 22–30 (2011). https://doi.org/10.1109/MCSE.2011.37
29. Villani, C.: Optimal Transport: Old and New. Springer Science & Business Media, Berlin (2008)
30. Zahalak, G.I.: A distribution-moment approximation for kinetic theories of muscular contraction. Math. Biosci. **55**(1–2), 89–114 (1981). https://doi.org/10.1016/0025-5564(81)90014-6
31. Zahalak, G.I.: Non-axial muscle stress and stiffness. J. Theor. Biol. **182**(1), 59–84 (1996). https://doi.org/10.1006/jtbi.1996.0143

Generalized Elements for a Structural Analysis of Circuits

Idoia Cortes Garcia, Sebastian Schöps, Christian Strohm,
and Caren Tischendorf

Abstract The structural analysis, i.e., the investigation of the differential-algebraic
nature, of circuits containing simple elements, i.e., resistances, inductances and
capacitances is well established. However, nowadays circuits contain all sorts of
elements, e.g. behavioral models or partial differential equations stemming from
refined device modelling. This paper proposes the definition of generalized circuit
elements which may for example contain additional internal degrees of freedom,
such that those elements still behave structurally like resistances, inductances
and capacitances. Hereby, a classification of more evolved circuit elements is
enabled. The structural analysis of circuits is expanded to systems containing such
generalized elements. Several complex examples demonstrate the relevance of those
definitions.

Keywords Differential algebraic equations · Differential index · Modified nodal
analysis · Maxwell's equations

Supported by Deutsche Forschungsgemeinschaft (DFG, German Research Foundation) under
Germany's Excellence Strategy: the Berlin Mathematics Research Center MATH+ (EXC-2046/1,
project ID: 390685689) and the Graduate School of Computational Engineering at TU Darmstadt
(GSC 233) as well as the DFG grant SCHO1562/1-2 and the Bundesministerium für Wirtschaft
und Energie (BMWi, Federal Ministry for Economic Affairs and Energy) grant 0324019E.

I. Cortes Garcia (✉)
Computational Electromagnetics Group, Technische Universität Darmstadt, Darmstadt, Germany
e-mail: idoia.cortes@tu-darmstadt.de

S. Schöps
Technische Universität Darmstadt, Darmstadt, Germany
e-mail: sebastian.schoeps@tu-darmstadt.de

C. Strohm · C. Tischendorf
Humboldt Universität zu Berlin, Berlin, Germany
e-mail: strohmch@math.hu-berlin.de; tischendorf@math.hu-berlin.de

T. Reis et al. (eds.), *Progress in Differential-Algebraic Equations II*,
Differential-Algebraic Equations Forum,
https://doi.org/10.1007/978-3-030-53905-4_13

397

Mathematics Subject Classification (2010) 34A09, 35Q61, 65M20, 78A20, 94C99

1 Introduction

Circuits or electric networks are a common modeling technique to describe the electrotechnical behavior of large systems. Their structural analysis, i.e., the investigation of the properties of the underlying differential-algebraic equations (DAEs), has a long tradition. For example Bill Gear studied in 1971 'the mixed differential and algebraic equations of the type that commonly occur in the transient analysis of large networks' in [21]. At that time several competing formulations were used in the circuit simulation community, for example the sparse tableau analysis (STA) was popular. This changed with the introduction of the modified nodal analysis (MNA) by Ho et. al in [26] and the subsequent development of the code SPICE [33]. Nowadays all major circuit simulation tools are using some dialect of MNA, e.g. the traditional formulation or the flux/charge oriented one [20]. The mathematical structure has been very well understood in the case of simple elements, i.e., resistances, inductances and capacitances as well as sources [19, 25].

However, the complexity of element models has increased quickly. For example, the semiconductor community develops various phenomenological and physical models, which are standardized e.g. in the BSIM (Berkeley Short-channel IGFET Model) family, [40]. The development of *mixed-mode device simulation* has become popular, which is mathematically speaking the coupling of DAEs with partial differential equations (PDEs), e.g. [22, 24, 30, 36]. Even earlier, low frequency engineers have established *field-circuit-coupling*, i.e., the interconnection of finite element machine models with circuits, first based on loop analysis, later (modified) nodal analysis e.g. [16, 35].

Until now, the structural DAE analysis of circuits which are based on complex ('refined') elements has mainly been carried out on a case by case basis, e.g. for elliptic semiconductor models in [1], parabolic-elliptic models of electrical machines in [2, 13, 42] and hyperbolic models stemming from the full set of Maxwell's equations in [4]. Based on the analysis made in [13], this contribution aims for a more systematic analysis: we consider each element as an arbitrary (smooth) function of voltages, currents, internal variables and their derivatives. Then, we formulate sets of assumptions ('generalized elements') on these functions, e.g. which quantity is derived or which DAE-index does the function have. Based on these assumptions we proof a DAE index result that generalizes [19]. Not surprisingly, it turns out that our generalized elements are natural generalizations of the classical elements, i.e., resistances, inductances and capacitances. The classification of complex circuit elements as well as the generalized index result eases the index analysis of the obtained DAE, as, analogously to the case of circuits with classic elements (see [19]), it depends on topological properties of the underlying circuit. All results are formulated in the context of electrical engineering but the presented approach is also of interest for the analysis and simulation of other networks such as gas transport networks [5, 23, 28] or power networks [31].

The paper is structured as follows: we start with a few basic mathematical definitions and results in Sect. 2, then we give the definitions of our generalized elements and some simple examples in Sect. 3. In Sect. 4 we discuss the mathematical modeling of circuits by modified nodal analysis. Finally, Sect. 5 proves the new DAE index results which are then applied to several very complex refined models in Sect. 6.

2 Mathematical Preliminaries

Let us collect some basic notations and definitions. The following preliminaries collect standard results from functional analysis, e.g. [45].

Definition 2.1 A function $f : \mathbb{R}^m \to \mathbb{R}^m$ is called strongly monotone if and only if there is a constant $c > 0$ such that

$$\forall x, \bar{x} \in \mathbb{R}^m : \quad \langle f(x) - f(\bar{x}), x - \bar{x} \rangle \geq c \|x - \bar{x}\|^2.$$

Lemma 2.1 *Let* $M \in \mathbb{R}^{m \times m}$ *be a matrix. Then, the linear function* $f(x) := Mx$ *is strongly monotone if and only if* M *is positive definite.*

Proof If $f(x) := Mx$ is strongly monotone we find a constant $c > 0$ such that for all $x \in \mathbb{R}^m$ with $x \neq 0$

$$\langle Mx, x \rangle = \langle f(x) - f(0), x - 0 \rangle \geq c \|x - 0\| > 0,$$

that means M is positive definite. Next, we show the opposite direction. Let M be positive definite. We split M into its symmetric and non-symmetric part

$$M = M_s + M_n, \quad M_s = \frac{1}{2}(M + M^\top), \quad M_n = \frac{1}{2}(M - M^\top).$$

Consequently, for all $x \in \mathbb{R}^m$ with $x \neq 0$,

$$\langle M_s x, x \rangle = \langle M x, x \rangle > 0.$$

Since M_s is symmetric, we find a unitary matrix T and a diagonal matrix D such that $M_s = T^{-1} D T$. We get that

$$0 < \langle M_s x, x \rangle = \langle T^{-1} D T x, x \rangle = \langle D T x, T x \rangle = \sum_{j=1}^{m} d_{jj} y_j^2 \quad \text{with} \quad y := Tx.$$

Choosing the unit vectors $y := e_i$, we find that $d_{ii} > 0$ for all $i = 1, \ldots, m$. Defining $c := \min_{i=1,\ldots,m} d_{ii}$, we see that for all $x \in \mathbb{R}^m$

$$\langle Mx, x \rangle = \langle M_s x, x \rangle \geq \sum_{j=1}^{m} c y_j^2 = c \|Tx\|^2 = c \|x\|^2.$$

Finally, we obtain, for any $x, \bar{x} \in \mathbb{R}^m$

$$\langle f(x) - f(\bar{x}), x - \bar{x} \rangle = \langle M(x - \bar{x}), x - \bar{x} \rangle \geq c \|x - \bar{x}\|^2. \qquad \square$$

Definition 2.2 A function $f : \mathbb{R}^m \times \mathbb{R}^n \to \mathbb{R}^m$ is called strongly monotone with respect to x if and only if there is a constant $c > 0$ such that

$$\forall y \in \mathbb{R}^n \, \forall x, \bar{x} \in \mathbb{R}^m : \quad \langle f(x, y) - f(\bar{x}, y), x - \bar{x} \rangle \geq c \|x - \bar{x}\|^2.$$

Remark 2.1 In case of variable matrix functions $M(y)$, the function $f(x, y) := M(y)x$ might be not strongly monotone with respect to x even if $M(y)$ is positive definite for each y. For strong monotony, one has to ensure that the eigenvalues of the symmetric part of $M(y)$ can be bounded from below by a constant $c > 0$ independent of y.

In the following we present a specific variant of the Theorem of Browder–Minty, see e.g. [34, 45].

Lemma 2.2 *Let $f = f(x, y) : \mathbb{R}^m \times \mathbb{R}^n \to \mathbb{R}^m$ be strongly monotone with respect to x and continuous. Then, there is a uniquely defined continuous function $g : \mathbb{R}^n \to \mathbb{R}^m$ such that $f(g(y), y) = 0$ for all $y \in \mathbb{R}^n$.*

Proof For fixed $y \in \mathbb{R}^n$ we define $F_y : \mathbb{R}^m \to \mathbb{R}^m$ by

$$F_y(x) := f(x, y) \quad \forall x \in \mathbb{R}^m.$$

Since f is strongly monotone with respect to x, the function F_y is strongly monotone. The Theorem of Browder–Minty, e.g. [45] and [34], provides a unique $z_y \in \mathbb{R}^m$ such that $F_y(z_y) = 0$ and, hence, $f(z_y, y) = 0$. We define $g : \mathbb{R}^n \to \mathbb{R}^m$ by

$$g(y) := z_y.$$

Obviously, $f(g(y), y) = 0$ for all $y \in \mathbb{R}^n$. It remains to show that g is continuous. Let (y_k) be a convergent series in $\mathbb{R}^n$ with $y_k \to y_* \in \mathbb{R}^n$ for $k \to \infty$. Since f is strongly monotone with respect to x, there is a constant $c > 0$ such that

$$\|g(y_k) - g(y_*)\|^2 \leq \frac{1}{c} \langle f(g(y_k), y_k) - f(g(y_*), y_k), g(y_k) - g(y_*) \rangle$$

$$\leq \frac{1}{c} \|f(g(y_k), y_k) - f(g(y_*), y_k)\| \, \|g(y_k) - g(y_*)\|$$

$$= \frac{1}{c} \| f(g(y_*), y_k) \| \, \| g(y_k) - g(y_*) \|$$

$$= \frac{1}{c} \| f(g(y_*), y_k) - f(g(y_*), y_*) \| \, \| g(y_k) - g(y_*) \|.$$

Since f is continuous, we may conclude that $g(y_k) \to g(y_*)$ for $k \to \infty$. $\qquad \square$

The next Lemma is new. It bases on the solvability statements of the previous Lemma but also provides structural properties of the solution for equations with a particular structure that is relevant for circuit equations.

Lemma 2.3 *Let $M \in \mathbb{R}^{m \times k}$ be a matrix and $P \in \mathbb{R}^{k \times k}$ be a projector along $\ker M$. Additionally, let $f = f(x, y) : \mathbb{R}^m \times \mathbb{R}^n \to \mathbb{R}^m$ be strongly monotone with respect to x and continuous as well as $r : \mathbb{R}^n \to \mathbb{R}^m$ be a continuous function. Then, there is a continuous function $g : \mathbb{R}^n \to \mathbb{R}^k$ such that*

$$M^\top f(Mz, y) + P^\top r(y) = 0 \quad \text{if and only if} \quad Pz = g(y). \tag{2.1}$$

Proof In the degenerated case that $M = 0$ we have $P = 0$ and the zero function $g(y) \equiv 0$ fulfills obviously the equivalence (2.1). Let be $M \neq 0$ for the further considerations. We chose a basis B of $\operatorname{im} P$. For $r := \operatorname{rank} P$, we form the full-column rank matrix $\widetilde{P} \in \mathbb{R}^{k \times r}$ as a matrix whose columns consist of all basis vectors of B. By construction, $\ker M\widetilde{P} = \{0\}$ and, hence, the matrix $(M\widetilde{P})^\top M\widetilde{P}$ is non-singular. Next, we introduce a function $F : \mathbb{R}^r \times \mathbb{R}^n \to \mathbb{R}^r$ by

$$F(u, y) := (M\widetilde{P})^\top f(M\widetilde{P}u, y) + \widetilde{P}^\top P^\top r(y).$$

Since f is continuous, also F is continuous. From the strong monotony of f with respect to x we can also conclude the strong monotony of F with respect to u since there is a constant $c > 0$ such that, for all $y \in \mathbb{R}^n$ and for all $u, \bar{u} \in \mathbb{R}^r$,

$$\langle F(u, y) - F(\bar{u}, y), u - \bar{u} \rangle = \langle (M\widetilde{P})^\top f(M\widetilde{P}u, y) - (M\widetilde{P})^\top f(M\widetilde{P}\bar{u}, y), u - \bar{u} \rangle$$

$$= \langle f(M\widetilde{P}u, y) - f(M\widetilde{P}\bar{u}, y), M\widetilde{P}u - M\widetilde{P}\bar{u} \rangle$$

$$\geq c \| M\widetilde{P}u - M\widetilde{P}\bar{u} \|^2$$

and

$$\| u - \bar{u} \| = \| ((M\widetilde{P})^\top M\widetilde{P})^{-1} (M\widetilde{P})^\top M\widetilde{P}(u - \bar{u}) \|$$

$$\leq \| ((M\widetilde{P})^\top M\widetilde{P})^{-1} (M\widetilde{P})^\top \| \, \| M\widetilde{P}(u - \bar{u}) \|$$

which implies

$$\langle F(u, y) - F(\bar{u}, y), u - \bar{u} \rangle \geq \frac{c}{c_1} \| u - \bar{u} \|^2$$

for $c_1 := \|((M\widetilde{P})^\top M\widetilde{P})^{-1}(M\widetilde{P})^\top\|^2 > 0$ since M is a non-zero matrix. From Lemma 2.2 we know that there is a unique continuous function $G : \mathbb{R}^n \to \mathbb{R}^r$ such that

$$F(G(y), y) = 0 \quad \forall y \in \mathbb{R}^n.$$

It means that $F(u, y) = 0$ if and only if $u = G(y)$. Next, we show that the function $g : \mathbb{R}^n \to \mathbb{R}^k$ defined by

$$g(y) := \widetilde{P}G(y)$$

satisfies the equivalence (2.1). First, we see that

$$\widetilde{P}^\top M^\top f(Mg(y), y) + \widetilde{P}^\top P^\top r(y) = F(G(y), y) = 0.$$

By construction of $\widetilde{P}$, we know that $\ker \widetilde{P}^\top = \ker P^\top$ and, therefore,

$$P^\top M^\top f(Mg(y), y) + P^\top P^\top r(y) = 0.$$

Since P is a projector along $\ker M$, we see that $M = MP$ and, hence,

$$M^\top f(Mg(y), y) + P^\top r(y) = 0.$$

From here, we can directly conclude the following direction of the equivalence (2.1). If $Pz = g(y)$ then $M^\top f(Mz, y) + P^\top r(y) = 0$. Finally, we show the opposite direction. If $M^\top f(Mz, y) + P^\top r(y) = 0$ then we again exploit the monotony of f in order to obtain

$$0 = \langle M^\top f(Mz, y) + P^\top r(y) - M^\top f(Mg(y), y) - P^\top r(y), z - g(y)\rangle$$
$$= \langle f(Mz, y) - f(Mg(y), y), Mz - Mg(y)\rangle \geq c\|Mz - Mg(y)\|,$$

that means $M(z - g(y)) = 0$. By assumption we have $\ker M = \ker P$ and, hence, $P(z - g(y)) = 0$. It follows $Pz = Pg(y) = P\widetilde{P}G(y) = \widetilde{P}G(y) = g(y)$. □

Corollary 2.4 *Let $M \in \mathbb{R}^{k\times m}$ be a matrix and $P \in \mathbb{R}^{k\times k}$ be a projector along $\ker M$. Additionally, let $r : \mathbb{R}^n \to \mathbb{R}^m$ be continuous. Then, there is a continuous function $g : \mathbb{R}^n \to \mathbb{R}^k$ such that*

$$M^\top Mz + P^\top r(y) = 0 \quad \text{if and only if} \quad Pz = g(y). \tag{2.2}$$

Proof It follows directly from Lemma 2.3 using the function $F : \mathbb{R}^m \times \mathbb{R}^n \to \mathbb{R}^m$ defined by

$$F(x, y) := x.$$

□

Following [8] we call a function $x(t)$ the solution of a general nonlinear DAE

$$f(x', x, t) = 0 \qquad (2.3)$$

on an interval $\mathcal{I} \subset \mathbb{R}$, if x is continuously differentiable on I and satisfies (2.3) for all $t \in \mathcal{I}$.

Assumption 2.1 *We assume solvability of (2.3), see e.g. [8, Definition 2.2.1], and that all functions involved are sufficiently smooth.*

Definition 2.3 ([8]) The minimum number of times that all or part of (2.3) must be differentiated with respect to t in order to determine x' as a continuous function of x, t, is the index of the DAE.

3 Generalized Circuit Elements

In this section we define new classes of generalized circuit elements motivated by the classical ones, i.e., resistances, inductances and capacitances. The first inductance-like element is based on the definition in [13]. The original version was designed to represent a specific class of models but also to be minimally invasive in the sense that the proofs in [19] could still be used. The following definition is more general and a new proof of the corresponding index results is given in Sect. 5.

Before presenting the definitions, we introduce the quantities that are required to describe the generalized circuit elements. We consider the functions

$$f_\star : \mathbb{R}^{n_{\mathrm{m},\star}} \times \mathbb{R}^{n_{\mathrm{x},\star}} \times \mathbb{R}^{n_{\mathrm{i},\star}} \times \mathbb{R}^{n_{\mathrm{v},\star}} \times \mathbb{R} \longrightarrow \mathbb{R}^{n_{\mathrm{f},\star}} \quad \text{and}$$

$$m_\star : \mathbb{R}^{n_{\mathrm{x},\star}} \times \mathbb{R}^{n_{\mathrm{i},\star}} \times \mathbb{R}^{n_{\mathrm{v},\star}} \times \mathbb{R} \longrightarrow \mathbb{R}^{n_{\mathrm{m},\star}}$$

and the time dependent vectors

$$x_\star : \mathbb{R} \longrightarrow \mathbb{R}^{n_{\mathrm{x},\star}}, \quad i_\star : \mathbb{R} \longrightarrow \mathbb{R}^{n_{\mathrm{i},\star}} \quad \text{and} \quad v_\star : \mathbb{R} \longrightarrow \mathbb{R}^{n_{\mathrm{v},\star}},$$

for $\star \in \{L, C, R\}$. Here, $f_\star$ represents the constitutive law, $x_\star$ the internal degrees of freedom, $i_\star$ the vector of branch currents across the element and $v_\star$ the vector of branch voltages through the element.

Definition 3.1 We define an **inductance-like** element as one element described by

$$f_{\mathrm{L}}\left(\frac{\mathrm{d}}{\mathrm{d}t} m_{\mathrm{L}}(x_{\mathrm{L}}, i_{\mathrm{L}}, v_{\mathrm{L}}, t), x_{\mathrm{L}}, i_{\mathrm{L}}, v_{\mathrm{L}}, t\right) = 0$$

where there is at most one differentiation $\frac{d}{dt}$ needed to obtain a model description of the form

$$\frac{d}{dt}x_L = \chi_L(\frac{d}{dt}v_L, x_L, i_L, v_L, t) \qquad (3.1)$$

$$\frac{d}{dt}i_L = g_L(x_L, i_L, v_L, t) \qquad (3.2)$$

We call it a **strongly inductance-like** element if, additionally, the function

$$F_L(v'_L, x_L, i_L, v_L, t) := \partial_{x_L} g_L(x_L, i_L, v_L, t)\chi_L(v'_L, x_L, i_L, v_L, t)$$
$$+ \partial_{v_L} g_L(x_L, i_L, v_L, t)v'_L \qquad (3.3)$$

is continuous and strongly monotone with respect to v'_L.

Proposition 3.1 *Linear inductances defined as*

$$v_L - L\frac{d}{dt}i_L = 0 ,$$

with L being positive definite, are strongly inductance-like elements.

Proof By inverting L we obtain without the need of any differentiation a model description as required in (3.2). Furthermore, $F_L(v_L') = L^{-1}v'_L$ is strongly monotone with respect to v'_L due to L^{-1} being positive definite and by using Lemma 2.1 in Definition 2.2. $\qquad\square$

Proposition 3.2 *Flux formulated inductances defined as*

$$v_L = \frac{d}{dt}\Phi_L ,$$

$$\Phi_L = \phi(i_L, t) ,$$

with $\partial_{i_L}\phi(i_L, t)$ being positive definite, are strongly inductance-like elements.

Proof we chose $x_L = \Phi_L$. Then, one time differentiation of the second equation yields $\frac{d}{dt}x_L = \partial_{i_L}\phi(i_L, t)\frac{d}{dt}i_L + \partial_t\phi(i_L, t)$ and exploiting the positive definiteness we write $\frac{d}{dt}i_L$ as in (3.2), for

$$g_L(i_L, v_L, t) := \partial_{i_L}\phi(i_L, t)^{-1}\frac{d}{dt}x_L - \partial_t\phi(i_L, t) = \partial_{i_L}\phi(i_L, t)^{-1}v_L - \partial_t\phi(i_L, t) .$$

Consequently, $F_L(v_L', i_L, t) = \partial_{i_L}\phi(i_L, t)^{-1}v'_L$ and $F_L(v_L', i_L, t)$ is strongly monotone with respect to v'_L. The latter follows again from $\partial_{i_L}\phi(i_L, t)^{-1}$ being positive definite and by using Lemma 2.1 in Definition 2.2. $\qquad\square$

Remark 3.1 Please note that the flux formulated inductances in Proposition 3.2 can be vector-valued such as for example in the case of a mutual inductance

$$\begin{pmatrix} v_1 \\ v_2 \end{pmatrix} = \frac{d}{dt} \begin{pmatrix} \Phi_1 \\ \Phi_2 \end{pmatrix}, \quad \begin{pmatrix} \Phi_1 \\ \Phi_2 \end{pmatrix} = \begin{pmatrix} L_{11} & L_{12} \\ L_{12} & L_{22} \end{pmatrix} \begin{pmatrix} i_1 \\ i_2 \end{pmatrix},$$

with $x_L = (\phi_1 \ \phi_2)^\top$. In this specific example,

$$g_L = \begin{pmatrix} L_{11} & L_{12} \\ L_{12} & L_{22} \end{pmatrix}^{-1} \begin{pmatrix} v_1 \\ v_2 \end{pmatrix}, \quad \chi_L = \begin{pmatrix} v_1 \\ v_2 \end{pmatrix}$$

and thus

$$F_L = \begin{pmatrix} L_{11} & L_{12} \\ L_{12} & L_{22} \end{pmatrix}^{-1} \begin{pmatrix} v_1 \\ v_2 \end{pmatrix}',$$

which is strongly monotone for a positive definite mutual inductance matrix L.

A more complex application of an electromagnetic element complying with this definition can be found in Sect. 6.1.

Definition 3.2 We define a **capacitance-like** element as one element described by

$$f_C \left(\frac{d}{dt} m_C(x_C, i_C, v_C, t), x_C, i_C, v_C, t \right) = 0$$

where there is at most one differentiation $\frac{d}{dt}$ needed to obtain a model description of the form

$$\frac{d}{dt} x_C = \chi_C(\frac{d}{dt} i_C, x_C, i_C, v_C, t) \tag{3.4}$$

$$\frac{d}{dt} v_C = g_C(x_C, i_C, v_C, t) \tag{3.5}$$

We call it a **strongly capacitance-like** element if, additionally, the function

$$F_C(i_C', x_C, i_C, v_C, t) := \partial_{x_C} g_C(x_C, i_C, v_C, t) \chi_C(i_C', x_C, i_C, v_C, t)$$
$$+ \partial_{i_C} g_C(x_C, i_C, v_C, t) i_C' \tag{3.6}$$

is continuous and strongly monotone with respect to i_C'.

Proposition 3.3 *Linear capacitances defined as*

$$C \frac{\mathrm{d}}{\mathrm{d}t} v_{\mathrm{C}} - i_{\mathrm{C}} = 0 \,,$$

with C being positive definite, are strongly capacitance-like elements.

Proof Analogous to the proof in Proposition 3.1, we exploit the fact that C is positive definite and here, $F_{\mathrm{C}}(i_{\mathrm{C}}') = C^{-1} i_{\mathrm{C}}'$ is shown to be strongly monote with respect to i_{C}' by using by using Lemma 2.1 and Definition 2.2.   □

Proposition 3.4 *Charge formulated capacitances defined as*

$$i_{\mathrm{C}} = \frac{\mathrm{d}}{\mathrm{d}t} q_{\mathrm{C}} \,,$$

$$q_{\mathrm{C}} = q(v_{\mathrm{C}}, t) \,,$$

with $\partial_{v_{\mathrm{C}}} q(v_{\mathrm{C}}, t)$ being positive definite, are strongly capacitance-like elements.

Proof There proof is analogous to the one of Proposition 3.2 by setting $x_{\mathrm{C}} = q_{\mathrm{C}}$ and $F_{\mathrm{C}}(i_{\mathrm{C}}', v_{\mathrm{C}}, t) = \partial_t q(v_{\mathrm{C}}, t)^{-1} i_{\mathrm{C}}'$.   □

Definition 3.3 We define a **resistance-like** element as one element described by

$$f_{\mathrm{R}} \left(\frac{\mathrm{d}}{\mathrm{d}t} m_{\mathrm{R}}(x_{\mathrm{R}}, i_{\mathrm{R}}, v_{\mathrm{R}}, t), x_{\mathrm{R}}, i_{\mathrm{R}}, v_{\mathrm{R}}, t \right) = 0$$

where there is at most one differentiation $\frac{\mathrm{d}}{\mathrm{d}t}$ needed to obtain a model description of the form

$$\frac{\mathrm{d}}{\mathrm{d}t} x_{\mathrm{R}} = \chi_{\mathrm{R}}(x_{\mathrm{R}}, i_{\mathrm{R}}, v_{\mathrm{R}}, t) \tag{3.7}$$

$$\frac{\mathrm{d}}{\mathrm{d}t} i_{\mathrm{R}} = g_{\mathrm{R}}(\frac{\mathrm{d}}{\mathrm{d}t} v_{\mathrm{R}}, x_{\mathrm{R}}, i_{\mathrm{R}}, v_{\mathrm{R}}, t) \tag{3.8}$$

We call it a **strongly resistance-like** element if, additionally, the function

$$g_{\mathrm{R}}(v_{\mathrm{R}}', x_{\mathrm{R}}, i_{\mathrm{R}}, v_{\mathrm{R}}, t) \tag{3.9}$$

is continuous and strongly monotone with respect to v_{R}'.

Proposition 3.5 *Linear resistances defined as*

$$v_{\mathrm{R}} - R i_{\mathrm{R}} = 0 \,,$$

with R being positive definite, are strongly resistance-like elements.

Proof Here, the equation is differentiated once to obtain

$$\frac{\mathrm{d}}{\mathrm{d}t} v_{\mathrm{R}} - R \frac{\mathrm{d}}{\mathrm{d}t} i_{\mathrm{R}} = 0.$$

Now, analogously to the proof in 3.1, we exploit the positive definiteness of R to invert it and obtain a function $g_{\mathrm{R}}(v_{\mathrm{R}}') = R^{-1} v_{\mathrm{R}}'$, which is strongly monote with respect to v_{R}'.   □

Remark 3.2 Definitions 3.1–3.3 are made for one-port elements or multi-port elements which are structurally identically for each port and do not change their structure, e.g. depending on state, time (or frequency). However, in practice an inductance-like device may turn into a capacitance-like device depending on its working point. Also, a two-port element may simply consist of an inductance and a capacitance. Those examples are not covered by our generalizations.

4 Circuit Structures and Circuit Graph Describing Matrices

In this section we provide common ingredients for the analysis of circuits, see e.g. [19, 37]. Note that the classical elements are replaced by their corresponding newly introduced generalization, see the following assumption.

Assumption 4.1 *Let a connected circuit be given whose elements belong to the set of capacitance-like devices (C), inductance-like devices (I), resistance-like devices (R), voltage sources (V) and current sources (I).*

   We consider the element related incidence matrices A_{C}, A_{L}, A_{R}, A_{V} and A_{I} whose entries a_{ij} are defined by

$$a_{ij} = \begin{cases} +1 & \text{if branch } j \text{ directs from node } i \\ -1 & \text{if branch } j \text{ directs to node } i \\ 0 & \text{else} \end{cases}$$

where the index i refers to a node (except the mass node) and the index j refers to branches of capacitance-like devices (A_{C}), inductance-like devices (A_{L}), resistance-like devices (A_{R}), voltage sources (A_{V}) and current sources (A_{I}).

Remark 4.1 If Assumption 4.1 is fulfilled then the incidence matrix A of the circuit is given by $A = [A_{\mathrm{C}} \, A_{\mathrm{L}} \, A_{\mathrm{R}} \, A_{\mathrm{V}} \, A_{\mathrm{I}}]$ and has full row rank (see [9]).

Lemma 4.1 (c.f. [19]) *Let a connected circuit be given and A_X be the incidence matrix of all branches of type X. All other branches shall be collected in the incidence matrix A_Y such that the incidence matrix of the circuit is given by $A = [A_X\, A_Y]$. Then,*

1. *the circuit contains no loops of only X-type branches if and only if A_X has full column rank,*
2. *the circuit contains no cutsets of only X-type branches if and only if A_Y has full row rank.*

Proof The incidence matrix of a subset S of branches of a circuit is non-singular if and only if S forms a spanning tree [9]. From this we can conclude the following statements.

1. The circuit contains no loops of only X-type branches if and only if there is a spanning tree containing all X-type branches. The latter condition is equivalent to the condition that A_X has full column rank.
2. The circuit contains no cutsets of only X-type branches if and only if there is a spanning tree containing only Y-type branches. The latter condition is equivalent to the condition that A_Y has full row rank. □

Corollary 4.2 ([19]) *Let Assumption 4.1 be fulfilled. Then,*

1. *the circuit contains no loops of only voltage sources if and only if A_V has full column rank,*
2. *the circuit contains no cutsets of only current sources if and only if $[A_C\, A_L\, A_R\, A_V]$ has full row rank.*

Since loops of only voltage sources and cutsets of only current sources are electrically forbidden, we suppose the following assumption to be fulfilled.

Assumption 4.2 ([19]) *The matrix A_V has full column rank and the matrix $[A_C\, A_L\, A_R\, A_V]$ has full row rank.*

Definition 4.1 We call a loop of branches of a circuit a **CV-loop** if it contains only capacitance-like devices and voltage sources. We call a cutset of branches of a circuit an **LI-cutset** if it contains only inductance-like devices and current sources.

Corollary 4.3 ([19]) *Let Assumption 4.1 be fulfilled. Then,*

1. *the circuit contains no CV-loops if and only if $[A_C\, A_V]$ has full column rank,*
2. *the circuit contains no LI-cutsets if and only if $[A_C\, A_R\, A_V]$ has full row rank.*

5 DAE Index for Circuits with Generalized Lumped Models

Let Assumptions 4.1 and 4.2 be fulfilled. Following the idea of the modified nodal analysis for circuits, we introduce the nodal potentials e, with which we can express the generalized elements' branch voltages as $v_\star = A_\star^\top e$, for $\star \in \{L, C, R\}$, and form the circuit equations as

$$A_C i_C + A_R i_R + A_V i_V + A_L i_L + A_I i_S = 0, \qquad (5.1a)$$

$$A_V^\top e = v_S, \qquad (5.1b)$$

$$f_L \left(\frac{d}{dt} m_L(x_L, i_L, A_L^\top e, t), x_L, i_L, A_L^\top e, t \right) = 0, \qquad (5.1c)$$

$$f_C \left(\frac{d}{dt} m_C(x_C, i_C, A_C^\top e, t), x_C, i_C, A_C^\top e, t \right) = 0, \qquad (5.1d)$$

$$f_R \left(\frac{d}{dt} m_R(x_R, i_R, A_R^\top e, t), x_R, i_R, A_R^\top e, t \right) = 0 \qquad (5.1e)$$

with given source functions $i_S(t)$ for current sources and $v_S(t)$ for voltage sources.

Remark 5.1 Please note that the currents i_C and i_R are variables of the system (5.1). This is in contrast to the traditional modified nodal analysis which is only based on simple lumped elements such that these variables can be eliminated by explicitly solving (5.1e) and (5.1d) for the currents i_C and i_R, respectively.

Theorem 5.1 *Let Assumption 4.1 be fulfilled. Furthermore, let all resistance-like devices be strongly resistance-like devices. If the circuit has no CV-loops and no LI-cutsets then the differentiation index of the system (5.1) is at most index 1.*

Proof Let Q_{cv} be a projector onto $\ker [A_C \ A_V]^\top$ and $P_{cv} := I - Q_{cv}$. It allows us to split

$$e = P_{cv} e + Q_{cv} e.$$

For the capacitance-like devices and the voltage sources we find after at most one differentiation of the device equations (5.1d) and (5.1b) that

$$A_C^\top \frac{d}{dt} e = g_C(x_C, i_C, A_C^\top e, t) \quad \text{and} \quad A_V^\top \frac{d}{dt} e = \frac{d}{dt} v_S. \qquad (5.2)$$

It implies

$$[A_C \ A_V] \left(\begin{bmatrix} A_C^\top \\ A_V^\top \end{bmatrix} \frac{d}{dt} e - \begin{bmatrix} g_C(x_C, i_C, A_C^\top e, t) \\ \frac{d}{dt} v_S \end{bmatrix} \right) = 0.$$

Applying Corollary 2.4 for $M := [A_C \ A_V]^\top$, $P := P_{cv}$,

$$z := \frac{d}{dt} e, \ y := (x_C, i_C, e, t), \ r(y) := -[A_C \ A_V] \begin{bmatrix} g_C(x_C, i_C, A_C^\top e, t) \\ \frac{d}{dt} v_S(t) \end{bmatrix}$$

we find a continuous function f_1 such that

$$P_{\text{CV}} \frac{\mathrm{d}}{\mathrm{d}t} e = f_1(x_{\text{C}}, i_{\text{C}}, e, t). \tag{5.3}$$

Next we exploit the nodal equations (5.1a). Multiplication by $Q_{\text{CV}}^{\top}$ and one differentiation yields

$$Q_{\text{CV}}^{\top} (A_{\text{R}} \frac{\mathrm{d}}{\mathrm{d}t} i_{\text{R}} + A_{\text{L}} \frac{\mathrm{d}}{\mathrm{d}t} i_{\text{L}} + A_{\text{I}} \frac{\mathrm{d}}{\mathrm{d}t} i_{\text{S}}) = 0. \tag{5.4}$$

For the resistance-like and inductance-like devices we get after at most one differentiation of the device equations (5.1e) and (5.1c) that

$$\frac{\mathrm{d}}{\mathrm{d}t} i_{\text{R}} = g_{\text{R}}(\frac{\mathrm{d}}{\mathrm{d}t} A_{\text{R}}^{\top} e, x_{\text{R}}, i_{\text{R}}, A_{\text{R}}^{\top} e, t) = g_{\text{R}}(A_{\text{R}}^{\top} Q_{\text{CV}} \frac{\mathrm{d}}{\mathrm{d}t} e + A_{\text{R}}^{\top} P_{\text{CV}} \frac{\mathrm{d}}{\mathrm{d}t} e, x_{\text{R}}, i_{\text{R}}, A_{\text{R}}^{\top} e, t) \tag{5.5}$$

and

$$\frac{\mathrm{d}}{\mathrm{d}t} i_{\text{L}} = g_{\text{L}}(x_{\text{L}}, i_{\text{L}}, A_{\text{L}}^{\top} e, t). \tag{5.6}$$

Together with (5.3) and (5.4) we obtain

$$Q_{\text{CV}}^{\top} \left(A_{\text{R}} g_{\text{R}}(A_{\text{R}}^{\top} Q_{\text{CV}} \frac{\mathrm{d}}{\mathrm{d}t} e + A_{\text{R}}^{\top} f_1(x_{\text{C}}, i_{\text{C}}, e, t), x_{\text{R}}, i_{\text{R}}, A_{\text{R}}^{\top} e, t) \right.$$
$$\left. + A_{\text{L}} g_{\text{L}}(x_{\text{L}}, i_{\text{L}}, A_{\text{L}}^{\top} e, t) + A_{\text{I}} \frac{\mathrm{d}}{\mathrm{d}t} i_{\text{S}} \right) = 0. \tag{5.7}$$

We choose a projector $P_{\text{R}-\text{CV}}$ along $\ker A_{\text{R}}^{\top} Q_{\text{CV}}$. Then, multiplication of (5.7) by $P_{\text{R}-\text{CV}}^{\top}$ yields

$$Q_{\text{CV}}^{\top} A_{\text{R}} g_{\text{R}} \left(A_{\text{R}}^{\top} Q_{\text{CV}} \frac{\mathrm{d}}{\mathrm{d}t} e + A_{\text{R}}^{\top} f_1(x_{\text{C}}, i_{\text{C}}, e, t), x_{\text{R}}, i_{\text{R}}, A_{\text{R}}^{\top} e, t \right)$$
$$+ P_{\text{R}-\text{CV}}^{\top} Q_{\text{CV}}^{\top} \left(A_{\text{L}} g_{\text{L}}(x_{\text{L}}, i_{\text{L}}, A_{\text{L}}^{\top} e, t) + A_{\text{I}} \frac{\mathrm{d}}{\mathrm{d}t} i_{\text{S}} \right) = 0. \tag{5.8}$$

It allows us to apply Lemma 2.3 for

$$M := A_{\text{R}}^{\top} Q_{\text{CV}}, \quad P := P_{\text{R}-\text{CV}}, \quad z := \frac{\mathrm{d}}{\mathrm{d}t} e, \quad y := (x_{\text{C}}, i_{\text{C}}, x_{\text{R}}, i_{\text{R}}, x_{\text{L}}, i_{\text{L}}, e, t),$$

and

$$f(x, y) := g_R(x + A_R^\top f_1(x_C, i_C, e, t), x_R, i_R, A_R^\top e, t),$$

$$r(y) := Q_{CV}^\top(A_L g_L(x_L, i_L, A_L^\top e, t) + A_I \frac{d}{dt} i_S).$$

Thus, we find a continuous function f_2 such that

$$P_{R-CV} \frac{d}{dt} e = f_2(x_C, i_C, x_R, i_R, x_L, i_L, e, t). \tag{5.9}$$

Since the circuit does not contain LI-cutsets, the matrix $[A_C \ A_R \ A_V]^\top$ has full column rank (see Corollary 4.3). It implies for $Q_{R-CV} := I - P_{R-CV}$ that

$$\ker Q_{CV} = \ker A_R^\top Q_{CV} = \ker P_{R-CV} = \text{im } Q_{R-CV}$$

and, therefore, $Q_{CV} = Q_{CV} P_{R-CV}$. Consequently,

$$Q_{CV} \frac{d}{dt} e = Q_{CV} f_2(x_C, i_C, x_R, i_R, x_L, i_L, e, t). \tag{5.10}$$

Regarding (5.3), (5.10), and (5.5), we find continuous functions f_3 and f_4 such that

$$\frac{d}{dt} e = f_3(x_C, i_C, x_R, i_R, x_L, i_L, e, t) \quad \text{and} \quad \frac{d}{dt} i_R = f_4(x_C, i_C, x_R, i_R, x_L, i_L, e, t). \tag{5.11}$$

Using again (5.1a), we get

$$[A_C \ A_V] \begin{bmatrix} \frac{d}{dt} i_C \\ \frac{d}{dt} i_V \end{bmatrix} + A_R \frac{d}{dt} i_R + A_L \frac{d}{dt} i_L + A_I \frac{d}{dt} i_S = 0.$$

Together with (5.11) and (5.6) we have

$$[A_C \ A_V] \begin{bmatrix} \frac{d}{dt} i_C \\ \frac{d}{dt} i_V \end{bmatrix} + A_R f_4(x_C, i_C, x_R, i_R, x_L, i_L, e, t) + A_L g_L(x_L, i_L, A_L^\top e, t) + A_I \frac{d}{dt} i_S = 0. \tag{5.12}$$

Since the circuit does not contain CV-loops, the matrix $[A_C \ A_V]$ has full column rank and, hence, $\ker[A_C \ A_V] = 0$. Multiplying (5.12) by $[A_C \ A_V]^\top$ allows us to apply Corollary 2.4 for

$$M := [A_C \ A_V], \quad P := I, \quad z := \begin{bmatrix} \frac{d}{dt} i_C \\ \frac{d}{dt} i_V \end{bmatrix}, \quad y := (x_C, i_C, x_R, i_R, x_L, i_L, e, t)$$

and

$$f(y) := A_R f_4(x_C, i_C, x_R, i_R, x_L, i_L, e, t) + A_L g_L(x_L, i_L, A_L^\top e, t) + A_I \frac{d}{dt} i_s(t).$$

Consequently, we find a continuous function f_5 such that

$$\begin{bmatrix} \frac{d}{dt} i_C \\ \frac{d}{dt} i_v \end{bmatrix} = f_5(x_C, i_C, x_R, i_R, x_L, i_L, e, t). \tag{5.13}$$

Finally, we obtain from (3.1) and (5.11) that

$$\frac{d}{dt} x_L = \chi_L(A_L^\top f_3(x_C, i_C, x_R, i_R, x_L, i_L, e, t), x_L, i_L, A_L^\top e, t) \tag{5.14}$$

and from (3.4) and (5.13) that

$$\frac{d}{dt} x_C = \chi_C([I \; 0] f_5(x_C, i_C, x_R, i_R, x_L, i_L, e, t), x_C, i_C, A_C^\top e, t). \tag{5.15}$$

Consequently, Eqs. (5.11), (5.6), (5.13), and (5.14), (5.15), (3.7) represent an explicit ordinary differential equation system. That means the differentiation index of the circuit system (5.1) is at most 1. ☐

Theorem 5.2 *Let Assumption 4.1 and Assumption 4.2 be fulfilled. Furthermore, let all resistance-like devices be strongly resistance-like devices. Additionally, let all inductance-like devices belonging to LI-cutsets be strongly inductance-like devices and all capacitance-like devices belonging to CV-loops be strongly capacitance-like devices. Then, the differentiation index of the system (5.1) is at most index 2.*

Proof First, we follow the proof of Theorem 5.1 and derive equations (5.2)–(5.9). Secondly, we perform the following splitting

$$\frac{d}{dt} e = P_{CV} \frac{d}{dt} e + Q_{CV} P_{R-CV} \frac{d}{dt} e + Q_{CV} Q_{R-CV} \frac{d}{dt} e, \tag{5.16}$$

$$\begin{bmatrix} \frac{d}{dt} i_C \\ \frac{d}{dt} i_v \end{bmatrix} = P_{CV-loop} \begin{bmatrix} \frac{d}{dt} i_C \\ \frac{d}{dt} i_v \end{bmatrix} + Q_{CV-loop} \begin{bmatrix} \frac{d}{dt} i_C \\ \frac{d}{dt} i_v \end{bmatrix}, \tag{5.17}$$

with projector $P_{CV-loop}$ along $\ker [A_C \; A_V]$ and $Q_{CV-loop} = I - P_{CV-loop}$.

Due to (5.3) and (5.9), the explicit ordinary differential equations for $P_{CV} \frac{d}{dt} e$ and $Q_{CV} P_{R-CV} \frac{d}{dt} e$ are already obtained with the need of at most one time differentiation of the original system. Now we derive the expression for $Q_{CV} Q_{R-CV} \frac{d}{dt} e$. Multiplication of (5.7) by $Q_{R-CV}^\top := I - P_{R-CV}^\top$ yields

$$Q_{R-CV}^\top Q_{CV}^\top (A_L g_L(x_L, i_L, A_L^\top e, t) + A_I \frac{d}{dt} i_s) = 0. \tag{5.18}$$

Differentiating (5.18) once again, we obtain

$$Q_{R-CV}^\top Q_{CV}^\top \left(A_L \partial_{x_L} g_L(x_L, i_L, A_L^\top e, t) \frac{d}{dt} x_L + A_L \partial_{i_L} g_L(x_L, i_L, A_L^\top e, t) \frac{d}{dt} i_L \right.$$

$$\left. + A_L \partial_{v_L} g_L(x_L, i_L, A_L^\top e, t) \frac{d}{dt} A_L^\top e + A_L \partial_t g_L(x_L, i_L, A_L^\top e, t) + A_I \frac{d^2}{dt^2} i_s \right) = 0.$$

Next, we plug in (3.1) and (5.6). Hence,

$$Q_{R-CV}^\top Q_{CV}^\top \left(A_L \partial_{x_L} g_L(x_L, i_L, A_L^\top e, t) \chi_L(\frac{d}{dt} A_L^\top e, x_L, i_L, v_L, t) + A_L \partial_{v_L} g_L(x_L, i_L, A_L^\top e, t) \frac{d}{dt} A_L^\top e \right.$$

$$\left. + A_L \partial_{i_L} g_L(x_L, i_L, A_L^\top e, t) g_L(x_L, i_L, A_L^\top e, t) + A_L \partial_t g_L(x_L, i_L, A_L^\top e, t) + A_I \frac{d^2}{dt^2} i_s \right) = 0.$$

Using (3.3), we see that

$$Q_{R-CV}^\top Q_{CV}^\top \left(A_L F_L(\frac{d}{dt} A_L^\top e, x_L, i_L, A_L^\top e, t) + A_L \partial_t g_L(x_L, i_L, A_L^\top e, t) \right.$$

$$\left. + A_L \partial_{i_L} g_L(x_L, i_L, A_L^\top e, t) g_L(x_L, i_L, A_L^\top e, t) + A_I \frac{d^2}{dt^2} i_s \right) = 0. \qquad (5.19)$$

Regarding (5.9) and (5.3), we can split

$$\frac{d}{dt} A_L^\top e = A_L^\top Q_{CV} Q_{R-CV} \frac{d}{dt} e + A_L^\top Q_{CV} P_{R-CV} \frac{d}{dt} e + A_L^\top P_{CV} \frac{d}{dt} e$$

$$= A_L^\top Q_{CV} Q_{R-CV} \frac{d}{dt} e + A_L^\top Q_{CV} f_2(x_C, i_C, x_R, i_R, x_L, i_L, e, t) + A_L^\top f_1(x_C, i_C, e, t).$$

We choose a projector P_{LI-cut} along ker $A_L^\top Q_{CV} Q_{R-CV}$. Since the circuit does not contain I-cutsets, the matrix $[A_C\ A_R\ A_V\ A_L]^\top$ has full column rank (see Corollary 4.2). It implies, for $Q_{LI-cut} := I - P_{LI-cut}$, that

$$\text{ker } Q_{CV} Q_{R-CV} = \text{ker } A_L^\top Q_{CV} Q_{R-CV} = \text{ker } P_{LI-cut} = \text{im } Q_{LI-cut}$$

and, therefore, $Q_{CV} Q_{R-CV} = Q_{CV} Q_{R-CV} P_{LI-cut}$ as well as $Q_{R-CV}^\top Q_{CV}^\top = P_{LI-cut}^\top Q_{R-CV}^\top Q_{CV}^\top$. Consequently, we can apply Lemma 2.3 onto (5.19) with

$$M := A_L^\top Q_{CV} Q_{R-CV}, \quad P := P_{LI-cut}, \quad z := \frac{d}{dt} e, \quad y := (x_C, i_C, x_R, i_R, x_L, i_L, e, t),$$

and

$$f(x, y) := F_L(x + A_L^\top Q_{cv} f_2(x_C, i_C, x_R, i_R, x_L, i_L, e, t) + A_L^\top f_1(x_C, i_C, e, t), x_L, i_L, A_L^\top e, t),$$

$$r(y) := Q_{R-cv}^\top Q_{cv}^\top A_L \partial_{i_L} g_L(x_L, i_L, A_L^\top e, t) g_L(x_L, i_L, A_L^\top e, t)$$

$$+ Q_{R-cv}^\top Q_{cv}^\top \left(A_L \partial_t g_L(x_L, i_L, A_L^\top e, t) + A_I \frac{d^2}{dt^2} i_s(t) \right).$$

Thus, we find a continuous function f_6 such that

$$P_{LI-cut} \frac{d}{dt} e = f_6(x_C, i_C, x_R, i_R, x_L, i_L, e, t). \tag{5.20}$$

implying

$$Q_{cv} Q_{R-cv} \frac{d}{dt} e = Q_{cv} Q_{R-cv} f_6(x_C, i_C, x_R, i_R, x_L, i_L, e, t). \tag{5.21}$$

Please note that two time differentiations were required to obtain this expression and, due to (5.19), it depends on the second time derivative of the current source function, i.e. $\frac{d^2}{dt^2} i_s$.

Regarding (5.9) and (5.3) again, we obtain

$$\frac{d}{dt} e = f_7(x_C, i_C, x_R, i_R, x_L, i_L, e, t) \tag{5.22}$$

for

$$f_7(x_C, i_C, x_R, i_R, x_L, i_L, e, t) := Q_{cv} Q_{R-cv} f_6(x_C, i_C, x_R, i_R, x_L, i_L, e, t)$$

$$+ Q_{cv} f_2(x_C, i_C, x_R, i_R, x_L, i_L, e, t) + f_1(x_C, i_C, e, t).$$

Regarding (5.5), (5.9), and (5.3) we get a continuous function f_8 such that

$$\frac{d}{dt} i_R = f_8(x_C, i_C, x_R, i_R, x_L, i_L, e, t), \tag{5.23}$$

without requiring a second time derivative of the original system.

So far the ordinary differential system equations for $\frac{d}{dt} e$, $\frac{d}{dt} i_R$ and $\frac{d}{dt} i_L$ have been obtained. In the following we derive the expressions for $\frac{d}{dt} i_C$ and $\frac{d}{dt} i_V$. Using again (5.1a), we get

$$[A_C\ A_V] \begin{bmatrix} \frac{d}{dt} i_C \\ \frac{d}{dt} i_V \end{bmatrix} + A_R \frac{d}{dt} i_R + A_L \frac{d}{dt} i_L + A_I \frac{d}{dt} i_s = 0.$$

Together with (5.23) and (5.6) we have

$$[A_C\ A_V]\begin{bmatrix}\frac{d}{dt}i_C\\\frac{d}{dt}i_V\end{bmatrix}+A_R f_8(x_C, i_C, x_R, i_R, x_L, i_L, e, t)+A_L g_L(x_L, i_L, A_L^\top e, t)+A_I\frac{d}{dt}i_s = 0.$$

$$(5.24)$$

Multiplying (5.24) by $[A_C\ A_V]^\top$ allows us to apply Corollary 2.4 for

$$M := [A_C\ A_V],\ P := P_{CV-loop},\ z := \begin{bmatrix}\frac{d}{dt}i_C\\\frac{d}{dt}i_V\end{bmatrix},\ y := (x_C, i_C, x_R, i_R, x_L, i_L, e, t)$$

and

$$r(y) := [A_C\ A_V]^\top\left(A_R f_7(x_C, i_C, x_R, i_R, x_L, i_L, e, t)+A_L g_L(x_L, i_L, A_L^\top e, t)+A_I\frac{d}{dt}i_s(t)\right).$$

Consequently, we find a continuous function f_9 such that

$$P_{CV-loop}\begin{bmatrix}\frac{d}{dt}i_C\\\frac{d}{dt}i_V\end{bmatrix} = f_9(x_C, i_C, x_R, i_R, x_L, i_L, e, t).$$

$$(5.25)$$

Rewriting (5.2) as equation system in column form and multiplication by $Q_{CV-loop}^\top$ yields

$$Q_{CV-loop}^\top\begin{bmatrix}g_C(x_C, i_C, A_C^\top e, t)\\\frac{d}{dt}v_s\end{bmatrix} = 0.$$

Differentiating this equation and regarding (3.4) and (5.22) as well as (3.6), we obtain

$$Q_{CV-loop}^\top\begin{bmatrix}\frac{d}{dt}g_C(x_C, i_C, A_C^\top e, t)\\\frac{d^2}{dt^2}v_s\end{bmatrix} = 0$$

$$(5.26)$$

with

$$\frac{d}{dt}g_C(x_C, i_C, A_C^\top e, t)$$

$$= \partial_{x_C}g_C(x_C, i_C, A_C^\top e, t)\chi_C(\frac{d}{dt}i_C, x_C, i_C, A_C^\top e, t)+\partial_{i_C}g_C(x_C, i_C, A_C^\top e, t)\frac{d}{dt}i_C$$

$$+ \partial_{v_C}g_C(x_C, i_C, A_C^\top e, t)A_C^\top f_7(x_C, i_C, x_R, i_R, x_L, i_L, e, t)+\partial_t g_C(x_C, i_C, A_C^\top e, t)$$

$$= F_C(\frac{d}{dt}i_C, x_C, i_C, A_C^\top e, t)$$

$$+ \partial_{v_C} g_C(x_C, i_C, A_C^\top e, t) A_C^\top f_7(x_C, i_C, x_R, i_R, x_L, i_L, e, t) + \partial_t g_C(x_C, i_C, A_C^\top e, t).$$

Using (5.25), we can split

$$\frac{d}{dt} i_C = \begin{bmatrix} I & 0 \end{bmatrix} \begin{bmatrix} \frac{d}{dt} i_C \\ \frac{d}{dt} i_V \end{bmatrix} = \begin{bmatrix} I & 0 \end{bmatrix} Q_{CV-loop} \begin{bmatrix} \frac{d}{dt} i_C \\ \frac{d}{dt} i_V \end{bmatrix} + \begin{bmatrix} I & 0 \end{bmatrix} P_{CV-loop} \begin{bmatrix} \frac{d}{dt} i_C \\ \frac{d}{dt} i_V \end{bmatrix}$$

$$= \begin{bmatrix} I & 0 \end{bmatrix} Q_{CV-loop} \begin{bmatrix} \frac{d}{dt} i_C \\ \frac{d}{dt} i_V \end{bmatrix} + \begin{bmatrix} I & 0 \end{bmatrix} f_8(x_C, i_C, x_R, i_R, x_L, i_L, e, t).$$

Since the circuit has no V-loop, the matrix A_V has full column rank, see Corollary 4.2. It implies

$$\ker \begin{bmatrix} I & 0 \end{bmatrix} Q_{CV-loop} = \ker \begin{bmatrix} I & 0 \\ A_C & A_V \end{bmatrix} Q_{CV-loop} = \ker Q_{CV-loop}.$$

Rewriting (5.26) as

$$Q_{CV-loop}^\top \begin{bmatrix} I \\ 0 \end{bmatrix} \frac{d}{dt} g_C(x_C, i_C, A_C^\top e, t) + Q_{CV-loop}^\top \begin{bmatrix} 0 \\ I \end{bmatrix} \frac{d^2}{dt^2} v_s = 0$$

allows us to apply Lemma 2.3 with

$$M := \begin{bmatrix} I & 0 \end{bmatrix} Q_{CV-loop}, \quad P := Q_{CV-loop}, \quad z := \begin{bmatrix} \frac{d}{dt} i_C \\ \frac{d}{dt} i_V \end{bmatrix}, \quad y := (x_C, i_C, x_R, i_R, x_L, i_L, e, t)$$

and

$$f(x, y) := F_C(x + \begin{bmatrix} I & 0 \end{bmatrix} f_8(x_C, i_C, x_R, i_R, x_L, i_L, e, t), x_C, i_C, A_C^\top e, t)$$

$$+ \partial_{v_C} g_C(x_C, i_C, A_C^\top e, t) A_C^\top f_7(x_C, i_C, x_R, i_R, x_L, i_L, e, t) + \partial_t g_C(x_C, i_C, A_C^\top e, t),$$

$$r(y) := \begin{bmatrix} 0 \\ I \end{bmatrix} \frac{d^2}{dt^2} v_s(t).$$

It means that we find a continuous function f_9 such that

$$Q_{CV-loop} \begin{bmatrix} \frac{d}{dt} i_C \\ \frac{d}{dt} i_V \end{bmatrix} = f_9(x_C, i_C, x_R, i_R, x_L, i_L, e, t).$$

Please note that two time differentations of the original system were required to obtain this expression and due to (5.26) depends on $\frac{d^2}{dt^2} v_s(t)$. Combining it with (5.25) we get

$$
\begin{bmatrix} \frac{d}{dt} i_C \\ \frac{d}{dt} i_v \end{bmatrix} = f_8(x_C, i_C, x_R, i_R, x_L, i_L, e, t) + f_9(x_C, i_C, x_R, i_R, x_L, i_L, e, t). \tag{5.27}
$$

Finally, we obtain from (3.1) and (5.22) that

$$
\frac{d}{dt} x_L = \chi_L (A_L^\top f_7(x_C, i_C, x_R, i_R, x_L, i_L, e, t), x_L, i_L, v_L, t) \tag{5.28}
$$

and from (3.4) and (5.27) that

$$
\frac{d}{dt} x_C = \chi_C([I\ 0](f_8 + f_9)(x_C, i_C, x_R, i_R, x_L, i_L, e, t), x_C, i_C, v_C, t). \tag{5.29}
$$

Consequently, Eqs. (5.22), (5.6), (5.27) and (5.28), (5.29), (3.7) represent an explicit ordinary differential equation system. That means the differentiation index of the circuit system (5.1) is at most 2. □

Remark 5.2 Note that, for simplicty, the analysed system (5.1) only contains time dependent sources. However, the authors see no reason that the results cannot be generalized to include controlled sources as in [19].

Theorems 5.1 and 5.2 contain the results of [19] in the case of circuits that only contain simple lumped elements in either traditional, i.e., Propositions 3.1, 3.5 and 3.3, or flux/charge formulation, i.e. Propositions. 3.2 and 3.4. Some minor differences arise due to Remark 5.1, e.g., loops of capacitances lead to index-2 systems since the corresponding current i_C is not eliminated from the system (5.1). Similarly, results for many refined models, for example when considering [2, 13, 42] as inductance-like elements, are included in Theorems 5.1 and 5.2. The next section discusses a few challenging examples.

6 Refined Models

We present examples for refined models based on PDEs describing electromagnetic fields, that are coupled to the circuit system of DAEs and can be categorized with the generalized elements of Sect. 3.

All models appearing in this section arise from Maxwell's equations [27, 29]. Those can be written in differential form for a system at rest as

$$\nabla \times \vec{E} = -\partial_t \vec{B} \ , \tag{6.1a}$$

$$\nabla \times \vec{H} = \partial_t \vec{D} + \vec{J} \ , \tag{6.1b}$$

$$\nabla \cdot \vec{D} = \rho \ , \tag{6.1c}$$

$$\nabla \cdot \vec{B} = 0 \ , \tag{6.1d}$$

where $\vec{E}$ is the electric field strength, $\vec{B}$ the magnetic flux density, $\vec{H}$ the magnetic field strength, $\vec{D}$ the electric flux density and $\vec{J}$ the electric current density. All these quantities are vector fields $\Omega \times \mathcal{I} \to \mathbb{R}^3$ defined in a domain $\Omega \subset \mathbb{R}^3$ and time interval $\mathcal{I} \subset \mathbb{R}$. The electric charge density ρ is a scalar field $\Omega \times \mathcal{I} \to \mathbb{R}$.

The field quantities are related to each other through the material equations

$$\vec{D} = \varepsilon \vec{E} \ , \qquad\qquad \vec{J}_{\mathrm{c}} = \sigma \vec{E} \ , \qquad\qquad \vec{H} = \mu \vec{B} \ , \tag{6.2}$$

where ε is the electric permittivity, σ the electric conductivity and μ the magnetic permeability. They are rank-2 tensor fields $\Omega \to \mathbb{R}^{3\times3}$. The current density in (6.1b) can be divided into the conduction current density $\vec{J}_{\mathrm{c}}$ of (6.2) and the source current density $\vec{J}_{\mathrm{s}}$

$$\vec{J} = \vec{J}_{\mathrm{c}} + \vec{J}_{\mathrm{s}} \ . \tag{6.3}$$

The inverse of the material relations in (6.2) is defined through the electric resistivity $\rho : \Omega \to \mathbb{R}^{3\times3}$ and the magnetic reluctivity $\nu : \Omega \to \mathbb{R}^{3\times3}$ such that

$$\vec{E} = \rho \vec{J}_{\mathrm{c}} \ , \qquad\qquad \vec{B} = \nu \vec{H} \ . \tag{6.4}$$

Assumption 6.1 ([15]) *We divide the space domain Ω into three disjoint subdomains Ω_{c} (the conducting domain), Ω_{s} (the source domain) and Ω_0 (the excitation-free domain) such that*

- *the material tensors ε, μ and ν are positive definite on the whole subdomain Ω.*
- *the material tensors ρ and σ are positive definite in Ω_{c} and zero everywhere else.*
- *the source current density is only nonzero in Ω_{s}.*

In order to simulate Maxwell's equations and its approximations, often potentials are defined, that allow to rewrite the equations as systems of PDEs that can be resolved. For the examples that are presented next, the magnetic vector potential $\vec{A} : \Omega \times \mathcal{I} \to \mathbb{R}^3$ and the electric scalar potential $\phi : \Omega \times \mathcal{I} \to \mathbb{R}$ are relevant. They are defined such that

$$\vec{B} = \nabla \times \vec{A} \qquad \text{and} \qquad \vec{E} = -\partial_t \vec{A} - \nabla\phi \ . \tag{6.5}$$

Following the *finite integration technique* (FIT), originally introduced in 1977 by Thomas Weiland [44], the discrete version of (6.1) is obtained as Maxwell's grid equations [39]

$$Ce = -\frac{d}{dt}b \quad \widetilde{C}h = \frac{d}{dt}d + j \quad \widetilde{S}d = q \quad Sb = 0 , \quad (6.6)$$

here C, $\widetilde{C} = C^\top$ (see [39]) and S, $\widetilde{S}$ are the discrete curl, dual curl, divergence and dual divergence operators, respectively. The discrete field vectors e, b, h, d, j and q are integrated quantities over points, edges, facets and volumes of two dual grids. Also, the material relations (6.2) and (6.4) can be formulated through the material matrices $M_\star$ as

$$d = M_\varepsilon e \quad j_c = M_\sigma e \quad h = M_\mu b \quad e = M_\rho j_c \quad b = M_\nu h . \quad (6.7)$$

Analogous to the continuous case, discrete potentials can be defined, which lead to the relation

$$b = Ca \quad e = -\frac{d}{dt}a - G\bar{\Phi} , \quad (6.8)$$

where a and $\bar{\Phi}$ are the discrete magnetic vector potential and electric scalar potential, respectively and $G = -\widetilde{S}^\top$ (see [39]) is the discrete gradient operator.

Assumption 6.2 *The boundary of the domain $\Gamma = \partial\Omega$ is divided into three disjoint sets $\Gamma_{neu,0}$, $\Gamma_{dir,0}$ and Γ_s, with*

$$\Gamma = \Gamma_{neu,0} \cup \Gamma_{dir,0} \cup \Gamma_s .$$

Here, $\Gamma_{neu,0}$ and $\Gamma_{dir,0}$ are the parts where homogeneous Neumann and Dirichlet boundary conditions are imposed and Γ_s where the field equation is excited.

In case of a device described by Maxwell's equations and coupled to a circuit through boundary conditions, Γ_s represents the area where the device is connected to the surrounding network.

Assumption 6.3 ([3, 15]) *We assume that at least the homogeneous Dirichlet boundary conditions of $\Gamma_{dir,0}$ are already incorporated into the discrete operator matrices, such that the gradient operator matrix $G = -\widetilde{S}^\top$ has full column rank.*

This is a standard assumption and has already been shown and used e.g. in [3, 15].

Remark 6.1 Both material as well as operator matrices with similar properties are also obtained with a finite element (FE) discretization of the partial differential equations obtained from Maxwell's equations, whenever appropriate basis and test functions are used, that fulfil the discrete de Rham sequence [7, 13]. Therefore, the subsequent analysis of the discretized systems is also valid for FE discretizations.

6.1   Inductance-Like Element

In the following we give an example of an electromagnetic (EM) device, with its formulation taken from [4], based upon full wave Maxwell's equation, that fits the form of a strong inductance-like element.

In the absence of source terms and Neumann boundary conditions, i.e., Ω_s, $\Gamma_{neu,0} = \emptyset$, one possibility to rewrite Maxwell's equations in terms of potentials is given by the following second order PDE system (see [3])

$$\varepsilon \nabla \partial_t \phi + \zeta \nabla \left[\xi \nabla \cdot \left(\zeta \vec{A} \right) \right] = 0 \quad \text{in } \Omega \, , \qquad (6.9a)$$

$$\nabla \times (\nu \nabla \times \vec{A}) + \partial_t \left[\varepsilon \left(\nabla \phi + \partial_t \vec{A} \right) \right] + \sigma \left(\nabla \phi + \partial_t \vec{A} \right) = 0 \quad \text{in } \Omega \, , \qquad (6.9b)$$

where ζ and ξ are artificial material tensors whose choice is discussed for example in [12] and [10]. We refer to system (6.9) as the $\vec{A} - \phi$ *formulation* which makes use of a *grad-type Lorenz gauge condition* in order to avoid ambiguity of the potentials, see [3, 12]. Let v_L and i_L be the time-dependent branch voltages and currents of the element, respectively. With Assumption 6.2 given, we complete (6.9) with the boundary conditions

$$\nabla \times \vec{A} = 0 \qquad\qquad \text{in } \Gamma_{dir,0} \, , \qquad (6.10a)$$

$$\phi = 0 \qquad\qquad \text{in } \Gamma_{dir,0} \, , \qquad (6.10b)$$

$$\phi = v_L \qquad\qquad \text{in } \Gamma_s \, . \qquad (6.10c)$$

The branch currents i_L shall comply with the model

$$\int_{\Gamma_s} \nabla \times \left(\nu \nabla \times \vec{A} \right) \cdot d\vec{S} = i_L \, . \qquad (6.11)$$

In order to apply the method of lines, we spatially discretize the system (6.9) using e.g. the finite integration technique. Since most of the required matrices and quantities were already introduced in this section's preliminaries, we proceed with the circuit coupling which is archived via the boundaries only ($\Omega_s = \emptyset$).

Given Assumption 6.3, the homogeneous Dirichlet boundaries (6.10a) and (6.10b) are already incorporated into the discrete operator matrices, e.g. G or $\widetilde{C}$. To incorporate the inhomogeneous Dirichlet boundary conditions, we split $\bar{\Phi}$ into Φ_s and Φ, belonging to the degrees of freedom in Γ_s and the rest, as follows

$$\bar{\Phi} = Q_s \Phi + P_s \Phi_s \, , \qquad (6.12)$$

with basis matrices Q_s and P_s of full column rank. The boundary voltage excitation (6.10c) is then obtained by setting $\Phi_s = \Lambda_s v_L$ with the element's terminal to $\Gamma_s^{(j)}$'s degrees of freedom mapping

$$(\Lambda_s)_{ij} = \begin{cases} 1, & \text{if } (\Phi_s)_i \text{ belongs to the } j\text{-th terminal } \Gamma_s^{(j)} \\ 0, & \text{otherwise.} \end{cases}$$

Here, $\Gamma_s = \Gamma_s^{(1)} \cup \ldots \cup \Gamma_s^{(k)}$, for a k-port device, where

$$\Gamma_s^i \cap \Gamma_s^j = \emptyset, \quad \text{for } i \neq j \ .$$

With the junction $Y_s = P_s \Lambda_s$ the discrete gradient in (6.8) reads:

$$G\bar{\Phi} = GQ_s\Phi + GY_s v_L \ .$$

Remark 6.2 Note that, as the different terminals $\Gamma_s^{(j)}$ are disjoint, per construction, Λ_s, and therefore also Y_s, have full column rank.

The spatially discretized version of (6.9) with incorporated boundary conditions (6.10) is then given by

$$Q_s^\top \tilde{S} M_\varepsilon GQ_s \frac{d}{dt}\Phi + Q_s^\top \tilde{S} M_\zeta GM_\xi \tilde{S} M_\zeta a = 0 \ ,$$

$$(6.13)$$

$$\check{C}M_\nu Ca + \frac{d}{dt}\left[M_\varepsilon \left(GQ_s\Phi + GY_s v_L + \pi\right)\right] + M_\sigma \left(GQ_s\Phi + GY_s v_L + \frac{d}{dt}a\right) = 0 \ ,$$

$$(6.14)$$

$$\frac{d}{dt}a - \pi = 0 \ ,$$

$$(6.15)$$

where π is a discrete quasi-canonical momentum introduced in order to avoid second order derivatives. The discretized current coupling model of (6.11) reads

$$i_L = Y_s^\top \tilde{S}\check{C}M_\nu Ca \ . \tag{6.16}$$

For $x_L = (\Phi, a, \pi)$, we define the system matrices

$$M := \begin{bmatrix} Q_s^\top \tilde{S} M_\varepsilon GQ_s & 0 & 0 \\ M_\varepsilon GQ_s & M_\sigma & M_\varepsilon \\ 0 & I & 0 \end{bmatrix}, \quad A := \begin{bmatrix} 0 & Q_s^\top \tilde{S} M_\zeta GM_\xi \tilde{S} M_\zeta & 0 \\ M_\sigma GQ_s & \check{C}M_\nu C & 0 \\ 0 & 0 & -I \end{bmatrix},$$

$$N := \begin{bmatrix} 0 \\ M_\varepsilon G Y_s \\ 0 \end{bmatrix}, \qquad\qquad B := \begin{bmatrix} 0 \\ M_\sigma G Y_s \\ 0 \end{bmatrix},$$

$$F := \begin{bmatrix} 0 & Y_s^\top \widetilde{S} \widetilde{C} M_\nu C & 0 \end{bmatrix}$$

from which we conclude the EM device's element description

$$f_L\left(\frac{\mathrm{d}}{\mathrm{d}t}m_L(x_L, i_L, v_L, t), x_L, i_L, v_L, t\right) := \begin{pmatrix} M\frac{\mathrm{d}}{\mathrm{d}t}x_L + Ax_L + Bv_L + N\frac{\mathrm{d}}{\mathrm{d}t}v_L \\ i_L - Fx_L \end{pmatrix} = 0 .$$

$$(6.17)$$

Proposition 6.1 *Provided Assumptions 6.1, 6.2 and 6.3 are fulfilled and the absence of inner sources and Neumann boundary conditions, the EM device, whose model is given by the element description* (6.17), *is a strongly inductance-like element.*

Proof The discrete gradient operator G and basis matrix Q_s have full column rank by Assumption 6.3 and construction (6.12). Further it is $G = -\widetilde{S}^\top$ and together with M_ε being positive definite, as of Assumption 6.1, we deduce that the Laplace-operator $L_Q := Q_s^\top \widetilde{S} M_\varepsilon G Q_s$ is non-singular. Hence, we find

$$M^{-1} = \begin{bmatrix} L_Q^{-1} & 0 & 0 \\ 0 & 0 & I \\ -G Q_s L_Q^{-1} & M_\varepsilon^{-1} & -M_\varepsilon^{-1} M_\sigma \end{bmatrix}.$$

Therefore, we can define the following matrices

$$\widetilde{A} := M^{-1}A = \begin{bmatrix} 0 & L_Q^{-1}H & 0 \\ 0 & 0 & -I \\ M_\varepsilon^{-1}M_\sigma G Q_s & -G Q_s L^{-1}H + M_\varepsilon^{-1}\widetilde{C}M_\nu C & M_\varepsilon^{-1}M_\sigma \end{bmatrix},$$

$$\widetilde{B} := M^{-1}B = \begin{bmatrix} 0 \\ 0 \\ M_\varepsilon^{-1}M_\sigma G Y_s \end{bmatrix}, \quad \widetilde{N} := M^{-1}N = \begin{bmatrix} 0 \\ 0 \\ G Y_s \end{bmatrix}$$

with $H := Q_s^\top \widetilde{S} M_\zeta G M_\xi \widetilde{S} M_\zeta$ and deduce from (6.17) a description for $\frac{\mathrm{d}}{\mathrm{d}t}x_L$ of the form (3.1)

$$\frac{\mathrm{d}}{\mathrm{d}t}x_L = -\widetilde{A}x_L - \widetilde{B}v_L - \widetilde{N}\frac{\mathrm{d}}{\mathrm{d}t}v_L =: \chi_L(\frac{\mathrm{d}}{\mathrm{d}t}v_L, x_L, i_L, v_L, t) .$$

$$(6.18)$$

Next, we differentiate (6.17) once, in particular the second part, and insert the expression for $\frac{d}{dt}x_L$ from (6.18) yielding

$$\frac{d}{dt}i_L = F(-\tilde{A}x_L - \tilde{B}v_L - \tilde{N}\frac{d}{dt}v_L) = -F\tilde{A}x_L - F\tilde{B}v_L =: g_L(x_L, i_L, v_L, t).$$

Thus, we found an expression of $\frac{d}{dt}i_L$ fitting (3.2). Finally, we observe that

$$F_L(v'_L, x_L, i_L, v_L, t) := \partial_{x_L}g_L(x_L, i_L, v_L, t)\chi_L(v'_L, x_L, i_L, v_L, t)$$
$$+ \partial_{v_L}g_L(x_L, i_L, v_L, t)v'_L$$
$$= F\tilde{A}\tilde{A}x_L + F\tilde{A}\tilde{B}v_L + F\tilde{A}\tilde{N}v'_L - \underbrace{F\tilde{B}}_{=0}v'_L$$

is continuous and strongly monotone with respect to v'_L, see Lemma 2.1 using that $F\tilde{A}\tilde{N} = -Y_s^\top \tilde{S}\tilde{C}M_\nu CGY_s = Y_s^\top G^\top C^\top M_\nu CGY_s$ is positive definite by construction. We conclude that this model for an EM device fulfills the strongly inductance-like property. □

Remark 6.3 The fact that $FM^{-1}N$ vanishes, as obtained by elemental matrix operations, plays a key role in the EM device's model fitting the inductance-like element description.

For two different field approximations of Maxwell's equations that result in strongly inductance-like elements, see [13]. In contrast to our example, there the strongly inductance-like element is given by the term $\partial_{v_L}g_L(x_L, i_L, v_L, t)v'_L$ in (3.3), like in the case of classical and flux-formulated inductances.

6.2 Capacitance-Like Element

We consider the electroquasistatic field approximation of Maxwell's equations [11, 15]. As in this approximation, the electric field $\vec{E}$ is rotation free, we can write it in terms of only the electric scalar potential ϕ [15].

Given a time-dependent excitation v_C, we can write the following boundary value problem to describe an electroquasistatic field

$$\nabla \cdot \sigma\nabla\phi + \frac{d}{dt}\nabla \cdot \varepsilon\nabla\phi = 0 \qquad \text{in } \Omega , \qquad (6.19a)$$

$$\phi = 0 \qquad \text{in } \Gamma_{\text{dir},0} , \qquad (6.19b)$$

$$\partial_{\vec{n}}\phi = 0 \qquad \text{in } \Gamma_{\text{neu},0} , \qquad (6.19c)$$

$$\phi = v_C \qquad \text{in } \Gamma_s , \qquad (6.19d)$$

with $\vec{n}$ being the outer normal vector to $\Gamma_{\text{neu},0}$. To couple the electroquasistatic system (6.19) to a circuit, the extraction of a current is necessary, so as to obtain an implicit voltage-to-current relation. For that we integrate the current density (6.19a) over the boundary, where the connections to the circuit are located (Γ_s), i.e.

$$\int_{\Gamma_s} \left(\nabla \cdot \sigma \nabla \phi + \frac{\mathrm{d}}{\mathrm{d}t} \nabla \cdot \varepsilon \nabla \phi \right) \cdot \mathrm{d}\vec{S} = i_c . \tag{6.20}$$

We assume first a spatial discretization of the PDEs (6.19a) and (6.20) has been applied, with only the boundary conditions

$$\phi = 0 \quad \text{in } \Gamma_{\text{dir},0} \qquad \text{and} \qquad \partial_{\vec{n}} \phi = 0 \quad \text{in } \Gamma_{\text{neu},0} . \tag{6.21}$$

Analogously to the previous examples and given the homogeneous boundary conditions of (6.21) are incorporated in the operator matrices, i.e., Assumption 6.3 holds, the spatially discretized electroquasitatic field equation with circuit coupling equation is obtained as [15]

$$Q_s^\top L_\sigma Q_s \Phi + Q_s^\top L_\varepsilon Q_s \frac{\mathrm{d}}{\mathrm{d}t} \Phi + Q_s^\top L_\sigma Y_s v_C + Q_s^\top L_\varepsilon Y_s \frac{\mathrm{d}}{\mathrm{d}t} v_C = 0 , \tag{6.22a}$$

$$Y_s^\top L_\sigma Q_s \Phi + Y_s^\top L_\varepsilon Q_s \frac{\mathrm{d}}{\mathrm{d}t} \Phi + Y_s^\top L_\sigma Y_s v_C + Y_s^\top L_\varepsilon Y_s \frac{\mathrm{d}}{\mathrm{d}t} v_C = i_C , \tag{6.22b}$$

where $L_\sigma = \widetilde{S} M_\sigma \widetilde{S}^\top$ and $L_\varepsilon = \widetilde{S} M_\varepsilon \widetilde{S}^\top$ are two Laplace matrices.

Proposition 6.2 *For Q_s and Y_s, we have that*

$$Q_s x_1 \neq Y_s x_2, \text{ for } x_1, x_2 \neq 0 .$$

Proof This property follows directly from the definition of both matrices. We have $Y_s x_2 = P_s y_2$ and, by construction, the image of P_s are the discrete elements living in Γ_s, while the image of Q_s are the rest. Also, by construction, both matrices have full column rank and thus a trivial kernel. □

Proposition 6.3 *Provided Assumptions 6.1, 6.2 and 6.3 are fulfilled, then the semidiscrete eletroquasistatic system of equations with circuit coupling equation (6.22) is a strongly capacitance-like element.*

Proof Due to Assumptions 6.1 and 6.3, and the fact that Q_s has full column rank, we start by rewriting (6.22a) as

$$\frac{\mathrm{d}}{\mathrm{d}t} \Phi = - (Q_s^\top L_\varepsilon Q_s)^{-1} Q_s^\top L_\sigma Q_s \Phi$$

$$- (Q_s^\top L_\varepsilon Q_s)^{-1} \left(Q_s^\top L_\varepsilon Y_s \frac{\mathrm{d}}{\mathrm{d}t} v_C + Q_s^\top L_\sigma Y_s v_C \right) . \tag{6.23}$$

Inserting this into (6.22b) yields

$$i_C = Y_s^\top \left(I - L_\varepsilon Q_s (Q_s^\top L_\varepsilon Q_s)^{-1} Q_s^\top\right) L_\sigma Q_s \Phi$$

$$+ Y_s^\top \left(I - L_\varepsilon Q_s (Q_s^\top L_\varepsilon Q_s)^{-1} Q_s^\top\right) L_\sigma Y_s v_C$$

$$+ Y_s^\top \left(L_\varepsilon - L_\varepsilon Q_s (Q_s^\top L_\varepsilon Q_s)^{-1} Q_s^\top L_\varepsilon\right) Y_s \frac{d}{dt} v_C . \tag{6.24}$$

Now we want to see that $C = Y_s^\top \left(L_\varepsilon - L_\varepsilon Q_s (Q_s^\top L_\varepsilon Q_s)^{-1} Q_s^\top L_\varepsilon\right) Y_s$ is positive definite. For that, using again that L_ε is symmetric positive definite (Assumptions 6.1 and 6.3) and thus its square root exists and is also symmetric positive definite, we rewrite

$$C = Y_s^\top L_\varepsilon^{\frac{1}{2}} \left(I - L_\varepsilon^{\frac{1}{2}} Q_s (Q_s^\top L_\varepsilon Q_s)^{-1} Q_s^\top L_\varepsilon^{\frac{1}{2}}\right) L_\varepsilon^{\frac{1}{2}} Y_s.$$

It can easily be seen that $\left(I - L_\varepsilon^{\frac{1}{2}} Q_s (Q_s^\top L_\varepsilon Q_s)^{-1} Q_s^\top L_\varepsilon^{\frac{1}{2}}\right)$ is a symmetric projector and thus positive semidefinite. Therefore we have that C is positive semidefinite. Let's assume that there exists a vector x such that $x^\top C x = 0$, then,

$$\left(I - L_\varepsilon^{\frac{1}{2}} Q_s (Q_s^\top L_\varepsilon Q_s)^{-1} Q_s^\top L_\varepsilon^{\frac{1}{2}}\right) L_\varepsilon^{\frac{1}{2}} Y_s x = 0.$$

However, this implies that

$$L_\varepsilon^{\frac{1}{2}} Y_s x = L_\varepsilon^{\frac{1}{2}} Q_s (Q_s^\top L_\varepsilon Q_s)^{-1} Q_s^\top L_\varepsilon Y_s x$$

and multiplying this by $L_\varepsilon^{-\frac{1}{2}}$ would yield $Y_s x = Q_s y$, with $y = (Q_s^\top L_\varepsilon Q_s)^{-1} Q_s^\top L_\varepsilon Y_s x$. Due to Proposition 6.2 this, however, is only possible if $Y_s x = 0$ and, as Y_s has full column rank (see Remark 6.2), $x = 0$. Therefore C has full rank and is positive definite.

According to Definition 3.2, we need to show that $\frac{d}{dt}\Phi$ can be written, with at most one differentiation, as a function depending only on $\frac{d}{dt}i_C$, Φ, v_C, i_C and t (see (3.4)). For that we invert C in (6.24) to obtain

$$\frac{d}{dt} v_C = g_C(\Phi, i_C, v_C) . \tag{6.25}$$

This can now be inserted into (6.23) to obtain a function

$$\frac{d}{dt} \Phi = \chi_C(\Phi, i_C, v_C) , \tag{6.26}$$

without having required any differentiation of the original system.

Due to (6.25), we have already shown that we obtain a capacitance-like element. Furthermore, as $\partial_{i_C} g_C(\Phi, i_C, v_C) = C$, is positive definite, using Lemma 2.1 and Definition 2.2, the system is shown to be strongly capacitance-like.   □

6.3   Resistance-Like Element

The last refined model we study is the eddy current equation for the simulation of magnets with superconducting coils. For that we consider a magnetoquasistatic approximation of Maxwell's equations [27] in terms of the $\vec{A}^*$ formulation [18]. Here, the gauging freedom of the magnetoquasistatic setting allows to choose a special magnetic vector potential $\vec{A}$, such that the electric scalar potential ϕ vanishes from the PDE. The governing equation reads

$$\nabla \times \nu \tau_{eq} \nabla \times \frac{\mathrm{d}}{\mathrm{d}t} \vec{A} - \nabla \times \nu \nabla \times \vec{A} = \vec{J}_s \ .$$

The non-standard expression $\nabla \times \nu \tau_{eq} \nabla \times \frac{\mathrm{d}}{\mathrm{d}t} \vec{A}$ is an homogenization model accounting for the cable magnetization, that represents the eddy current effects of the superconducting coils [17]. It contains the cable time constant τ_{eq}, which depends on certain properties of the cable [43]. This formulation is coupled to a circuit in order to simulate the superconducting magnet's protection system of the LHC at CERN [6, 14]. For the boundary value problem we also set the boundary conditions

$$\vec{n} \times \vec{A} = 0, \quad \text{on } \Gamma_{\mathrm{dir},0} \qquad \text{and} \qquad \vec{n} \times (\nu \nabla \times \vec{A}) = 0, \quad \text{on } \Gamma_{\mathrm{neu},0} \ , \qquad (6.27)$$

where $\vec{n}$ is again the outer normal vector to the boundary Γ. Please note that here, no boundary conditions where set on Γ_s, as for this example $\Gamma_s = \emptyset$.

In this case, as $\Gamma_s = \emptyset$, the circuit coupling is not performed through the boundary but by a characteristic function (winding density function) [38], that discributes the zero dimensional current i_R on the two or three dimensional domain of the PDE. For the excitation of the coil's cross-section we define a $\chi_s : \Omega \to \mathbb{R}^3$, such that

$$\vec{J}_s = \chi_s i_R \ .$$

This also allows to extract the voltage across the coil as

$$v_R = -\int_\Omega \chi_s \cdot \vec{E} \, \mathrm{d}V \ .$$

Assumption 6.4 *As the magnet is excited through the superconducting coils, we assume that the domain, where the source current density is nonzero also corresponds to the domain, where the cable time constant is positive, that is*

$$\sup \tau_{\text{eq}} = \sup \chi_s = \Omega_s \, .$$

After spatial discretisation of the eddy current PDE with coupling equation, we obtain the DAE

$$C^\top M_{\nu,\tau_{\text{eq}}} C \frac{\mathrm{d}}{\mathrm{d}t} a + C^\top M_\nu C a = X i_{\text{R}} \tag{6.28a}$$

$$X^\top \frac{\mathrm{d}}{\mathrm{d}t} a = v_{\text{R}} \, , \tag{6.28b}$$

where X is a vector, containing the discretisation of the winding density function. We define the orthogonal projector Q_τ onto $\ker C^\top M_{\nu,\tau_{\text{eq}}} C$ and its complementary $P_\tau = I - Q_\tau$.

Assumption 6.5 *We assume that*

- *the curl matrix C and the discrete magnetic vector potential a are gauged and contain homogeneous Dirichlet boundary conditions, such that C has full column rank.*
- *there is no excitation outside of the coils, i.e., $Q_\tau^\top X = 0$.*

The first part of the assumption is necessary, such that the DAE system (6.28) is uniquely solvable. This is possible by for example using a tree-cotree gauge [32], where the degrees of freedom of a belonging to a gradient field are eliminated. The second part of the assumption is motivated by the fact that the source current density has to be divergence-free, together with Assumption 6.4.

Proposition 6.4 *Provided Assumptions 6.1, 6.4–6.5 are fulfilled, then the semidiscrete homogenized eddy current system of equations with circuit coupling equation (6.28) is a strongly resistance-like element.*

Proof We start by multiplying equation (6.28a) by $Q_\tau^\top$ and $P_\tau^\top$ and obtain

$$C^\top M_{\nu,\tau_{\text{eq}}} C \frac{\mathrm{d}}{\mathrm{d}t} a + P_\tau^\top C^\top M_\nu C a = P_\tau^\top X i_{\text{R}} \tag{6.29a}$$

$$Q_\tau^\top C^\top M_\nu C a = Q_\tau^\top X i_{\text{R}} \tag{6.29b}$$

From (6.29a) we obtain without the need of any differentiation

$$P_\tau \frac{\mathrm{d}}{\mathrm{d}t} a = (C^\top M_{\nu,\tau_{\text{eq}}} C + Q_\tau^\top Q_\tau)^{-1} (P_\tau^\top X i_{\text{R}} - P_\tau^\top C^\top M_\nu C a) \, . \tag{6.30}$$

Differentiating (6.29b) once and using Assumption 6.5 we have

$$Q_\tau \frac{d}{dt} a = -(Q_\tau^\top C^\top M_\nu C Q_\tau + P_\tau^\top P_\tau)^{-1} Q_\tau^\top C^\top M_\nu C P_\tau \frac{d}{dt} a \qquad (6.31)$$

Inserting (6.30) into (6.31) we obtain an ODE with the structure

$$\frac{d}{dt} x_R = \chi_R(x_R, i_R) \,,$$

where $x_R = P_\tau a + Q_\tau a$. Now we use Assumption 6.5 and insert (6.30) into the circuit coupling equation to obtain

$$v_R = X^\top \frac{d}{dt}(P_\tau a + Q_\tau a) = X^\top \frac{d}{dt} P_\tau a$$
$$= X^\top P_\tau (C^\top M_{\nu,\tau_{eq}} C + Q_\tau^\top Q_\tau)^{-1} (P_\tau^\top X i_R - P_\tau^\top C^\top M_\nu C a) \,.$$

To obtain this expression no differentiation was needed, thus if we differentiate it once, according to Definition 3.3 and using Lemma 2.1 and Definition 2.2, we now only need to show that $G = \partial_{v_R'} g_R(v_R', x_R, i_R, v_R, t)$, with

$$G^{-1} = X^\top P_\tau (C^\top M_{\nu,\tau_{eq}} C + Q_\tau^\top Q_\tau)^{-1} P_\tau^\top X \,,$$

is positive definite to obtain that (6.28) is a strongly resistance-like element. This follows immediately by the fact that $M_{\nu,\tau_{eq}}$ is positive semidefinite (Assumptions 6.1 and 6.4) and X has full column rank, as it is only a vector. □

7   Conclusions

This paper has demonstrated that even very complicated refined models with internal degrees of freedom can be characterized by generalizations of the basic circuit elements, i.e., resistance, inductance and capacitance. This knowledge significantly simplifies the structural analysis of future networks consisting of refined models. Structural properties of the network, e.g. the differential algebraic index, can easily be deduced if all elements are identified in terms of the proposed generalized elements.

Already known index results of different approximations of Maxwell's equations coupled to circuits could be confirmed within this framework. This includes the results of circuits containing elements described with the classic eddy current equation in [2] as well as the $\vec{A} - \phi$ formulation of full Maxwell's equations with Lorenz gauge in [3]. In addition, new index results for other field approximations have been obtained. For instance, the structural analysis of circuits containing elements described with an electroquasistatic setting or the eddy current effects on superconducting coils is achieved. In an analogous manner, the classification of further refined models or complicated circuit elements can be performed in future works to further unify the analysis of such coupled systems, see e. g. [41].

References

1. Alì, G., Bartel, A., Günther, M., Tischendorf, C.: Elliptic partial differential-algebraic multiphysics models in electrical network design. Math. Models Methods Appl. Sci. **13**(9), 1261–1278 (2003)
2. Bartel, A., Baumanns, S., Schöps, S.: Structural analysis of electrical circuits including magnetoquasistatic devices. Appl. Numer. Math. **61**, 1257–1270 (2011)
3. Baumanns, S.: Coupled electromagnetic field/circuit simulation: modeling and numerical analysis. Ph.D. thesis, Universität zu Köln, Köln (2012)
4. Baumanns, S., Clemens, M., Schöps, S.: Structural aspects of regularized full Maxwell electrodynamic potential formulations using FIT. In: Manara, G. (ed.) Proceedings of 2013 URSI International Symposium on Electromagnetic Theory (EMTS), pp. 1007–1010. IEEE, Piscataway (2013)
5. Benner, P., Grundel, S., Himpe, C., Huck, C., Streubel, T., Tischendorf, C.: Gas network benchmark models. In: Campbell, S., Ilchmann, A., Mehrmann, V., Reis, T. (eds.) Applications of Differential-Algebraic Equations: Examples and Benchmarks, Differential-Algebraic Equations Forum. Springer, Cham (2018)
6. Bortot, L., Auchmann, B., Cortes Garcia, I., Fernando Navarro, A.M., Maciejewski, M., Mentink, M., Prioli, M., Ravaioli, E., Schöps, S., Verweij, A.: STEAM: a hierarchical co-simulation framework for superconducting accelerator magnet circuits. IEEE Trans. Appl. Super. **28**(3) (2018). https://doi.org/10.1109/TASC.2017.2787665
7. Bossavit, A.: Computational Electromagnetism: Variational Formulations, Complementarity, Edge Elements. Academic Press, San Diego (1998). https://doi.org/10.1016/B978-0-12-118710-1.X5000-4
8. Brenan, K.E., Campbell, S.L., Petzold, L.R.: Numerical Solution of Initial-Value Problems in Differential-Algebraic Equations. Society for Industrial and Applied Mathematics, Philadelphia (1995). https://doi.org/10.1137/1.9781611971224
9. Chua, L.O., Desoer, C.A., Kuh, E.S.: Linear and nonlinear Circuits. McGraw-Hill, Singapore (1987)
10. Clemens, M.: Large systems of equations in a discrete electromagnetism: formulations and numerical algorithms. IEE Proc. Sci. Meas. Technol. **152**(2), 50–72 (2005). https://doi.org/10.1049/ip-smt:20050849
11. Clemens, M., Steinmetz, T., Weida, D., Hinrichsen, V.: Coupled thermal-electroquasistatic 3D field simulation of high-voltage surge arrester structures. In: Computational Electromagnetics Conference (CEM) (2006)
12. Clemens, M., Weiland, T.: Regularization of eddy-current formulations using discrete grad-div operators. IEEE Trans. Magn. **38**(2), 569–572 (2002)
13. Cortes Garcia, I., De Gersem, H., Schöps, S.: A structural analysis of field/circuit coupled problems based on a generalised circuit element. Numer. Algorithm 83, 373–394 (2020). https://doi.org/10.1007/s11075-019-00686-x
14. Cortes Garcia, I., Schöps, S., Bortot, L., Maciejewski, M., Prioli, M., Fernandez Navarro, A.M., Auchmann, B., Verweij, A.P.: Optimized field/circuit coupling for the simulation of quenches in superconducting magnets. IEEE J. Multiscale Multiphys. Comput. Tech. **2**(1), 97–104 (2017). https://doi.org/10.1109/JMMCT.2017.2710128

15. Cortes Garcia, I., Schöps, S., De Gersem, H., Baumanns, S.: Systems of Differential Algebraic Equations in Computational Electromagnetics. Differential-Algebraic Equations Forum. Springer, Heidelberg (2018)
16. Costa, M.C., Nabeta, S.I., Cardoso, J.R.: Modified nodal analysis applied to electric circuits coupled with FEM in the simulation of a universal motor. IEEE Trans. Magn. **36**(4), 1431–1434 (2000)
17. De Gersem, H., Weiland, T.: Finite-element models for superconductive cables with finite interwire resistance. IEEE Trans. Magn. **40**(2), 667–670 (2004). https://doi.org/10.1109/TMAG.2004.825454
18. Emson, C.R.I., Trowbridge, C.W.: Transient 3D eddy currents using modified magnetic vector potentials and magnetic scalar potentials. IEEE Trans. Magn. **24**(1), 86–89 (1988). https://doi.org/10.1109/20.43862
19. Estévez Schwarz, D., Tischendorf, C.: Structural analysis of electric circuits and consequences for MNA. Int. J. Circuit Theory Appl. **28**(2), 131–162 (2000). https://doi.org/10.1002/(SICI)1097-007X(200003/04)28:2<131::AID-CTA100>3.0.CO;2-W
20. Feldmann, U., Günther, M.: CAD-based electric-circuit modeling in industry I: mathematical structure and index of network equations. Surv. Math. Ind. **8**(2), 97–129 (1999)
21. Gear, C.: Simultaneous numerical solution of differential-algebraic equations. IEEE Trans. Circuit Theory **18**(1), 89–95 (1971)
22. Grasser, K.T., Selberherr, S.: Mixed-mode device simulation. Microelectron. J. **31**(11), 873–881 (2000)
23. Grundel, S., Jansen, L., Hornung, N., Clees, T., Tischendorf, C., Benner, P.: Model order reduction of differential algebraic equations arising from the simulation of gas transport networks. In: Schöps, S., Bartel, A., Günther, M., ter Maten, E.J.W., Müller, P.C. (eds.) Progress in Differential-Algebraic Equations, Differential-Algebraic Equations Forum, pp. 183–205. Springer Berlin (2014). https://doi.org/10.1007/978-3-662-44926-4_9
24. Günther, M.: A joint DAE/PDE model for interconnected electrical networks. Math. Model. Systems **6**(2), 114–128 (2000)
25. Günther, M., Feldmann, U.: The DAE-index in electric circuit simulation. Math. Comput. Simul. **39**, 573–582 (1995)
26. Ho, C.W., Ruehli, A.E., Brennan, P.A.: The modified nodal approach to network analysis. IEEE Trans. Circuits Syst. **22**(6), 504–509 (1975)
27. Jackson, J.D.: Classical Electrodynamics, 3rd edn. Wiley, New York (1998). https://doi.org/10.1017/CBO9780511760396
28. Jansen, L., Tischendorf, C.: A unified (P)DAE modeling approach for flow networks. In: Schöps, S., Bartel, A., Günther, M., ter Maten, E.J.W., Müller, P.C. (eds.) Progress in Differential-Algebraic Equations, Differential-Algebraic Equations Forum, pp. 127–151. Springer Berlin (2014). https://doi.org/10.1007/978-3-662-44926-4_7
29. Maxwell, J.C.: A dynamical theory of the electromagnetic field. R. Soc. Trans. **CLV**, 459–512 (1864)
30. Mayaram, K., Pederson, D.O.: Coupling algorithms for mixed-level circuit and device simulation. IEEE Trans. Comput. Aided Des. Integr. Circuits Syst. **11**(8), 1003–1012 (1992)
31. Mehrmann, V., Morandin, R., Olmi, S., Schöll, E.: Qualitative stability and synchronicity analysis of power network models in port-hamiltonian form. Chaos: Interdisc. J. Nonlinear Sci. **28**(10), 101,102 (2018). https://doi.org/10.1063/1.5054850
32. Munteanu, I.: Tree-cotree condensation properties. ICS Newslett. **9**, 10–14 (2002). http://www.compumag.org/jsite/images/stories/newsletter/ICS-02-09-1-Munteanu.pdf
33. Nagel, L.W.: SPICE2: a computer program to simulate semiconductor circuits. Tech. rep., University of Berkeley (1975)
34. Ortega, J.M., Rheinboldt, W.C.: Iterative solution of nonlinear equations in several variables, vol. 30. SIAM, Philadelphia (1970)
35. Potter, P.G., Cambrell, G.K.: A combined finite element and loop analysis for nonlinearly interacting magnetic fields and circuits. IEEE Trans. Magn. **19**(6), 2352–2355 (1983)

36. Rollins, J.G., Choma, J.: Mixed-mode PISCES-SPICE coupled circuit and device solver. IEEE Trans. Comput. Aided Des. Integr. Circuits Syst. **7**(8), 862–867 (1988)
37. Ruehli, A.E., Antonini, G., Jiang, L.: Circuit Oriented Electromagnetic Modeling Using the PEEC Techniques, 1st edn. Wiley, London (2017)
38. Schöps, S., De Gersem, H., Weiland, T.: Winding functions in transient magnetoquasistatic field-circuit coupled simulations. COMPEL **32**(6), 2063–2083 (2013). https://doi.org/10.1108/COMPEL-01-2013-0004
39. Schuhmann, R., Weiland, T.: Conservation of discrete energy and related laws in the finite integration technique. Prog. Electromagn. Res. **32**, 301–316 (2001). https://doi.org/10.2528/PIER00080112
40. Sheu, B.J., Scharfetter, D.L., Ko, P.K., M.-C., J.: BSIM: Berkeley short-channel IGFET model for MOS transistors. IEEE J. Solid-State Circuits **22**(4), 558–566 (1987)
41. Strohm, C.: Circuit simulation including full-wave Maxwell's equations: Modeling aspeects and numerical analysis. Ph.D. thesis, Humboldt-Universiät zu Berlin, Berlin, Germany (2020). Submitted
42. Tsukerman, I.A.: Finite element differential-algebraic systems for eddy current problems. Numer. Algorithms **31**(1), 319–335 (2002)
43. Verweij, A.P.: Electrodynamics of superconducting cables in accelerator magnets. Ph.D. thesis, Universiteit Twente, Twente, The Netherlands (1995)
44. Weiland, T.: A discretization method for the solution of Maxwell's equations for six-component fields. AEÜ **31**, 116–120 (1977)
45. Zeidler, E.: Nonlinear Functional Analysis and Its Applications: II/B: Nonlinear Monotone Operators. Springer, Berlin (2013)

Singularities of the Robotic Arm DAE

Diana Estévez Schwarz, René Lamour, and Roswitha März

Abstract One of the benchmarks for higher-index DAEs is the so-called robotic arm, which results from a tracking problem in robotics. Testing this benchmark, we became aware of the singularities that appear and started to analyze them thoroughly. To our knowledge, there is no comprehensive description of the singularities appearing in this example in the DAE literature so far. For our analysis, we use different methodologies, which are elaborated in this article. This detailed inspection results from two different index concepts, namely the projector based analysis of the derivative array and the direct projector based DAE analysis associated with the tractability index. As a result, with both approaches we identify the same kinds of singularities. Some of them are obvious, but others are unexpected.

Keywords Robotic arm problem · Singularity · Differential-algebraic equation · DAE · Projector based analysis · Derivative array · Differentiation index · Tractability index

Mathematics Subject Classification (2010) 34A09, 34C05, 65L05, 65L80, 93C85

R. Lamour (✉) · R. März
Humboldt University of Berlin, Department of Mathematics, Berlin, Germany
e-mail: lamour@math.hu-berlin.de; maerz@math.hu-berlin.de

D. Estévez Schwarz
Beuth Hochschule für Technik, Berlin, Germany
e-mail: estevez@beuth-hochschule.de

T. Reis et al. (eds.), *Progress in Differential-Algebraic Equations II*,
Differential-Algebraic Equations Forum,
https://doi.org/10.1007/978-3-030-53905-4_14

433

1 Introduction

The diagnosis of singularities of differential-algebraic equations (DAEs) is necessary to evaluate the reliability of numerical results. Nevertheless, this aspect has not
been considered sufficiently in practice up to now. Therefore, in the last couple of
years we developed some tools that provided indications for numerical difficulties,
in particular the code InitDAE, [6, 10]. The Taylor-coefficients computed with
InitDAE can be used for an integration method described in [11]. Altogether, in
this way we obtain detailed information while integrating DAEs.

Looking for an ambitious higher-index test example, we recalled the path control
of a two-link, flexible joint, planar Robotic Arm from Campbell [2] and Campbell
and Griepentrog [3], which based on a more general model by de Luca and Isidori
[14] and De Luca [5].

To our surprise, first tests suggested the existence of various singularities. In
particular, we wondered that integrating over the interval [0, 2], InitDAE finds a
singularity at ∼1.5 and the integration stops (see Fig. 1), where in [17] a successful
integration up to 1.7 is reported. This motivated a deeper theoretical analysis of this
particular higher-index DAE.

We aim at a comprehensive description of this famous test example and analyze
the DAE in great detail from different points of view. The equations of the DAE are
presented in Sect. 2, where also a difference between two particular versions of the
Robotic Arm DAE is discussed. This difference seems to result from a sign error
and is discussed in Remark 7.1 (Sect. 7). In Sect. 3 we investigate the properties of
the DAE by direct consideration of the model equations. This direct analysis shows
also a way to represent the solution of the Robotic Arm equations and permits a first
characterization of the singularities.

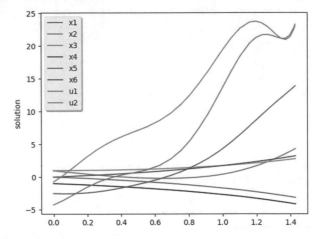

Fig. 1 Numerical solution of Robotic Arm problem using Eqs. (2.1)–(2.2) with (2.3), obtained
with the Taylor methods described in [11]

From a more general point of view, we use two approaches to analyze the DAE:

- In Sect. 4 we apply the algorithm used in InitDAE, which is based on a projector based analysis of the derivative array.
- An admissible matrix function sequence and associated admissible projector functions in the context of the direct projector based DAE-analysis are developed in Sect. 5.

Both concepts are supported by certain constant-rank conditions and, following [18, p. 137], "... singular points of DAEs will be locally defined as those for which the assumptions supporting an index notion fail." In both cases, the same singularities are indicated by corresponding rank drops. Note that singularities in the DAE context are more multifarious than those of a vector field given on manifolds. We refer to [18, Chapter 4] for a careful analysis of the state of the art and further discussions concerning singularities in the DAE context—including impasse points, bifurcations, singularity-crossing phenomena, harmless critical points and others. In [18] it is pointed out that, for instance, impasse points arise in the last reduction step of the so-called geometric index reduction in problems with positive dynamical degree. In contrast, the problem to be considered here has the dynamical degree of freedom zero.

Our numerical results are reported in Sect. 6. The paper closes with an investigation of a more general formulation for the Robotic Arm problem, in dependence of several parameters. The critical constellations for these parameters are described in Sect. 7.

2 Equations of the Robotic Arm

The problem we will consider is a semi-explicit DAE of dimension 8 with two constraints, which is constructed to model a Robotic Arm, see Fig. 2. The variables (x_1, x_2, x_3) are angular coordinates that describe the robot's configuration and

Fig. 2 Drawing of the Robotic Arm problem. (Modification of a graphic from [5])

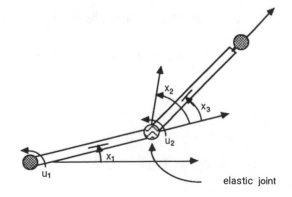

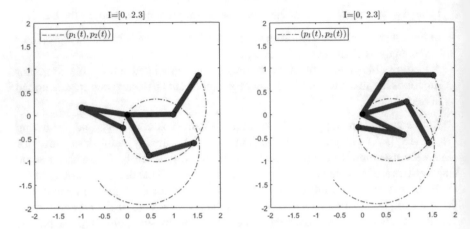

Fig. 3 Path prescribed by (2.2) for the endpoint of the outer arm. In both images, the Robotic Arm is represented at different positions for $\sin x_3 > 0$ (left) and $\sin x_3 < 0$ (right), the black colored joint is fixed at the origin and the elastic joint corresponds to the blue marker

(x_4, x_5, x_6) are their derivatives. Finally, the variables u_1 and u_2 are rotational torques and $(p_1(t), p_2(t))$ is the prescribed endpoint of the outer arm in Cartesian coordinates.

The structure of the Robotic Arm model we consider is illustrated in Fig. 2. It describes a two-link Robotic Arm with an elastic joint moving on a horizontal plane. In fact, x_1 corresponds to the rotation of the first link with respect to the base frame, x_2 to the rotation of the motor at the second joint and x_3 to the rotation of the second link with respect to the first link. Analogously to the well known simple model for the inverse kinematics in a two-joint robotic arm, there is usually more than one solution, cf. Fig. 3. The additional challenge of this example results from the model of the elastic joint. For more details we refer to [5, 14]. The description of the more general DAE in dependence of all parameters can be found in [4] and will be discussed in Sect. 7.

2.1   The Equations from the DAE-Literature

First, in Sects. 2–6, we will analyze the equations

$$x_1' - x_4 = 0,$$

$$x_2' - x_5 = 0,$$

$$x_3' - x_6 = 0,$$

$$x_4' - 2c(x_3)(x_4 + x_6)^2 - x_4^2 d(x_3) - (2x_3 - x_2)(a(x_3) + 2b(x_3))$$
$$-a(x_3)(u_1 - u_2) = 0,$$

$$x_5' + 2c(x_3)(x_4 + x_6)^2 + x_4^2 d(x_3) - (2x_3 - x_2)(1 - 3a(x_3) - 2b(x_3))$$

$$+ a(x_3)(u_1 - u_2) - u_2 = 0,$$

$$(2.1)$$

$$x_6' + 2c(x_3)(x_4 + x_6)^2 + x_4^2 d(x_3) - (2x_3 - x_2)e(x_3)$$

$$+ (a(x_3) + b(x_3))(u_1 - u_2) + d(x_3)(x_4 + x_6)^2 + 2x_4^2 c(x_3) = 0,$$

$$\cos x_1 + \cos(x_1 + x_3) - p_1(t) = 0,$$

$$\sin x_1 + \sin(x_1 + x_3) - p_2(t) = 0,$$

with given

$$p_1(t) = \cos(e^t - 1) + \cos(t - 1), \quad p_2(t) = \sin(1 - e^t) + \sin(1 - t), \quad (2.2)$$

and

$$a(z) = \frac{2}{2 - \cos^2 z}, \quad b(z) = \frac{\cos z}{2 - \cos^2 z},$$

$$c(z) = \frac{\sin z}{2 - \cos^2 z}, \quad d(z) = \frac{\cos z \sin z}{2 - \cos^2 z}.$$

For

$$e(x_3) = e_{Ca88}(x_3) = a(x_3) - 9b(x_3) \qquad (2.3)$$

the resulting DAE (2.1) coincides with that introduced by Campbell in [2] (but better available, e.g., in [3]), frequently been used to illustrate DAE procedures (e.g., [1–3, 9, 10, 17]). However, writing this article, we noticed that (2.3) does not correspond to the more general model described by Campbell and Kunkel in [4] 2019, since we could not fit the parameters described there and in de Luca and Isidori [14] and De Luca [5] to obtain e_{Ca88}.

Assuming that all masses and lengths are one and the remaining parameters are chosen in the general model as described in Remark 7.1 in Sect. 7, we obtain the DAE (2.1)–(2.2) for

$$e(x_3) = e_{DeLu87}(x_3) = -3a(x_3) + 5b(x_3). \qquad (2.4)$$

Therefore, we think that the equations with (2.3) do not correspond, in fact, to the model. A possible source for the discrepancy between (2.3) and (2.4) is a sign error described also in Remark 7.1 in Sect. 7.

For the intention of this article, this mistake complicates the setting. On the one hand, the equations with e_{Ca88} were discussed in several publications over the last decades, such that we want to compare our results with those discussed there. On the other hand, we should stop the circle of citing equations that do not correspond to the model, such that considering e_{DeLu87} is mandatory. Although this complicates slightly the overall exposition, we therefore discuss below all properties in dependence of a general function e and distinguish between e_{Ca88} and e_{DeLu87} whenever it becomes reasonable.

As will be confirmed below, both DAEs have analogous mathematical properties. In particular, their index is 5, the degree of freedom is zero. However, there are several singular points that do not coincide.

2.2 Structural Properties

In order to present a more intelligible representation of some structural properties, we reformulate the equations with new variables $x_7 := u_1 - u_2$, $x_8 = u_2$, as already done in [1], and discuss the corresponding DAE, which will be considered in the form

$$\left(\begin{pmatrix} I_6 & \\ & 0_2 \end{pmatrix} x \right)' + b(x, t) = 0. \tag{2.5}$$

For this notation, we have then

$b_1(x, t) = -x_4,$

$b_2(x, t) = -x_5,$

$b_3(x, t) = -x_6,$

$b_4(x, t) = -2c(x_3)(x_4 + x_6)^2 - x_4^2 d(x_3) - (2x_3 - x_2)(a(x_3) + 2b(x_3)) - a(x_3)x_7,$

$b_5(x, t) = +2c(x_3)(x_4 + x_6)^2 + x_4^2 d(x_3) - (2x_3 - x_2)(1 - 3a(x_3) - 2b(x_3))$

$$+ a(x_3)x_7 - x_8,$$

$b_6(x, t) = +2c(x_3)(x_4 + x_6)^2 + x_4^2 d(x_3) - (2x_3 - x_2)e(x_3)$

$$+ (a(x_3) + b(x_3))x_7 + d(x_3)(x_4 + x_6)^2 + 2x_4^2 c(x_3),$$

$b_7(x, t) = \cos x_1 + \cos(x_1 + x_3) - p_1(t),$

$b_8(x, t) = \sin x_1 + \sin(x_1 + x_3) - p_2(t),$

resulting in $x \in \mathbb{R}^8$. The function $b : \mathbb{R}^8 \times \mathbb{R} \to \mathbb{R}^8$ is continuously differentiable and the partial Jacobian matrix $b_x(x, t)$ reads

$$
b_x(x, t) = \begin{pmatrix}
0 & 0 & 0 & -1 & 0 & 0 & 0 & 0 \\
0 & 0 & 0 & 0 & -1 & 0 & 0 & 0 \\
0 & 0 & 0 & 0 & 0 & -1 & 0 & 0 \\
0 & b_{42} & b_{43} & b_{44} & 0 & b_{46} & b_{47} & 0 \\
0 & b_{52} & b_{53} & b_{54} & 0 & b_{56} & b_{57} & -1 \\
0 & b_{62} & b_{63} & b_{64} & 0 & b_{66} & b_{67} & 0 \\
b_{71} & 0 & b_{73} & 0 & 0 & 0 & 0 & 0 \\
b_{81} & 0 & b_{83} & 0 & 0 & 0 & 0 & 0
\end{pmatrix},
$$

for entries $b_{ik} = \frac{\partial b_i}{\partial x_k}$ that are again smooth functions of x. In particular, $b_{42}, b_{47}, b_{57}, b_{62}, b_{67}$ depend only on x_3, and $b_{71}, b_{73}, b_{81}, b_{83}$ depend only on x_1, x_3. We drop the arguments in the majority of cases.

Since the particular form of several coefficients b_{ik} does not matter, we present only those coefficients which will actually play a role later on:

$$
b_{42} = a(x_3) + 2b(x_3) = \frac{2 + 2\cos x_3}{2 - \cos^2 x_3},
$$

$$
b_{47} = -a(x_3) = -\frac{2}{2 - \cos^2 x_3},
$$

$$
b_{62} = e(x_3),
$$

$$
b_{67} = a(x_3) + b(x_3) = \frac{2 + \cos x_3}{2 - \cos^2 x_3},
$$

$$
b_{71} = -\sin x_1 - \sin(x_1 + x_3),
$$

$$
b_{73} = -\sin(x_1 + x_3),
$$

$$
b_{81} = \cos x_1 + \cos(x_1 + x_3),
$$

$$
b_{83} = \cos(x_1 + x_3).
$$

We also want to emphasize some special relations.

Lemma 2.1

(a) *The function b_{67} is smooth and has no zeros. It depends on x_3 only.*
(b) *The functions*

$$
p := \frac{b_{47}}{b_{67}} \quad \text{and} \quad r := \frac{1}{b_{67}}
$$

are smooth and depend on x_3 only. They have no zeros.

(c) *The matrix function*

$$M(x_3) := \begin{pmatrix} b_{42} \; b_{47} \\ b_{62} \; b_{67} \end{pmatrix} = \begin{pmatrix} a(x_3) + 2b(x_3) & -a(x_3) \\ e(x_3) & a(x_3) + b(x_3) \end{pmatrix} \tag{2.6}$$

has smooth entries depending on x_3 only. $M(x_3)$ is nonsingular precisely if

$$b_{42}(x_3) - p(x_3)b_{62}(x_3) \neq 0.$$

(d) – *If $e(x_3) = e_{Ca88}(x_3) = a(x_3) - 9b(x_3)$ according to (2.3), for $z_\star = 3 - \sqrt{5}$*
     it holds that $\cos x_3 = z_\star$ implies $b_{42}(x_3) - p(x_3)b_{62}(x_3) = 0$ and vice versa.
   – *If $e(x_3) = e_{DeLu87}(x_3) = -3a(x_3) + 5b(x_3)$ according to (2.4), for $z_\star = 2\sqrt{5} - 4$ it holds that $\cos x_3 = z_\star$ implies $b_{42}(x_3) - p(x_3)b_{62}(x_3) = 0$ and*
     vice versa.

(e) *The function*

$$S(x_3) = \frac{1}{b_{42}(x_3) - p(x_3)b_{62}(x_3)}, \qquad x_3 \in \operatorname{dom} S = \{\tau \in \mathbb{R} : \cos \tau \neq z_\star\},$$

is smooth on its definition domain, and so is

$$M^{-1} = \begin{pmatrix} S & -pS \\ -rb_{62}S \; r + rpb_{62}S \end{pmatrix}.$$

(f) *The matrix function*

$$N(x_1, x_3) := \begin{pmatrix} b_{71} \; b_{73} \\ b_{81} \; b_{83} \end{pmatrix} = \begin{pmatrix} -\sin x_1 - \sin(x_1 + x_3) & -\sin(x_1 + x_3) \\ \cos x_1 + \cos(x_1 + x_3) & \cos(x_1 + x_3) \end{pmatrix} \tag{2.7}$$

depends only on x_1 and x_3. $N(x_1, x_3)$ is nonsingular precisely if
$\det N(x_1, x_3) = \sin x_3 \neq 0$.

Proof Assertion (d):

– By definition, for $e = e_{Ca88} = a - 9b$ one has

$$b_{42}(x_3) - p(x_3)b_{62}(x_3) = \frac{2}{2 - \cos^2 x_3} \frac{4 + \cos^2 x_3 - 6\cos x_3}{2 + \cos x_3}.$$

This expression becomes zero exactly if $4 + \cos^2 x_3 - 6\cos x_3 = 0$. Next, $z_\star = 3 - \sqrt{5}$ is the only zero of the polynomial $z^2 - 6z + 4$ that belongs to the interval $[-1, 1]$. This proves the assertion for (2.3).

– Analogously, for $e = e_{\text{DeLu87}} = -3a + 5b$ according to (2.4), $b_{42}(x_3) - p(x_3)b_{62}(x_3)$ becomes zero iff $-4 + \cos^2 x_3 + 8 \cos x_3 = 0$. In this case, $z_\star = 2\sqrt{5} - 4$ is the only zero of the polynomial $z^2 + 8z - 4$ that belongs to the interval $[-1, 1]$.

The other assertions are now evident.   □

A corresponding generalization can be found in Lemma 7.1 from Sect. 7.

Consider the function $h : \mathbb{R}^2 \to \mathbb{R}^2$, given by

$$h_1(x_1, x_3) = \cos x_1 + \cos(x_1 + x_3),$$

$$h_2(x_1, x_3) = \sin x_1 + \sin(x_1 + x_3), \; x_1, x_3 \in \mathbb{R},$$

which is closely related to the derivative-free equations in (2.5), ($h_1 = b_7 + p_1$, $h_2 = b_8 + p_2$), and $\mathcal{N}$ from Lemma 2.1, which is at the same time the Jacobian matrix of h. Recall that $\det \mathcal{N}(x_1, x_3) = \sin x_3$.

Using addition theorems we represent

$$h_1(x_1, x_3) = \cos x_1(1 + \cos x_3) - \sin x_1 \sin x_3,$$

$$h_2(x_1, x_3) = \sin x_1(1 + \cos x_3) + \cos x_1 \sin x_3, \; x_1, x_3 \in \mathbb{R}.$$

For arbitrary $x_1, x_3 \in \mathbb{R}$ and corresponding $y_1 = h_1(x_1, x_3)$, $y_2 = h_2(x_1, x_3)$, it holds that

$$(1 + \cos x_3)^2 + \sin^2 x_3 = 2(1 + \cos x_3) = y_1^2 + y_2^2,$$

and therefore $y_1^2 + y_2^2 \le 4$.

Denote $\mathcal{C}_r = \{y \in \mathbb{R}^2 : y_1^2 + y_2^2 = r^2\}$, $\mathcal{B}_r = \{y \in \mathbb{R}^2 : y_1^2 + y_2^2 \le r^2\}$, $r \ge 0$. The function h maps $\mathbb{R}^2$ to the closed ball $\mathcal{B}_2$. The points outside $\mathcal{B}_2$ are out of reach. Since $\cos 2\pi k = 1$, $\cos(2\pi k + \pi) = -1$, $k \in \mathbb{Z}$, h maps the

lines $\left\{ \begin{pmatrix} x_1 \\ 2\pi k \end{pmatrix} \in \mathbb{R}^2 : x_1 \in \mathbb{R} \right\}$ and $\left\{ \begin{pmatrix} x_1 \\ 2\pi k + \pi \end{pmatrix} \in \mathbb{R}^2 : x_1 \in \mathbb{R} \right\}$ to $\mathcal{C}_2$ and $\mathcal{C}_0$,

respectively. Note that the Jacobian matrix $\mathcal{N}$ is singular on these lines. Each point on $\mathcal{C}_2$ can be reached so that $\text{im} \, h = \mathcal{B}_2$. Therefore, a consistent tracking path has always to remain within $\mathcal{B}_2$. The border points of $\mathcal{B}_2$, i.e., the points on $\mathcal{C}_2$, are reachable but impervious.

3   Inspection by Hand Method

In this section, we explain how the properties characterized in general terms in the following sections can be determined by an intuitive analysis for this particular DAE. However, we want to emphasize that, in general, such a direct manual analysis can only be conducted if structural information is given a priori.

One result of this analysis is that the dynamical degree of freedom is zero. This means that we cannot prescribe any initial values. In terms of the two approaches considered below, this means that the projectors commonly used to describe the inherent dynamics consist of zeros only.

The other result is that we can characterize the singular points using Lemma 2.1. Rearranging the equations (similar as in [3]) we obtain

$$
\begin{pmatrix} x_1' \\ x_3' \end{pmatrix} - \begin{pmatrix} x_4 \\ x_6 \end{pmatrix} = 0,
$$

(3.1)

$$
\begin{pmatrix} x_4' \\ x_6' \end{pmatrix} + \begin{pmatrix} -2c(x_3)(x_4 + x_6)^2 - x_4^2 d(x_3) - 2x_3(a(x_3) + 2b(x_3)) \\ 2c(x_3)(x_4 + x_6)^2 + x_4^2 d(x_3) - 2x_3(e(x_3) \ \ldots \\ +d(x_3)(x_4 + x_6)^2 + 2x_4^2 c(x_3)) \end{pmatrix}
$$
$$
+ \underbrace{\begin{pmatrix} a(x_3) + 2b(x_3) & -a(x_3) \\ e(x_3) & a(x_3) + b(x_3) \end{pmatrix}}_{\mathcal{M}(x_3)} \begin{pmatrix} x_2 \\ x_7 \end{pmatrix} = 0,
$$

(3.2)

$$
x_2' - x_5 = 0,
$$

(3.3)

$$
x_5' + 2c(x_3)(x_4 + x_6)^2 + x_4^2 d(x_3) - (2x_3 - x_2)(1 - 3a(x_3)
$$
$$
-2b(x_3)) + a(x_3)x_7 - x_8 = 0,
$$

(3.4)

$$
\cos x_1 + \cos(x_1 + x_3) - p_1(t) = 0,
$$

(3.5)

$$
\sin x_1 + \sin(x_1 + x_3) - p_2(t) = 0.
$$

(3.6)

In this rearranged form, we can recognize the following:

1. x_1 and x_3 are uniquely determined by Eqs. (3.5)–(3.6) whenever the Jacobian matrix with respect to x_1 and x_3, i.e., the matrix $\mathcal{N}$ from Lemma 2.1, is nonsingular. Therefore, if $\det \mathcal{N} = \sin x_3 \neq 0$, Eqs. (3.5) and (3.6) provide

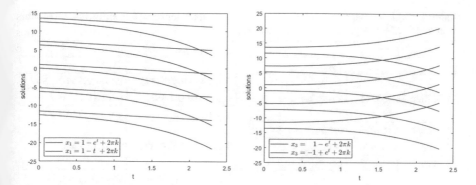

Fig. 4 Different solutions x_1 and x_3 for $k = -2, \ldots, 2$

expressions for x_1, x_3. For the particular choice of p_1, p_2, cf. (2.2), we obtain the solutions

$$\begin{pmatrix} x_1(t) \\ x_3(t) \end{pmatrix} = \begin{pmatrix} 1 - e^t + 2\pi k_1 \\ e^t - t + 2\pi k_2 \end{pmatrix} \text{ or } \begin{pmatrix} x_1(t) \\ x_3(t) \end{pmatrix} = \begin{pmatrix} 1 - t + 2\pi k_3 \\ -(e^t - t) + 2\pi k_4 \end{pmatrix}, \quad (3.7)$$

for $k_1, k_2, k_3, k_4 \in \mathbb{Z}$, which may intersect, see Fig. 4.

At points $t_\star$ with $x_3(t_\star) = k\pi, k \in \mathbb{Z}$, singularities are indicated.

2. By differentiation of Eq. (3.7) according to Eq. (3.1) we further obtain

$$\begin{pmatrix} x_1'(t) \\ x_3'(t) \end{pmatrix} = \begin{pmatrix} x_4(t) \\ x_6(t) \end{pmatrix} = \begin{pmatrix} -e^t \\ e^t - 1 \end{pmatrix} \text{ or } \begin{pmatrix} x_1'(t) \\ x_3'(t) \end{pmatrix} = \begin{pmatrix} x_4(t) \\ x_6(t) \end{pmatrix} = \begin{pmatrix} -1 \\ -e^t + 1 \end{pmatrix}.$$

3. Differentiating one of this latter expression and inserting it into Eq. (3.2) provide expressions for (x_2, x_7) everywhere where the matrix $\mathcal{M}$ is nonsingular, cf. Lemma 2.1. At points $t_\star$ with $\cos x_3(t_\star) = z_\star$ singularities are indicated, too.
4. Differentiating the expressions obtained for x_2, Eq. (3.3) provides expressions for x_5.
5. A final differentiation of x_5 provides, with (3.4), an expression for x_8.

The number of differentiations indicates that the classical differentiation index is 5.

We call a point $x_\star \in \mathbb{R}^8$ *regular* if both matrices $\mathcal{M}$ and $\mathcal{N}$ are nonsingular, which means that $\sin x_{\star,3} \neq 0$ and $\cos x_{\star,3} \neq z_\star$ and, otherwise, $x_\star$ is said to be a *singular* point.

At regular points there are explicit representations of the solutions of (2.1). One of these solutions was also described in [1], but without mentioning the possible singularities.

Using Mathematica, we succeeded in computing explicit formulas for these explicit representations, although they turned out to be rather extensive. In fact, we

Table 1 Critical timepoints $t_\star$ at which singularities appear for $x_1 = 1 - e^t$, $x_3 = e^t - t$ in the interval $[0, 2.3]$ in dependence of e_{Ca88} and e_{DeLu87}

$e = e_{Ca88}$		$e = e_{DeLu87}$	
$t_\star$	Type	$t_\star$	Type
		0.372999460773573	$\cos x_3(t_\star) = 2\sqrt{5} - 4$
1.544626000035211	$\sin x_3(t_\star) = 0$	1.544626000035211	$\sin x_3(t_\star) = 0$
		1.970554820677743	$\cos x_3(t_\star) = 2\sqrt{5} - 4$
2.029650268169820	$\cos x_3(t_\star) = 3 - \sqrt{5}$		
2.129773196104886	$\sin x_3(t_\star) = 0$	2.129773196104886	$\sin x_3(t_\star) = 0$
2.219666830742805	$\cos x_3(t_\star) = 3 - \sqrt{5}$		
		2.264553226360753	$\cos x_3(t_\star) = 2\sqrt{5} - 4$

Fig. 5 Graph of $\sin x_3(t)$

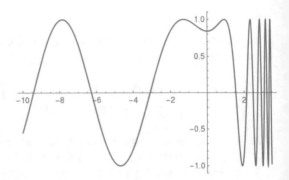

fixed $x_1 = 1 - e^t$, $x_3 = e^t - t$ and computed the remaining components accordingly. For this solution, we listed the timepoints $t_\star$ at which singularities appear in Table 1, that can also be appreciated in Figs. 5 and 6.

A Python code to evaluate the solutions using these formulas for $e = e_{Ca88}$ and $e = e_{DeLu87}$ is available online in [12]. The obtained results are represented in Figs. 7 and 8. The singularities of type $\sin x_3 = 0$ are not really visible, e.g., in Fig. 7 (right), because they do not lead to a pole in any component. This explains why the integration was successful up to 1.7 in [17]. But these singularities influence the numerical computation because of the large condition number as observed in Fig. 1. This will be explained in more detail in Sect. 6.

Now, it is easy to see that $\sin x_3 = 0$ may lead to two kinds of singularities:

- $x_3 = 2\pi k$, $k \in \mathbb{Z}$, which corresponds to the completely extended arm with the tip at a point of C_2. In this case, $\cos x_3 = 1$ and, hence, also x_1 is locally determined. However, in Fig. 4 we can guess that different solutions for x_1 and x_3 intersect. This means that, if no sufficient smoothness is assumed for the solution, it can switch from a solution with $\sin x_3 > 0$ to a solution with $\sin x_3 < 0$ or vice versa, cf. Fig. 3.
- $x_3 = \pi + 2\pi k$, $k \in \mathbb{Z}$, which corresponds to the completely folded arm with the tip at the origin, i.e., at C_0. In this case, $\cos x_3 = -1$ and x_1 is not determined

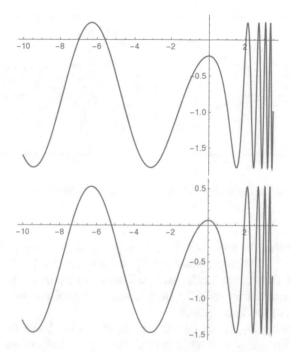

Fig. 6 Graphs of $\cos x_3(t) - z_\star$ for $z_\star = 3 - \sqrt{5}$ (top) and $z_\star = 2\sqrt{5} - 4$ (bottom), cf. Lemma 2.1

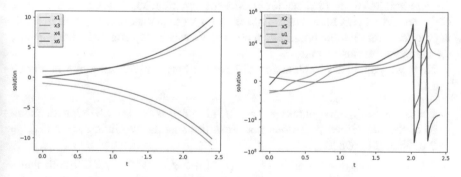

Fig. 7 Solution obtained by Estévez Schwarz and Lamour [12] for $e = e_{\text{Ca88}}$ with $x_1 = 1 - e^t$, $x_3 = e^t - t$

by the constraints. Indeed, the first column from the matrix $\mathcal{N}$ from Lemma 2.1 consists of zeros only. Moreover, in this case we cannot jump from a solution with $\sin x_3 < 0$ to a solution with $\sin x_3 > 0$ or vice versa in practice, because this would imply a discontinuity in x_1, cf. Fig. 4.

We emphasize that the singularities $\sin x_3 = 0$ arise from the constraints equation of the model.

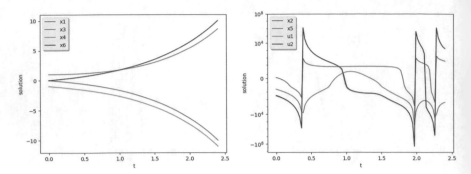

Fig. 8 Solution obtained by Estévez Schwarz and Lamour [12] for $e = e_{\text{DeLu87}}$ with $x_1 = 1 - e^t$, $x_3 = e^t - t$

Having said that the singularities $\cos x_3 = z_\star$ result from the considered model of the elastic joint, they correspond to a tip at the circle $C_{r^\star}$ for $r^\star = \sqrt{2(1 + z_\star)}$. Although x_1, x_3, x_4, x_6 are locally well defined at such points, the change of sign of the determinant of the matrix $\mathcal{M}$ from Lemma 2.1 implies a pole with change of sign for x_2, u_1, u_2, cf. Figs. 7 and 8.

In the following, C_0, $C_{r^\star}$, C_2 denote the *singularity circles* of the Robotic Arm. Obviously, a singularity appears if the trajectory intersects or touches one of the singularity circles.

Due to the singularities arising from the elastic joint, the inner workspace limit is further restricted to the disk inside $C_{r^\star}$. Moreover, a trajectory towards the proximity of $C_{r^\star}$ requires a very high rotation of the motor x_2 such that it may be unachievable in practice. From a more pragmatical point of view, we may conclude that reaching $C_{r^\star}$ damages the elastic joint.

Now we can better explain the differences between the solutions in Figs. 7 and 8:

- For $e = e_{\text{Ca88}}$, we obtain $r^\star = \sqrt{2(1 + 3 - \sqrt{5})} = 1.8783$ and, therefore, the prescribed path does not cross the singularity circle $C_{r^\star}$ until $t_\star = 2.029650268169820$.
- For $e = e_{\text{DeLu87}}$, we obtain $r^\star = \sqrt{2(1 + 2\sqrt{5} - 4)} = 1.7159$ and, therefore, the prescribed path crosses the singularity circle $C_{r^\star}$ already at $t_\star = 0.3729994607735725$.

The existence of the singularities of the type $\cos x_3 = z_\star$ was probably not noticed earlier because of this difference, since, for $e = e_{\text{Ca88}}$, the numerical behavior near $C_{r^\star}$ could easily be confounded with the proximity to C_2 at first glance (Fig. 9). We focus now on the mathematical tools and concepts we used to discover these singularities.

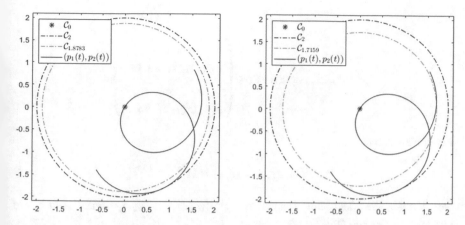

Fig. 9 Path prescribed by (2.2) and singularity circles for $e = e_{Ca88}$ (left) and $e = e_{DeLu87}$ (right). Since the radius of C_{r^*} is smaller for $e = e_{DeLu87}$, singularities appear earlier along the solution, cf. Table 1 and Figs. 7 and 8

4 Projector Based Derivative Array Procedures

In this section, we show how the steps from Sect. 3 are described in terms of the approach presented in [7] and [8].

The Robotic Arm problem is a semi-explicit DAE

$$f(x', x, t) := f((Px)', x, t) = (Px)' + b(x(t), t) = 0,$$

with the constant projectors P and, for later use, $Q := I - P$

$$P = \begin{pmatrix} I_6 & \\ & 0_2 \end{pmatrix}, \quad Q = \begin{pmatrix} 0_6 & \\ & I_2 \end{pmatrix}.$$

This means that $x_1, \ldots, x_6$ (the Px-component) is the differentiated component, while x_7 and x_8 (or, in the original formulation, u_1 and u_2) (the Qx-component) is the undifferentiated component.
For vectors $z_j \in \mathbb{R}^8$, $j = 0, \ldots, k$, we define

$$F_j(x^{(j+1)}, x^{(j)}, \ldots, \dot{x}, x, t) := \frac{d^j}{dt^j} f(\dot{x}, x, t)$$

and

$$
g^{[k]}(z_0, z_1, \ldots, z_k, t) := \begin{pmatrix} F_0(z_1, z_0, t) \\ F_1(z_2, z_1, z_0, t) \\ \vdots \\ F_{k-1}(z_k, \ldots, z_0, t) \end{pmatrix}.
$$

Furthermore, by

$$
G^{[k]}(z_0, z_1, \ldots, z_k, t) \in \mathbb{R}^{8k \times 8(k+1)}
$$

we denote the Jacobian matrix of $g^{[k]}(z_0, z_1, \ldots, z_k, t)$ with respect to $(z_0, z_1, \ldots, z_k)$ and split it into

$$
G^{[k]} = \left(G_L^{[k]} \ G_R^{[k]} \right),
$$

$G_L^{[k]} \in \mathbb{R}^{8 \cdot k \times 8}$, $G_R^{[k]} \in \mathbb{R}^{8 \cdot k \times 8 \cdot k}$ (note that L and R stand for left-hand side and right-hand side, respectively).

Let us now consider the matrices

$$
\mathcal{B}^{[k]} := \begin{pmatrix} P & 0 \\ G_L^{[k]} & G_R^{[k]} \end{pmatrix} \in \mathbb{R}^{8(k+1) \times 8(k+1)}. \tag{4.1}
$$

According to [7], to determine the index we check whether the matrices $\mathcal{B}^{[k]}$ are 1-full with respect to the first 8 columns for $k = 1, 2, \ldots$, i.e., whether

$$
\ker \mathcal{B}^{[k]} \subseteq \left\{ \begin{pmatrix} s_0 \\ s_1 \end{pmatrix} : s_0 \in \mathbb{R}^8, \ s_0 = 0, \ s_1 \in \mathbb{R}^{8k} \right\}. \tag{4.2}
$$

If $k = \mu$ is the smallest integer for which $\mathcal{B}^{[k]}$ is 1-full, then the index is μ.

For the Robotic Arm equations, $k = 1, 2, 3, 4$ do not lead to the required 1-fullness. We illustrate the 1-fullness of $\mathcal{B}^{[5]}$ by the patterns of a transformation into a block diagonal form, see Fig. 10. The orange dots represent 1, the blue dots -1, and the brown ones other nonzero elements. For the transformation we use rows with one nonzero entry only. The used rows are marked by small ellipses and arrows. In this procedure we have to exclude the singularities of the Jacobian matrix $\mathcal{N}$ from (2.7) first and later the singularities of the Jacobian matrix $\mathcal{M}$ from (2.6), too.

In order to characterize the different components of the solution we further analyze the matrices $G^{[k]}$ for $k = 1, 2, 3, 4, 5$.

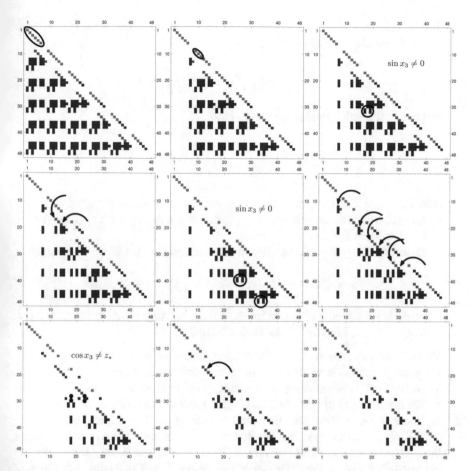

Fig. 10 Illustration of the 1-fullness of (4.1) for Eqs. (2.5) and $k = 5$. Therefore, the index results to be 5

- To decouple the undifferentiated component Q, for each k we consider a basis $W_R^{[k]}$ along im $G_R^{[k]}$ and define T_k as the orthogonal projector onto

$$\ker \begin{pmatrix} P \\ W_R^{[k]} G_L^{[k]} \end{pmatrix} =: \text{im } T_k.$$

Consequently, $T_k x$ corresponds to the part of the undifferentiated component Qx that cannot be represented as a function of (Px, t) after k-1 differentiations. Note that, by definition, $T_\mu = 0$, cf. [7].

If we define further $U_k := Q - T_k$, then we obtain the following decoupling for the Q-component:

$$Qx = QU_1 x + T_1 U_2 x + \cdots T_{\mu-2} U_{\mu-1} x + T_{\mu-1} x.$$

– To characterize the different parts of the differentiated component Px, in each step k we consider a basis $W^{[k]}_{LQ-R}$ along

$$\text{im} \left(G^{[k]}_L Q \quad G^{[k]}_R \right)$$

and define the orthogonal projector V_k onto

$$\ker \left(W^{[k]}_{LQ-R} G^{[k]}_L \right) =: \text{im } V_k.$$

then $V_k x$ represents the part of the differentiated components Px not determined by the constraints that result after $k-1$ differentiations. By definition, the degree of freedom is rank $V_{\mu-1}$.

Defining $Z_k := P - V_k$, we also obtain a decoupling for the P-component:

$$Px = PZ_1 x + V_1 Z_2 x + \cdots V_{\mu-2} Z_{\mu-1} x + V_{\mu-1} x.$$

We summarize the results obtained for the considered Robotic Arm DAE for $x_3 \neq k\pi$, $\cos x_3 \neq z_\star$ (cf. Lemma 2.1) in Tables 2 and 3:

– Table 2 corresponds to the reformulated equations with $(x_1, \ldots, x_6, x_7, x_8)$. Here, all projectors have diagonal form with only ones or zeros in the diagonal. Therefore, the different components correspond to particular rows of the vector x. The obtained splitting corresponds to the representations deduced in Sect. 3.
– In Table 3 we present the structural properties of the original formulation with $(x_1, \ldots, x_6, u_1, u_2)$. For the first steps, we obtain identical results as for the reformulated equations. Hence, in Table 3, we present only the projectors obtained for $k = 3, 4$. There we can recognize that T_3 and T_4 do not have diagonal form, since we cannot assign this higher-index-property to a particular row, i.e., to either u_1 nor u_2. Indeed, the description of $u_1 + u_2$ by the corresponding projector results to be adequate.

Since the diagnosis procedures of InitDAE are conceived for general DAEs, consistent initial values and the corresponding projectors can be computed for both formulations at regular points. We further observe that, for singular timepoints InitDAE cannot solve the minimization problem, i.e., no consistent values can be computed. Summarizing, the computed differentiation index and the detection of singularities correlates,[1] independent of the chosen variables. This is a crucial difference to the structural index, where the introduction of the variable $x_7 = u_1 - u_2$ is essential for the correct index determination, cf. [17].

---

[1]Recall that the definition of a singularity in [7] precisely bases on the successful computation of consistent initial values.

Table 2 Projectors associated with the derivative array analysis for the reformulated Eqs. (2.5)

$(x_1, x_2, x_3, x_4, x_5, x_6, x_7, x_8)$

A	$P = \begin{pmatrix} 1 & & & & & & & \\ & 1 & & & & & & \\ & & 1 & & & & & \\ & & & 1 & & & & \\ & & & & 1 & & & \\ & & & & & 1 & & \\ & & & & & & 0 & \\ & & & & & & & 0 \end{pmatrix}$, $Q = \begin{pmatrix} 0 & & & & & & & \\ & 0 & & & & & & \\ & & 0 & & & & & \\ & & & 0 & & & & \\ & & & & 0 & & & \\ & & & & & 0 & & \\ & & & & & & 1 & \\ & & & & & & & 1 \end{pmatrix}$

$G^{[1]}$	$V_1 = \begin{pmatrix} 0 & & & & & & & \\ & 1 & & & & & & \\ & & 0 & & & & & \\ & & & 1 & & & & \\ & & & & 1 & & & \\ & & & & & 1 & & \\ & & & & & & 0 & \\ & & & & & & & 0 \end{pmatrix}$, $T_1 = \begin{pmatrix} 0 & & & & & & & \\ & 0 & & & & & & \\ & & 0 & & & & & \\ & & & 0 & & & & \\ & & & & 0 & & & \\ & & & & & 0 & & \\ & & & & & & 1 & \\ & & & & & & & 1 \end{pmatrix}$

$G^{[2]}$	$V_2 = \begin{pmatrix} 0 & & & & & & & \\ & 1 & & & & & & \\ & & 0 & & & & & \\ & & & 0 & & & & \\ & & & & 1 & & & \\ & & & & & 0 & & \\ & & & & & & 0 & \\ & & & & & & & 0 \end{pmatrix}$, $T_2 = \begin{pmatrix} 0 & & & & & & & \\ & 0 & & & & & & \\ & & 0 & & & & & \\ & & & 0 & & & & \\ & & & & 0 & & & \\ & & & & & 0 & & \\ & & & & & & 1 & \\ & & & & & & & 1 \end{pmatrix}$

$G^{[3]}$	$V_3 = \begin{pmatrix} 0 & & & & & & & \\ & 0 & & & & & & \\ & & 0 & & & & & \\ & & & 0 & & & & \\ & & & & 1 & & & \\ & & & & & 0 & & \\ & & & & & & 0 & \\ & & & & & & & 0 \end{pmatrix}$, $T_3 = \begin{pmatrix} 0 & & & & & & & \\ & 0 & & & & & & \\ & & 0 & & & & & \\ & & & 0 & & & & \\ & & & & 0 & & & \\ & & & & & 0 & & \\ & & & & & & 0 & \\ & & & & & & & 1 \end{pmatrix}$

(continued)

Table 2 (continued)

	$(x_1, x_2, x_3, x_4, x_5, x_6, x_7, x_8)$
$G^{[4]}$	$V_4 = \begin{pmatrix} 0 & & & & & & & \\ & 0 & & & & & & \\ & & 0 & & & & & \\ & & & 0 & & & & \\ & & & & 0 & & & \\ & & & & & 0 & & \\ & & & & & & 0 & \\ & & & & & & & 0 \end{pmatrix}, \ T_4 = \begin{pmatrix} 0 & & & & & & & \\ & 0 & & & & & & \\ & & 0 & & & & & \\ & & & 0 & & & & \\ & & & & 0 & & & \\ & & & & & 0 & & \\ & & & & & & 0 & \\ & & & & & & & 1 \end{pmatrix}$
$G^{[5]}$	$V_5 = \begin{pmatrix} 0 & & & & & & & \\ & 0 & & & & & & \\ & & 0 & & & & & \\ & & & 0 & & & & \\ & & & & 0 & & & \\ & & & & & 0 & & \\ & & & & & & 0 & \\ & & & & & & & 0 \end{pmatrix}, \ T_5 = \begin{pmatrix} 0 & & & & & & & \\ & 0 & & & & & & \\ & & 0 & & & & & \\ & & & 0 & & & & \\ & & & & 0 & & & \\ & & & & & 0 & & \\ & & & & & & 0 & \\ & & & & & & & 0 \end{pmatrix}$

Table 3 Projectors associated with the derivative array analysis: differences to Table 2 when using the original formulation (2.1) independent of the term e

	$(x_1, x_2, x_3, x_4, x_5, x_6, u_1, u_2)$
$G^{[3]}$	$V_3 = \begin{pmatrix} 0 & & & & & & & \\ & 0 & & & & & & \\ & & 0 & & & & & \\ & & & 0 & & & & \\ & & & & 1 & & & \\ & & & & & 0 & & \\ & & & & & & 0 & \\ & & & & & & & 0 \end{pmatrix}, \ T_3 = \begin{pmatrix} 0 & & & & & & & \\ & 0 & & & & & & \\ & & 0 & & & & & \\ & & & 0 & & & & \\ & & & & 0 & & & \\ & & & & & 0 & & \\ & & & & & & 0.5 & 0.5 \\ & & & & & & 0.5 & 0.5 \end{pmatrix}$
$G^{[4]}$	$V_4 = \begin{pmatrix} 0 & & & & & & & \\ & 0 & & & & & & \\ & & 0 & & & & & \\ & & & 0 & & & & \\ & & & & 0 & & & \\ & & & & & 0 & & \\ & & & & & & 0 & \\ & & & & & & & 0 \end{pmatrix}, \ T_4 = \begin{pmatrix} 0 & & & & & & & \\ & 0 & & & & & & \\ & & 0 & & & & & \\ & & & 0 & & & & \\ & & & & 0 & & & \\ & & & & & 0 & & \\ & & & & & & 0.5 & 0.5 \\ & & & & & & 0.5 & 0.5 \end{pmatrix}$

5   Direct Projector Based DAE Analysis and Tractability Index

Here we provide an admissible sequence of matrix functions and describe the regularity regions with their characteristic values, including the tractability index (cf. [13]). For this purpose, we rewrite the DAE in the proper form

$$A(Dx)'(t) + b(x(t), t) = 0, \tag{5.1}$$

where

$$A = \begin{pmatrix} 1\,0\,0\,0\,0\,0 \\ 0\,1\,0\,0\,0\,0 \\ 0\,0\,1\,0\,0\,0 \\ 0\,0\,0\,1\,0\,0 \\ 0\,0\,0\,0\,1\,0 \\ 0\,0\,0\,0\,0\,1 \\ 0\,0\,0\,0\,0\,0 \\ 0\,0\,0\,0\,0\,0 \end{pmatrix}, \quad D = \begin{pmatrix} 1\,0\,0\,0\,0\,0\,0\,0 \\ 0\,1\,0\,0\,0\,0\,0\,0 \\ 0\,0\,1\,0\,0\,0\,0\,0 \\ 0\,0\,0\,1\,0\,0\,0\,0 \\ 0\,0\,0\,0\,1\,0\,0\,0 \\ 0\,0\,0\,0\,0\,1\,0\,0 \end{pmatrix}, \quad D^- = A.$$

Following the projector based approach (e.g., [13]) we construct an admissible matrix function sequence to analyze the DAE. The matrix function sequence to be built pointwise for x, t comes from the given matrix functions

$$G_0 = AD, \quad B_0 = b_x, \quad P_0 = D^- D, \quad Q_0 = I - P_0, \quad \Pi_0 = P_0.$$

First we obtain the matrix function

$$G_1 = G_0 + B_0 Q_0 = \begin{pmatrix} 1\,0\,0\,0\,0\,0\ \ 0\ \ 0 \\ 0\,1\,0\,0\,0\,0\ \ 0\ \ 0 \\ 0\,0\,1\,0\,0\,0\ \ 0\ \ 0 \\ 0\,0\,0\,1\,0\,0\ b_{47}\ 0 \\ 0\,0\,0\,0\,1\,0\ b_{57}\ {-}1 \\ 0\,0\,0\,0\,0\,1\ b_{67}\ 0 \\ 0\,0\,0\,0\,0\,0\ \ 0\ \ 0 \\ 0\,0\,0\,0\,0\,0\ \ 0\ \ 0 \end{pmatrix}$$

and its nullspace

$$N_1 = \{z \in \mathbb{R}^8 : z_1 = 0, z_2 = 0, z_3 = 0, z_4 + b_{47}z_7 = 0, z_5 + b_{57}z_7z_8 = 0,$$

$$z_6 + b_{67}z_7 = 0\}.$$

Furthermore, the intersection $N_1 \cap \ker \Pi_0$ is trivial, thus there is a projector function Q_1 onto N_1 such that $\ker \Pi_0 \subseteq \ker Q_1$. It is evident that

$$Q_1 = \begin{pmatrix} 0 & 0 & 0 & 0 & 0 & 0 & 0 & 0 \\ 0 & 0 & 0 & 0 & 0 & 0 & 0 & 0 \\ 0 & 0 & 0 & 0 & 0 & 0 & 0 & 0 \\ 0 & 0 & 0 & 0 & 0 & p & 0 & 0 \\ 0 & 0 & 0 & 0 & 1 & 0 & 0 & 0 \\ 0 & 0 & 0 & 0 & 0 & 1 & 0 & 0 \\ 0 & 0 & 0 & 0 & 0 & -r & 0 & 0 \\ 0 & 0 & 0 & 0 & 1 & -rb_{57} & 0 & 0 \end{pmatrix}$$

is such a projector function. We also derive

$$\Pi_0 Q_1 = \begin{pmatrix} 0 & 0 & 0 & 0 & 0 & 0 & 0 & 0 \\ 0 & 0 & 0 & 0 & 0 & 0 & 0 & 0 \\ 0 & 0 & 0 & 0 & 0 & 0 & 0 & 0 \\ 0 & 0 & 0 & 0 & 0 & p & 0 & 0 \\ 0 & 0 & 0 & 0 & 1 & 0 & 0 & 0 \\ 0 & 0 & 0 & 0 & 0 & 1 & 0 & 0 \\ 0 & 0 & 0 & 0 & 0 & 0 & 0 & 0 \\ 0 & 0 & 0 & 0 & 0 & 0 & 0 & 0 \end{pmatrix}, \quad \Pi_1 = \Pi_0 - \Pi_0 Q_1 = \begin{pmatrix} 1 & 0 & 0 & 0 & 0 & 0 & 0 & 0 \\ 0 & 1 & 0 & 0 & 0 & 0 & 0 & 0 \\ 0 & 0 & 1 & 0 & 0 & 0 & 0 & 0 \\ 0 & 0 & 0 & 1 & 0 & -p & 0 & 0 \\ 0 & 0 & 0 & 0 & 0 & 0 & 0 & 0 \\ 0 & 0 & 0 & 0 & 0 & 0 & 0 & 0 \\ 0 & 0 & 0 & 0 & 0 & 0 & 0 & 0 \\ 0 & 0 & 0 & 0 & 0 & 0 & 0 & 0 \end{pmatrix},$$

and

$$B_1 = B_0 \Pi_0 - G_1 D^- (D\Pi_1 D^-)' D = \begin{pmatrix} 0 & 0 & 0 & -1 & 0 & 0 & 0 & 0 \\ 0 & 0 & 0 & 0 & -1 & 0 & 0 & 0 \\ 0 & 0 & 0 & 0 & 0 & -1 & 0 & 0 \\ 0 & b_{42} & b_{43} & b_{44} & 0 & b_{46} + p' & 0 & 0 \\ 0 & b_{52} & b_{53} & b_{54} & 0 & b_{56} & 0 & 0 \\ 0 & b_{62} & b_{63} & b_{64} & 0 & b_{66} & 0 & 0 \\ b_{71} & 0 & b_{73} & 0 & 0 & 0 & 0 & 0 \\ b_{81} & 0 & b_{83} & 0 & 0 & 0 & 0 & 0 \end{pmatrix},$$

where the sign prime indicates the total derivative in jet variables. In particular, p' stands for the function $p'(x_3, x_3^1) = p'(x_3)x_3^1$ of $x_3 \in \mathbb{R}$ and the jet variable $x_3^1 \in \mathbb{R}$, see e.g., [13, Section 3.2].

Next we compute the matrix function

$$G_2 = G_1 + B_1 Q_1 = \begin{pmatrix} 1 & 0 & 0 & 0 & 0 & -p & 0 & 0 \\ 0 & 1 & 0 & 0 & -1 & 0 & 0 & 0 \\ 0 & 0 & 1 & 0 & 0 & -1 & 0 & 0 \\ 0 & 0 & 0 & 1 & 0 & b_{44}p + b_{46} + p' & b_{47} & 0 \\ 0 & 0 & 0 & 0 & 1 & b_{54}p + b_{56} & b_{57} & -1 \\ 0 & 0 & 0 & 0 & 0 & 1 + b_{64}p + b_{66} & b_{67} & 0 \\ 0 & 0 & 0 & 0 & 0 & 0 & 0 & 0 \\ 0 & 0 & 0 & 0 & 0 & 0 & 0 & 0 \end{pmatrix},$$

its nullspace

$$N_2 = \{ z \in \mathbb{R}^8 : z_1 - pz_6 = 0, \ z_2 - z_5 = 0, \ z_3 - z_6 = 0,$$
$$z_4 + (b_{44}p + b_{46} + p')z_6 + b_{47}z_7 = 0,$$
$$z_5 + (b_{54}p + b_{56})z_6 + b_{57}z_7 - z_8 = 0,$$
$$(1 + b_{64}p + b_{66})z_6 + b_{67}z_7 = 0 \},$$

and the intersection

$$N_2 \cap \ker \Pi_1 = N_2 \cap \{ z \in \mathbb{R}^8 : z_1 = 0, \ z_2 = 0, \ z_3 = 0, z_4 - pz_6 = 0 \} = \{0\}.$$

With

$$Q_2 = \begin{pmatrix} 0 & 0 & p & 0 & 0 & 0 & 0 & 0 \\ 0 & 1 & 0 & 0 & 0 & 0 & 0 & 0 \\ 0 & 0 & 1 & 0 & 0 & 0 & 0 & 0 \\ 0 & 0 & p + \mathcal{A} & 0 & 0 & 0 & 0 & 0 \\ 0 & 1 & 0 & 0 & 0 & 0 & 0 & 0 \\ 0 & 0 & 1 & 0 & 0 & 0 & 0 & 0 \\ 0 & 0 & -r(1 + pb_{64} + b_{66}) & 0 & 0 & 0 & 0 & 0 \\ 0 & 1 & pb_{54} + b_{56} - rb_{57}(1 + pb_{64} + b_{66}) & 0 & 0 & 0 & 0 & 0 \end{pmatrix},$$

$$\mathcal{A} = p^2 b_{64} + p(b_{66} - b_{44}) - b_{46} - p',$$

we find an admissible projector function onto N_2 such that $\ker \Pi_1 \subseteq \ker Q_2$. Then it results that

$$
\Pi_1 Q_2 = \begin{pmatrix}
0 & 0 & p & 0 & 0 & 0 & 0 & 0 \\
0 & 1 & 0 & 0 & 0 & 0 & 0 & 0 \\
0 & 0 & 1 & 0 & 0 & 0 & 0 & 0 \\
0 & 0 & \mathcal{A} & 0 & 0 & 0 & 0 & 0 \\
0 & 0 & 0 & 0 & 0 & 0 & 0 & 0 \\
0 & 0 & 0 & 0 & 0 & 0 & 0 & 0 \\
0 & 0 & 0 & 0 & 0 & 0 & 0 & 0 \\
0 & 0 & 0 & 0 & 0 & 0 & 0 & 0
\end{pmatrix},
\quad
\Pi_2 = \Pi_1 - \Pi_1 Q_2 = \begin{pmatrix}
1 & 0 & -p & 0 & 0 & 0 & 0 & 0 \\
0 & 0 & 0 & 0 & 0 & 0 & 0 & 0 \\
0 & 0 & 0 & 0 & 0 & 0 & 0 & 0 \\
0 & 0 & -\mathcal{A} & 1 & 0 & -p & 0 & 0 \\
0 & 0 & 0 & 0 & 0 & 0 & 0 & 0 \\
0 & 0 & 0 & 0 & 0 & 0 & 0 & 0 \\
0 & 0 & 0 & 0 & 0 & 0 & 0 & 0 \\
0 & 0 & 0 & 0 & 0 & 0 & 0 & 0
\end{pmatrix},
$$

as well as

$$
B_2 = B_1 \Pi_1 - G_2 D^-(D\Pi_2 D^-)'D\Pi_1 = \begin{pmatrix}
0 & 0 & +p' & -1 & 0 & p & 0 & 0 \\
0 & 0 & 0 & 0 & 0 & 0 & 0 & 0 \\
0 & 0 & 0 & 0 & 0 & 0 & 0 & 0 \\
0 & b_{42} & b_{43}+\mathcal{A}' & b_{44} & 0 & -pb_{44} & 0 & 0 \\
0 & b_{52} & b_{53} & b_{54} & 0 & -pb_{54} & 0 & 0 \\
0 & b_{62} & b_{63} & b_{64} & 0 & -pb_{64} & 0 & 0 \\
b_{71} & 0 & b_{73} & 0 & 0 & 0 & 0 & 0 \\
b_{81} & 0 & b_{83} & 0 & 0 & 0 & 0 & 0
\end{pmatrix},
$$

$$
B_2 Q_2 = \begin{pmatrix}
0 & 0 & p'-\mathcal{A} & 0 & 0 & 0 & 0 & 0 \\
0 & 0 & 0 & 0 & 0 & 0 & 0 & 0 \\
0 & 0 & 0 & 0 & 0 & 0 & 0 & 0 \\
0 & b_{42} & b_{43}+\mathcal{A}'+b_{44}\mathcal{A} & 0 & 0 & 0 & 0 & 0 \\
0 & b_{52} & b_{53}+b_{54}\mathcal{A} & 0 & 0 & 0 & 0 & 0 \\
0 & b_{62} & b_{63}+b_{64}\mathcal{A} & 0 & 0 & 0 & 0 & 0 \\
0 & 0 & b_{73}+pb_{71} & 0 & 0 & 0 & 0 & 0 \\
0 & 0 & b_{83}+pb_{81} & 0 & 0 & 0 & 0 & 0
\end{pmatrix},
$$

$$
G_3 = G_2 + B_2 Q_2 = \begin{pmatrix}
1 & 0 & p'-\mathcal{A} & 0 & 0 & -p & 0 & 0 \\
0 & 1 & 0 & 0 & -1 & 0 & 0 & 0 \\
0 & 0 & 1 & 0 & 0 & -1 & 0 & 0 \\
0 & b_{42} & b_{43}+\mathcal{A}'+b_{44}\mathcal{A} & 1 & 0 & b_{44}p+b_{46}+p' & b_{47} & 0 \\
0 & b_{52} & b_{53}+b_{54}\mathcal{A} & 0 & 1 & b_{54}p+b_{56} & b_{57} & -1 \\
0 & b_{62} & b_{63}+b_{64}\mathcal{A} & 0 & 0 & 1+b_{64}p+b_{66} & b_{67} & 0 \\
0 & 0 & b_{73}+pb_{71} & 0 & 0 & 0 & 0 & 0 \\
0 & 0 & b_{83}+pb_{81} & 0 & 0 & 0 & 0 & 0
\end{pmatrix}.
$$

Therefore, the nullspace of G_3 is

$$N_3 = \{z \in \mathbb{R}^8 : z_3 = 0, \ z_2 - z_5 = 0, \ z_6 = 0, \ z_1 = 0, \ b_{42}z_2 + z_4 + b_{47}z_7 = 0,$$
$$b_{52}z_2 + z_5 + b_{57}z_7 - z_8 = 0, \ b_{62}z_2 + b_{67}z_7 = 0\}$$
$$= \{z \in \mathbb{R}^8 : z_1 = 0, \ z_3 = 0, \ z_6 = 0, \ z_5 = z_2, \ z_7 = -rb_{62}z_2,$$
$$z_4 = (-b_{42} + pb_{62})z_2, \ z_8 = (b_{52} + 1 - b_{57}rb_{62})z_2 \}.$$

The intersection

$$N_3 \cap \ker \Pi_2 = N_3 \cap \{z \in \mathbb{R}^8 : z_1 - pz_3 = 0, \ -\mathcal{A}z_3 + z_4 - pz_6 = 0\}$$
$$= \{z \in \mathbb{R}^8 : z_3 = 0, \ z_6 = 0, \ z_1 = 0, \ z_4 = 0, \ z_5 = z_2,$$
$$z_8 = (1 + b_{52})z_2 + b_{57}z_7, b_{42}z_2 + b_{47}z_7 = 0, \ b_{62}z_2 + b_{67}z_7 = 0\}$$

is trivial, precisely where the matrix $\mathcal{M}$ (see Lemma 2.1) is nonsingular, that means,

$$N_3(x) \cap \ker \Pi_2(x) = \{0\} \Leftrightarrow \cos x_3 \neq z_\star.$$

The hyperplanes in $\mathbb{R}^8$ described by $\cos x_3 = z_\star$ indicate critical points of the DAE. Denote the set of critical points arising at this level (cf. [13, Definition 2.75], also [18, Def. 42] for linear DAEs) by

$$S_{crit}^{3-B} = \{x \in \mathbb{R}^8 : \cos x_3 = z_\star\}. \tag{5.2}$$

The function S (see Lemma 2.1) will play its role when constructing the next projector function Q_3 onto N_3 such that $\ker \Pi_2 \subseteq \ker Q_3$ for arguments outside the critical point set. We observe that there

$$N_3 = \mathrm{im} \begin{pmatrix} 0 \\ 1 \\ 0 \\ -b_{42} + pb_{62} \\ 1 \\ 0 \\ -rb_{62} \\ B \end{pmatrix} = \mathrm{im} \begin{pmatrix} 0 \\ -S \\ 0 \\ 1 \\ -S \\ 0 \\ Srb_{62} \\ -SB \end{pmatrix}, \quad B = 1 + b_{52} - rb_{57}b_{62},$$

leading to

$$
Q_3 = \begin{pmatrix}
0 & 0\,0 & 0 & 0 & 0 & 0\,0 \\
\mathcal{S}\frac{1}{p}\mathcal{A} & 0\,0 & -\mathcal{S} & 0 & p\mathcal{S} & 0\,0 \\
0 & 0\,0 & 0 & 0 & 0 & 0\,0 \\
-\frac{1}{p}\mathcal{A} & 0\,0 & 1 & 0 & -p & 0\,0 \\
\mathcal{S}\frac{1}{p}\mathcal{A} & 0\,0 & -\mathcal{S} & 0 & p\mathcal{S} & 0\,0 \\
0 & 0\,0 & 0 & 0 & 0 & 0\,0 \\
-rb_{62}\mathcal{S}\frac{1}{p}\mathcal{A} & 0\,0 & rb_{62}\mathcal{S} & 0 & -rb_{62}p\mathcal{S} & 0\,0 \\
\mathcal{S}\frac{1}{p}\mathcal{A}\mathcal{B} & 0\,0 & -\mathcal{S}\mathcal{B} & 0 & p\mathcal{S}\mathcal{B} & 0\,0
\end{pmatrix},
$$

$$
\Pi_2 Q_3 = \begin{pmatrix}
0 & 0\,0\,0\,0 & 0 & 0\,0 \\
0 & 0\,0\,0\,0 & 0 & 0\,0 \\
0 & 0\,0\,0\,0 & 0 & 0\,0 \\
-\frac{1}{p}\mathcal{A} & 0\,0\,1\,0 & -p & 0\,0 \\
0 & 0\,0\,0\,0 & 0 & 0\,0 \\
0 & 0\,0\,0\,0 & 0 & 0\,0 \\
0 & 0\,0\,0\,0 & 0 & 0\,0 \\
0 & 0\,0\,0\,0 & 0 & 0\,0
\end{pmatrix}, \quad
\Pi_3 = \Pi_2 - \Pi_2 Q_3 = \begin{pmatrix}
1 & 0 & -p & 0\,0\,0\,0\,0 \\
0 & 0 & 0 & 0\,0\,0\,0\,0 \\
0 & 0 & 0 & 0\,0\,0\,0\,0 \\
\frac{1}{p}\mathcal{A} & 0 & -\mathcal{A} & 0\,0\,0\,0\,0 \\
0 & 0 & 0 & 0\,0\,0\,0\,0 \\
0 & 0 & 0 & 0\,0\,0\,0\,0 \\
0 & 0 & 0 & 0\,0\,0\,0\,0 \\
0 & 0 & 0 & 0\,0\,0\,0\,0
\end{pmatrix},
$$

and

$$
B_3 = B_2\Pi_2 - G_3 D^-(D\Pi_3 D^-)'D\Pi_2
$$

$$
= \begin{pmatrix}
0 & 0 & \mathcal{A} & -1 & 0 & p & 0\,0 \\
0 & 0 & 0 & 0 & 0 & 0 & 0\,0 \\
0 & 0 & 0 & 0 & 0 & 0 & 0\,0 \\
0 & 0 & -b_{44}\mathcal{A} & b_{44} & 0 & -pb_{44} & 0\,0 \\
0 & 0 & -b_{54}\mathcal{A} & b_{54} & 0 & -pb_{54} & 0\,0 \\
0 & 0 & -b_{64}\mathcal{A} & b_{64} & 0 & -pb_{64} & 0\,0 \\
b_{71} & 0 & -pb_{71} & 0 & 0 & 0 & 0\,0 \\
b_{81} & 0 & -pb_{81} & 0 & 0 & 0 & 0\,0
\end{pmatrix}
- \begin{pmatrix}
0 & 0 & 0 & 0\,0\,0\,0\,0 \\
0 & 0 & 0 & 0\,0\,0\,0\,0 \\
0 & 0 & 0 & 0\,0\,0\,0\,0 \\
(\frac{1}{p}\mathcal{A})' & 0 & -p(\frac{1}{p}\mathcal{A})' & 0\,0\,0\,0\,0 \\
0 & 0 & 0 & 0\,0\,0\,0\,0 \\
0 & 0 & 0 & 0\,0\,0\,0\,0 \\
0 & 0 & 0 & 0\,0\,0\,0\,0 \\
0 & 0 & 0 & 0\,0\,0\,0\,0
\end{pmatrix},
$$

$$= \begin{pmatrix}
0 & 0 & \mathcal{A} & -1\,0 & p & 0\,0 \\
0 & 0 & 0 & 0\,0 & 0 & 0\,0 \\
0 & 0 & 0 & 0\,0 & 0 & 0\,0 \\
-(\frac{1}{p}\mathcal{A})' & 0 & -b_{44}\mathcal{A}+p(\frac{1}{p}\mathcal{A})' & b_{44}\,0 & -pb_{44} & 0\,0 \\
0 & 0 & -b_{54}\mathcal{A} & b_{54}\,0 & -pb_{54} & 0\,0 \\
0 & 0 & -b_{64}\mathcal{A} & b_{64}\,0 & -pb_{64} & 0\,0 \\
b_{71} & 0 & -pb_{71} & 0\,0 & 0 & 0\,0 \\
b_{81} & 0 & -pb_{81} & 0\,0 & 0 & 0\,0
\end{pmatrix}.$$

Next we obtain

$$B_3 Q_3 = \begin{pmatrix}
\frac{1}{p}\mathcal{A} & 0\,0 & -1\,0 & p & 0\,0 \\
0 & 0\,0 & 0\,0 & 0 & 0\,0 \\
0 & 0\,0 & 0\,0 & 0 & 0\,0 \\
-b_{44}\frac{1}{p}\mathcal{A} & 0\,0 & b_{44}\,0 & -pb_{44} & 0\,0 \\
-b_{54}\frac{1}{p}\mathcal{A} & 0\,0 & b_{54}\,0 & -pb_{54} & 0\,0 \\
-b_{64}\frac{1}{p}\mathcal{A} & 0\,0 & b_{64}\,0 & -pb_{64} & 0\,0 \\
0 & 0\,0 & 0\,0 & 0 & 0\,0 \\
0 & 0\,0 & 0\,0 & 0 & 0\,0
\end{pmatrix}$$

and

$$G_4 = G_3 + B_3 Q_3 = \begin{pmatrix}
1+\frac{1}{p}\mathcal{A} & 0 & p'-\mathcal{A} & -1 & 0 & 0 & 0 & 0 \\
0 & 1 & 0 & 0 & -1 & 0 & 0 & 0 \\
0 & 0 & 1 & 0 & 0 & -1 & 0 & 0 \\
-b_{44}\frac{1}{p}\mathcal{A} & b_{42} & b_{43}+\mathcal{A}'+b_{44}\mathcal{A} & 1+b_{44} & 0 & b_{46}+p' & b_{47} & 0 \\
-b_{54}\frac{1}{p}\mathcal{A} & b_{52} & b_{53}+b_{54}\mathcal{A} & b_{54} & 1 & b_{56} & b_{57} & -1 \\
-b_{64}\frac{1}{p}\mathcal{A} & b_{62} & b_{63}+b_{64}\mathcal{A} & b_{64} & 0 & 1+b_{66} & b_{67} & 0 \\
0 & 0 & b_{73}+pb_{71} & 0 & 0 & 0 & 0 & 0 \\
0 & 0 & b_{83}+pb_{81} & 0 & 0 & 0 & 0 & 0
\end{pmatrix},$$

as well as the nullspace

$$N_4 = \{z \in \mathbb{R}^8 : z_3 = 0, z_6 = 0, (1+\frac{\mathcal{A}}{p})z_1 - z_4 = 0, z_2 - z_5 = 0,$$

$$-b_{44}\frac{\mathcal{A}}{p}z_1 + b_{42}z_2 + (1+b_{44})z_4 + b_{47}z_7 = 0,$$

$$-b_{54}\frac{\mathcal{A}}{p}z_1 + b_{52}z_2 + b_{54}z_4 + z_5 + b_{57}z_7 - z_8 = 0,$$

$$-b_{64}\frac{\mathcal{A}}{p}z_1 + b_{62}z_2 + b_{64}z_4 + b_{67}z_7 = 0\}.$$

The intersection

$$N_4 \cap \ker \Pi_3 = \{z \in \mathbb{R}^8 : z_1 = 0, z_3 = 0, z_6 = 0, z_4 = 0, z_2 = z_5,$$

$$b_{42}z_2 + b_{47}z_7 = 0, b_{62}z_2 + b_{67}z_7 = 0, (b_{52} + 1)z_2 + b_{57}z_7 - z_8 = 0\}$$

becomes trivial exactly where the matrix $\mathcal{M}$ (see Lemma 2.1) is nonsingular. Thus, rank $G_4 = 7$ and $N_4 \cap \ker \Pi_3 = \{0\}$ on $\{x \in \mathbb{R}^8 : x \notin S_{crit}^{3-B}\}$, and we find a projector matrix Q_4 onto N_4 such that $\ker \Pi_3 \subseteq \ker Q_4$.
For $z \in N_4$, it holds, in particular, that $z_4 = (1 + \frac{A}{p})z_1$ and

$$b_{42}z_2 + b_{47}z_7 = b_{44}\frac{A}{p}z_1 - (1 + b_{44})(1 + \frac{A}{p})z_1,$$

$$b_{62}z_2 + b_{67}z_7 = b_{64}\frac{A}{p}z_1 - b_{64}(1 + \frac{A}{p})z_1.$$

Since here the coefficient matrix $\mathcal{M}$ is nonsingular, we obtain the expressions[2]

$$z_2 = g\, z_1, \quad z_7 = h\, z_1,$$

with functions g and h being continuous outside of S_{crit}^{3-B}. Denoting further

$$f = b_{54} + (1 + b_{52})g + b_{57}h,$$

we arrive at

$$N_4 = \mathrm{im} \begin{pmatrix} 1 \\ g \\ 0 \\ 1 + \frac{A}{p} \\ g \\ 0 \\ h \\ f \end{pmatrix}.$$

---

[2]In detail $g = -\mathcal{S}(1 + \frac{1}{p}A + b_{44}) + p\mathcal{S}b_{64}$ and $h = rb_{62}\mathcal{S}(1 + \frac{1}{p}A + b_{44}) - (r + rb_{62}p\mathcal{S})b_{64}$.

Regarding that $\ker \Pi_3 = \{z \in \mathbb{R}^8 : z_1 - pz_3 = 0\}$, we choose

$$Q_4 = \begin{pmatrix} 1 & 0 & -p & 0 & 0 & 0 & 0 & 0 \\ g & 0 & -pg & 0 & 0 & 0 & 0 & 0 \\ 0 & 0 & 0 & 0 & 0 & 0 & 0 & 0 \\ 1+\frac{A}{p} & 0 & -p-A & 0 & 0 & 0 & 0 & 0 \\ g & 0 & -pg & 0 & 0 & 0 & 0 & 0 \\ 0 & 0 & 0 & 0 & 0 & 0 & 0 & 0 \\ h & 0 & -ph & 0 & 0 & 0 & 0 & 0 \\ f & 0 & -pf & 0 & 0 & 0 & 0 & 0 \end{pmatrix}.$$

This yields

$$\Pi_3 Q_4 = \begin{pmatrix} 1 & 0 & -p & 0 & 0 & 0 & 0 & 0 \\ 0 & 0 & 0 & 0 & 0 & 0 & 0 & 0 \\ 0 & 0 & 0 & 0 & 0 & 0 & 0 & 0 \\ \frac{1}{p}A & 0 & -A & 0 & 0 & 0 & 0 & 0 \\ 0 & 0 & 0 & 0 & 0 & 0 & 0 & 0 \\ 0 & 0 & 0 & 0 & 0 & 0 & 0 & 0 \\ 0 & 0 & 0 & 0 & 0 & 0 & 0 & 0 \\ 0 & 0 & 0 & 0 & 0 & 0 & 0 & 0 \end{pmatrix}, \quad \Pi_4 = \Pi_3 - \Pi_3 Q_4 = 0,$$

and

$$B_4 = B_3 \Pi_3 = \begin{pmatrix} -\frac{1}{p}A & 0 & A & 0 & 0 & 0 & 0 & 0 \\ 0 & 0 & 0 & 0 & 0 & 0 & 0 & 0 \\ 0 & 0 & 0 & 0 & 0 & 0 & 0 & 0 \\ b_{44}\frac{1}{p}A-(\frac{1}{p}A)' & 0 & -Ab_{44}+p(\frac{1}{p}A)' & 0 & 0 & 0 & 0 & 0 \\ b_{54}\frac{1}{p}A & 0 & -Ab_{54} & 0 & 0 & 0 & 0 & 0 \\ b_{64}\frac{1}{p}A & 0 & -Ab_{64} & 0 & 0 & 0 & 0 & 0 \\ b_{71} & 0 & -pb_{73} & 0 & 0 & 0 & 0 & 0 \\ b_{81} & 0 & -pb_{83} & 0 & 0 & 0 & 0 & 0 \end{pmatrix},$$

$$B_4 Q_4 = B_3 \Pi_3 Q_4 = B_3 \Pi_3,$$

$$G_5 = G_4 + B_4 Q_4 = \begin{pmatrix} 1 & 0 & p' & -1 & 0 & 0 & 0 & 0 \\ 0 & 1 & 0 & 0 & -1 & 0 & 0 & 0 \\ 0 & 0 & 1 & 0 & 0 & -1 & 0 & 0 \\ -(\frac{1}{p}\mathcal{A})' & b_{42} & b_{43} + \mathcal{A}' + p(\frac{1}{p}\mathcal{A})' & 1 + b_{44} & 0 & b_{46} + p' & b_{47} & 0 \\ 0 & b_{52} & b_{53} & b_{54} & 1 & b_{56} & b_{57} & -1 \\ 0 & b_{62} & b_{63} & b_{64} & 0 & 1 + b_{66} & b_{67} & 0 \\ b_{71} & 0 & b_{73} & 0 & 0 & 0 & 0 & 0 \\ b_{81} & 0 & b_{83} & 0 & 0 & 0 & 0 & 0 \end{pmatrix}.$$

It remains to check if G_5 is nonsingular. $z \in N_5 = \ker G_5$ implies

$$z_5 = z_2,$$

$$z_6 = z_3,$$

$$z_4 = z_1 + p' z_3,$$

$$\begin{pmatrix} b_{42} & b_{47} \\ b_{62} & b_{67} \end{pmatrix} \begin{pmatrix} z_2 \\ z_7 \end{pmatrix} = -\begin{pmatrix} 1 + b_{44} - (\frac{1}{p}\mathcal{A})' & b_{43} + \mathcal{A}' + p(\frac{1}{p}\mathcal{A})' + (1 + b_{44})p' + b_{46} + p' \\ b_{64} & 1 + b_{66} + b_{64}p' \end{pmatrix} \begin{pmatrix} z_1 \\ z_3 \end{pmatrix},$$

$$z_8 = \begin{pmatrix} b_{52} + 1 & b_{57} \end{pmatrix} \begin{pmatrix} z_2 \\ z_7 \end{pmatrix} + \begin{pmatrix} b_{54} & b_{53} + b_{54}p' + b_{56} \end{pmatrix} \begin{pmatrix} z_1 \\ z_3 \end{pmatrix},$$

and

$$\begin{pmatrix} b_{71} & b_{73} \\ b_{81} & b_{83} \end{pmatrix} \begin{pmatrix} z_1 \\ z_3 \end{pmatrix} = 0.$$

This shows that G_5 becomes nonsingular precisely if the matrix functions $\mathcal{M}$ and $\mathcal{N}$ are nonsingular (see Lemma 2.1). The matrix function $\mathcal{N}$ depends only on x_1 and x_3. However, since $\det \mathcal{N}(x_1, x_3) = \sin x_3$, the DAE features also critical points of type 5-A, (cf. [13, Definition 2.75])

$$S_{crit}^{5-A} = \{x \in \mathbb{R}^8 : \sin x_3 = 0\}.$$

We summarize the results as a proposition:

Proposition 5.1 *The definition domain $\mathbb{R}^8 \times \mathbb{R}$ of the data of the given DAE (5.1) decomposes into an infinite number of regularity regions $\mathfrak{G}$, each of which is an open connected set determined by*

$$\mathfrak{G} = \{(x, t) \in \mathbb{R}^8 \times \mathbb{R} : \cos x_3 \neq z_\star, \ \sin x_3 \neq 0\},$$

whereas z_* fulfills the condition described in Lemma 2.1. On all these regularity regions the DAE has the tractability index 5 and the characteristic values $r_0 = 6$, $r_1 = 6$, $r_2 = 6$, $r_3 = 7$, $r_4 = 7$, $r_5 = 8$.
The regularity regions are separated by hyperplanes corresponding to the sets of critical points S_{crit}^{3-B} and S_{crit}^{5-A}, respectively.

The given DAE (5.1) has no dynamics owing to the fact that $\Pi_4 = 0$. Note that this is confirmed already by the observations in Sect. 3. The solution $x_*(\cdot)$ decomposes according to $I = Q_0 + \Pi_0 Q_1 + \Pi_1 Q_2 + \Pi_2 Q_3 + \Pi_3 Q_4$,

$$Q_0 x_* = \begin{pmatrix} 0 \\ 0 \\ 0 \\ 0 \\ 0 \\ 0 \\ x_{*7} \\ x_{*8} \end{pmatrix}, \quad \Pi_0 Q_1 x_* = \begin{pmatrix} 0 \\ 0 \\ 0 \\ px_{*6} \\ x_{*5} \\ x_{*6} \\ 0 \\ 0 \end{pmatrix}, \quad \Pi_1 Q_2 x_* = \begin{pmatrix} px_{*3} \\ x_{*2} \\ x_{*3} \\ Ax_{*3} \\ 0 \\ 0 \\ 0 \\ 0 \end{pmatrix},$$

$$\Pi_2 Q_3 x_* = \begin{pmatrix} 0 \\ 0 \\ 0 \\ x_{*4} - \frac{1}{p}Ax_{*1} - px_{*6} \\ 0 \\ 0 \\ 0 \\ 0 \end{pmatrix}, \quad \Pi_3 Q_4 x_* = \begin{pmatrix} x_{*1} - px_{*3} \\ 0 \\ 0 \\ \frac{1}{p}Ax_{*1} - Ax_{*3} \\ 0 \\ 0 \\ 0 \\ 0 \end{pmatrix},$$

It is worth noting that the projector functions Π_0, Π_1, Π_2 are continuous, but Π_3, Π_4 have continuous extensions through the critical points. Therefore, also the matrix functions G_i, B_i are continuous or have continuous extensions. This fact is closely related to the approach in [16] and seems to be helpful for further critical point studies.

We want to emphasize that, according to [13, Theorem 3.39], the characteristic values $r_0, \ldots, r_5$ are invariant under regular transformations. Therefore, analogous results are obtained if the original equation (2.1) is used instead of (2.5).

6 Types of Singularities and Numerical Experiments

The different ways of investigation of the Robotic Arm problem in Sects. 3–5 discover the same two types of singularities. This underlines that the singularities belong to the problem and are not owed to the used technical procedure.

– In [15, 16, 18] a classification of singularities is introduced. Using the corresponding nomenclature, we concluded in Sect. 5 that two singularity sets appear:

1. $S_{crit}^{3-B} = \{x \in \mathbb{R}^8 : \cos x_3 = z_\star\}$
2. $S_{crit}^{5-A} = \{x \in \mathbb{R}^8 : \sin x_3 = 0\}$.

– In terms of the nomenclature used in Sect. 4, we observed that

1. the projector V_1 can only be obtained if x_3 is not in the set of critical points $\{x \in \mathbb{R}^8 : \sin x_3 = 0\}$.
2. the projectors V_3 and T_3 can only be obtained if, furthermore, x_3 is not in the set of critical points $\{x \in \mathbb{R}^8 : \cos x_3 = z_\star\}$.

According to the concept of regularity regions developed in [13, Section 3.3], the original definition domain of the DAE decomposes into several maximal regularity regions whose borders consist of critical points. It may well happen that the characteristic values including the index on different regions are different. Related to Proposition 5.1, the Robotic Arm DAE has the same characteristics on all regularity regions. Different kinds of problems may happen if a solution of the DAE approaches or crosses such a critical point set. For the Robotic Arm, singularities arise if the component $x_3(t)$ touches or crosses a singularity set. This happens if the prescribed path (p_1, p_2) touches or crosses a singularity circle.

According to [7], the condition number of $\mathcal{B}^{[5]}$ is an indicator for singularities of DAEs, whereas a more detailed analysis can be obtained observing the ranks of the associated projectors.

We monitored the condition number of solutions obtained in [12] (see also Figs. 7 and 8). This numerical experiment perfectly showed the position of the singularities, see Figs. 11 and 12. Therefore, the theoretical analysis presented in Sect. 3–5 confirmed, indeed, the existence of the singularities noticed monitoring this condition number. Moreover, the results we obtained numerically can now perfectly be explained:

– For Eqs. (2.1)–(2.2) with (2.3), i.e., using $e = e_{Ca88}$, we have both types of singularities in the interval $[-6, 2.21967]$, as shown in Fig. 11. In particular, we now understand why the integration stopped at $t \approx 1.5$, as illustrated in Fig. 1. The reason is $\sin(x_3) \approx 0$.
– For Eqs. (2.1)–(2.2) with (2.4), i.e., using $e = e_{DeLu87}$, we also have both types of singularities in the interval $[-6, 2.21967]$, as shown in Fig. 12. Now, the integration stops already at ≈ 0.3730, because then $\cos x_3 \approx 0.4721$, cf. Fig. 13.

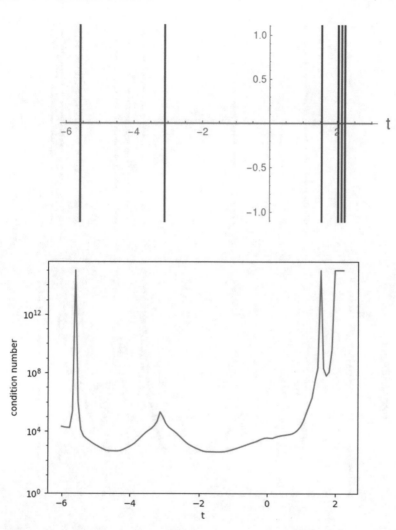

Fig. 11 The singularities in $[-6, 2.21967]$ (top) and the condition number of $\mathcal{B}^{[5]}$ (bottom) for $e = e_{\text{Ca88}}$

- For Eqs. (2.1)–(2.2) with (2.4), i.e. using $e = e_{\text{DeLu87}}$, we furthermore computed successfully the numerical solution piecewise for the part of the trajectory inside the circle $\mathcal{C}_{r*}$, see Figs. 14 and 15. At the singularities, InitDAE stopped the integration as expected, according to the high condition number.

Since (2.1)–(2.2) with (2.3) (i.e., $e = e_{\text{Ca88}}$) has no physical background, in the following we focus on Eqs. (2.1)–(2.2) with (2.4) only (i.e., $e = e_{\text{DeLu87}}$).

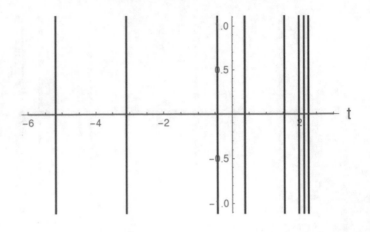

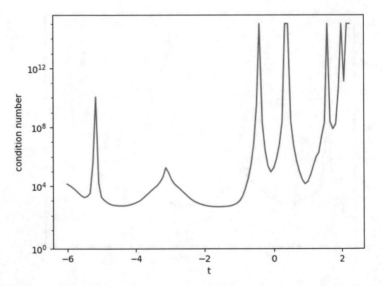

Fig. 12 The singularities in $[-6, 2.21967]$ (top) and the condition number of $\mathcal{B}^{[5]}$ (bottom) for $e = e_{\text{DeLu87}}$

If we choose a trajectory $\bar{p}_1$, $\bar{p}_2$ that does not touch or cross a singularity circle from Sect. 3, then the resulting solution component x_3 never crosses a singularity hyperplane and we obtain a singularity-free solution. We illustrate that by choosing the path

$$\begin{aligned} \bar{p}_1(t) &= \cos(1 - t) + \cos(3 + \tfrac{\sin t}{2} - t), \\ \bar{p}_2(t) &= \sin(1 - t) + \sin(3 + \tfrac{\sin t}{2} - t), \end{aligned} \tag{6.1}$$

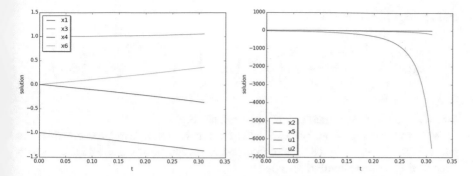

Fig. 13 Numerical solution of the Robotic Arm problem using Eqs. (2.1)–(2.2) with $e = e_{\text{DeLu87}}$, obtained with the Taylor methods described in [11]. These numerical results coincide with those obtained with Mathematica in Sect. 3 up to an absolute error of $1e - 10$

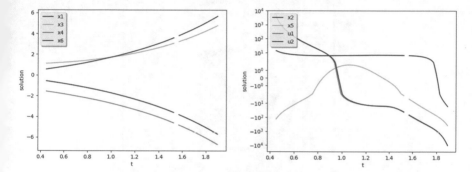

Fig. 14 Piecewise computed numerical solution of the Robotic Arm problem using Eqs. (2.1)–(2.2) with $e = e_{\text{DeLu87}}$, obtained with the Taylor methods described in [11] for $t \in [0.4500, 1.5262]$ and $t \in [1.5700, 1.8991]$. These numerical results coincide with those obtained with Mathematica in Sect. 3 up to an absolute error of $1e - 8$

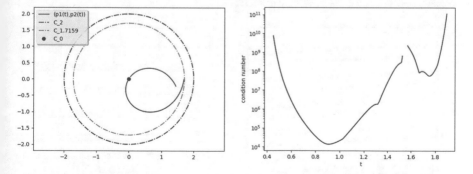

Fig. 15 Piecewise computed path and condition for the Robotic Arm problem using the equations according to Fig. 14

cf. Fig. 16, which leads to

$$\begin{pmatrix} x_1(t) \\ x_3(t) \end{pmatrix} = \begin{pmatrix} 1 - t + 2\pi k_1 \\ 2 + \frac{\sin t}{2} + 2\pi k_2 \end{pmatrix} \text{ or } \begin{pmatrix} x_1(t) \\ x_3(t) \end{pmatrix} = \begin{pmatrix} 3 + \frac{\sin t}{2} - t + 2\pi k_3 \\ -(2 + \frac{\sin t}{2}) + 2\pi k_4 \end{pmatrix},$$

(6.2)

for $k_1, k_2, k_3, k_4 \in \mathbb{Z}$, which do not intersect, see Fig. 17.

With this modification, we can integrate the Robotic Arm problem over the interval $[0, 10]$, see Fig. 18. The computed condition number, see Fig. 19, varies uncritically, i.e., no singularities appear.

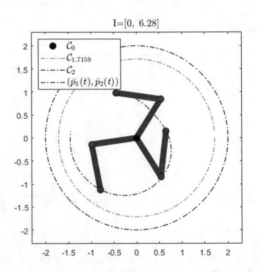

Fig. 16 Path prescribed by (6.1) for the endpoint of the outer arm. Due to the particular choice of the path, $\cos x_3 < z_\star$ and $\sin x_3 \neq 0$ are given for all t

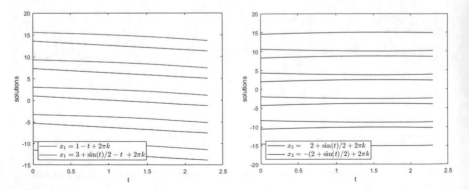

Fig. 17 Different solutions x_1 and x_3 for $k = -2, \ldots, 2$ for the trajectory $(\bar{p}_1, \bar{p}_2)$ from (6.1)

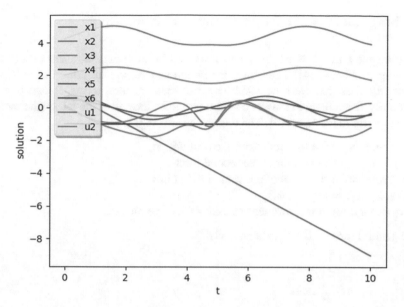

Fig. 18 Singularity free solution of modified Robotic Arm problem using Eqs. (2.1) and (6.1) with $e = e_{\text{DeLu87}}$, obtained with the Taylor methods described in [11]

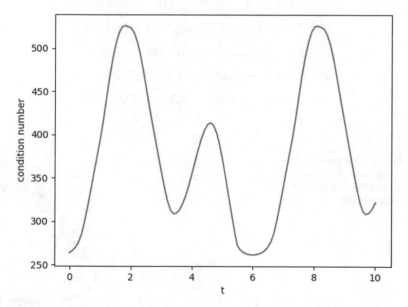

Fig. 19 Condition number of the modified Robotic Arm using Eqs. (2.1) with (6.1) and $e = e_{\text{DeLu87}}$, obtained with the Taylor methods described in [11]. In this case, no singularities are detected, as expected

7 Singularities of a More General Formulation

Only recently, the Robotic Arm was described in more general terms as a test for tracking problems, [4]. Therefore, for completeness, we describe the two types of singularities that were detected above in these general terms. Recall that, for simplicity, the centers of masses are assumed to be located at the joints (the motors) and at the tip (a load), cf. [5]. The unknowns are (Fig. 20)

- x_1: rotation of the first link w.r.t. the base frame,
- x_2: rotation of the motor at the second joint,
- x_3: rotation of the second link w.r.t. the first link.
- $(x_4, x_5, x_6) = (x_1', x_2', x_3')$,
- u_1 and u_2: rotational torques caused by the drive motors.

The general form of the equations reads

$$x_1' = x_4,$$
$$x_2' = x_5,$$
$$x_3' = x_6,$$
$$x_4' = f_4(x_2, x_3, x_4, x_6) + g_{41}(x_3)u_1 - g_{41}(x_3)u_2,$$
$$x_5' = f_5(x_2, x_3, x_4, x_6) - g_{41}(x_3)u_1 + g_{52}(x_3)u_2, \tag{7.1}$$
$$x_6' = f_6(x_2, x_3, x_4, x_6) + g_{61}(x_3)u_1 - g_{61}(x_3)u_2,$$
$$0 = l_1 \cos x_1 + l_2 \cos(x_1 + x_3) - p_1(t),$$
$$0 = l_1 \sin x_1 + l_2 \sin(x_1 + x_3) - p_2(t),$$

Fig. 20 Two-link planar Robot Arm with the second joint elastic. (Modification of a graphic from [5])

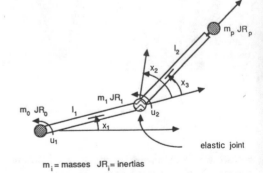

m_i = masses JR$_i$ = inertias

where $(p_1(t), p_2(t))$ is again the endpoint of the outer arm in Cartesian coordinates, cf. (2.2), l_1, l_2 are possibly different lengths of the two links and f_i, g_{ij} are suitable functions resulting from the dynamic model of the Robotic Arm, and read

$$g_{41}(x_3) = \frac{A_2}{A_3(A_4 - A_3 \cos^2 x_3)}$$

$$g_{52}(x_3) = g_{41}(x_3) + \frac{1}{JR_1}$$

$$g_{61}(x_3) = -g_{41}(x_3) - \frac{\cos x_3}{A_4 - A_3 \cos^2 x_3}$$

$$f_4(x_2, x_3, x_4, x_6) = \frac{A_2 \sin x_3 (x_4 + x_6)^2 + A_3 x_4^2 \sin x_3 \cos x_3}{A_4 - A_3 \cos^2 x_3}$$

$$+ \frac{K \left(x_3 - \frac{x_2}{NT}\right)\left(\frac{A_2}{A_3}\left(\frac{NT-1}{NT}\right) + \cos x_3\right)}{A_4 - A_3 \cos^2 x_3}$$

$$f_5(x_2, x_3, x_4, x_6) = -f_4(x_2, x_3, x_4, x_6) + \frac{K}{NT}\left(x_3 - \frac{x_2}{NT}\right)\left(\frac{1}{JR_1} - 2g_{41}(x_3)\right),$$

$$f_6(x_2, x_3, x_4, x_6) = -f_4(x_2, x_3, x_4, x_6) - \frac{K\left(x_3 - \frac{x_2}{NT}\right)\left(\frac{A_5}{A_3} - \left(\frac{3NT+1}{NT}\right)\cos x_3\right)}{A_4 - A_3 \cos^2 x_3}$$

$$- \frac{A_5 x_4^2 \sin x_3 + A_3 \sin x_3 \cos x_3 (x_4 + x_6)^2}{A_4 - A_3 \cos^2 x_3},$$

where

- K is the coefficient of elasticity of the second joint,
- NT is the transmission ratio at the second joint,
- m_p is the mass of the object being held,
- m_0 and m_1 are the masses of the motors and arms viewed as concentrated at the corresponding joints,
- JR_1 and JR_p are corresponding rotor inertias, and
- the constants are defined by

$$A_2 = JR_p + m_p l_2^2,$$

$$A_3 = m_p l_1 l_2,$$

$$A_4 = (m_1 + m_p) l_1 l_2,$$

$$A_5 = (m_1 + m_p) l_1^2,$$

whereas $A_4 > A_3 \geq A_3 \cos x_3$ is always given by definition.

Remark 7.1 We noticed that there is something wrong in Eqs. (2.1)–(2.2) for $e = e_{Ca88}$ from (2.3), since we could not fit parameters to obtain that specific equation. We highly appreciate that it was confirmed by the author of [2] that there is an error in one sign. In fact, comparing Eqs. (2.1)–(2.2) with the general equation (7.1), it is easy to deduce that in (2.1)–(2.2) $m_1 = 1$, $m_p = 1$, $l_1 = 1$, $l_2 = 1$, $JR_p = 1$ have to be given. Under this assumption, comparing the fourth and fifth equations of both DAEs, we obtain $NT = 2$ and $K = 4$. However, inserting these values leads to e_{DeLu87} in the sixth equation of (2.1). In contrast, e_{Ca88} is obtained in the sixth equation if, in (7.1), the corresponding function f_6 is computed considering a different sign for the second summand, i.e.,

$$f_6(x_2, x_3, x_4, x_6) = -f_4(x_2, x_3, x_4, x_6) + \frac{K \ldots}{A_4 - A_3 \cos^2 x_3} - \ldots$$

Therefore, it seems likely that e_{Ca88} has its origin in this sign error.

Analogously to the results from the previous sections, in [4], $x_3 \neq k\pi$ is identified to be a necessary condition for reasonable rank properties in (7.1). The second type of singularities that we described in the previous sections depends, in general, on several parameters.

In terms of the notation used before, for the general equations we obtain

$$b_{42} = -\frac{\partial f_4}{\partial x_2} = \frac{K \left(\cos x_3 + \frac{A_2 (NT-1)}{A_3 NT}\right)}{NT \left(A_4 - A_3 \cos^2 x_3\right)},$$

$$b_{47} = -g_{41} = -\frac{A_2}{A_3 \left(A_4 - A_3 \cos^2 x_3\right)},$$

$$b_{62} = -\frac{\partial f_6}{\partial x_2} = \frac{K \left(A_2 + A_3 \cos x_3 - A_2 NT - A_5 NT + 2 A_3 NT \cos x_3\right)}{A_3 NT^2 \left(A_4 - A_3 \cos^2 x_3\right)},$$

$$b_{67} = -g_{61} = \frac{A_2 + A_3 \cos x_3}{A_3 \left(A_4 - A_3 \cos^2 x_3\right)},$$

$$b_{71} = -l_1 \sin x_1 - l_2 \sin(x_1 + x_3),$$

$$b_{73} = -l_2 \sin(x_1 + x_3),$$

$$b_{81} = l_1 \cos x_1 + l_2 \cos(x_1 + x_3),$$

$$b_{83} = l_2 \cos(x_1 + x_3),$$

such that analogous structural properties result directly.

Lemma 7.1

(a) *The function $g_{61}(x_3) = -b_{67}$ is smooth and has no zeros. It depends on x_3 only.*

(b) *The functions*

$$p := \frac{g_{41}(x_3)}{g_{61}(x_3)} = \frac{b_{47}}{b_{67}} \quad and \quad r := -\frac{1}{g_{61}(x_3)} = \frac{1}{b_{67}}$$

are smooth and depend on x_3 only. They have no zeros.

(c) *The matrix function*

$$M(x_3) := -\begin{pmatrix} \frac{\partial f_4}{\partial x_2}(x_3) & g_{41}(x_3) \\ \frac{\partial f_6}{\partial x_2}(x_3) & g_{61}(x_3) \end{pmatrix} = \begin{pmatrix} b_{42} & b_{47} \\ b_{62} & b_{67} \end{pmatrix}$$

$$= -\begin{pmatrix} -\dfrac{K\left(\cos x_3 + \frac{A_2(NT-1)}{A_3 NT}\right)}{NT\left(A_4 - A_3 \cos^2 x_3\right)} & \dfrac{A_2}{A_3\left(A_4 - A_3 \cos^2 x_3\right)} \\ -\dfrac{K\left(A_2 + A_3\cos x_3 - A_2 NT - A_5 NT + 2 A_3 NT \cos x_3\right)}{A_3 NT^2\left(A_4 - A_3 \cos^2 x_3\right)} & -\dfrac{A_2 + A_3\cos x_3}{A_3\left(A_4 - A_3 \cos^2 x_3\right)} \end{pmatrix},$$

has smooth entries depending on x_3 only. $M(x_3)$ is nonsingular precisely if

$$b_{42}(x_3) - p(x_3)b_{62}(x_3) = \frac{\partial f_4}{\partial x_2}(x_3) - \frac{g_{41}(x_3)}{g_{61}(x_3)}\frac{\partial f_6}{\partial x_2}(x_3) \neq 0.$$

(d) *For*

$$\begin{pmatrix} z_1 \\ z_2 \end{pmatrix} = \begin{pmatrix} -\dfrac{2A_2 + \sqrt{A_2(4A_2 + A_5)}}{A_3} \\ -\dfrac{2A_2 - \sqrt{A_2(4A_2 + A_5)}}{A_3} \end{pmatrix}$$

$$= \begin{pmatrix} -\dfrac{2JR_p + 2l_2^2 m_p + \sqrt{\left(m_p l_2^2 + JR_p\right)\left(4JR_p + l_1^2 m_1 + l_1^2 m_p + 4l_2^2 m_p\right)}}{l_1 l_2 m_p} \\ -\dfrac{2JR_p + 2l_2^2 m_p - \sqrt{\left(m_p l_2^2 + JR_p\right)\left(4JR_p + l_1^2 m_1 + l_1^2 m_p + 4l_2^2 m_p\right)}}{l_1 l_2 m_p} \end{pmatrix}.$$

it holds that $\cos x_3 = z_1$ or $\cos x_3 = z_2$ imply $b_{42}(x_3) - p(x_3)b_{62}(x_3) = 0$ and vice versa.

(e) *The function*

$$S(x_3) = \frac{1}{b_{42}(x_3) - p(x_3)b_{62}(x_3)}, \quad x_3 \in \operatorname{dom} S = \{\tau \in \mathbb{R} : \cos\tau \neq z_{1,2}\},$$

is smooth on its definition domain, and so is

$$M^{-1} = \begin{pmatrix} S & -pS \\ -rb_{62}S & r + rpb_{62}S \end{pmatrix}.$$

(f) *The matrix function*

$$N = \begin{pmatrix} b_{71} & b_{73} \\ b_{81} & b_{83} \end{pmatrix} = \begin{pmatrix} -l_1 \sin x_1 - l_2 \sin(x_1 + x_3) & -l_2 \sin(x_1 + x_3) \\ l_1 \cos x_1 + l_2 \cos(x_1 + x_3) & l_2 \cos(x_1 + x_3) \end{pmatrix} \quad (7.2)$$

depends only on x_1 and x_3. $N(x_1, x_3)$ is nonsingular precisely if $\det N(x_1, x_3) = l_1 l_2 \sin x_3 \neq 0$.

Proof Assertion (d): Since

$$A_4 - A_3 \cos^2 x_3 = m_1 l_1 l_2 + m_p l_1 l_2 (1 - \cos^2 x_3),$$

we can assume that all denominators are nonzero and, analogously to Lemma 2.1, focus on the singularities of the matrix $H(x_3) := -(A_4 - A_3 \cos^2 x_3)M$, i.e.,

$$H(x_3) = \begin{pmatrix} -\dfrac{K\left(\cos x_3 + \frac{A_2 (NT-1)}{A_3 NT}\right)}{NT} & \dfrac{A_2}{A_3} \\ -\dfrac{K (A_2 + A_3 \cos x_3 - A_2 NT - A_5 NT + 2 A_3 NT \cos x_3)}{A_3 NT^2} & -\dfrac{A_2 + A_3 \cos x_3}{A_3} \end{pmatrix},$$

with the determinant

$$\det(H(x_3)) = (A_3{}^2 \cos^2 x_3 + 4 A_2 A_3 \cos x_3 - A_2 A_5) \underbrace{K/(A_3^2 NT)}_{\neq 0}.$$

For the substitution $z = \cos x_3$ we obtain $A_3{}^2 z^2 + 4 A_2 A_3 z - A_2 A_5$ with the roots

$$\begin{pmatrix} z_1 \\ z_2 \end{pmatrix} = \begin{pmatrix} -\dfrac{2 A_2 + \sqrt{A_2 (4 A_2 + A_5)}}{A_3} \\ -\dfrac{2 A_2 - \sqrt{A_2 (4 A_2 + A_5)}}{A_3} \end{pmatrix}$$

$$= \begin{pmatrix} -\dfrac{2 JR_p + 2 l_2{}^2 m_p + \sqrt{\left(m_p l_2{}^2 + JR_p\right)\left(4 JR_p + l_1{}^2 m_1 + l_1{}^2 m_p + 4 l_2{}^2 m_p\right)}}{l_1 l_2 m_p} \\ -\dfrac{2 JR_p + 2 l_2{}^2 m_p - \sqrt{\left(m_p l_2{}^2 + JR_p\right)\left(4 JR_p + l_1{}^2 m_1 + l_1{}^2 m_p + 4 l_2{}^2 m_p\right)}}{l_1 l_2 m_p} \end{pmatrix}.$$

This proves the assertion. All further assertions follow straightforward.   □

At regular points, these structural properties imply that all the results from the previous sections can also be applied to the general equations. In particular, the shape of all the described projectors is analogous.

Let us now discuss the two types of singularities. For the general case, using addition theorems we obtain

$$p_1 = l_1 \cos x_1 + l_2 \cos x_1 \cos x_3 - l_2 \sin x_1 \sin x_3$$
$$= \cos x_1 (l_1 + l_2 \cos x_3) - l_2 \sin x_1 \sin x_3,$$
$$p_2 = l_1 \sin x_1 + l_2 \sin x_1 \cos x_3 + l_2 \cos x_1 \sin x_3$$
$$= \sin x_1 (l_1 + l_2 \cos x_3) + l_2 \cos x_1 \sin x_3,$$

and

$$p_1^2 + p_2^2 = (l_1 \cos x_1 + l_2 \cos(x_1 + x_3))^2 + (l_1 \sin x_1 + l_2 \sin(x_1 + x_3))^2$$
$$= l_1^2 + l_2^2 + 2l_1 l_2 \cos x_1 \cos(x_1 + x_3) + 2l_1 l_2 \sin x_1 \sin(x_1 + x_3)$$
$$= l_1^2 + l_2^2 + 2l_1 l_2 \cos x_3.$$

For $l_1 \geq l_2$, the circles of singularity are therefore

$$C_{l_1 \, l_2}, \quad C_{r^*}, \quad C_{l_1 + l_2}$$

for $r^* = \sqrt{l_1^2 + l_2^2 + 2l_1 l_2 z_*}$.

Again, for $\sin x_3 = 0$ the corresponding Jacobian matrix $\mathcal{N}$ is singular for $\sin x_3 = 0$. Also in this case, this is given if the arm is fully extended or completely folded.

The singularities of the type $\cos x_3 = z_*$ are considerably more difficult to understand. If z_1 or z_2 belongs to the interval $[-1, 1]$, then the corresponding singularities appear, defining a corresponding singularity circle C_{r^*}. In Table 4 we present some critical z_* in dependence of some values for the parameters. This means that, in general, there actually appear singularities in configurations, depending on the particular values for JR_p, m_p, m_1, l_1, l_2.

On the one hand, if such a value z_* is given, singularities can only be avoided restricting the allowed values for x_3. This means that p_1, p_2 have to be chosen correspondingly, see e.g. (6.1).

On the other hand, also a condition on the parameters can be formulated to characterize the existence of these singularities for arbitrary x_3. For this purpose, we realize that, in terms of the introduced constants, these singularities appear iff

$$A_2 \cdot (A_5 - 4 A_3 \cos x_3) = A_3^2 \cos^2 x_3. \tag{7.3}$$

Table 4 For the specified parameters, $\cos x_3 = z_\star$ leads to a singularity

Parameters			Critical values
$m_1 = 1, \; m_p = 1$	$l_1 = 1, \; l_2 = 1$	$JR_p = 1$	$z_\star = 2\sqrt{5} - 4 = 0.4721$
$m_1 = 1, \; m_p = 10$	$l_1 = 1, \; l_2 = 1$	$JR_p = 1$	$z_\star = \frac{11\sqrt{5}}{10} - \frac{11}{5} = 0.2597$
$m_1 = 10, \; m_p = 10$	$l_1 = 1, \; l_2 = 1$	$JR_p = 1$	$z_\star = \frac{4\sqrt{11}}{5} - \frac{11}{5} = 0.4533$
$m_1 = 10, \; m_p = 1$	$l_1 = 1, \; l_2 = 1$	$JR_p = 1$	–
$m_1 = 10, \; m_p = 1$	$l_1 = 1, \; l_2 = 1$	$JR_p = 0.5$	–
$m_1 = 10, \; m_p = 3$	$l_1 = 1, \; l_2 = 1$	$JR_p = 0.5$	$z_\star = \frac{\sqrt{42}}{2} - \frac{7}{3} = 0.4346$
$m_1 = 1, \; m_p = 1$	$l_1 = 1, \; l_2 = 2$	$JR_p = 1$	$z_\star = \frac{\sqrt{110}}{2} - 5 = 0.2440$
$m_1 = 1, \; m_p = 1$	$l_1 = 1, \; l_2 = 0.5$	$JR_p = 1$	$z_\star = \sqrt{35} - 5 = 0.9161$
$m_1 = 1, \; m_p = 1$	$l_1 = 1, \; l_2 = 0.3$	$JR_p = 1$	–
$m_1 = 1, \; m_p = 1$	$l_1 = 1, \; l_2 = 1$	$JR_p = 0.9$	$z_\star = \frac{2\sqrt{114}}{5} - \frac{19}{5} = 0.4708$
$m_1 = 1, \; m_p = 1$	$l_1 = 1, \; l_2 = 1$	$JR_p = 1.1$	$z_\star = \frac{\sqrt{546}}{5} - \frac{21}{5} = 0.4733$

Therefore, we consider the following cases:

- If we suppose that $A_5 - 4\,A_3\,\cos x_3 = 0$, i.e.,

$$\cos x_3 = \frac{A_5}{4A_3} = \frac{(m_1 + m_p)l_2^2}{4m_p l_1 l_2} = \frac{1}{4}\left(\frac{m_1}{m_p} - 1\right)\frac{l_1}{l_2},$$

we obtain

$$A_3{}^2 \cos^2 x_3 = \left(\frac{A_5}{4}\right)^2 > 0$$

such that (7.3) cannot be given.
- For $A_5 - 4\,A_3\,\cos x_3 \neq 0$, the condition (7.3) can be represented as

$$A_2 = \frac{A_3{}^2 \cos^2 x_3}{A_5 - 4\,A_3\,\cos x_3}.$$

Therefore, singularities appear iff

$$JR_p + l_2{}^2 m_p = \frac{l_1{}^2 l_2{}^2 m_p{}^2 \cos^2 x_3}{l_1{}^2 \left(m_1 + m_p\right) - 4 l_1 l_2 m_p \cos x_3}.$$

Focusing on the denominator of the latter two expressions, we consider now the following cases:

- If $A_5 \leq 4 A_3$, i.e., $l_1{}^2 \left(m_1 + m_p\right) \leq 4 l_1 l_2 m_p$, or, equivalently

$$\frac{m_1}{m_p} + 1 \leq 4\frac{l_2}{l_1},$$

there are singularities for all values of JR_p.

- In case $A_5 - 4\,A_3 = l_1{}^2\left(m_1 + m_p\right) - 4\,l_1\,l_2\,m_p > 0$, i.e.,

$$\frac{m_1}{m_p} + 1 > 4\,\frac{l_2}{l_1} \tag{7.4}$$

is given, then it holds for all x_3 that

$$\left|\frac{l_1{}^2\,l_2{}^2\,m_p{}^2\,\cos^2 x_3}{l_{21}{}^2\left(m_1 + m_p\right) - 4\,l_1\,l_2\,m_p\,\cos x_3}\right| \le \frac{l_1{}^2\,l_2{}^2\,m_p{}^2}{l_1{}^2\left(m_1 + m_p\right) - 4\,l_1\,l_2\,m_p}\,,$$

and we can conclude that, at least for

$$JR_p + l_2{}^2\,m_p > \frac{l_1{}^2\,l_2{}^2\,m_p{}^2}{l_1{}^2\left(m_1 + m_p\right) - 4\,l_1\,l_2\,m_p}\,, \tag{7.5}$$

no singularities may appear. To obtain a rule of thumb applicable for all values of JR_p, we use $JR_p + l_2{}^2\,m_p > l_2{}^2\,m_p$, and, hence, that it suffices to assume

$$1 > \frac{l_1{}^2\,m_p}{l_1{}^2\left(m_1 + m_p\right) - 4\,l_1\,l_2\,m_p} = \frac{1}{\left(\frac{m_1}{m_p} + 1\right) - 4\,\frac{l_2}{l_1}}\,,$$

i.e.,

$$l_2\,m_p < \frac{1}{4}\,l_1\,m_1.$$

We summarize the results as a corollary:

Corollary 7.2 *With respect to the singularities arising from* $\cos x_3 = z_\star$ *for the DAE (7.1) it holds that:*

- *If*

$$\frac{m_1}{m_p} + 1 \le 4\,\frac{l_2}{l_1},$$

there are critical values $z_\star$ *and therefore singularities for all values of* JR_p.
- *If*

$$\frac{m_1}{m_p} + 1 > 4\,\frac{l_2}{l_1}$$

at least for

$$JR_p > \frac{l_1{}^2 \, l_2{}^2 \, m_p{}^2}{l_1{}^2 \, (m_1 + m_p) - 4 \, l_1 \, l_2 \, m_p} - l_2{}^2 \, m_p,$$

no critical values $z_\star$ exist and, hence, no singularities may appear.
– *If*

$$l_2 \, m_p < \frac{1}{4} l_1 \, m_1,$$

then no critical values $z_\star$ exist and, therefore, no singularities may appear, independent of the value of JR_p.

Let us finally have a closer look at the case if there exists a singularity of this type. If we are at the singularity at a time-point $t_\star$, we can assume that there is a value

$$A_6 = \pm\sqrt{A_2 \, (4 \, A_2 + A_5)}$$

such that

$$\cos x_3 = -\frac{2 A_2 + A_6}{A_3} = z_\star.$$

In that case, we obtain for the matrix $\mathcal{M}$ from Lemma 7.1 that

$$\mathcal{M} = - \begin{pmatrix} \frac{K}{A_3 NT^2} (A_2 + NT \, A_6 + A_2 \, NT) & \frac{A_2}{A_3} \\[2ex] -\dfrac{K\left(A_4 - \frac{(2 A_2 + A_6)^2}{A_3}\right)(A_2 + A_6 + 2 NT \, A_6 + 5 A_2 \, NT + A_5 \, NT)}{NT^2 \left(8 A_2{}^2 + 4 A_2 \, A_6 + A_2 \, A_5 - A_3 \, A_4\right)} & -\dfrac{\left(A_4 - \frac{(2 A_2 + A_6)^2}{A_3}\right)(A_2 + A_6)}{8 A_2{}^2 + 4 A_2 \, A_6 + A_2 \, A_5 - A_3 \, A_4} \end{pmatrix}$$

and, therefore,

$$\ker \mathcal{M} = \mathrm{im} \begin{pmatrix} -A_2 \\[1ex] \frac{K}{NT^2} (A_2 + NT \, A_6 + A_2 \, NT) \end{pmatrix}.$$

For a better understanding of the type of singularity we are dealing with, we emphasize that the setting is analogous to the singularity of nonlinear systems of equations for unknown functions $(y_1(t), y_2(t), y_3(t))$ of the form

$$\begin{pmatrix} \mathcal{M}(y_3) & \\ & 1 \end{pmatrix} \begin{pmatrix} y_1 \\ y_2 \\ y_3 \end{pmatrix} = \begin{pmatrix} r_1(t) \\ r_2(t) \\ r_3(t) \end{pmatrix},$$

which are considered for all values t from a given interval. If, at value $t_\star$, the right-hand side $r_3(t_\star)$ leads to a value for y_3 at which $\mathcal{M}(y_3)$ is singular, then no unique solvability is given at $t_\star$. For the Robotic Arm, we deal with (x_2, x_7) instead of (y_1, y_2), or, more precisely, with $(x_2, u_1 - u_2)$ if the original unknowns are considered.

Finally, we notice that, according to Lemma 7.1, also the singularity circle $C_{r^\star}$ corresponds to poles for x_2, u_1, u_2 in the general case.

8 Conclusions

In this article, we applied two different methodologies to characterize singular points of higher-index DAEs considering the Robotic Arm equations, the well-known benchmark from literature.

The two methodologies, which are related to the projector based differentiation index and the tractability index, are based on rank consideration of matrices that are constructed in accordance to each of the two index concepts. For the differentiation index, the 1-fullness of the expanded derivative array is considered. For the tractability index, the corresponding matrix sequence has to deliver a nonsingular matrix.

Although the matrices considered in both approaches are constructed in very different ways, both give us, in the end, hints to the same two types of singular points. The existence of these singularities depends on the particular values of the variable x_3, which describes the angular coordinate of the outer arm.

The detected singularity for $\cos x_3 = z_\star$ means that, for the Robotic Arm, there are singular configurations that, to our knowledge, have not been described so far in the DAE literature. These particular values are influenced by several model parameters.

Now that we finally understand these very illustrative singularities, in future work they can be analyzed in more detail following different mathematical concepts of the theory of DAEs. For instance, the interpretation of $C_{l_1-l_2}$ for $l_2 \to l_1$, $C_{r^\star}$ for $r^\star \to l_1 - l_2$ or $r^\star \to l_1 + l_2$ in dependence of the masses and/or the combination with other equations with dynamical degree greater than zero can be considered.

References

1. Barrlund, A.: Constrained least squares methods for non-linear differential algebraic systems. Technical report, University of Umeå Sweden, Institute of Information Processing (1991)
2. Campbell, S.L.: A general method for nonlinear descriptor systems: an example from robotic path control. Technical Report CRSC 090588-01, North Carolina State University Raleigh (1988)
3. Campbell, S.L., Griepentrog, E.: Solvability of general differential algebraic equations. SIAM J. Sci. Comput. **16**(2), 257–270 (1995)

4. Campbell, S.L., Kunkel, P.: Applications of Differential-Algebraic Equations: Examples and Benchmarks, Chapter General Nonlinear Differential Algebraic Equations and Tracking Problems: A Robotics Example. Differential-Algebraic Equations Forum. Springer, Cham (2019)
5. De Luca, A.: Control properties of robot arms with joint elasticity. Analysis and control of nonlinear systems. In: 8th International Symposium on Mathematical Networks System. Selected Paper Phoenix, 1987, pp. 61–70 (1988)
6. Estévez Schwarz, D., Lamour, R.: InitDAE's documentation. https://www.mathematik.hu-berlin.de/~lamour/software/python/InitDAE/html/
7. Estévez Schwarz, D., Lamour, R.: A new projector based decoupling of linear DAEs for monitoring singularities. Numer. Algorithms **73**(2), 535–565 (2016)
8. Estévez Schwarz, D., Lamour, R.: Consistent initialization for higher-index DAEs using a projector based minimum-norm specification. Technical Report 1, Institut für Mathematik, Humboldt-Universität zu Berlin (2016)
9. Estévez Schwarz, D., Lamour, R.: A new approach for computing consistent initial values and Taylor coefficients for DAEs using projector-based constrained optimization. Numer. Algorithms **78**(2), 355–377 (2018)
10. Estévez Schwarz, D., Lamour, R.: InitDAE: Computation of consistent values, index determination and diagnosis of singularities of DAEs using automatic differentiation in Python. J. Comput. Appl. Math. **2019**, 112486 (2019). https://doi.org/10.1016/j.cam.2019.112486
11. Estévez Schwarz, D., Lamour, R.: Projected explicit and implicit Taylor series methods for DAEs. Technical Report 8, Institut für Mathematik, Humboldt-Universität zu Berlin (2019)
12. Estévez Schwarz, D., Lamour, R.: Python code of the solution of the Robotic Arm problem (2019). https://www.mathematik.hu-berlin.de/~lamour/software/python/
13. Lamour, R., März, R., Tischendorf, C.: Differential-Algebraic Equations: A Projector Based Analysis. In: Ilchman, A., Reis, T. (eds.) Differential-Algebraic Equations Forum. Springer, Berlin (2013)
14. De Luca, A., Isidori, A.: Feedback linearization of invertible systems. In: Second Duisburger Kolloquium Automation und Robotik (1987)
15. März, R., Riaza, R.: Linear differential-algebraic equations with properly stated leading term: A-critical points. Math. Comput. Model. Dyn. Syst. **13**(3), 291–314 (2007).
16. März, R., Riaza, R.: Linear differential-algebraic equations with properly stated leading term: B-critical points. Dyn. Syst. **23**(4), 505–522 (2008)
17. Pryce, J.: Solving high-index DAEs by Taylor series. Numer. Algorithms **19**, 195–211 (1998)
18. Riaza, R.: Differential-Algebraic Systems. Analytical Aspects and Circuit Applications. World Scientific, Hackensack (2008)

Index

© The Editor(s) (if applicable) and The Author(s), under exclusive licence
to Springer Nature Switzerland AG 2020
T. Reis et al. (eds.), *Progress in Differential-Algebraic Equations II*,
Differential-Algebraic Equations Forum,
https://doi.org/10.1007/978-3-030-53905-4

Printed in the United States
By Bookmasters